Pro/ENGINEER Wildfire for Designers
Release 2.0

Sham Tickoo
Professor
Department of Mechanical Engineering Technology
Purdue University Calumet
Hammond, Indiana
U.S.A.

Contributing Authors
Deepak Maini
Sr. CADD Engineer
Sanjib Sahu
CADD Engineer
CADCIM Technologies

CADCIM Technologies
(www.cadcim.com)
U.S.A.

 **CADCIM Technologies**

Pro/ENGINEER Wildfire for Designers Release 2.0
Sham Tickoo

Published by CADCIM Technologies, 525 St Andrews Drive, Schererville, IN 46375 USA. Copyright © CADCIM Technologies. All rights reserved. No part of this publication may be reproduced or distributed in any form or by any means, or stored in the database or retrieval system without the prior permission of CADCIM Technologies.

ISBN 1-932709-03-7

Cover designer: *CADCIM Technologies*
Cover illustration: *Created by Deepak Maini and Gurpreet Singh using Pro/ENGINEER and SolidWorks*
Technical editors: *Gurpreet Singh*
Typeface: *10/12 New Baskerville Bt*

NOTICE TO THE READER

Publisher does not warrant or guarantee any of the products described in the text or perform any independent analysis in connection with any of the product information contained in the text. Publisher does not assume, and expressly disclaims, any obligation to obtain and include information other than that provided to it by the manufacturer.

The reader is expressly warned to consider and adopt all safety precautions that might be indicated by the activities herein and to avoid all potential hazards. By following the instructions contained herein, the reader willingly assumes all risks in connection with such instructions.

The Publisher makes no representation or warranties of any kind, including but not limited to, the warranties of fitness for particular purpose or merchantability, nor are any such representations implied with respect to the material set forth herein, and the publisher takes no responsibility with respect to such material. The publisher shall not be liable for any special, consequential, or exemplary damages resulting, in whole or part, from the reader's use of, or reliance upon, this material.

www.cadcim.com

DEDICATION

*To teachers, who make it possible to disseminate knowledge
to enlighten the young and curious minds
of our future generations*

*To students, who are dedicated to learning new technologies
and making the world a better place to live*

THANKS

*To the faculty and students of the MET department of
Purdue University Calumet for their cooperation*

*To Santosh Tickoo, Gurpreet Singh, and Monalisa Bhol
of CADCIM/CADSoft Technologies for their valuable help*

Free Teaching Aids for Faculty

The following teaching aids are available for free to faculty.

1. Free online technical support by contacting techsupport@cadcim.com.
2. All Part files, Assembly files, and Drawing files used for theory, tutorials and exercises in this book.
3. PowerPoint presentations for every chapter of the book.
4. Instructor's Guide with answers to review questions and solution to exercises.
5. Course outlines.
6. Students projects.
7. Class Test

To access the web site that contains these teaching aids, please contact the author, Prof. Sham Tickoo, at the following address.

        stickoo@calumet.purdue.edu
        or
        tickoo@cadcim.com

For Students: You can download solid modeling exercises, tutorials, and special topics by accessing the author's Web site **http://technology.calumet.purdue.edu/met/tickoo/students/students.htm** or the publisher's Web site **www.cadcim.com**

        For more information, please visit www.cadcim.com

Table of Contents

Dedication iii
Preface xiii
Introduction xvii

Chapter 1: Creating Sketches in the Sketch Mode-I

The Sketch Mode	1-2
Using the Sketch Mode	1-2
Invoking the Sketch Mode	1-3
The Sketcher Environment	1-3
Working With the Sketch in the Sketch Mode	1-4
Drawing a Sketch Using the Sketcher Tools Toolbar	1-5
Placing a Point	1-5
Drawing a Line	1-5
Drawing a Rectangle	1-7
Drawing a Circle	1-7
Drawing an Arc	1-10
Dimensioning the Sketch	1-13
Dimensioning the Sketch using the Create defining dimension button or the Normal option	1-13
Dimensioning the Basic Sketched Entities	1-14
Linear Dimensioning a Line	1-14
Angular Dimensioning an Arc	1-14
Diameter Dimensioning	1-14
Radial Dimensioning	1-15
Dimensioning Revolved Sections	1-15
Working With Constraints	1-16
Explain Option	1-17
Disabling the Constraints	1-17
Converting a Weak Constraint into a Strong Constraint	1-17
Modifying the Dimensions of a Sketch	1-18
Using the Modify the values of dimensions, geometry of splines, or text entities Tool	1-18
Using the Edit Menu	1-19
Modifying a Dimension by Double-Clicking	1-19
Modifying Dimensions Dynamically	1-19
Resolve Sketch Dialog Box	1-19
Deleting the Sketcher Entities	1-20
Trimming the Sketcher Entities	1-21
Mirroring the Sketcher Entities	1-22
Drawing Display Options	1-22
Tutorial 1	1-23

Tutorial 2	1-30
Tutorial 3	1-34
Self-Evaluation Test	1-37
Review Questions	1-38
Exercise 1	1-39
Exercise 2	1-39
Exercise 3	1-40
Exercise 4	1-40

Chapter 2: Creating Sketches in the Sketch Mode-II

Dimensioning the Sketch	2-2
Dimensioning a Sketch using the Baseline Option	2-2
Replacing the Dimensions of a Sketch using the Replace option	2-3
Creating Fillets	2-3
Creating Circular Fillets	2-3
Creating Elliptical Fillets	2-6
Creating a Reference Coordinate System	2-7
Working With Splines	2-7
Creating a Spline	2-7
Dimensioning of Splines	2-8
Modifying a Spline	2-9
Writing Text in Sketcher Environment	2-11
Scaling and Rotating Entities	2-12
Copying Sketched Entities in Sketch Mode	2-13
Importing 2D Drawings in the Sketch Mode	2-13
Tutorial 1	2-15
Tutorial 2	2-20
Tutorial 3	2-25
Self-Evaluation Test	2-29
Review Questions	2-30
Exercise 1	2-31
Exercise 2	2-31
Exercise 3	2-31

Chapter 3: Creating Base Features

Creating Base Features	3-2
Invoking the Part mode	3-2
The Default Datum Planes	3-3
Creating a Protrusion	3-5
Extruding a Sketch	3-5
Revolving a Sketch	3-16
Understanding the Orientation of Datum Planes	3-17
Parent-Child Relationship	3-23
Implicit Relationship	3-23
Explicit Relationship	3-23

Table of Contents

Nesting of Sketches	3-23
Tutorial 1	3-25
Tutorial 2	3-32
Tutorial 3	3-38
Tutorial 4	3-43
Self-Evaluation Test	3-48
Review Questions	3-49
Exercise 1	3-50
Exercise 2	3-50
Exercise 3	3-51
Exercise 4	3-51

Chapter 4: Datums

Datums	4-2
Default Datum Planes	4-2
Need for Datums in Modeling	4-2
Selection Method in Pro/ENGINEER Wildfire 2.0	4-3
Datum Options	4-5
Datum Planes	4-5
Creating Datum Planes	4-8
Datum Planes Created On-The-Fly	4-14
Datum Axes	4-15
Datum Points	4-21
Creating Cuts	4-28
Removing Material by Extruding a Sketch	4-28
Removing Material by Revolving a Sketch	4-29
Tutorial 1	4-30
Tutorial 2	4-40
Tutorial 3	4-48
Self-Evaluation Test	4-54
Review Questions	4-55
Exercise 1	4-56
Exercise 2	4-56
Exercise 3	4-57
Exercise 4	4-59

Chapter 5: Options Aiding Construction of Parts-I

Options Aiding Construction of Parts	5-2
Creating Holes	5-2
Important Points to Remember While Creating Hole	5-10
Creating Rounds	5-11
Creating Basic Rounds	5-11
Creating a Variable Radius Round	5-19
Points to Remember While Creating Rounds	5-20
Creating Chamfers	5-21

Corner Chamfer	5-21
Edge Chamfer	5-22
Understanding Ribs	5-27
Editing Features of a Model	5-29
Editing Definition or Redefining Features	5-29
Reordering Features	5-30
Rerouting Features	5-31
Suppressing Features	5-32
Deleting a Feature	5-33
Modifying a Feature	5-33
Tutorial 1	5-34
Tutorial 2	5-44
Tutorial 3	5-49
Tutorial 4	5-54
Self-Evaluation Test	5-62
Review Questions	5-63
Exercise 1	5-64
Exercise 2	5-65
Exercise 3	5-65

Chapter 6: Options Aiding Construction of Parts-II

Creating Feature Patterns	6-2
Uses of Patterns	6-2
Creating Patterns	6-2
Deleting a Pattern	6-13
Copying Features	6-13
Mirror	6-14
Move	6-15
Mirroring a Geometry	6-16
Creating a Section of the Solid Model	6-17
Work Region Method	6-18
Tutorial 1	6-20
Tutorial 2	6-26
Tutorial 3	6-37
Tutorial 4	6-43
Self-Evaluation Test	6-58
Review Questions	6-59
Exercise 1	6-60
Exercise 2	6-61
Exercise 3	6-61
Exercise 4	6-62

Chapter 7: Advanced Modeling Tools-I

Other Protrusion Options	7-2
Sweep Features	7-2

Creating Swept Protrusions	7-2
Sketching a Geometry Aligned to an Existing Geometry	7-4
Creating Sweep Feature by Selecting a Trajectory	7-5
Creating Thin Sweep Protrusion	7-7
Creating a Sweep Cut	7-7

Blend Features 7-8
    Parallel Option 7-8
    Rotational Blend 7-10
    General Option 7-12
Using Blend Vertex 7-12
Shell Option 7-12
    Creating Constant Thickness Shell 7-13
    Creating Variable Thickness Shell 7-14
Datum Curves 7-15
    Creating Datum Curve Using the Insert a datum curve button 7-15
    Creating Datum Curve by Sketching 7-18
    Intersect Option 7-18
    Project Option 7-19
    Wrap Option 7-23
Creating Draft Features 7-24
Tutorial 1 7-34
Tutorial 2 7-41
Tutorial 3 7-48
Tutorial 4 7-53
Self-Evaluation Test 7-58
Review Questions 7-59
Exercise 1 7-60
Exercise 2 7-60
Exercise 3 7-61
Exercise 4 7-63

Chapter 8: Advanced Modeling Tools-II

Advanced Feature Creation Tools 8-2
    Variable Section Sweep Option 8-2
    Swept Blend 8-7
    Helical Sweep 8-8
    Blend Section to Surfaces 8-11
    Blend Between Surfaces 8-11
    Toroidal Blend 8-12
Tutorial 1 8-14
Tutorial 2 8-20
Tutorial 3 8-32
Tutorial 4 8-39
Self-Evaluation Test 8-41
Review Questions 8-42
Exercise 1 8-43

Exercise 2   8-43

Chapter 9: Assembly Modeling

Assembly Modeling	9-2
Important Terms Related to Assembly Mode	9-2
Top-Down Approach	9-2
Bottom-Up Approach	9-3
Creating Top-Down Assemblies	9-4
Creating Components in the Assembly Mode	9-4
Creating Bottom-Up Assemblies	9-5
Inserting Components in the Assembly	9-5
Placement Constraints	9-5
Automatic	9-5
Mate	9-6
Align	9-7
Insert	9-9
Coord Sys	9-9
Tangent	9-9
Pnt On Line	9-9
Pnt On Srf	9-9
Edge On Srf	9-9
Default	9-9
Fix	9-10
Convert	9-10
Assembly Datum Planes	9-10
Assembling the Components	9-10
Packaging the Components	9-16
Creating Simplified Representations	9-17
Redefining the Components of the Assembly	9-22
Reordering the Components	9-22
Suppressing/Resuming the Components	9-22
Replacing Components	9-23
Assembling Repeated Copies of a Component	9-23
Modifying the Components of the Assembly	9-25
Modifying the Dimensions of a Feature of a Component	9-25
Redefining a Feature of a Component	9-25
Creating the Exploded State	9-25
Offset Lines	9-28
Creating the Offset Lines	9-29
Modifying the Offset Lines	9-29
The Bill Of Material	9-30
Tutorial 1	9-31
Tutorial 2	9-48
Self-Evaluation Test	9-58
Review Questions	9-59
Exercise 1	9-60

Chapter 10: Generating, Editing, and Modifying the Drawing Views

The Drawing Mode	10-2
Generating the Drawing Views	10-5
Generating the General View	10-5
Generating the Projection View	10-6
Generating the Detailed View	10-7
Generating the Auxiliary View	10-7
Generating the Revolved Section View	10-8
Generating the Copy and Aligned View	10-19
Editing the Drawing Views	10-20
Moving the Drawing View	10-20
Erasing the Drawing View	10-20
Deleting the Drawing Views	10-21
Adding New Parts or Assemblies to the Current Drawing	10-21
Modifying the Drawing Views	10-21
Changing the View Type	10-21
Changing the View Scale	10-22
Reorienting the Views	10-22
Modifying the Cross-sections	10-23
Modifying the Boundaries of Views	10-23
Adding or Removing the Cross-section Arrows	10-23
Modifying the Perspective Views	10-23
Modifying Other Parameters	10-23
Editing the Cross-section Hatching	10-24
Tutorial 1	10-24
Tutorial 2	10-33
Self-Evaluation Test	10-40
Review Questions	10-40
Exercise 1	10-41

Chapter 11: Dimensioning the Drawing Views

Dimensioning the Drawing Views	11-2
Show / Erase dialog box Options	11-2
Modifying and Editing Dimensions	11-6
Modifying the Dimensions Using the Dimension Properties Dialog Box	11-7
Modifying the Drawing Items Using the Shortcut Menu	11-11
Cleaning up the Dimensions	11-12
Adding Reference Datums to the Drawing Views	11-16
Adding Tolerances in the Drawing Views	11-17
Dimensional Tolerances	11-17
Geometric Tolerances	11-17
Editing the Geometric Tolerances	11-20
Adding Notes to the Drawing	11-21
Adding Balloons to the Assembly Views	11-21
Tutorial 1	11-23

Tutorial 2	11-29
Self-Evaluation Test	11-34
Review Questions	11-34
Exercise 1	11-35

Chapter 12: Other Drawing Options

Sketching in the Drawing Mode	12-2
Modifying the Sketched Entities	12-3
User-Defined Drawing Formats	12-8
Retrieving the User-Defined Formats in the Drawings	12-9
Adding and Removing Sheets in the Drawing	12-11
Creating Tables in the Drawing Mode	12-11
Generating the BOM and Balloons in Drawings	12-12
Tutorial 1	12-16
Tutorial 2	12-27
Self-Evaluation Test	12-36
Review Questions	12-37
Exercise 1	12-38

Index

I-1

Chapters for Free Download

The following chapters are available for free download on the author's and the publisher's Web site. To download the free chapters, logon to

www.cadcim.com or
http://technology.calumet.purdue.edu/met/tickoo/students/students.htm

Chapter 13: Projects   Chapter For Free Download

Chapter 14: Surface Modeling   Chapter For Free Download

Author's Web Sites

For Faculty: Please contact the author at stickoo@calumet.purdue.edu or tickoo@cadcim.com to access the web site that contain the PowerPoint presentations, solid models used in the text book, Instructor's Guide, and other related material.

For Students: You can download solid modeling exercises, tutorials, and special topics by accessing the author's web site at **http://technology.calumet.purdue.edu/met/tickoo/students/students.htm** or **www.cadcim.com**

Preface

Pro/ENGINEER WILDFIRE 2.0

Pro/ENGINEER, developed by Parametric Technology Corporation, is one of the world's fastest growing solid modeling softwares. It is a parametric feature-based solid modeling tool and it not only unites the 3D parametric features with 2D tools but also addresses every design-through-manufacturing process. Based mainly on the solid modeling users feedback, this solid modeling tool is remarkably user-friendly and it allows you to be productive from day one.

This solid modeling tool allows you to easily import the standard format files with an amazing compatibility.

The 2D drawing views of the components are automatically generated in the Drawing mode. The drawing views that can be generated include detailed, orthographic, isometric, auxiliary, section, and so on. You can use any predefined drawing standard files for generating the drawing views. You can display the model dimensions in the drawing views or add reference dimensions whenever you want. The bidirectional associative nature of this software ensures that any modification made in the model is automatically reflected in the drawing views and any modification made in the dimensions in drawing views automatically updates the model.

Pro/ENGINEER Wildfire for Designers Release 2.0 is a book written with an intention of helping the readers effectively use Pro/ENGINEER Wildfire 2.0. This book is written with the tutorial point of view, with learn-by-doing as the theme. The mechanical engineering industry examples and tutorials are used in this book to ensure that the user can relate his knowledge of this book with the actual mechanical industry designs. The salient features of the book are as follows:

- **Free Downloadable Chapters.**

    The author has provided chapters on Surface Modeling and Projects for free download. The chapters are available on the publisher's Web site www.cadcim.com.

- **Tutorial approach.**

    The author has adopted the tutorial point-of-view, with learn-by-doing as the theme, throughout the book. This approach will guide the users through the process of creating the model in the tutorial.

- **Real-World Projects as Tutorials.**

    The author has used the real-world mechanical engineering projects as tutorials in this book so that the reader can correlate the tutorials in this book with the real-time models in the mechanical engineering industry.

- **Tips and Notes.**

    The additional information related to the topics is provided to the users in the form of tips and notes.

- **Learning Objectives.**

    The first page of every chapter provides in brief the topics that will be covered in that chapter. This will help the users to easily refer to a topic.

- **Tools section.**

    Every chapter begins with the tools section that provides the detailed explanation of the Pro/ENGINEER tools.

- **Self-Evaluation Test, Review Questions, and Exercises.**

    Every chapter ends with a Self-Evaluation Test so that the users can assess their knowledge of the chapter. The author has given the answers of the Self-Evaluation Tests so that the users can compare their answers with the correct answers. The Review Questions and Exercises are also given at the end of each chapter and can be used by the Instructors as test questions and exercises in the classroom.

- **Heavily illustrated text.**

    The text in this book is heavily illustrated with the help of around 500 line diagrams and 700 photos that support the tools section and tutorials.

Features of the Text

Learning Objectives Each chapter begins with Learning Objectives. These are the points that in brief explain that tools and options that the reader will be able to work with after completing the chapter.

Tutorials These are real-world mechanical automobile engineering components whose modeling is explained in a step-by-step method.

Self-Evaluation Test These are the questions given at the end of each chapter. Answers to these questions are also given at the end of the chapter so that the user can check his answers with the correct answers.

Review Questions These are the questions that are also given at the end of the chapter. However, the answer to these questions are not given in the book and so the instructor can use these questions as test questions. The teaching faculty will get the answer to these questions in the Instructor's Guide that is provided by the Authors.

Note

Authors have provided a lot of additional information to the users about the topic being discussed in the form of notes.

Tip

Special information and techniques are provided in the form of tips that allow the users to increase their efficiency.

About the Author

Sham Tickoo is a professor of Manufacturing Engineering Technology at Purdue University Calumet. He has been in the education and training since 1981 and joined Purdue University in the year 1987. Since then, he has been teaching drafting and design, AutoCAD, AutoLISP, 3ds max, 3ds viz, Mechanical Desktop, Pro/ENGINEER, Autodesk Inventor, etc.

Prof. Tickoo is also one of the best known authors of CAD/CAM books in the world market. He has authored a number of books on the latest CAD/CAM technology. He is one of the very few authors in the world whose books are translated and sold in the countries like Italy, China, Philippines, and Russia. The list of some of his books is given below. Visit www.cadcim.com for the complete listing of Prof. Tickoo's textbooks.

- CATIA for Designers, CADCIM Technologies, USA
- Solid Edge for Designers, CADCIM Technologies, USA
- Autodesk Inventor for Designers, CADCIM Technologies, USA
- Solid Works for Designers, CADCIM Technologies, USA
- Pro/ENGINEER Wildfire for Designers, CADCIM Technologies, USA
- Pro/ENGINEER for Designers, CADCIM Technologies, USA
- AutoCAD: A Problem Solving Approach, Delmar Publishers, USA
- AutoCAD LT: A Problem Solving Approach, Delmar Publishers, USA
- Customizing AutoCAD, Delmar Publishers, USA
- 3D Studio VIZ, Goodheart-Willcox Publishers, USA
- 3D Studio MAX, Goodheart-Willcox Publishers, USA
- Mechanical Desktop Instructor, McGraw-Hill, USA
- AutoCAD Tecniche Avanzate, APOGeO education, Italy
- AutoCAD Fondamenti, APOGeO education, Italy
- AutoCAD: A Problem Solving Approach, China Machine Press, China
- Customizing AutoCAD 2000, China Machine Press, China
- AutoCAD 2004, Piter Press, Russia

In addition to his academic career, he has also been employed as a design engineer, quality control engineer, software developer, and has also worked as a consultant for a number of companies in US and Canada.

Prof. Tickoo has also developed a software package called "SMLayout" that has been registered by Autodesk Inc. as a third party software product. This software package is currently being used by some companies in Canada, United States, and Columbia. SMLayout is written in the AutoLISP programming language. It generates flat layout drawing of some complicated geometric shapes like: transitions, cones, cone-cylinder intersections, sphere-cylinder intersections, and cylinder-cylinder intersections. This software package addresses the needs of steel fabricators and manufacturers of sheet metal products. The programs also provide a dimensioning option for automatic dimensioning of the drawing.

In March 2000, Prof. Tickoo was issued a US patent for a product called "Self Adjusting Cargo Organizer" that he designed for vehicles.

Introduction

Pro/ENGINEER WILDFIRE 2.0

Welcome to Pro/ENGINEER Wildfire 2.0. If you are a new user of Pro/ENGINEER software package, you are going to join hands with thousands of users of this high-end CAD/CAM/CAE tool worldwide. If you are a user of the previous releases of this software, you are going to upgrade your designing skills with the tremendous improvement in this latest release.

Pro/ENGINEER Wildfire 2.0 is a powerful program used to create complex designs with a great precision. The design intent of any three-dimensional (3D) model or an assembly is defined by its specification and its use. You can use the powerful tools of Pro/ENGINEER Wildfire 2.0 to capture the design intent of any complex model by incorporating intelligence into the design.

To make the designing process simple and quick, this software package has divided the steps of designing into different modules. This means that each step of designing is completed in a different module. For example, generally a design process consists of the following steps:

- Sketching using the basic sketch entities.
- Converting the sketch into features and parts.
- Assembling different parts and analyzing them.
- Documentation of the parts and the assembly in terms of drawing views.
- Manufacturing the final part and assembly.

All these steps are divided into different modes of Pro/ENGINEER Wildfire 2.0, namely, the **Sketch** mode, **Part** mode, **Assembly** mode, **Drawing** mode, and **Manufacturing** mode.

In spite of making various modifications in a design, the parametric nature of this software helps to preserve the design intent of a model with tremendous ease. Once you understand the feature-based, associative, and parametric nature of Pro/ENGINEER Wildfire 2.0, you can appreciate its power as a solid modeler. It allows you to work in a 3D environment and calculates the mass properties directly from the created geometry. You can switch to various display modes like wireframe, shaded, hidden, and no hidden at any time with ease as it only changes the appearance of the model.

FEATURE-BASED NATURE OF Pro/ENGINEER WILDFIRE 2.0

Pro/ENGINEER Wildfire 2.0 is a feature-based solid modeling tool. A feature is defined as the smallest building block and any solid model created in Pro/ENGINEER Wildfire 2.0 is an integration of a number of these building blocks. Each feature can be edited individually to bring in any change in the solid model. The use of the feature-based property provides greater flexibility to the parts created. For example, consider the part shown in Figure I-1. It consists of a counterbore hole at the center and six counterbore holes around it at some Bolt Circle Diameter (BCD).

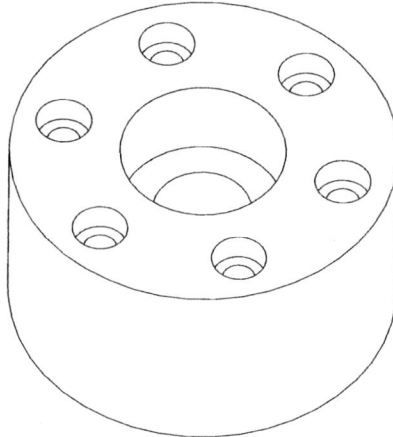

Figure I-1 *Model displaying the counterbore holes*

Now, consider a case where you have to change all outer counterbore holes to drill holes keeping the central counterbore hole and the BCD for the outer holes same. Also, you have to change the number of holes from six to eight. In a nonfeature-based software package, you will have to delete the entire part and then create a new part as per the new specifications; whereas, Pro/ENGINEER Wildfire 2.0 allows you to make this modification by just modifying some values in the same part, see Figure I-2. This shows that the solid parts created in Pro/ENGINEER Wildfire 2.0 are a combination of various features that can be modified individually at any time.

BIDIRECTIONAL ASSOCIATIVE NATURE OF Pro/ENGINEER WILDFIRE 2.0

There is a bidirectional associativity between all modes of Pro/ENGINEER Wildfire 2.0. The bidirectional associative nature of a software package is defined as its ability to ensure that if any modifications are made in a particular model in one mode, the corresponding modifications are also reflected in the same model in other modes. For example, if you make any change in a model in the **Part** mode and regenerate it, the changes will also be highlighted in the **Assembly** mode. Similarly, if you make any change in a part in the **Assembly** mode, after regeneration, the change will also be highlighted in the **Part** mode. This bidirectional associativity also correlates the two-dimensional (2D) drawing views generated in

Introduction

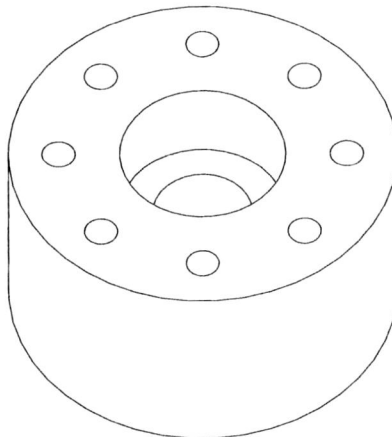

Figure I-2 Model after making the modifications

the **Drawing** mode and the solid model created in the **Part** mode of Pro/ENGINEER Wildfire 2.0. This means that if you modify the dimensions of the 2D drawing views in the **Drawing** mode, the change will be automatically reflected in the solid model and also in the assembly after regeneration. Likewise, if you modify the solid model in the **Part** mode, the changes will also be seen in the 2D drawing views of that model in the **Drawing** mode. Thus, bidirectional associativity means that if modification is made to any one application, it changes the output of all other modes related to the model. This nature relates the various modes available in Pro/ENGINEER.

Figure I-3 shows the drawing views of the part shown in Figure I-1 generated in the **Drawing** mode. The views show that the part consists of a counterbore hole at the center and six counterbore holes around it.

Now, when the part is modified in the **Part** mode, the modifications are automatically reflected in the **Drawing** mode, as shown in Figure I-4. The views in this figure show that all outer counterbore holes are converted into the drilled holes and the number of holes is increased from six to eight.

Figure I-5 shows the Crosshead assembly. It is clear from the assembly that the diameter of the hole is more than what is required (shown using dotted lines). In an ideal case, the diameter of the hole should be equal to the diameter of the bolt.

The diameter of the hole can be easily changed by opening the file in the **Part** mode and making the necessary modifications in the part. This modification is reflected in the assembly, as shown in Figure I-6. This is due to the bidirectional associative nature of Pro/ENGINEER.

Because all modes of Pro/ENGINEER Wildfire 2.0 are related to each other, it becomes very easy to modify your model at any time. This makes the application software more user-friendly.

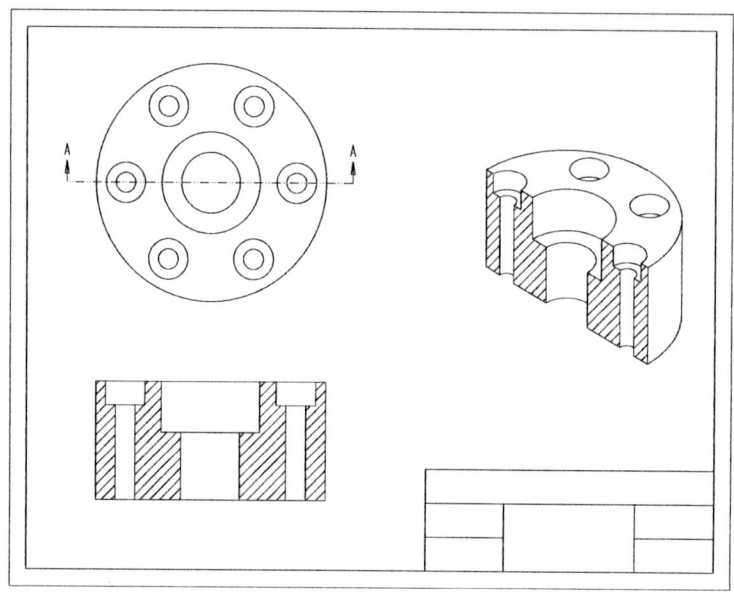

Figure I-3 Drawing views of the model before modifications

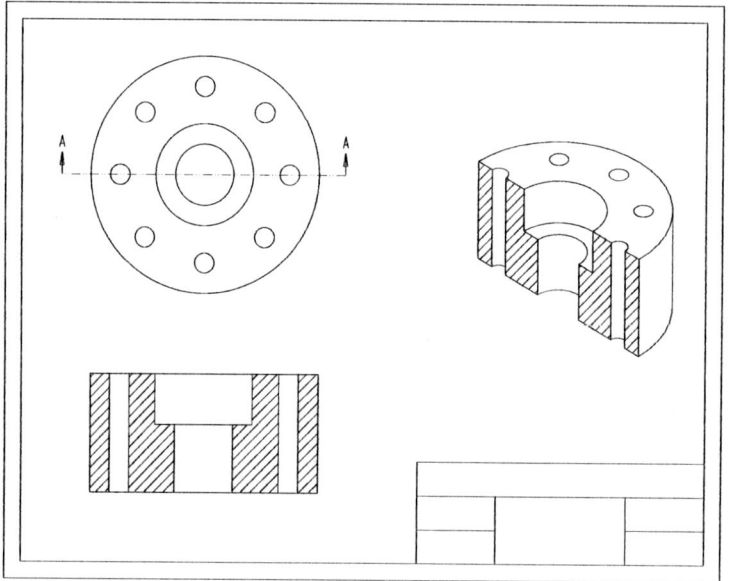

Figure I-4 Drawing views of the model after modifications

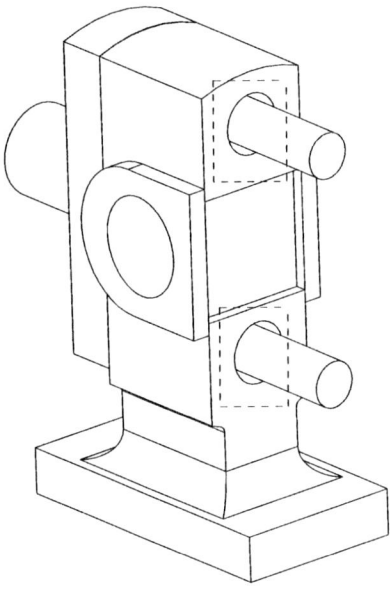

Figure I-5 *Diameter of the hole and the bolt*

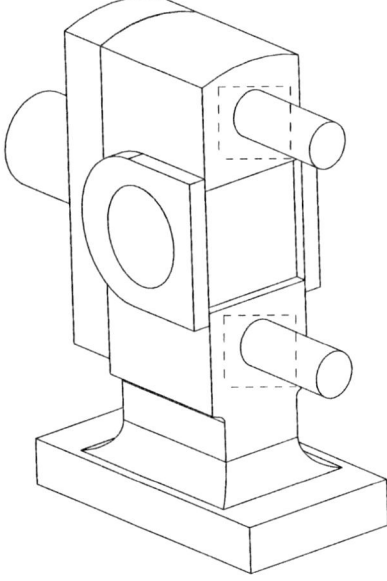

Figure I-6 *Figure after modifying the diameter of the hole*

PARAMETRIC NATURE OF Pro/ENGINEER WILDFIRE 2.0

Pro/ENGINEER Wildfire 2.0 is parametric in nature, which means that the features of a part become interrelated if they are drawn by taking the reference of each other. You can redefine the dimensions or the attributes of a feature at any time. The changes will propagate automatically throughout the model. Thus, they develop a relationship among themselves. This relationship is known as the parent-child relationship. So if you want to change the placement of the child feature, you can make alterations in the dimensions of the references and hence change the design as per your requirement. The parent-child relationship will be discussed in detail while discussing the datums in later chapters.

SYSTEM REQUIREMENTS FOR Pro/ENGINEER WILDFIRE 2.0

The system requirements for Pro/ENGINEER Wildfire 2.0 are given below.

1. Operating System: Windows NT 4.0 SP5, Windows 2000, Windows XP Home Edition, XP Professional Edition, UNIX.

2. Processor 233MHz or higher.

3. Memory: 128MB RAM minimum requirement (512MB or higher recommended).

4. Swap Memory: 256MB minimum (1024MB or higher recommended).

5. Hard disk space: 1.12GB.

6. An ethernet adapter interface card or network card.

7. Microsoft approved three button mouse.

8. Microsoft Internet Explorer 6.0.

GETTING STARTED WITH Pro/ENGINEER WILDFIRE 2.0

Once you have Pro/ENGINEER Wildfire 2.0 installed on your system, there are two options to start it. The first option is to choose the **Start** button at the lower left corner of the screen and select **Start > Programs > PTC > Pro ENGINEER > Pro ENGINEER,** as shown in Figure I-7.

The second option to start Pro/ENGINEER Wildfire 2.0 is by double-clicking on its shortcut icon on the desktop of the computer.

Figure I-8 shows the screen that appears when you start Pro/ENGINEER Wildfire 2.0.

IMPORTANT TERMS AND DEFINITIONS

Some important terms that will be used in this book while working with Pro/ENGINEER Wildfire 2.0 are discussed next.

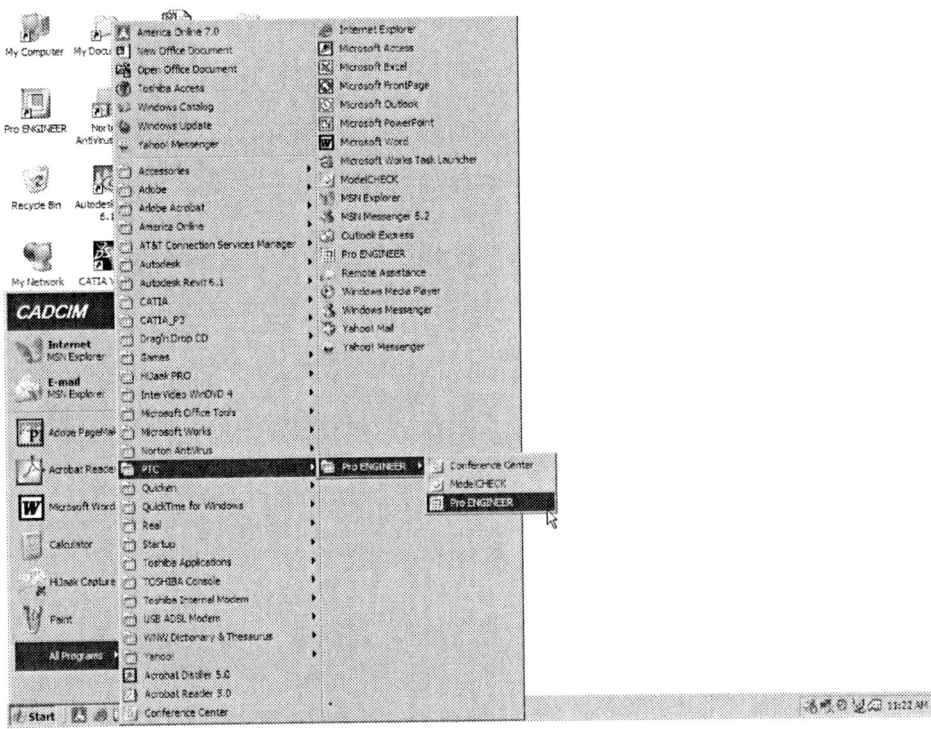

Figure I-7 Windows screen with task bar and application icons

Entity

An element of a section geometry is called an entity. The entity can be an arc, line, circle, point, conic, coordinate system, and so on. When one entity is divided at a point then the total number of entities is said to be two.

Dimension

It is the measurement of one or more entities.

Constraint

Constraints are logical operations that are performed on the selected geometry to make it more accurate in defining its position and size with respect to the other geometry.

Parameter

It is defined as a numeric value or any definition that defines a feature. For example, all dimensions in a sketch are parameters. The parameters can be modified at any time.

Relation

A relation is an equation that relates two entities.

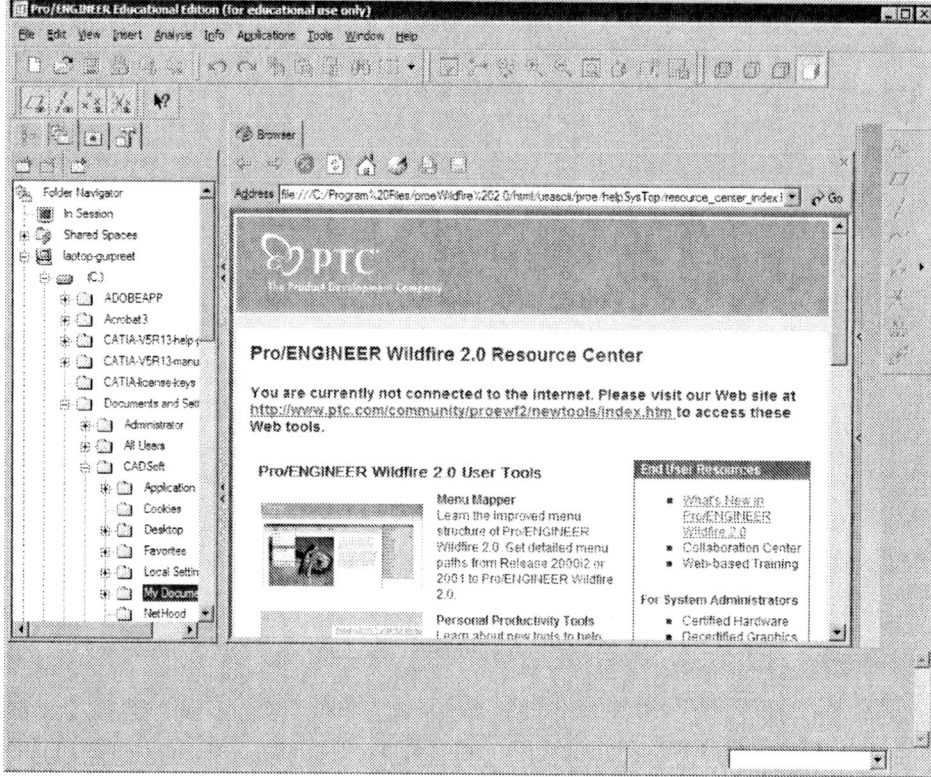

Figure I-8 Initial screen appearance after starting Pro/ENGINEER Wildfire

Weak Dimensions and Weak Constraints

Weak dimensions and weak constraints are temporary dimensions or constraints that appear in gray color. These are automatically applied to the sketch when it is drawn using the **Intent Manager**. They are removed from the sketch without any confirmation from the user. The weak dimensions or the weak constraints should be changed to strong dimensions or constraints if they seem to be useful for the sketch. This only saves an extra step of dimensioning the sketch or applying constraints to the sketch.

Strong Dimensions and Strong Constraints

Strong dimensions and strong constraints appear in yellow color. These dimensions and constraints are neither removed automatically nor applied automatically. All dimensions added manually to a sketch are strong dimensions.

FILE MENU OPTIONS

The options that are displayed when you choose **File** from the menu bar, are discussed next.

Working Directory

A working directory is a directory on your system where you can save the work done in the current session of Pro/ENGINEER Wildfire 2.0. You can set any directory existing on your system as the working directory. Before starting work in Pro/ENGINEER Wildfire 2.0, it is important to specify the working directory. If the working directory is not selected before saving an object file then the object file will be saved in a default directory. This default directory is set at the time of installing Pro/ENGINEER Wildfire 2.0. If the working directory is selected before saving the object files that you create, it becomes easy to organize them. In Pro/ENGINEER Wildfire 2.0, the working directory can be set in two ways:

Tip: *When several strong dimensions or constraints conflict, Pro/ENGINEER make the constraints and dimensions appear in red color, and prompts you to delete one or more.*

Using the Navigator

When you start a Pro/ENGINEER Wildfire 2.0 session, the **Navigator** is displayed on the left of the graphics window. This **Navigator** can be used to select a folder and set it as the working directory.

Right-click on the folder that you need to set as the working directory. The shortcut menu appears, as shown in Figure I-9. From this shortcut menu, choose the **Make Working Directory** option to set the selected folder as the working directory. To make a new folder, choose the **New Folder** option from the shortcut menu.

Figure I-9 Shortcut menu

Using the Select Working Directory Dialog Box

In order to specify a working directory, choose **File > Set Working Directory** from the menu bar. The **Select Working Directory** dialog box is displayed, as shown in Figure I-10. Using this dialog box you can set any directory as the working directory. The various options in this dialog box are discussed next.

Tip: *The Select Working Directory dialog box has some of the properties of the MS Windows operating system. You can set the working directory using this dialog box by browsing through directories and folders. However, you cannot rename a file, directory, or a folder name in this dialog box. You can create a new directory using this dialog box.*

Look In Drop-down List

The **Look In** drop-down list displays all drives present on your computer along with a **Favorites** folder, as shown in Figure I-10. When the **Select Working Directory** dialog box is invoked, by default, it displays the contents of a default directory. However, the default directory that appears every time you open this dialog box can be changed. This is discussed later. The **Favorites** folder contains all directories saved as favorites. The saving of the favorite directories will be discussed later.

Name

The **Name** edit box displays the name of the directory selected in the **Select Working Directory** dialog box. You can either select a directory from the **Look In** drop-down list or enter the path of any existing directory in this edit box.

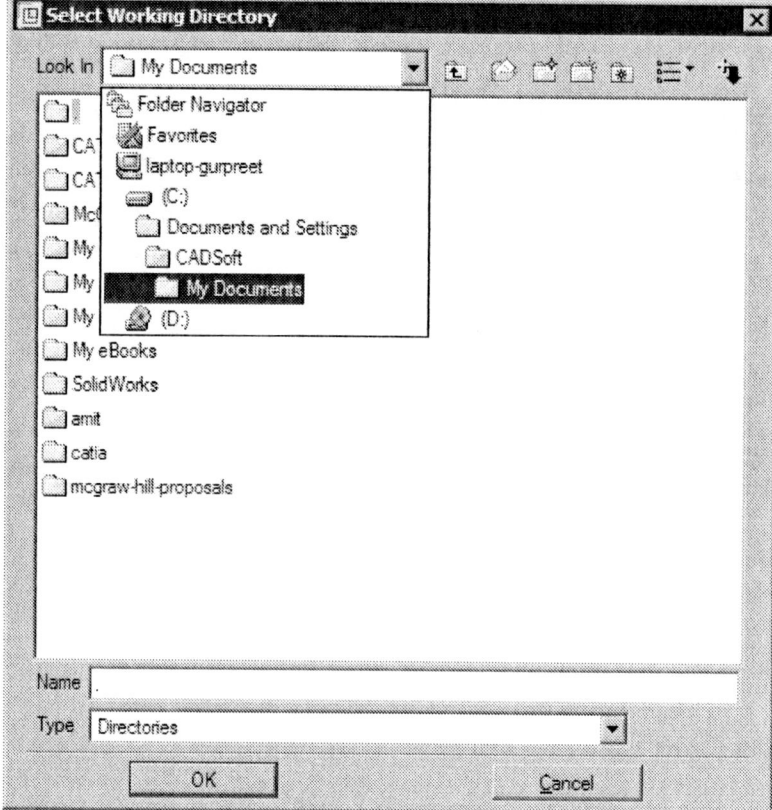

*Figure I-10 The **Select Working Directory** dialog box*

Type Drop-down List

The **Type** drop-down list has two options: **Directories** and **All Files**. If you select the **Directories** option, all directories present are listed and if you select **All Files** then all files along with the directories are listed in the dialog box.

Up One Level

The **Up One Level** button allows you to move one level up in the directory. When you choose this button, a directory is displayed that is one level above the current directory. This button is generally available in most of the dialog boxes of Windows operating system and has the same function.

Working Directory

This is used when you have already set the working directory. You may browse through the directories in the **Select Working Directory** dialog box, but when you choose this button, the directory selected previously as working directory is displayed in the **Look In** drop-down list.

New Directory

The **New Directory** button is used to create a new directory that can be selected as a working directory. When you choose the **New Directory** button, the **New Directory** dialog box is displayed. You are prompted to enter the name of the new directory you want to create.

Favorites

The **Favorites** button is used to save the location of the directories that are to be used frequently. You just have to specify the working directories to be used frequently and save the location of those directories by selecting the **Favorites** button.

When you want to select one of the favorite working directories, you can select the **Favorites** folder from the **Look In** drop-down list available in the **Select Working Directory** dialog box. In this folder there is a subfolder named **Personal Favorites**. When you double-click on this folder, all directories that were selected as favorites are displayed.

When you choose the **Favorites** button, a menu is displayed. The options in this menu are discussed next.

Save location
The **Save location** option is used to save the current directory in the **Favorites** folder. This option is available only when the directory selected is not already saved in the **Favorites** folder.

Remove location
The **Remove location** option removes the directory from the **Favorites** folder. This option is available only when the directory selected is already saved in **Favorites** folder.

Browse favorites
The **Browse favorites** option allows you to browse through your favorite directories that you saved using the **Save location** option.

Display Configuration

When you choose the **Display Configuration** button, a menu is displayed. The options in this menu are discussed next.

List
The **List** option is used to view the contents of the current directory or drive, which includes files and folders in the form of a list.

Details
The **Details** option is used to view the contents of the current folder or drive in the form of a table, which indicates the name, size, and date on which it is modified.

Commands and Settings

 The **Command and Settings** button can be used to customize the **Select Working Directory** dialog box. This button when chosen displays a menu. The options in this menu are discussed next.

'Look In' Default

The **'Look In' Default** option allows you to set a directory as a default directory. When you select this option, the **'Look In' Default** dialog box is displayed. Figure I-11 shows this dialog box with the options in the drop-down list. In the drop-down list, there are four options. If you select the **Default** option, whenever the **File Open** dialog box is invoked it displays the directory that is set as default. If you select the **Working Directory** option in this drop-down list, then whenever the **File Open** dialog box is invoked it displays the working directory that is set. If you select the **In Session** option, then whenever you select **Open** from the **File** menu in the **File Open** dialog box the **In Session** folder is selected by default. Similarly, the **Pro/Library** can be set as the working directory.

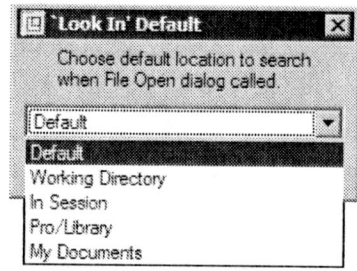

Figure I-11 The 'Look In' Default dialog box

All Versions

This option when selected displays all versions of an object file. In Pro/ENGINEER Wildfire 2.0, the file once saved will generate a new version of it with an extension 1. An object file is not copied on another object file but a new version of it is created. Therefore, every time you save an object using the **Save** option, a new version of it is created on the disk in the current working directory.

By default, in the **Select Working Directory** dialog box, the **Directories** option is displayed in the **Type** drop-down list. From this list if you select the **All Files** option and then select the **All Versions** option, all versions of the object files are displayed in the dialog box.

Note
An object in Pro/ENGINEER Wildfire 2.0 is defined as a file that is created using any of its modes like Part, Drawing, Sketch, and so on.

New

In order to create a new object, select **New** from the **File** menu or choose the **Create a new object** button from the **Top Toolchest**. The **New** dialog box is displayed, as shown in Figure I-12. The dialog box displays the various modes available in Pro/ENGINEER Wildfire 2.0. By default, the **Part** mode radio button is selected. A default name of the object file is displayed in the **Name** edit box. You can also enter the name of the object file as desired.

Introduction

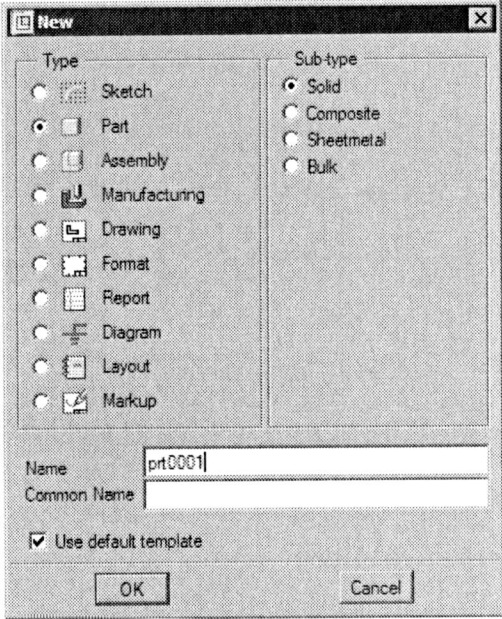

*Figure I-12 The **New** dialog box*

When you select the **Part**, **Assembly**, and **Manufacturing** mode radio buttons in this dialog box, their subtypes are displayed under the **Sub-type** area of this dialog box.

Accept the default name of the **Part** mode file, and choose the **OK** button in the **New** dialog box. The three default datum planes are displayed on the graphics window and some of the toolbars become active. Also, the **Model Tree** appears on the left of the screen, as shown in Figure I-13. The options available in the **New** dialog box are discussed next.

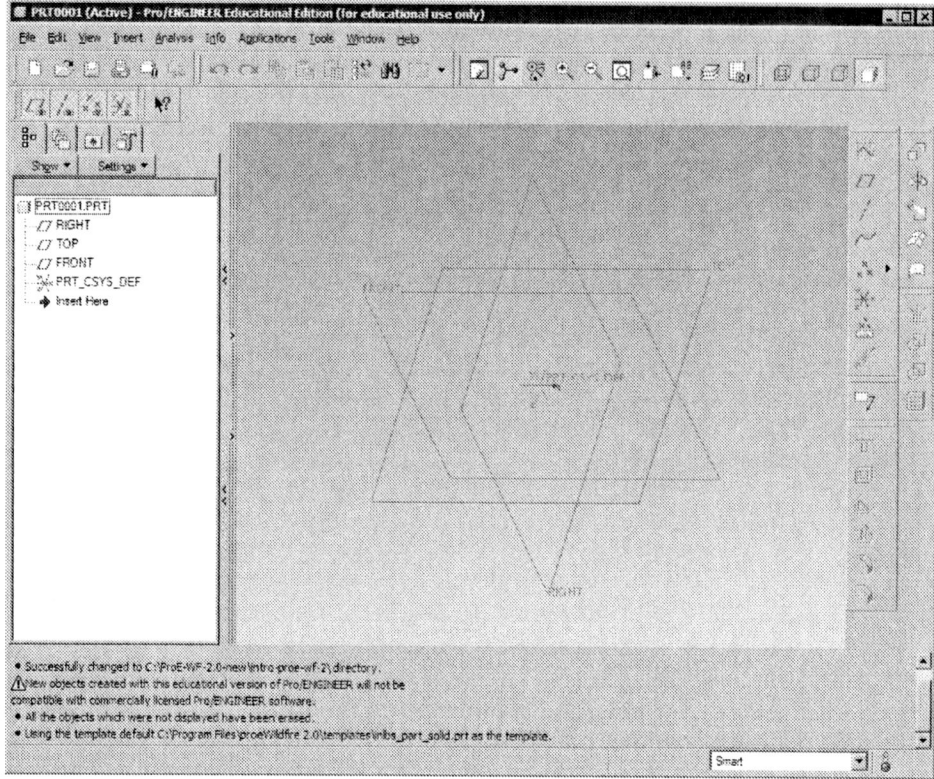

Figure I-13 The initial screen appearance after entering the Part mode

Use default template

The **Use default template** check box is selected to start a new file using an existing default template file that includes three default datum planes and a coordinate system. This default template file creates the model in inches. The check box is selected by default. If you clear this check box and choose the **OK** button from the **New** dialog box, the **New File Options** dialog box is displayed, as shown in Figure I-14. Using the **New File Options** dialog box you can select the predefined templates provided in Pro/ENGINEER Wildfire 2.0 or a user-defined template created and saved earlier. You can also open an empty template provided in the **New File Options** dialog box in which you have to create the datum planes and the coordinate system manually.

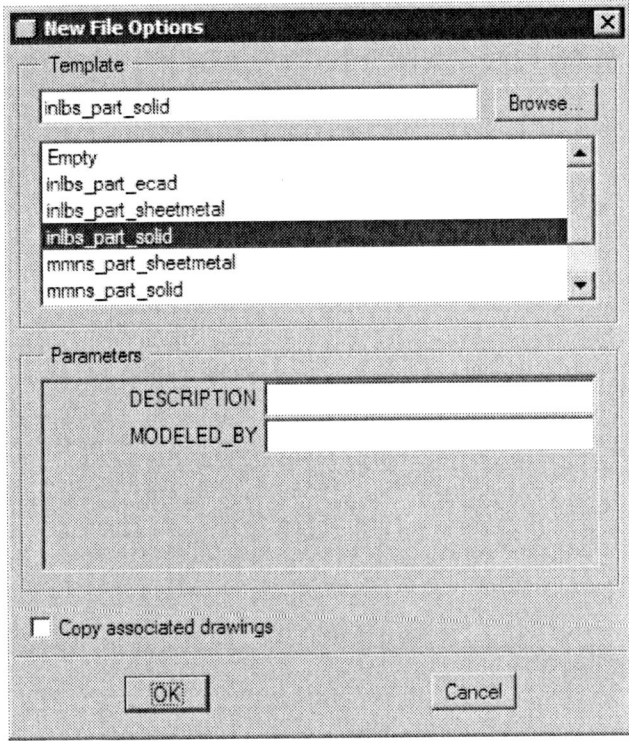

*Figure I-14 The **New File Options** dialog*

Open

The **Open** option is used to open an existing object file. When you choose the **Open** option from the **File** menu or choose the **Open an existing object** button from the **File** toolbar, the **File Open** dialog box is displayed, as shown in Figure I-15. The working directory you had selected is displayed in the **Look In** drop-down list. The various options in this dialog box are discussed next.

Look In

In the **Look In** drop-down list, various browsing options for selecting the directories are available. You can browse through these folders to search for the object file you want to open.

In Session

The **In Session** option displays all object files that are in the current session. The object files that you open in Pro/ENGINEER Wildfire 2.0 in the current session are stored in its temporary memory. This temporary memory is a folder named **In Session**. Once you exit Pro/ENGINEER Wildfire 2.0, the contents of this folder are deleted automatically. However, the original files are not removed from their actual location.

Commands and Settings

The **Commands and Settings** option can be used to customize the **File Open** dialog box.

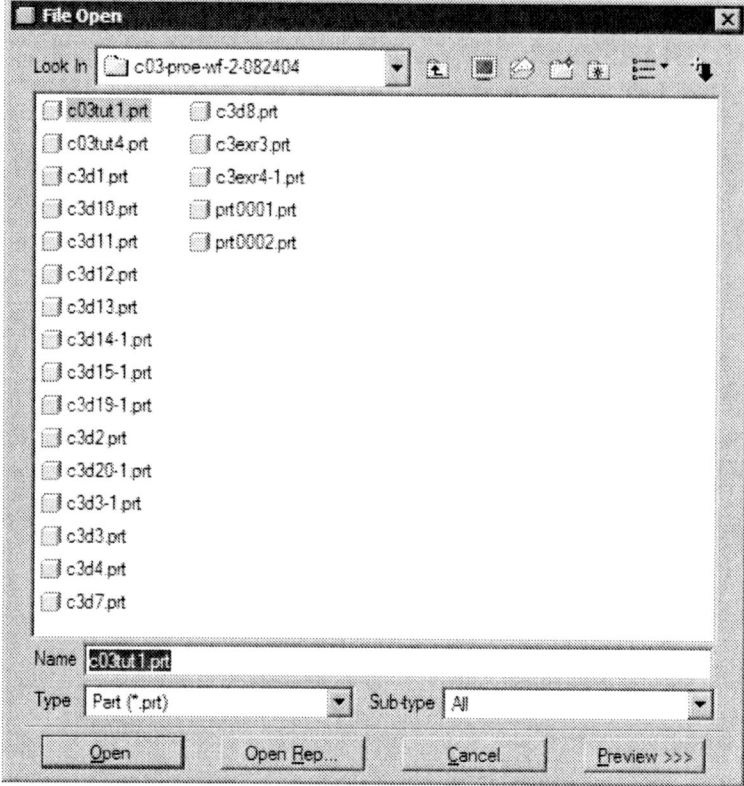

*Figure I-15 The **File Open** dialog box*

When you choose this button a shortcut menu is displayed. The options under this menu are similar to those available in the **Select Working Directory** dialog box.

Sort By

The **Sort By** option displays all files present in the directory in order so that it becomes easy to search a file. When you choose the **Sort By** option, a cascading menu is displayed. In the cascading menu, there are two options: **Model Name** and **Markup/Instance Name**. If you select the **Model Name** option, the file list will be sorted alphabetically by model name in the **File Open** dialog box. The **Markup/Instance Name** option sorts the file list by specific markups or instance names in the **File Open** dialog box.

Retrieve Drawing as View Only

The **Retrieve Drawing as View Only** option is used to open the drawing as view only. Generally when you open a drawing file, its part model is listed in the current session. You can verify this by opening the current session. But when you open a drawing file using this option, its part model is not listed in the current session.

The modification of an object in view only mode is not possible unless you choose **File > Retrieve Models** from the menu bar. The **Retrieve Drawing as View only** option is used to open drawings in the **Drawing** mode, that is, the object files having the extension *.drw*.

Name

In the **Name** edit box you can enter the name of the existing object file you want to open.

Type

The **Type** drop-down list contains the file formats of the various modes available in Pro/ENGINEER Wildfire 2.0. This drop-down list also has other file formats that can be imported in Pro/ENGINEER Wildfire 2.0. By default, in this drop-down list the **Pro/ENGINEER Files** option is selected, and hence the files created in any mode of Pro/ENGINEER can be opened. However, if you select a specific mode from this drop-down list, only the files created in that mode are displayed. For example, if you select **Part** in the drop-down list, then only the *.prt* files are displayed. This makes the selection and opening of the files easy.

Preview

The **Preview** button is used to see the preview of the model. When you choose this button, the **File Open** dialog box expands and a preview is displayed on the right of the dialog box. In this window, you can see the preview of the model selected. This feature of the **File Open** dialog box helps in previewing the model before actually opening it.

Note

*Assembly files with the file extension .asm can also be previewed using the **Preview** button. If you are not able to see the preview of the assembly files on the preview displayed on the right side of the **File Open** dialog box then choose the **Refit** button present on the top of the preview area to resize the assembly according to the preview area.*

*There is no command prompt in Pro/ENGINEER Wildfire 2.0. However, you are provided with prompts in the **Message Area**. Whenever you have to enter any numerical value or text, a **Message Input Window** is displayed in the **Message Area**.*

Erase

The **Erase** option is used to delete the files from the temporary memory known as the **In Session** folder. To invoke this option, choose **File > Erase** from the menu bar. The cascading menu is displayed with two options; **Current** and **Not Displayed**. As discussed earlier, all files opened in a session of Pro/ENGINEER Wildfire 2.0 are saved in the temporary memory. You can use the **Erase** option to erase files from the temporary memory. The options that are displayed in the cascading menu are discussed next.

Current

The **Current** option is used to erase the file that is opened and displayed on the graphics window. When the **Current** option of the cascading menu is chosen, the system prompts you to confirm erase file. The **Erase Confirm** dialog box is displayed, as shown in Figure I-16.

Not Displayed

This option is used to erase the files present in the **In Session** folder but not available presently on the screen. The **Erase Not Displayed** dialog box is displayed, as shown in Figure I-17. You can select the files from this dialog box to erase them.

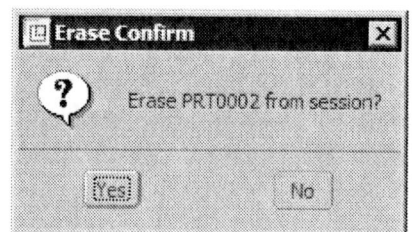

Figure I-16 The erase Confirm dialog box

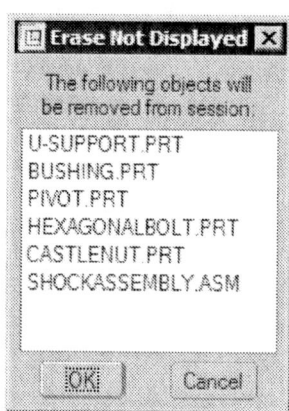

Figure I-17 The Erase Not Displayed dialog box

Delete

This option removes the selected file permanently from the hard disk. To invoke this option, choose **File > Delete** from the menu bar. The cascading menu is displayed with two options; **Old Versions** and **All Versions**.

Old Versions

This option is used to delete all old versions of the current file. When you choose the **Old Versions** option, you are prompted to enter the name of the object file of which the old versions have to be deleted. When the **Message Input Window** is displayed, enter the object file name in this window. All versions of that file will be deleted from the hard disk except the latest version.

All Version

This option is used to delete all versions including the current file from the hard disk. When you choose the **All Versions** option, a warning is displayed stating that performing this function can result in loss of data. This option is chosen when the file is opened and is displayed on the graphics window.

Save

The **Save** option is used to save the objects present in the **In Session** folder or an object on the graphics window. When you choose the **Save** option from the **File** menu or the **Save the active object** button on the **File** toolbar, you are prompted to specify the name of the object file. The name is displayed in the **Message Input Window** that is displayed when you choose this button.

Note
*Remember that all object files saved using the **Save** option in the **File** menu should be present in the **In Session** folder. If the file to be saved is not present in the **In Session** folder then Pro/ENGINEER Wildfire 2.0 will display an error informing you that the file does not exist in the current session.*

*It should be noted here that the file opened currently on the screen may not be in the **In Session** folder but is in the current session and hence will be saved without any error.*

Save a Copy

The **Save a Copy** option is used to save a copy of the current object in the same working directory or in some other directory. When you invoke this option from the **File** menu or select the **Save a copy of the active object** button from the **File** toolbar, the **Save a Copy** dialog box is displayed and you need to specify the new name of the object file to be saved as a copy and the name of the target directory in the **Save a Copy** dialog box. You can browse through the directories and select the target directory. The file will be saved in the selected directory.

Using this option you can also export the file in other file formats like IGES, STEP, STL, and so on. After specifying the name of the new file and the target directory choose the file format in which you want to export the file from the **Type** drop-down list in the **Save a Copy** dialog box.

Backup

The **Backup** option of the **File** menu creates a backup copy of an object file that is in memory. When you choose this option, the **Backup** dialog box is displayed, as shown in Figure I-18. In the **Model Name** edit box the name of the file that you want to take a backup of is displayed. In the **Backup To** edit box the name of the directory is specified where the object will be saved as a backup. If you backup an assembly or a drawing object, Pro/ENGINEER Wildfire 2.0 saves all its dependent files in the specified directory. For example, if you backup a drawing file in a different directory then the part file of the drawing will also be created automatically in the same directory where the drawing backup is created.

Rename

The **Rename** option in the **File** menu renames the active object that is on the screen. When you choose this option, the **Rename** dialog box is displayed, as shown in Figure I-19.

MANAGING FILES IN Pro/ENGINEER WILDFIRE 2.0

As discussed earlier, a new file is generated whenever you save an object. The number of files generated is directly proportional to the number of times you save that object. Hence, these files occupy a lot of disk space. The latest version of the object is of use and should be stored. Latest version implies to the highest number that is suffixed with the file name of that object. The rest of the files called old versions should be deleted from the hard disk if they are not required.

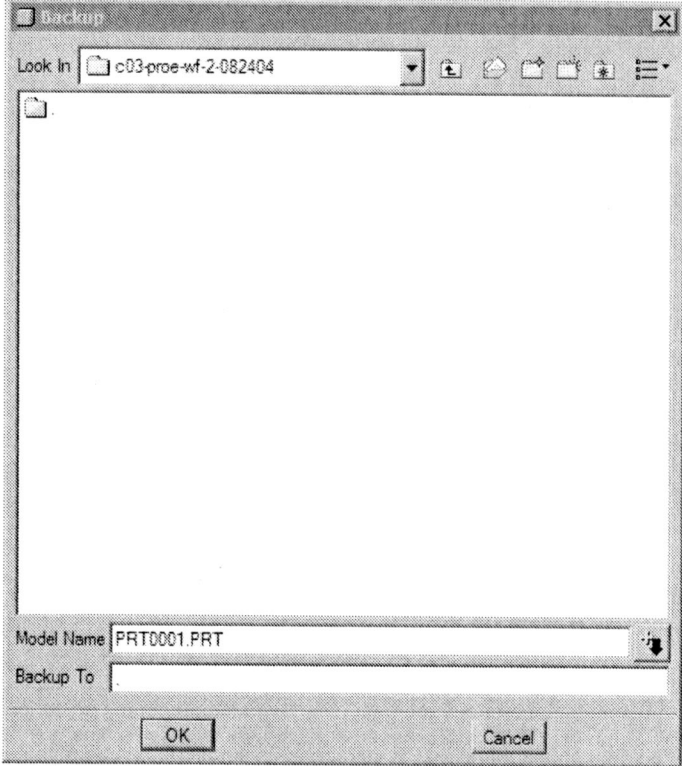

*Figure I-18 The **Backup** dialog box*

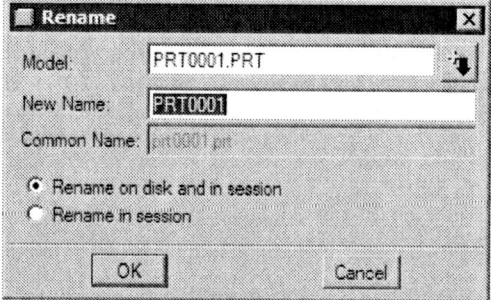

*Figure I-19 The **Rename** dialog box*

Note
*To save disk space you should keep deleting the old versions of a file. This is done using the **Delete** option in the **File** menu.*

Introduction

Tip: *Suppose you open an assembly that has a component named Nut. Close the assembly and now open another assembly that has a component named Nut. Now, there are chances that the second assembly that you choose to open may open with the Nut that was present in the previous assembly. This is because the component with the file name Nut was already present in the memory of Pro/ENGINEER (In session) before opening the second assembly.*

To avoid this error of assemblies, you should erase the In session memory of Pro/ENGINEER before opening the second assembly.

MENU MANAGER

From this release of Pro/ENGINEER the **Menu Manager** is not available when you enter the **Part** mode, **Assembly** mode, or the **Drawing** mode. The **Menu Manager** is displayed with some selected options of feature creation. There are menus and submenus cascaded in the **Menu Manager**. In the **Menu Manager**, all options are available to complete the desired task using Pro/ENGINEER Wildfire 2.0.

While using the **Menu Manager**, always complete the option selected by choosing **Done** or **Done Sel** after the current task is over. This is important when you are in the **Drawing** mode of Pro/ENGINEER. If you are directly selecting one option after another, then it is easy to loose track of commands or options in the **Menu Manager**.

Note
*From this release of Pro/ENGINEER most of the features are created using the **Dashboard**. The **Dashboard** is displayed above the **Message Area** and it contains all options to complete the selected operation on the model.*

MODEL TREE

The **Model Tree** stores and displays all features in a chronicle. You can select any desired feature of a model or an assembly from the model tree and apply different operations on the selected feature. You can select the feature by right-clicking on it. The shortcut menu will be displayed, as shown in Figure I-20. Move the cursor on the shortcut menu and choose an option from the menu pressing the right mouse button.

When you create a new object file, the **Model Tree** appears and is attached to the graphics window by default. But in this case with the **Model Tree** attached to the graphics window, the graphics window becomes small in size. You can hide the **Model Tree** by clicking the sash on the right edge of the **Navigator**, as shown in Figure I-20. The Model Tree can slide out or slide in, thus increasing the area on the graphics window. It can also be stretched horizontally to cover the graphics window.

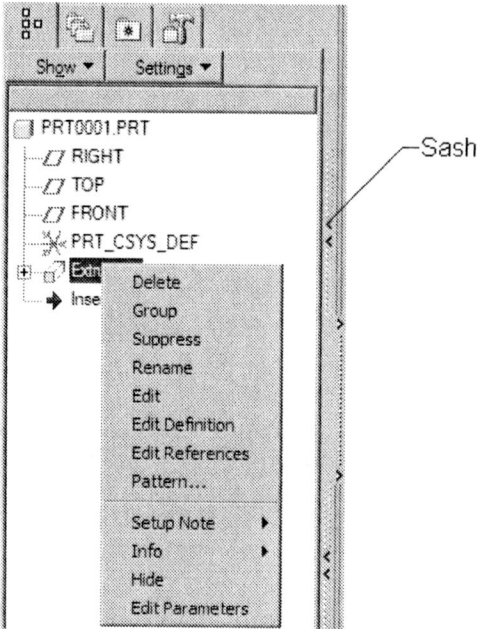

Figure I-20 The **Model Tree** with shortcut menu

Tip: *By looking at the **Model Tree** of a model, you can understand the method and approach used by the designer to create the model. If you have the **Model Tree** of a model you can modify the model. Generally when you import a model of a different file format in Pro/ENGINEER, you do not get the features of the model in the **Model Tree**, and therefore, you are not able to modify that model.*

UNDERSTANDING THE FUNCTION OF THE MOUSE BUTTONS

While working with Pro/ENGINEER Wildfire 2.0, it is important to understand the function of the three buttons of the mouse to make efficient use of this device. The various combinations of the keys and three buttons of the mouse are listed below.

1. Figure I-21 shows the function of the left mouse button. The left mouse button is used to make a selection. Using CTRL+left mouse button you can add or remove items from the selection set. Using SHIFT+left mouse button you can select a chain of entities or a set of surfaces.

2. Figure I-22 shows the function of the right mouse button. The right mouse button is used to invoke shortcut menus and to query select the items. When you bring the cursor on an item, it is highlighted in cyan color. Now if you hold down the right mouse button, a shortcut menu is displayed. Choose the **Pick From List** option from the shortcut menu. The

Introduction

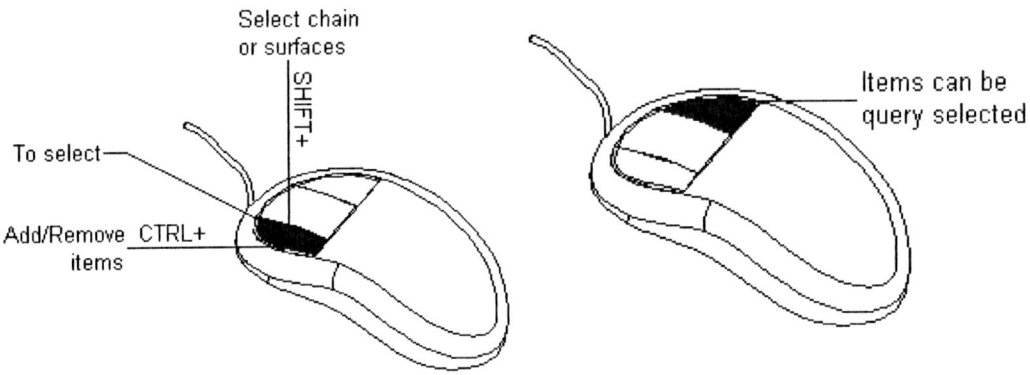

Figure I-21 Functions of the left mouse button *Figure I-22* Functions of the right mouse button

Pick From List dialog box is displayed. You can make selections from this dialog box.

3. Figure I-23 shows the function of the middle mouse button in the 3D mode. The middle mouse button is used to spin the model in the graphics window and view it from different directions.

   CTRL+middle mouse button is used to dynamically zoom in and zoom out by moving the mouse. When you press and hold the CTRL key and move the mouse up, the view is reduced, and thus you zoom out. When the mouse is moved down, the view is enlarged, and thus you zoom in.

   When you use CTRL+middle mouse button and move the mouse horizontally, the model is turned.

   Shift+middle mouse button is used to pan the object on the screen.

4. Figure I-24 shows the function of the middle mouse button in the 2D mode (sketcher environment and **Drawing** mode). The middle mouse button is used to place dimensions on the graphics window.

   It is also used to confirm an option or to abort the creation of an entity.

   The middle mouse button is used to pan in the **Sketch** mode and the **Drawing** mode.

 Note
*When you spin the model with the **Spin Center** on, the model is rotated about the spin center. If the spin center is turned off, then the model is rotated about the cursor.*

TOOLBARS

Before starting work on Pro/ENGINEER Wildfire 2.0, it is very important for you to understand the

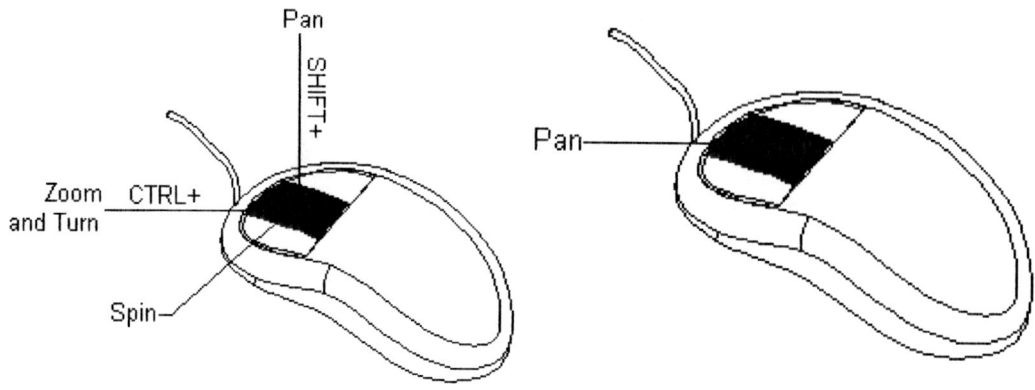

Figure I-23 *Functions of the middle mouse button in the 3D mode* **Figure I-24** *Functions of the middle mouse button in the 2D mode*

default toolbars and buttons on the main window. Figure I-25 shows various default screen components in Pro/ENGINEER Wildfire 2.0.

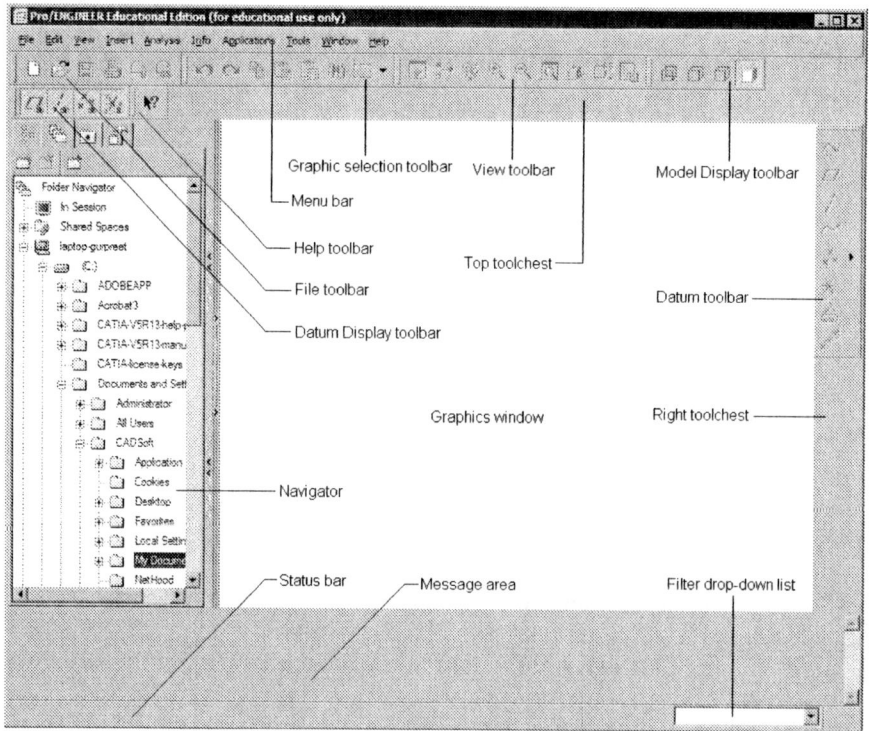

Figure I-25 *Items on the main window*

In Pro/ENGINEER Wildfire 2.0, the toolbars are at two locations: the **Top Toolchest** and the **Right Toolchest**. The area on the top of the graphics window where the toolbars exist is called the **Top Toolchest** and the area on the right of the graphics window where the toolbars exist is called the **Right Toolchest.** The toolbars that initially appear on the screen are highlighted in the figure shown. Here you will notice that all toolbar buttons are not available. These buttons are available after you open a new or existing file. Only those buttons are active that are required for the current session. As you proceed with the procedure to enter one of the modes provided by Pro/ENGINEER Wildfire 2.0, you will notice that the toolbar buttons required by that mode will be active. To make the designing easy and user-friendly, this software package has provided you with a number of toolbars. Different modes of Pro/ENGINEER Wildfire 2.0 display different toolbars. Some of the frequently used toolbars are shown in Figure I-25.

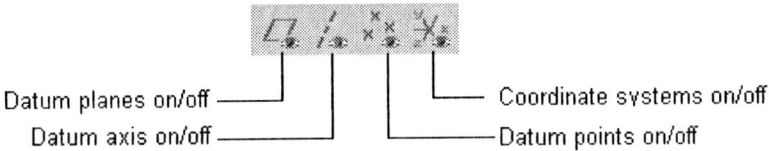

*Figure I-26 The **Datum Display** toolbar*

Tip: *To add buttons to the toolbar and add toolbars to the Toolchest, choose **Tools > Customize Screen** from the menu bar. The **Customize** dialog box is displayed. This dialog box is used to customize the main window of Pro/ENGINEER Wildfire 2.0.*

*Right-click on the **Right Toolchest** to invoke a shortcut menu that has all toolbar names that are available for the current session. Select any toolbar name to display that toolbar in the selected toolchest.*

Datum Display Toolbar

Figure I-26 shows the **Datum Display** toolbar. It has four buttons that are used to control the visibility of datum features like the datum planes, datum axes, and datum points. You can also use this toolbar to control the visibility of the coordinate systems.

File Toolbar

The **File** toolbar has six buttons. The buttons available in this toolbar are used to save a file, print the current file, open a new file, or open an existing file. The description of the buttons provided under this toolbar is shown in Figure I-27.

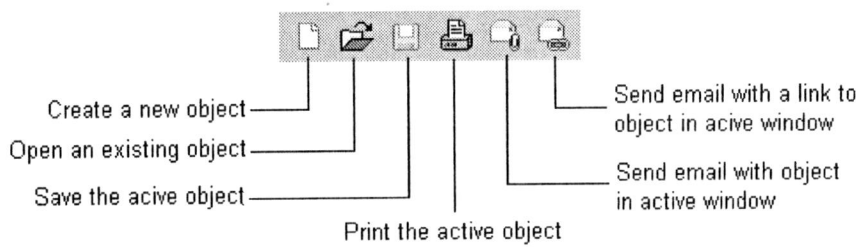

*Figure I-27 The **File** toolbar*

With this release of Pro/ENGINEER, there are two buttons added to the **File** toolbar. The **Send**

email with object in active window button is used to send an email with the current model attached with mail. The attachment may be zipped by using the **Send As Attachment** dialog box which is displayed when you choose this button. The **Send As Attachment** dialog box is shown in Figure I-28.

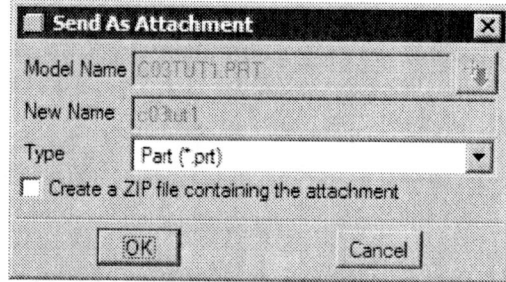

Figure I-28 The Send As Attachment dialog box

The **Send email with a link to object in active window** button is used to send an email with a link to the current model. This button is available to the users of Windchill.

Help Button

The **Help** button is known as the **Context Sensitive help** button, shown in Figure I-29. When you choose this button, the question mark symbol (?) is attached with the cursor. Depending on the location in the main window where you click, the **Pro/ENGINEER Help** window is displayed with the help topics.

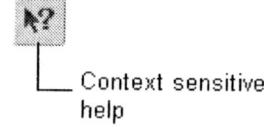

Figure I-29 Help button

Model Display Toolbar

Figure I-30 shows the **Model Display** toolbar. It has four buttons that are used to set the display mode for viewing a model. The **Wireframe** button is used to display the model as a wireframe model. The visible and the invisible edges of the model will be displayed in this display mode. The **Hidden Line** button is used to display the visible lines of the model in bright color and the hidden lines in dull color. However, if you plot the model in this display mode, you will notice that the hidden lines are shown by dashed lines in the plot. The **No Hidden** button displays only the visible edges of the model. The last one is the **Shading** button. In this type of display mode, the model is displayed as shaded using a default color. However, you have the flexibility to alter the color of your model according to your needs.

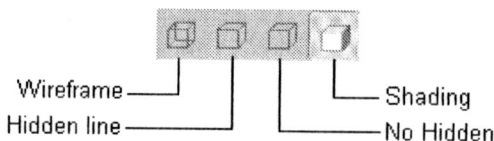

Figure I-30 The Model Display toolbar

Introduction

View Toolbar

The **View** toolbar shown in Figure I-31 has ten buttons. The first button **Redraw the current view** is used for repainting the screen, meaning to remove any temporary information from the graphics window. The second button **View Mode on/off** when chosen turns off the visibility of datums and allows you to spin the model using the middle mouse button. The third button toggles the visibility of the spin center. The fourth and fifth buttons are used to zoom in and zoom out the object. These options are discussed in Chapter 1. The **Refit object to fully display it on the screen** button is used to fit the model on the screen. The **Reorient view** button is used to orient the model. The next button is **Saved view list**. This button is used to display the saved views. The **Set layers, layer items and display states** button displays the layers in the navigation area.

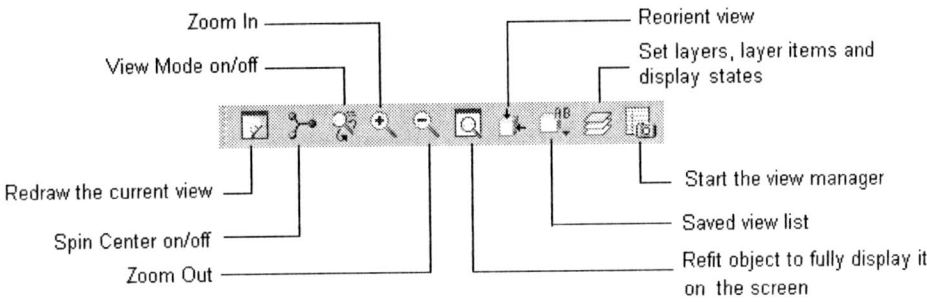

*Figure I-31 The **View** toolbar*

Note
*The default system of units in Pro/ENGINEER Wildfire 2.0 is in Inch Pound Seconds. To change the units choose **Edit > Setup** to display the **PART SETUP** menu in the **Menu Manager**. From this menu choose the **Units** option to invoke the **Units Manager** dialog box. Using this dialog box you can set the system of units available in Pro/ENGINEER Wildfire 2.0.*

NAVIGATOR

The **Navigator** is present on the left of the graphics window and can slide in or out on the graphics window. To make the **Navigator** slide in or out, you need to select the sash present on its right edge. The **Navigator** is shown in Figure I-32. It has the following functions:

1. When you browse files using the **Navigator**, the browser expands and the files in the selected folder are displayed in the browser.

2. When you open a model, the **Model Tree** is displayed in the navigation area.

3. The buttons on the top of the **Navigator** are used to display the different items in the navigation area. The **Model Tree** button is used to display the model tree in the navigation area. This button is available only when a model is opened on the screen. The **Folder Browser** button is used to display the folders that are in the local system of the user. The **Favorites** button is used to display the contents of the **Personal Favorites** folder. The

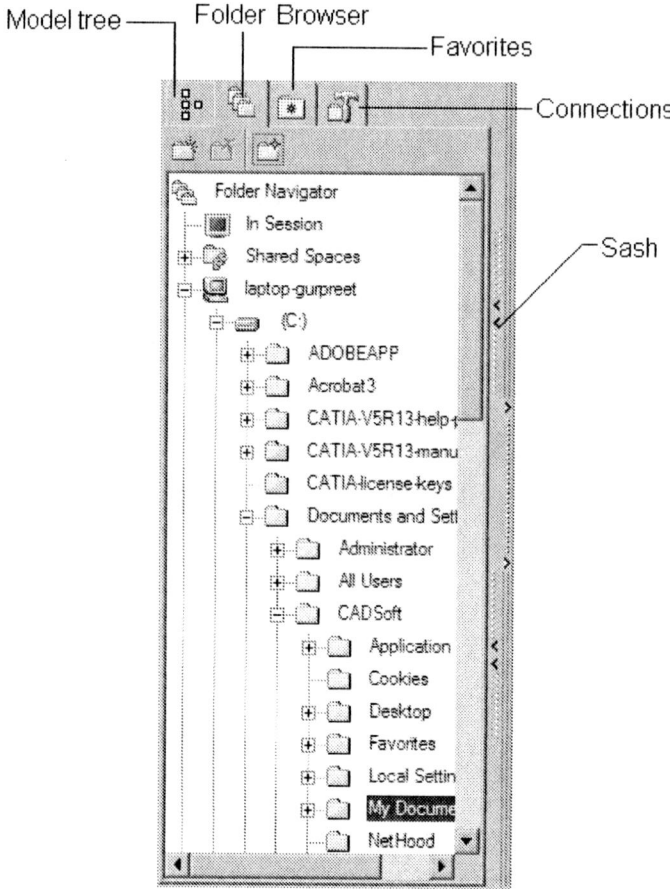

Figure I-32 Navigator

Connections button is used to connect you to the network. This button is used only when you are connected to the internal or the external network.

4. Any other locations if available to your system can also be accessed by using the **Navigator**.

BROWSER

The Browser is present on the right of the **Navigator**. It can slide in and out of the **Navigator** and can also be stretched horizontally. When the Pro/ENGINEER Wildfire 2.0 session is started, the browser is displayed on the screen. Figure I-33 shows the browser that displays some part files. The functions of the browser are listed next.

1. It is used to preview the Pro/ENGINEER Wildfire 2.0 files and browse the file system.

Introduction

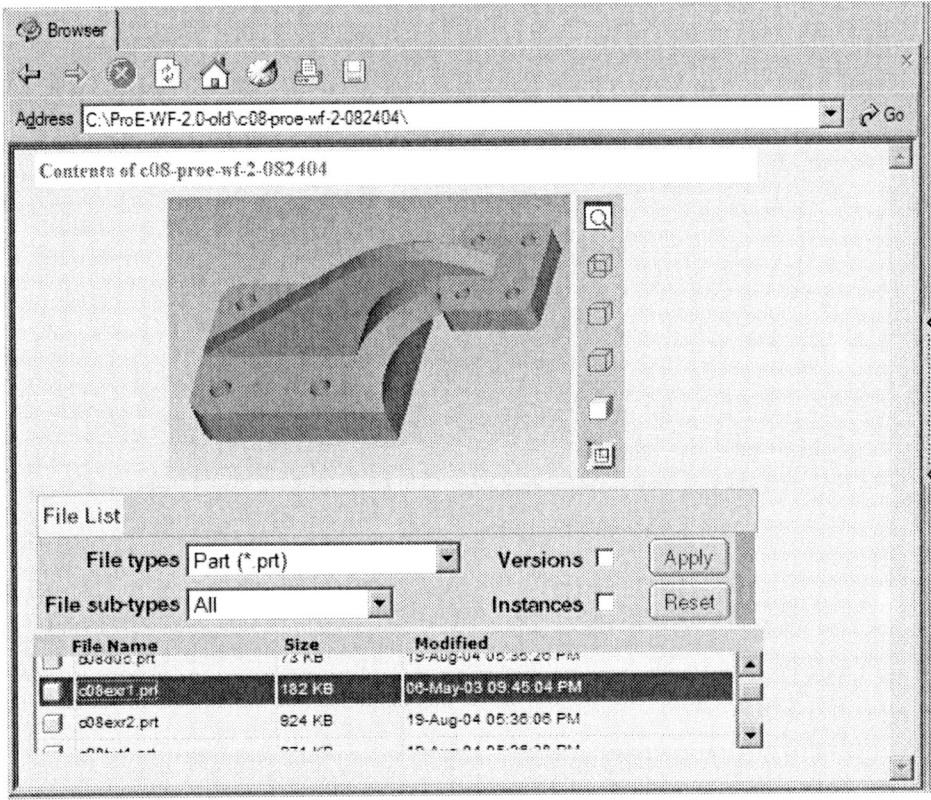

Figure I-33 Pro/ENGINEER Wildfire 2.0 Browser

2. A Pro/ENGINEER file can be opened by using the browser by double-clicking on the file or by dragging it to the graphics window. When you open a file using the browser, the model is displayed on the screen and the browser is closed automatically.

3. By default, the homepage of this browser is set to ptc.com. But you can view other Web sites and access ftp sites too.

4. You can connect to your client's computer and jointly work with them using the Browser.

Chapter 1

Creating Sketches in the Sketch Mode-I

Learning Objectives

After completing this chapter you will be able to:
- *Use various tools to create a geometry.*
- *Dimension a sketch.*
- *Apply constraints to a sketch.*
- *Modify a sketch.*
- *Use Modify Dimensions dialog box.*
- *Edit geometry of a sketch by trimming.*
- *Mirror a sketch.*
- *Use drawing display options.*

THE SKETCH MODE

Almost all models designed in Pro/ENGINEER Wildfire 2.0 consists of datums, sketched features, and placed features. For creating datums and placed features, you do not need to draw sketches. However, to create a three-dimensional (3D) feature, it is necessary to draw its two-dimensional (2D) sketch. When you enter the **Part** mode and select the options to create any sketched feature, the system automatically takes you to the sketcher environment. In the sketcher environment, the sketch of the feature is created, dimensioned, and constrained. The sketches created in the **Sketch** mode are stored in the *.sec* format. Then you return to the **Part** mode to create the required feature.

Note

You will learn about datums and placed features in later chapters.

In Pro/ENGINEER, a sketch can be drawn in the **Sketch** mode or in the sketcher environment. A designer can draw a 2D sketch of the product and assign the required dimensions and constraints to it. By assigning the dimensions, the designer can make sure that the 2D sketch of the product or model is satisfying the necessary conditions. He can then continue to create the 3D model of the product in the **Part** mode.

Using the Sketch Mode

To create any section in the **Sketch** mode of Pro/ENGINEER Wildfire 2.0, certain basic steps have to be followed. The following steps outline the procedure to use the **Sketch** mode:

1. Sketch the required section geometry

The different sketcher tools available in this mode can be used to sketch the required section geometry.

2. Add the constraints and dimensions to the sketched section

While sketching the section geometry, weak constraints and dimensions are automatically added to the section. The sketch can also be dimensioned and constrained manually. After adding the dimensions you can modify them as required.

3. Add relations to the sketch if needed

The geometry of the sketch can be controlled by adding relations.

4. Regenerate the section

If the sketch is fully dimensioned and constrained, the sketch is automatically regenerated. Throughout this book, it is assumed that you are sketching in the **Sketch** mode with the **Intent Manager** on. Pro/ENGINEER has the capability to analyze the section, and if the section is not complete for any reason, the section will not be regenerated. You will learn about these reasons as you go through this chapter.

Tip: *Throughout this book, sketcher environment is referred to the environment in Pro/ENGINEER where you can draw 2D geometries. Apart from the **Sketch** mode, the sketcher environment can be accessed in other modes of Pro/ENGINEER also.*

Invoking the Sketch Mode

To invoke the **Sketch** mode, choose **New** from the **File** menu or choose the **Create a new object** button from the **File** toolbar. The **New** dialog box is displayed with available Pro/ENGINEER modes. Select the **Sketch** radio button to start a new file in the **Sketch** mode, see Figure 1-1. When you select the **Sketch** radio button, a default name of the sketch file appears in the **Name** edit box. You can change the sketch name as required and then choose the **OK** button to enter the **Sketch** mode.

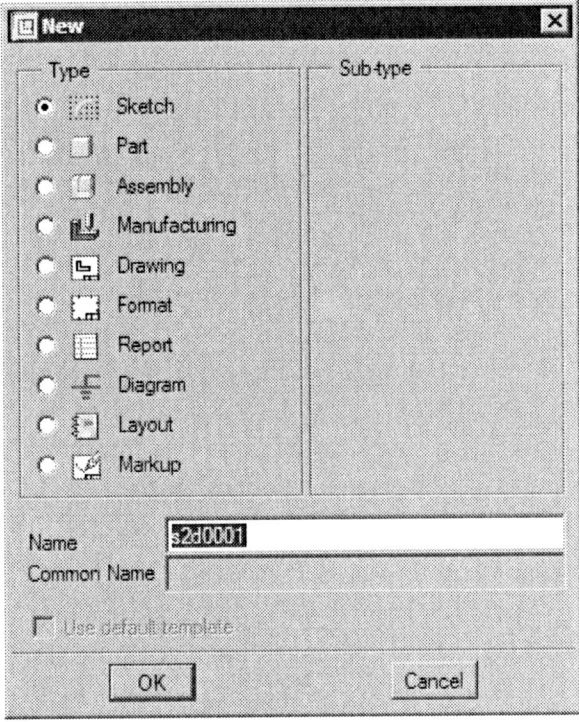

*Figure 1-1 The **New** dialog box*

THE SKETCHER ENVIRONMENT

When you invoke the **Sketch** mode, the initial screen appearance is similar to the one shown in Figure 1-2. This figure also shows the **Sketcher Tools** toolbar that is displayed on the right side of the graphics window. The buttons available in this toolbar are used to draw sketches. The drawing tools are also available in the **Sketch** menu in the menu bar. When you enter the sketcher environment, the **Intent Manager** is on by default. Also, when you are in the selection mode, shortcut menus can be invoked by holding down the right mouse button in the graphics window. The options in these shortcut menus vary depending on the item selected. These shortcut menus also contain the tools to draw sketches.

 Note
The selection mode in the sketcher environment is discussed later in this chapter.

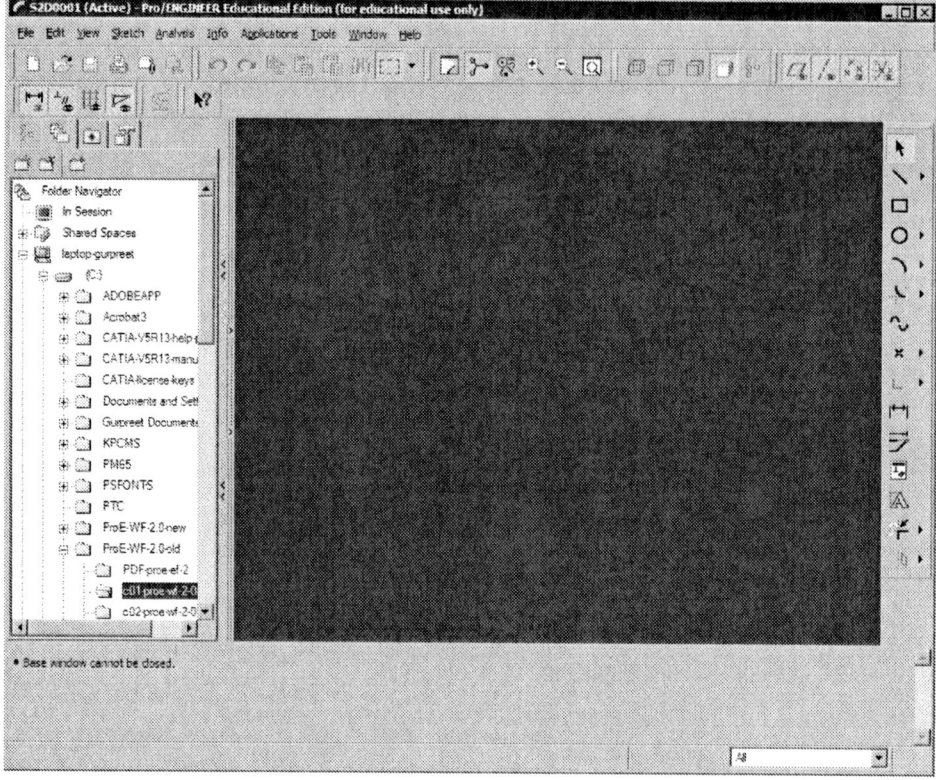

*Figure 1-2 Initial screen appearance in the **Sketch** mode*

The Navigator is displayed on the left side of the graphics window. It covers a part of the graphics window and therefore the drawing area is decreased. You can make the Navigator slide in by clicking the Navigator sash. Now, the area available for sketching is increased.

Note
The functions of Navigator are discussed in the Introduction chapter.

WORKING WITH THE SKETCH IN THE SKETCH MODE

When you invoke the sketcher environment, the **Select items** button is chosen by default. If this button is chosen, the sketcher environment is said to be in the selection mode. In the selection mode, you can select entities of the sketch to edit or to invoke the shortcut menu. The options in the shortcut menu can be used to apply various operations on the selected item.

Note
The sketch is saved with a .sec file extension. While drawing a sketch in the part file, if you save the sketch in the sketcher environment, this sketch is also saved with the .sec file extension.

Creating Sketches in the Sketch Mode-I

Note
You can create a simple sketch by using the options available in the shortcut menu. To invoke the shortcut menu, hold down the right mouse button in the graphics window. Note that, once the shortcut menu is displayed, the right mouse button can be released.

DRAWING A SKETCH USING THE SKETCHER TOOLS TOOLBAR

When you are in the sketcher environment, the **Sketcher Tools** toolbar that is available on the **Right Toolchest**, which contains the tools to draw a sketch, dimension it, and modify the dimensions. In this section, you will learn how to draw sketched entities using the tools available in the **Sketcher Tools** toolbar.

Placing a Point

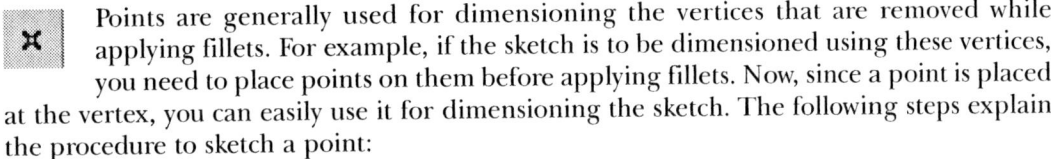

Points are generally used for dimensioning the vertices that are removed while applying fillets. For example, if the sketch is to be dimensioned using these vertices, you need to place points on them before applying fillets. Now, since a point is placed at the vertex, you can easily use it for dimensioning the sketch. The following steps explain the procedure to sketch a point:

1. Choose the **Create points** button from the **Sketcher Tools** toolbar; the system prompts you to select a location for the point.

2. As soon as you select the location of the point by clicking, the point is placed at the desired location in the graphics window.

Note
*To increase the number of visible command prompt lines in the **Message Area**, select the top sash of the **Message Area** using the left mouse button and drag it upwards, toward the screen.*

When you draw a single point no dimensions appear. But, when you draw two points, they are dimensioned with each other.

Drawing a Line

To draw lines, there are three tool buttons available in the **Sketcher Tools** toolbar. To view these buttons, choose the black arrow on the right of the **Create 2 point lines** button; the flyout appears with three tool buttons. The first button is **Create 2 point lines**. This button is used to create a line by selecting two points in the graphics window. The second button on the flyout is **Create lines tangent to 2 entities**. This button is used to create a tangent between two entities. The third button is **Create 2 point centerlines**. This button is used to create a centerline by selecting two points in the graphics window. The centerline is used for creating revolved features, mirroring, and so on.

The procedure to create lines using these tools are discussed next.

Drawing a Line Using the Create 2 point lines Tool

The following steps explain the procedure to create a line using the **Create 2 point lines** tool:

1. Choose the **Create 2 point lines** button. Click on graphics window to start the line; a rubber-band line appears from the selected point with the other end attached to the cursor. The symbols **V** and **H** that appear while drawing the vertical and horizontal lines are called constraints. Constraints are discussed in detail later in this chapter.

2. The system prompts you to specify the endpoint. Move the cursor in the graphics window to a desired location and click to specify the endpoint of the line; the line appears in yellow color. The rubber-band line continues and you can draw the second line.

3. Repeat step 2 until all lines are drawn. You can end the line creation by pressing the middle mouse button. To abort line creation, use the middle mouse button.

Note

When you draw a single line, the color of the line after you have drawn it is red. If you draw multiple lines, the color is yellow.

After drawing a line, when you press the middle mouse button to end the line creation, the line drawn is highlighted in red color. In the sketcher environment, the red color of an entity indicates that it is selected. If you press the DELETE key, the line will be erased from the graphics window. After drawing a line, weak dimensions in gray color are applied to the sketch. The weak dimensions are applied automatically to the sketched entities as you draw them.

Drawing a Line Using the Create lines tangent to 2 entities Tool

The **Create lines tangent to 2 entities** button is used to draw a tangent between two entities such as arcs, circles, splines, or a combination of these. The following steps explain the procedure to draw a tangent using this tool:

1. Choose the arrow on the right of the **Create 2 point lines** button and then choose the **Create lines tangent to 2 entities** button from the **Sketcher Tools** toolbar. You are prompted to select start location on arc or circle.

2. Select the first entity from where the tangent line will be drawn; the color of the selected entity changes to red. Now, you are prompted to select the end location on arc or circle. As soon as you select the second entity, a line that is tangent to both the selected entities is drawn.

Note

*Whenever you are prompted to select an entity in the sketcher environment, the **SELECT** dialog box is displayed. You can ignore this dialog box because it appears automatically and disappears without any confirmation.*

Creating Sketches in the Sketch Mode-I

Drawing a Centerline Using the Create 2 point centerlines Tool

You can draw horizontal, vertical, or inclined centerlines using the **Create 2 point centerlines** button. This button is available in the flyout that is displayed when you choose the black arrow on the right of the **Create 2 point lines** button. The centerline in a sketch is used as an axis of rotation, for mirroring, aligning, and dimensioning entities.

The following steps explain the procedure to draw a centerline:

1. In the **Sketcher Tools** toolbar, choose the black arrow on the right of the **Create 2 point lines** button and then choose the **Create 2 point centerlines** button; you are prompted to select the start point.

2. Click in the graphics window to specify the start point. Now, you are prompted to select the end point.

3. Click in the graphics window to specify the endpoint. A centerline of infinite length is drawn.

Drawing a Rectangle

The following steps explain the procedure to sketch a rectangle using the **Create rectangle** tool:

1. Choose the **Create rectangle** button from the **Sketcher Tools** toolbar; you are prompted to select two points to indicate the diagonal of box. Click to specify the first point; a yellow rubber-band box appears with the cursor attached to the opposite corner of the box.

2. Move the cursor to the desired location in the graphics window to size the diagonal of the rectangle. Click to specify the second point for the diagonal of the rectangle.

Drawing a Circle

In the **Sketcher Tools** toolbar, there are four tools to draw a circle and one tool to draw an ellipse. To view the tools available to draw circles and ellipses, choose the black arrow on the right of the **Create circle by picking the center and a point on the circle** button. The flyout appears with five buttons. The procedures to create circle and ellipse using the various tools are discussed next.

Drawing a Circle Using the Create circle by picking the center and a point on the circle Tool

As the name suggests, the **Create circle by picking the center and a point on the circle** tool is used to draw a circle by specifying the center of the circle and a point on it. The following steps explain the procedure to draw a circle using this tool.

1. Choose the **Create circle by picking the center and a point on the circle** button; you are prompted to select the center of a circle.

2. Click in the graphics window to specify the center point of the circle; you are prompted to select a point on the circle. A yellow rubber-band circle appears with the center at the specified point and the cursor attached to its circumference.

3. Move the cursor to size the circle. Click to complete the circle creation; you are again prompted to select the center of the circle.

4. Repeat steps 2 and 3 until you have drawn all required circles. If you want to abort circle creation, press the middle mouse button.

Drawing a Construction Circle

A construction circle is a circle that is used to align entities, create diametrical or radial dimensioning, and to reference entities. Figure 1-3 shows an application of construction circle. In the sketch of a flange, centers of the circles lie on a particular bolt circle diameter (BCD) that is defined using a construction circle.

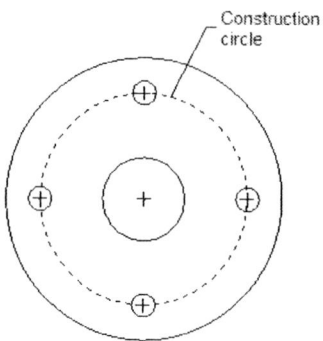

Figure 1-3 Sketch of a flange

To create a construction circle, draw a circle or select a previously drawn circle. Then, hold down the right mouse button on it to invoke the shortcut menu, as shown in Figure 1-4. Choose the **Construction** option from the shortcut menu. The circle appears yellow in color and with a dashed line style, indicating that it is a construction circle.

Drawing a Circle Using the Create concentric circle Tool

 The following steps explain the procedure to draw a concentric circle using the **Create concentric circle** tool:

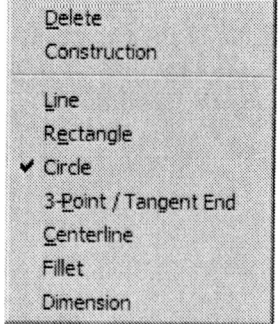

***Figure 1-4** Construction option in the shortcut menu*

Creating Sketches in the Sketch Mode-I   1-9

Tip: *To convert a construction circle back to a solid entity, select the construction circle and hold down the right mouse button to invoke the shortcut menu. Choose the **Solid** option from this shortcut menu.*

1. Choose the **Create concentric circle** button from the flyout in the **Sketcher Tools** toolbar; you are prompted to select an arc to determine the center. You can select an arc or a circle to specify the center point.

2. Click on an arc or a circle to determine the concentricity of the circle to be drawn. Move the mouse to size the circle.

3. After sizing the circle, finish circle creation by clicking and then pressing the middle mouse button to end circle creation.

Drawing a Circle Using the Create circle by picking its 3 points Tool

The following steps explain the procedure to draw a circle using the **Create circle by picking its 3 points** tool:

1. Choose the **Create circle by picking its 3 points** button from the flyout in the **Sketcher Tools** toolbar; you are prompted to specify the first point on the circle.

2. Click for the first point at the desired location in the graphics window; you are prompted to select the second point on the circle. Move the cursor and click to select the second point in the graphics window.

3. As soon as you select the second point, a yellow rubber-band circle appears with the cursor attached to it. You are prompted to select the third point. Move the mouse to size the circle and click to specify the third point; a circle is drawn. You will be again prompted to select first point on the circle to draw the next circle.

4. You can draw another circle or press the middle mouse button to end circle creation or to abort circle creation before the next circle is completed.

Drawing a Circle Using the Create a circle tangent to 3 entities Tool

The **Create a circle tangent to 3 entities** tool is used to draw a circle that is tangent to three existing entities. This tool references other entities to draw a circle. The circle created using this tool is drawn irrespective of the points selected on the entities. The following steps explain the procedure to draw a circle using the **Create a circle tangent to 3 entities** tool:

1. Choose the **Create a circle tangent to 3 entities** button from the flyout in the **Sketcher Tools** toolbar; you are prompted to select the start location on an arc, circle, or line.

2. Click to select the first entity. The color of the entity changes to red and you are prompted to select an end location on an arc, circle, or line. Select the second tangent entity. Next,

you are prompted to select the third location on an arc, circle, or line. Select the third tangent entity.

As you select the three entities, a circle that is tangent to these three entities is drawn.

3. To end the creation of circle, press the middle mouse button.

Drawing an Ellipse Using the Create a full ellipse Tool

The following steps explain the procedure to draw an ellipse using the **Create a full ellipse** tool:

1. Choose the **Create a full ellipse** button from the flyout in the **Sketcher Tools** toolbar; you are prompted to specify the center of an ellipse.

2. Click at the desired location in the graphics window to specify the center point; a yellow rubber-band ellipse appears with the cursor attached to the ellipse. Move the cursor in the graphics window to size the ellipse.

3. An ellipse is drawn when you click to specify the second point.

Drawing an Arc

To draw an arc there are five tools in the **Sketcher Tools** toolbar. To view them, choose the black arrow on the right of the **Create an arc by 3 points or tangent to an entity at its endpoint** button; the flyout appears with five buttons. The procedures to draw arcs using various tool buttons in the flyout are discussed next.

Drawing an Arc Using the Create an arc by 3 points or tangent to an entity at its endpoint Tool

The **Create an arc by 3 points or tangent to an entity at its endpoint** tool is used to draw arcs tangent from the endpoint of an existing entity or by defining three points in the graphics window.

When you choose this button to draw an arc from an endpoint, the **Target** symbol is displayed as soon as you select the endpoint. This **Target** symbol is in the form of a circle that is divided into four quadrants. The following steps explain the procedure to draw an arc from the endpoint of an existing entity by using this tool:

1. Choose the **Create an arc by 3 points or tangent to an entity at its endpoint** button from the **Sketcher Tools** toolbar; you are prompted to select the start point of the arc.

2. Specify three points in the graphics window to draw an arc. If you want to draw an arc from the endpoint of an existing entity, select the endpoint of that entity. As soon as you select the endpoint, the **Target** symbol appears at the endpoint of the entity. Move the cursor along the tangent direction through a small distance, a yellow rubber-band arc appears with one end attached to the endpoint of the entity and the other end attached to the cursor. Note that when you move the cursor out of the **Target** symbol perpendicular

Creating Sketches in the Sketch Mode-I   1-11

to the endpoint, the arc is drawn by specifying three points. In this case the rubber-band arc does not appear, as shown in Figure 1-5.

On the other hand, if you move the cursor out horizontally from one of the quadrants of the **Target** symbol, the arc is drawn tangent to the endpoint, as shown in Figure 1-6.

3. Move the cursor to the desired position in the graphics window to size the arc. Use the left mouse button to complete the arc.

Tip: *If you do not want to draw a tangent arc, move the cursor out of the **Target** symbol perpendicular to the endpoint.*

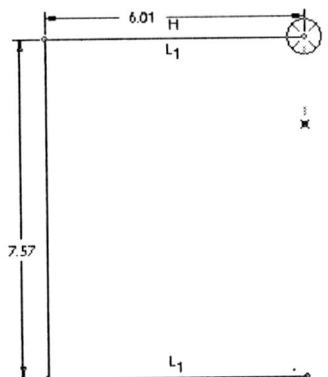
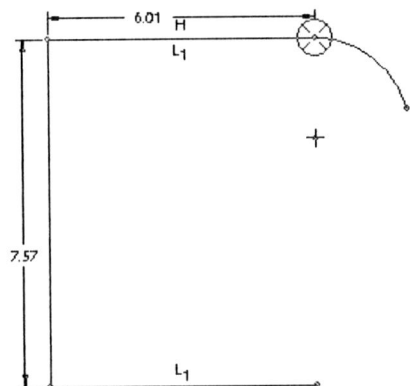

*Figure 1-5 Cursor moved out of **Target** symbol perpendicular to the endpoint*

*Figure 1-6 Cursor moved out of the **Target** symbol along the tangent direction.*

Drawing an Arc Using the Create concentric arc Tool

The **Create concentric arc** tool is used to draw an arc concentric to an existing arc. You will have to select an entity to which the arc will be concentric. The entity selected must be an arc or a circle. The following steps explain the procedure to draw an arc using this tool:

1. Choose the **Create concentric arc** button from the flyout in the **Sketcher Tools** toolbar; you are prompted to select an arc to determine the center of the arc to be created.

2. As soon as you select an entity, a dotted circle appears on the screen and you are prompted to select the start point of the arc. After doing so, a yellow rubber-band arc will appear with one end attached to the start point. The length of the arc will change as you move the cursor. Next, you are prompted to select the endpoint of the arc.

3. Click to specify the endpoint; the arc is created.

4. You can continue drawing another arc or end arc creation by pressing the middle mouse button.

Drawing an Arc Using the Create an arc by picking its center and endpoints Tool

 The following steps explain the procedure to draw an arc using the **Create an arc by picking its center and endpoints** tool:

1. Choose the **Create an arc by picking its center and endpoints** button from the flyout in the **Sketcher Tools** toolbar; you are prompted to select the center of the arc.

2. Click to specify a center point for the arc in the graphics window; a yellow colored center mark appears at that point. Now, you are prompted to select the start point of the arc. As you move the cursor, a dotted circle appears and is attached to the cursor.

3. Select the start point of the arc on the circumference of the dotted circle; a yellow rubber-band arc appears from the start point. The length of this arc changes dynamically as you move the cursor.

4. You are prompted to select the endpoint of the arc. Move the cursor to size the arc, and then click to select the endpoint of the arc. An arc is drawn between the two selected points.

Note
You can draw only one arc with one center. If you want to draw another arc you will have to select the center again.

Drawing an Arc Using the Create an arc tangent to 3 entities Tool

 The **Create an arc tangent to 3 entities** tool is used to draw an arc that is tangent to three selected entities. The following steps explain the procedure to draw an arc using this tool:

1. Choose the **Create an arc tangent to 3 entities** button from the flyout in the **Sketcher Tools** toolbar; you are prompted to select the start location on an arc, circle, or line.

2. As soon as you select the first entity, the color of the entity changes to red. Now, you are prompted to select the end location on an arc, circle, or line.

3. After doing so, you are prompted to select a third location on an arc, circle, or line. Select a third entity. An arc is drawn tangent to the three entities selected.

4. You can continue drawing arcs or press the middle mouse button to abort arc creation.

Drawing an Arc Using the Create a conic arc Tool

 The **Create a conic arc** tool is used to draw a conic arc. The following steps explain the procedure to draw a conic arc using this tool:

1. Choose the **Create a conic arc** button from the flyout in the **Sketcher Tools** toolbar; you are prompted to specify the first endpoint of the conic entity.

Creating Sketches in the Sketch Mode-I

2. Specify a point in the graphics window; you are prompted to specify the second endpoint of the conic entity.

3. Specify the second endpoint; a centerline is drawn between the two points. Now, you are prompted to specify the shoulder point of the conic. Specify a point on the screen; the conic arc is drawn.

Note
If you delete the centerline of the conic arc, the arc will not be deleted.

Remember that if the conic arc is the only entity in the graphics window, then you cannot delete its centerline.

DIMENSIONING THE SKETCH

After you draw a sketch, the next step involves the dimensioning of the sketch. The basic purpose of dimensioning in Pro/ENGINEER is to control the size of the sketch and to locate it with some reference. In Pro/ENGINEER, a sketch cannot be regenerated unless it is fully dimensioned and constrained. The phrase "the sketch cannot be regenerated" means that the sketch cannot be accepted by Pro/ENGINEER.

By default, the sketched entities are dimensioned and constrained automatically while sketching or as soon as you are done with the sketch. However, sometimes you need to add additional dimensions to the sketch. The **Create defining dimension** button in the **Sketcher Tools** toolbar is used to manually dimension the entities. You can also choose **Sketch > Dimension** from the menu bar to display a cascading menu, as shown in Figure 1-7. This menu contains different dimensioning options.

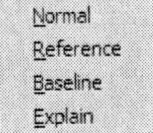
Figure 1-7 Dimensioning options

Note
*When the intent manager is off, the dimensions and constraints are not automatically applied. You need to manually add the dimensions by choosing the **Dimension** option form the **SKECHER** menu.*

Dimensioning a Sketch Using the Create defining dimension button or the Normal Option

The **Create defining dimension** button or the **Normal** option are used for normal dimensioning of the sketch. The following steps explain the procedure to dimension a sketch:

1. Choose the **Create defining dimension** button from the **Sketcher Tools** toolbar. Click on the entity you want to dimension; the color of the entity changes from yellow to red.

2. Move the cursor and place the dimension at the desired place by pressing the middle mouse button. You can modify the dimension values using the modifying options discussed later in this chapter.

The remaining options in the cascading menu are not used while sketching and therefore, they are discussed in Chapter 2.

DIMENSIONING THE BASIC SKETCHED ENTITIES

Choose the **Create defining dimension** button and follow the procedures given below to dimension the sketched entities.

Linear Dimensioning a Line

You can dimension a line by selecting its endpoints or by selecting the line. After selecting the two endpoints or the line, press the middle mouse button to place the dimension. If the line is inclined and you select the two endpoints to dimension, then the location where you press the middle mouse button is important, because it defines the orientation of the dimension that is displayed on the screen.

Figure 1-8 explains the three possible orientations of dimension that can be displayed when you dimension a line.

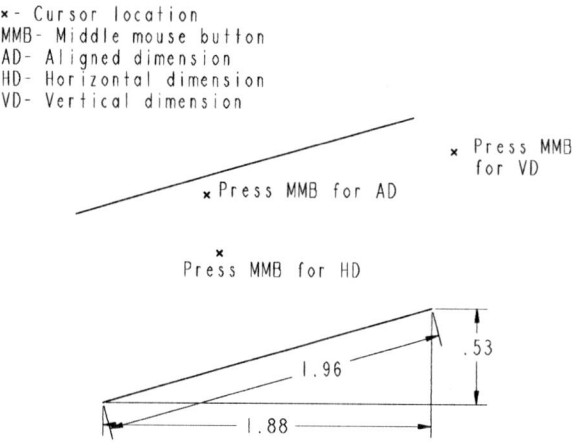

Figure 1-8 Approximate locations of the cursor to achieve different dimensions

Note
It is not possible to dimension a line in the three orientations at the same time in the sketcher environment. The dimensions in Figure 1-8 are only for explanation.

Angular Dimensioning of an Arc

To add angular dimension to an arc, select both ends of the arc by pressing the left mouse button and then select a point on the arc. Next, place the dimension at the desired point by pressing the middle mouse button. The dimension appears, as shown in Figure 1-9. You can modify the dimension using tools that are discussed later.

Diameter Dimensioning

For diameter dimensioning, click on a circle twice. Then place the dimension at the desired location by pressing the middle mouse button. The diameter dimension is displayed, as shown in Figure 1-10. The same diameter dimensioning technique can also be used for arcs.

Creating Sketches in the Sketch Mode-I

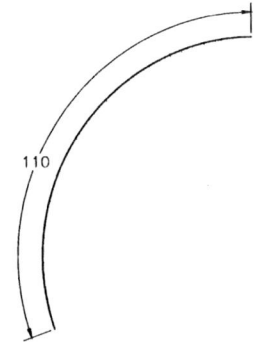

Figure 1-9 Angular dimensioning

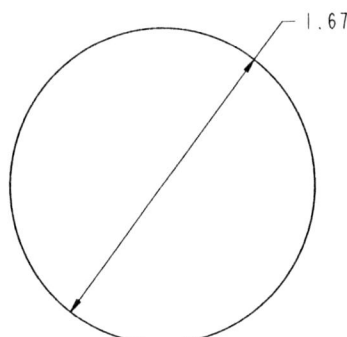

Figure 1-10 Diameter dimensioning

Radial Dimensioning

For radial dimensioning, click on the entity once. Then place the dimension at the desired location by pressing the middle mouse button. The radial dimension is displayed, as shown in Figure 1-11.

Dimensioning Revolved Sections

Revolved sections are used to create revolved features such as flanges, couplings, and so on. To dimension a revolved section, click on the entity to be dimensioned. Next, select the centerline about which you want the section to be revolved. Once again select the original entity that you want to dimension. Now, place the dimension at the desired location by pressing the middle mouse button. The dimension is displayed, as shown in Figure 1-12. This dimension represents the diameter of a revolved section.

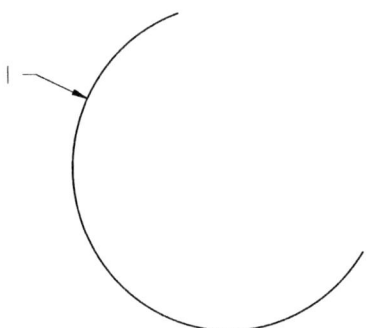

Figure 1-11 Radial dimensioning

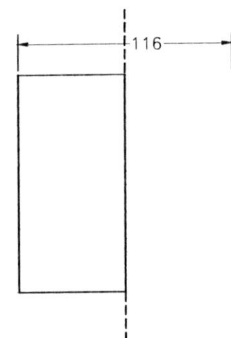

Figure 1-12 Dimensioning for revolved sections

Tip: *To add dimension to a revolved section, you can also first select the centerline, next the entity to dimension, and then again the centerline.*

WORKING WITH CONSTRAINTS

In Pro/ENGINEER, the entities in a sketch have to be fully specified in terms of size, shape, orientation, and location. This is achieved by setting constraints. Using constraints in the sketch reduces the number of dimensions in that sketch.

Constraints are the logical operations that are performed on the selected geometry to make it more accurate in defining its position with respect to the other geometry. For example, if a line is nearly parallel to another line, Pro/ENGINEER snaps the parallel line and displays the parallel constraint symbol. Now, if you confirm the line creation, the line is drawn parallel to the other line. You can also apply constraints manually.

There are two types of constraints in Pro/ENGINEER, **Geometry** constraints and **Assembly** constraints. Here, you will learn about the **Geometry** constraints and the **Assembly** constraints will be discussed in later chapters.

To apply constraints manually, choose the **Impose sketcher constraints on the section** button from the **Sketcher Tools** toolbar to display the **Constraints** dialog box. This dialog box is shown in Figure 1-13.

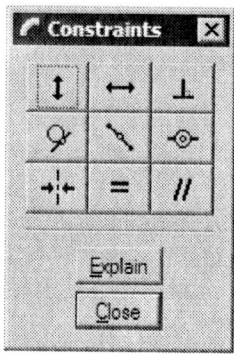

Figure 1-13 The Constraints dialog box

This dialog box is used to apply constraints manually. Although the constraints are applied automatically as you draw the sketch, you can use this dialog box if you want to manually apply additional constraints. The constraints that are applied automatically are weak constraints and they appear in gray color. Weak constraints can be made strong and this is discussed later in this chapter. The constraints in the **Constraints** dialog box are discussed next.

Make a line or two vertices vertical

This constraint forces the selected line segment to become a vertical line. This constraint also forces the two vertices to be placed along a vertical line.

Make a line or two vertices horizontal

This constraint forces the selected line segment or two vertices that are apart by some distance to become horizontal or to lie in a horizontal line.

Make two entities perpendicular

This constraint forces the selected entity to become normal to another selected entity.

Make two entities tangent

This constraint forces the two selected entities to become tangent to each other.

Place point on the middle of the line

This constraint forces a selected point or vertex to lie on the middle of a line.

Creating Sketches in the Sketch Mode-I

Create same points, points on entity or collinear constraint

This constraint performs three functions. This constraint can be used to force the two selected points to become coincident to constrain a point on the selected entity, and to make two selected entities collinear, so that they lie on the same line. This constraint aligns two vertices or entities.

Make two points or vertices symmetric about a centerline

This constraint makes a section symmetrical about the centerline. When you select this constraint, you are prompted to select a centerline and two vertices to make them symmetrical.

Create Equal Lengths, Equal Radii, or Same Curvature constraint

This constraint forces any two selected entities to become equal in dimension. When you select this constraint, you are prompted to select two lines to make their lengths equal, or you are prompted to select two arcs, circles, or ellipses to make their radii equal.

Make two lines parallel

This constraint is used to force two lines to become parallel. When selected, this constraint prompts you to select two entities that you want to make parallel.

Explain Option

The **Explain** option of the **Constraints** dialog box provides information about the constraints that are applied to a sketch. The constraints in the sketch are displayed as symbols. When you choose the **Explain** button, you are prompted to select the constraint or dimension on which you want the explanation. Select the symbol using the left mouse button. The information about the selected constraint is displayed in the **Message Area**.

Note
This option is generally helpful when you view a sketch drawn by some other person. By using the Explain option you can obtain information about the various constraints applied in the sketch.

Disabling the Constraints

The need to disable a constraint arises when you are drawing an entity. For example, if you draw a circle at some distance apart from a circle. While drawing it, the system tends to apply the equal radius constraint when the sizes of the two circles become equal. If at this moment you do not want to apply the equal radius constraint, right-click to disable the equal radius constraint. When you right-click to disable a constraint, an orange line / appears across the symbol. To enable the constraint, right-click once again.

Converting a Weak Constraint into a Strong Constraint

As discussed earlier, when you draw a sketch, some weak dimensions are automatically applied to the sketch. As you proceed to complete the sketch, these dimensions are automatically deleted from the sketch without any confirmation.

Select a weak dimension or a weak constraint from the graphics window. The selected dimension or constraint is highlighted in red. Press and hold the right mouse button to invoke the shortcut menu, as shown in Figure 1-14. Choose the **Strong** option from the shortcut menu.

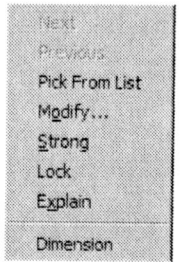

Figure 1-14 Shortcut menu to convert the weak dimensions to strong

You can also choose **Edit > Convert To > Strong** from the menu bar. The color of the selected dimension is changed from gray to yellow, indicating that the selected constraint or dimension is made permanent.

MODIFYING THE DIMENSIONS OF A SKETCH

There are four ways to modify the dimensions of a sketch. These methods are discussed next.

Using the Modify the values of dimensions, geometry of splines, or text entities Tool

You can select one or more dimensions from the sketch to modify. When you select dimension(s) from a sketch, they are highlighted in red. If you want to select more than one dimension, hold down the CTRL key and select the dimensions by clicking on them. You can also use CTRL+ALT+A or define a window to select the dimensions in the sketch. Choose the **Modify the values of dimensions, geometry of splines, or text entities** button from the **Sketcher Tools** toolbar to modify the dimensions; the **Modify Dimensions** dialog box is displayed, as shown in Figure 1-15.

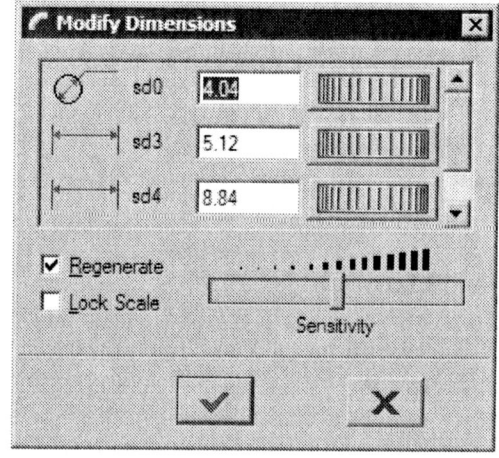

Figure 1-15 The **Modify Dimensions** dialog box

To modify dimensions using this dialog box, you can either enter a value in the edit box or use

Creating Sketches in the Sketch Mode-I

the thumbwheel that is available on the right of the edit box. The **Sensitivity** slider is used to set the sensitivity of the thumbwheel.

By default, the **Regenerate** check box is selected and any modifications in the dimensions are automatically updated in the sketch. If you want to delay the modification process of the sketch based on the new value of the selected dimension, you need to clear this check box. If this check box is cleared, the dimensions will not be modified until you exit this dialog box. This means that Pro/ENGINEER allows you to make multiple modifications before updating the sketch.

Note
*It is recommended that you clear the **Regenerate** check box and then modify the dimensions if you have to modify more than one dimension.*

The **Lock Scale** check box is used to lock the scale of the selected dimensions. After locking the scale, if you modify any dimension, all other dimensions will also be modified by the same scale.

Using the Edit Menu
The **Modify** option available in the **Edit** menu in the menu bar can also be used to modify the dimensions. When you choose the **Modify** option from the **Edit** menu, a check mark appears to the left of the **Modify** option in the **Edit** menu. Now, you can select a dimension from the sketch to modify. When you select a dimension, the **Modify Dimensions** dialog box is displayed. By default, the **Regenerate** check box is selected. Therefore, the sketch will be regenerated dynamically as you modify the dimension.

Modifying a Dimension by Double-Clicking
You can also modify a dimension by double-clicking on it. When you double-click on a dimension, the pop-up text field appears. Enter a new dimension value in this field and press ENTER or use the middle mouse button. Remember that you can select a dimension only when you choose **Select items** button from the **Sketch Tools** toolbar.

Modifying Dimensions Dynamically
In the sketcher environment, Pro/ENGINEER is always in the selection mode, unless you have invoked some other tool. When you bring the cursor to an entity, the color of the entity changes to cyan. Now, if you hold down the left mouse button, you can modify the entity by dragging the mouse. You will notice that as the entity is modified, the dimensions referenced to the selected entity are also modified.

RESOLVE SKETCH DIALOG BOX
While applying constraints or dimensions, the system may sometimes prompt you to delete one or more highlighted dimensions or constraints. This is because while adding dimensions or constraints some strong dimensions or constraints conflict with the existing dimensions or constraints. As soon as the conflict occur, the **Resolve Sketch** dialog box is displayed, as shown in Figure 1-16 and the constraints or dimensions under conflict are displayed in red. When you select a dimension or constraint from the **Resolve Sketch** dialog box, the corresponding

dimension or constraint in the graphics window is enclosed in a yellow box. The buttons available in the **Resolve Sketch** dialog box are discussed next.

Undo

When you choose the **Undo** button, the section is brought back to the state that was just before the conflict occurred.

Delete

The **Delete** button is used to delete a selected dimension or constraint that is enclosed within the yellow box. Select the dimension or the constraint to delete before you choose the **Delete** button from the **Resolve Sketch** dialog box.

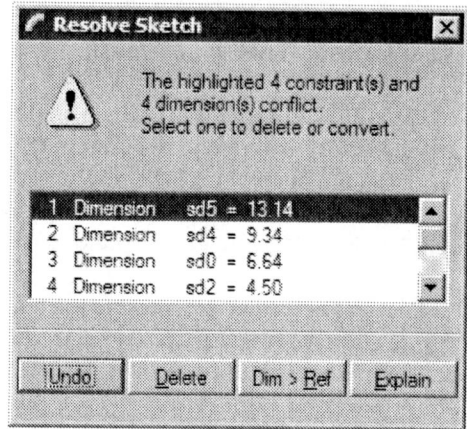

*Figure 1-16 The **Resolve Sketch** dialog box*

Dim > Ref

When you choose the **Dim > Ref** button, the selected dimension is converted to a reference dimension.

Note

The reference dimensions are used only for reference. They do not participate in feature creation.

Explain

When you choose the **Explain** button, the system provides you with information about the selected constraint or dimension. The information is displayed in the **Message Area**.

DELETING THE SKETCHER ENTITIES

To delete a sketched entity, select it by defining a window. You can specify a window by picking two points so that the entity or entities are enclosed in the window. After specifying the window, the color of the selected entity changes to red. Right-click in the graphics window and hold down the right mouse button until a shortcut menu appears. Now, choose the **Delete** option from this menu to delete the selected item.

You can also delete an item by selecting it and pressing the DELETE key when the selected item turns red in color.

To delete more than one item from the graphics window, press the CTRL key and click to select the entities to be deleted. Press the DELETE key to delete the selected entities. You can also specify a window to select the entities.

Creating Sketches in the Sketch Mode-I

Note
It is necessary to be in the selection mode while selecting the items. The term "items" used in this chapter refers to dimensions and entities.

*The **Geometry**, **Dimension**, and **Constraint** filters are available in the drop-down list located in the status bar. These filters allow you to select exactly the item that you need to select. This means, if you want to select all constraints in the sketch, choose the **Constraint** filter and specify a window to select. You will notice that only the constraints are selected.*

To restore the last deleted item, choose the **Undo** button. This button is available in the **Edit** toolbar on the **Top Toolchest**.

TRIMMING THE SKETCHER ENTITIES

When creating a design, there are a number of places where you need to remove the unwanted and extended entities. You can do this by using the tools available for trimming that are available in the **Sketcher Tools** toolbar. You can trim entities using three tools. These tools are discussed next.

Dynamically trim section entities Tool

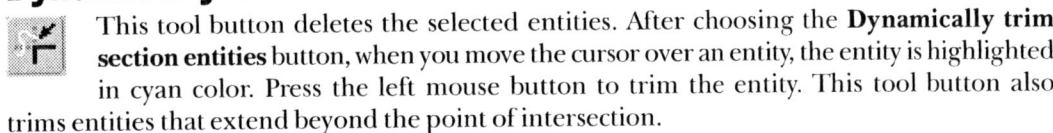
This tool button deletes the selected entities. After choosing the **Dynamically trim section entities** button, when you move the cursor over an entity, the entity is highlighted in cyan color. Press the left mouse button to trim the entity. This tool button also trims entities that extend beyond the point of intersection.

Trim entities (cut or extend) to other entities or geometry Tool

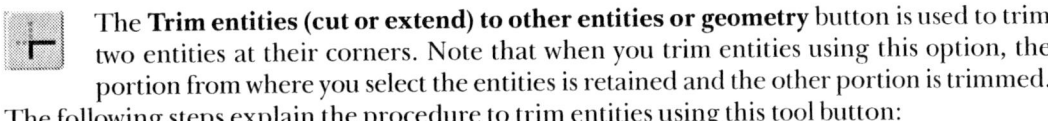
The **Trim entities (cut or extend) to other entities or geometry** button is used to trim two entities at their corners. Note that when you trim entities using this option, the portion from where you select the entities is retained and the other portion is trimmed. The following steps explain the procedure to trim entities using this tool button:

1. Choose the black arrow on the right of the **Dynamically trim section entities** button to display the flyout. From this flyout, choose the **Trim entities (cut or extend) to other entities or geometry** button; you are prompted to select two entities to be trimmed.

2. Click to select the two entities on the sides you want to keep after trimming, see Figure 1-17. These two entities must be intersecting entities. The entities are trimmed from the point of intersection.

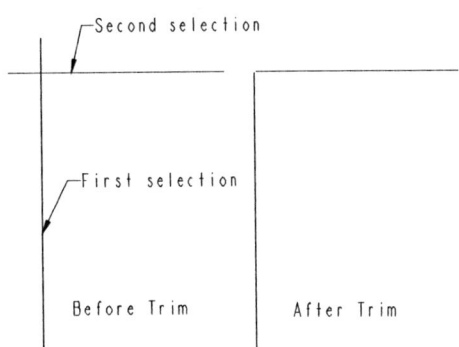

***Figure 1-17** Trimming the lines*

Divide an entity at the point of selection Tool

 The **Divide an entity at the point of selection** button is used to divide an entity into any number of parts or entities by specifying points on the entity.

This button is available on the flyout that is displayed when you choose the black arrow that is on the right side of the **Dynamically trim section entities** button.

The following steps explain the procedure to divide an entity:

1. Choose **Divide an entity at the point of selection** button from the flyout; you are prompted to select an entity to be divided.

2. Click to select the entity at the point where you want to divide it. The entity is divided into two different entities. They can now be treated as two separate entities.

3. Similarly, you can break other entities like circles or arcs into several smaller entities.

MIRRORING THE SKETCHER ENTITIES

 The **Mirror selected entities** button is used to mirror sketched geometries about a centerline. This tool helps to reduce the time used for creation of symmetrical geometries and dimensioning them.

The following steps explain the procedure to mirror a sketched geometry:

1. Sketch a geometry and then sketch a centerline about which you need to mirror the geometry.

2. Select the entities that you need to mirror. The selected entities turn red in color.

3. Choose the **Mirror selected entities** button from the **Sketcher Tools** toolbar. You are prompted to select the centerline about which you need to mirror, hence do so. The selected entities are mirrored about the centerline.

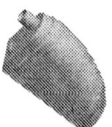 *Tip: In case of symmetrical parts, you can save time involved in dimensioning a sketch by dimensioning half of the section and then mirroring it. Pro/ENGINEER will assume that the mirrored half has the same dimensions as the sketched half.*

DRAWING DISPLAY OPTIONS

While working with complex sketches, sometimes you need to increase the display of a particular portion of a sketch so that you can work on the minute details of the sketch. For example, you are drawing a sketch of a piston and you have to work on the minute details of the grooves for the piston rings. To work on these minute details, you have to enlarge the display of these grooves. You can enlarge or reduce the drawing display using various drawing display tools provided in Pro/ENGINEER. These tools are available in the **View** toolbar. Some of these drawing display options are discussed next. The remaining drawing display options will be discussed in later chapters.

Zoom In

 This tool enlarges the view of the drawing on the screen. After choosing the **Zoom In** button, you will be prompted to define a box. The area that you will enclose inside the box will be enlarged and displayed in the graphics window. Note that when you enlarge the view of the drawing, the original size of the entities is not changed. To exit the zoom tool right-click in the graphics window.

Zoom Out

 This tool reduces the view of the drawing on the screen, thus increasing the drawing display area. Each time you choose this button to zoom out, the display of the sketch in the graphics window is reduced.

Refit object to fully display it on the screen

This option reduces or enlarges the display such that all entities that comprise the sketch are fitted inside the current display. Note that the dimensions may not necessarily be included in the current display.

Redraw the current view

 While working with complex sketches, some unwanted temporary information is retained on the screen. The unwanted information may include the shadows of the deleted sketched entities, dimensions, and so on. This unwanted information can be removed from the graphics window using the **Redraw the current view** button. This option is extensively used when designing in Pro/ENGINEER.

Note
*To remove the temporary information you can also choose **View** > **Repaint** from the menu bar or CTRL+R to repaint the screen.*

If you have a mouse that has a middle mouse button wheel, then scrolling the wheel will zoom in and out.

One more way to zoom in and out is to use the middle mouse button and the CTRL key. When you use CTRL+middle mouse button and drag the mouse upward the sketch is zoomed out and when you drag the mouse downward, the sketch is zoomed in.

*In the **Sketch** mode, you can pan the sketch using the middle mouse button but in the **Part** mode, use SHIFT+middle mouse button to pan the model.*

TUTORIALS

Tutorial 1

In this tutorial, you will draw the sketch for the model shown in Figure 1-18. The sketch is shown in Figure 1-19.   **(Expected time: 30 min)**

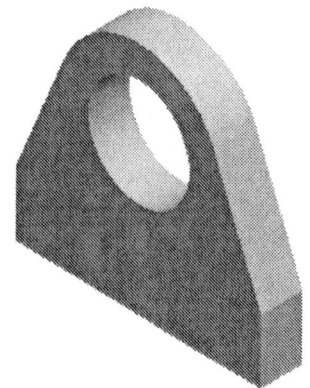

Figure 1-18 Model for Tutorial 1

Figure 1-19 Sketch of the model

The following steps are required to complete this tutorial:

a. Start Pro/ENGINEER Wildfire 2.0 session.
b. Set the working directory and create a new sketch file.
c. Draw lines using the line tool, refer to Figures 1-20 and 1-21.
d. Draw an arc and a circle, refer to Figures 1-22 and 1-23.
e. Dimension the sketch and then modify the dimensions of the sketch, refer to Figure 1-24
f. Save the sketch and close the file.

Starting Pro/ENGINEER

1. Start Pro/ENGINEER Wildfire 2.0 by double-clicking on the Pro/ENGINEER icon on the desktop of your computer or by using the **Start** menu.

Setting the Working Directory

When the Pro/ENGINEER session is started, the first task is to set the working directory. A working directory is a directory on your system where you can save the work done in the current session of Pro/ENGINEER. You can set any directory existing on your system as the working directory. Because this is the first tutorial of this chapter, you need to create a folder named *c01*, if it does not exist.

1. Choose the **Set Working Directory** option from the **File** menu. The **Select Working Directory** dialog box is displayed.

2. Browse and select *C:\ProE-WF-2.0*. If this folder is non existing then create it prior to setting the working directory.

3. Choose the **New Directory** button in the **Select Working Directory** dialog box; the **New Directory** dialog box is displayed.

4. Type **c01** in the **New Directory** edit box and choose **OK** from the dialog box. You have created a folder named *c01* in *C:\ProE-WF-2.0*.

Creating Sketches in the Sketch Mode-I

5. Choose **OK** from the **Select Working Directory** dialog box. You have set the working directory to *C:\ProE-WF-2.0\c01*. A message is displayed in the **Message Area** that the directory successfully changed to *C:\ProE-WF-2.0\c01* directory.

Starting a New Object File

Any sketch drawn in the **Sketch** mode is saved with the *.sec* file extension. This file format is one of the file formats available in Pro/ENGINEER.

1. Choose the **Create a new object** button from the **File** toolbar. The **New** dialog box is displayed. Select the **Sketch** radio button from the **Type** area of the **New** dialog box. A default name of the sketch appears in the **Name** edit box.

2. Enter *c01tut1* in the **Name** edit box and choose the **OK** button.

   You are in the sketcher environment of the **Sketch** mode. When you enter the sketcher environment, the Navigator is displayed to the left in the graphics window.

3. Slide in the Navigator to the left by clicking on the sash present on its right edge. Now, the drawing area is increased.

Drawing the Lines of the Sketch

Start drawing the sketch with the right vertical line.

1. Choose the **Create 2 point lines** button from the **Sketcher Tools** toolbar.

2. Specify the start point to the right in graphics window by clicking. One end of the line is attached to the cursor. Move the cursor down to get an approximate size of the line.

   Notice that when the cursor is moved vertically downward, a red colored constraint **V** appears in the graphics window next to the line. This indicates that if you draw a line now, the vertical constraint will be applied to the line.

3. Click to specify the endpoint of the line. The vertical constraint **V** is applied to the line and the symbol **V** appears in yellow. The color of the constraint indicates that this constraint is strong. This means that you cannot change the orientation of this line until you delete the constraint that is applied on the line.

   Another rubber-band line is attached to the cursor with its start point at the endpoint of the last line.

4. Move the cursor horizontally toward the left; a horizontal rubber-band line extends to the left as you move the mouse.

5. After you get the desired size of the line, click to end the line. Notice that a horizontal constraint **H** that is yellow in color is applied to the line.

6. Move the cursor upward in the graphics window; a vertical rubber-band line extends as

you move the mouse. As you move the cursor upward, notice that at a particular point where the length of the left vertical line is equal to the length of the right vertical line, L_1 symbol is displayed on both the vertical lines. This symbol suggests that the equal length constraint is applied to the two vertical lines.

7. When the L_1 constraint appears on the vertical line, click to specify the endpoint of the vertical line. Notice that the L_1 constraint is displayed in gray color, as shown in Figure 1-20. This suggests that it is a weak constraint. The rubber-band line is still attached to the cursor.

   You can also apply the constraints later. But to save an extra step of adding the constraints, you will use the constraints that are applied automatically while drawing.

8. Move the cursor to size the line and specify the endpoint of the left inclined line, see Figure 1-20.

9. Press the middle mouse button to end line creation. You will notice that gray colored dimensions are applied to the sketch, see Figure 1-20. The color of these dimensions indicates that these dimensions are weak dimensions. These dimensions are automatically deleted when you have completed the sketch or when you are adding dimensions and constraints manually. When system deletes weak dimensions, it does not confirm their deletion.

10. The line option is still active. Move the cursor close to the top end of the right vertical line. You will notice that as you bring the cursor close to the top end, the cursor snaps to that point. Select the point by clicking.

11. Size the inclined line and specify the endpoint of the right inclined line. Press the middle mouse button to end line creation.

    Figure 1-21 shows the lines that you have drawn. Now, the arc and circle will be drawn.

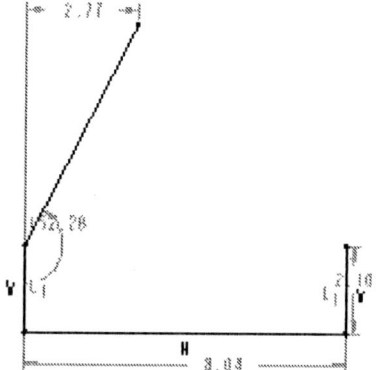

Figure 1-20 *Lines with weak dimensions*

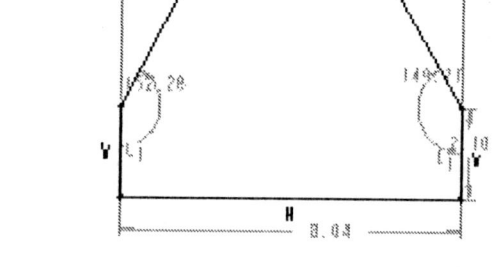

Figure 1-21 *Partial sketch with weak dimensions*

Creating Sketches in the Sketch Mode-I

Drawing the Arc

1. Choose the **Create an arc by 3 points or tangent to an entity at its endpoint** button from the **Sketcher Tools** toolbar. You are prompted to select the start point of the arc.

2. Select the endpoint of the left inclined line; the **Target** symbol appears in green color.

3. Move the cursor along the tangent direction through a small distance. A rubber-band arc that is tangent to the endpoint of the line appears. As you move the cursor to the endpoint of the right inclined line, at a particular point the tangent constraint is applied at both the ends of the arc. This is indicated by the symbol **T** that appears on the endpoints of the inclined lines.

4. As the tangent constraint appears, click to end arc creation. You will notice that the tangent constraint with a symbol **T** appears at the endpoints of the arc, as shown in Figure 1-22. Press the middle mouse button to end arc creation.

The tangent constraint **T** will appear in white, which suggests that it is a strong constraint and the tangency of the inclined line with the arc cannot be modified until you delete the tangent constraint.

Note that in Figure 1-21 there are some weak dimensions that are not displayed in Figure 1-22. This is because the weak dimensions are deleted without confirming their deletion. Hence, after drawing the arc some weak dimensions get deleted automatically.

Note
*If the tangent constraint symbol is not displayed on any of the inclined lines, apply the constraint manually using the **Constraints** dialog box that is displayed when you choose the **Impose sketcher constraints on the section** button from the **Sketcher Tools** toolbar (see page 1-16 for more information).*

Drawing the Circle

1. Choose the black arrow on the right of the **Create circle by picking the center and a point on the circle** button to display the flyout. From this flyout, choose the **Create concentric circle** button; you are prompted to select an arc.

2. Select the arc by clicking by clicking on it. Move the mouse and a circle appears.

3. To draw the circle, click to select a point inside the sketch.

4. Press the middle mouse button to end circle creation. The sketch is complete and appears similar to that shown in Figure 1-23.

Dimensioning the Sketch

The right vertical line, the bottom horizontal line, the arc, and the circle are dimensioned automatically and weak dimensions are applied to them. You will use these dimensions. Hence, there is no need to dimension these entities again.

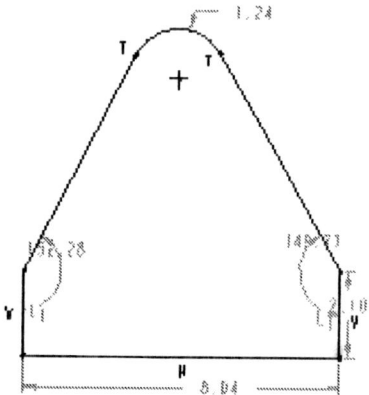 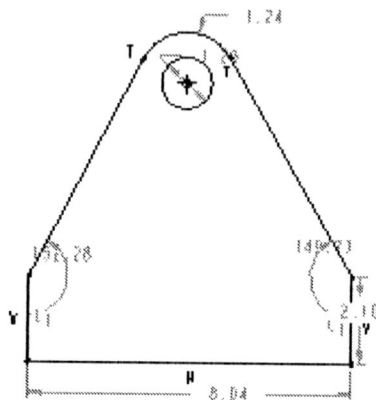

Figure 1-22 *Sketch with arc*   Figure 1-23 *Sketch with all entities, weak dimensions, and weak constraints*

1. Choose the **Create defining dimension** button from the **Sketcher Tools** toolbar.

2. Select the center of the circle and then select the bottom horizontal line by clicking on them. Both, the center and the line turn red in color.

3. Place the dimension on the right of the sketch by pressing the middle mouse button.

4. Select the center of the circle and then select the left vertical line. Both the center and the vertical line turn red in color.

5. Press the middle mouse button to place the dimension below the sketch, refer to Figure 1-24.

Modifying the Dimensions

The sketch is dimensioned with default values. You need to modify these values to the given values.

1. Choose the **Select items** button.

2. Select all dimensions by specifying a window around them.

Note
You can also use CTRL+ALT+A to select the entire sketch with dimensions.

3. When all dimensions turn red in color, choose the **Modify the values of dimensions, geometry of splines, or text entities** button. The **Modify Dimensions** dialog box is displayed.

All dimensions in the sketch are displayed in this dialog box and each dimension has a separate thumbwheel and an edit box. You can use the thumbwheel or the edit box to

Creating Sketches in the Sketch Mode-I

modify the dimensions. It is recommended that you use the edit boxes to modify the dimensions if the change in the dimension value is large.

4. Clear the **Regenerate** check box and then modify the values of the dimensions.

   When you clear this check box, any modification in a dimension value does not update the sketch. It is recommended that you clear the **Regenerate** check box when more than one dimension has to be modified.

   Notice that the dimension you select in the **Modify Dimensions** dialog box gets enclosed in a yellow box in the graphics window.

5. Modify all dimensions, as shown in Figure 1-24. After modifying the dimensions, choose the **Regenerate the section and close the dialog** button from the **Modify Dimensions** dialog box. A message **Dimension modifications successfully completed** is displayed in the **Message Area**.

The sketch is complete and is shown in Figure 1-24.

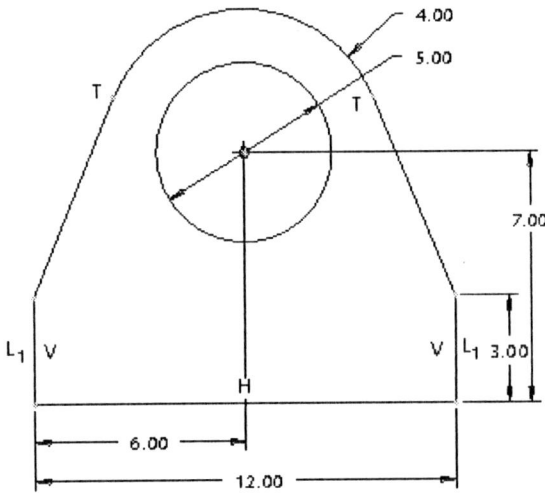

Figure 1-24 *The complete sketch with dimensions and constraints*

 **Tip**: *You can modify the location of the dimensions as they appear on the screen by selecting and dragging them to a new location.*

Saving the Sketch

The sketch will now be saved. You have to save the sketch because you may need the sketch later in the **Part** mode in order to create a 3D model.

1. Choose **Save the active object** button from the **File** toolbar. The **Save Object** dialog box is displayed with the name of the sketch that you had entered earlier.

2. Choose the **OK** button; the sketch is saved.

3. After saving the sketch, choose **Window > Close** from the menu bar.

Tutorial 2

In this tutorial, you will draw the sketch for the model shown in Figure 1-25. The sketch is shown in Figure 1-26. For your reference, all entities in the sketch are labeled alphabetically.

(Expected time: 30 min)

Figure 1-25 Model for Tutorial 2

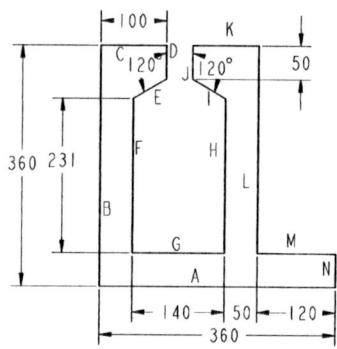

Figure 1-26 Sketch of the model

The following steps are required to complete this tutorial:

a. Set the Working directory and create a new object file.
b. Draw the sketch using the line tool, refer to Figure 1-27.
c. Dimension the required entities and then modify the dimensions of the sketch, refer to Figures 1-28 and 1-29.
d. Save the sketch and close the file.

Setting the Working Directory

The working directory was selected in Tutorial 1, and therefore there is no need to select the working directory again. But if a new session of Pro/Engineer is started, then you have to again set the working directory by following the steps given next.

1. Open the Navigator by clicking the top arrows on the left edge of the Pro/ENGINEER main window; the Navigator slides out.

2. Click on the plus symbol adjacent to the *ProE-WF-2.0* folder in the Navigator; the contents of the *ProE-WF-2.0* folder are displayed.

Creating Sketches in the Sketch Mode-I

3. Now right-click on the *c01* folder to display a shortcut menu. From this shortcut menu, choose the **Set Working Directory** option; the working directory is set to *c01*.

4. Close the Navigator by clicking on the sash located at the right edge of Navigator. The Navigator slides in.

Starting a New Object File

1. Choose the **Create a new object** button from the **File** toolbar; the **New** dialog box is displayed. Select the **Sketch** radio button from the **Type** area of the **New** dialog box; a default name of the sketch appears in the **Name** edit box.

2. Enter *c01tut2* in the **Name** edit box and choose **OK**, you are in the sketcher environment of the **Sketch** mode.

Drawing the Sketch

The sketch in Figure 1-26 consists of only lines. For ease of understanding, all lines in the sketch are labelled alphabetically.

1. Choose the **Create 2 point lines** button from the **Sketcher Tools** toolbar. Select a point close to the lower right corner of the graphics window by clicking and start drawing the horizontal line A. Here, you will notice that as you draw line A, the **H** symbol is displayed on the line. This indicates that the line is horizontally constrained. Move the cursor toward the left and specify the endpoint of the line.

2. Move the cursor vertically upwards so that the **V** constraint appears on the line. When you get the appropriate size of the line, click to specify the endpoint of line B; line B is completed.

3. Move the cursor to the right in the graphics window and click to specify the endpoint of line C.

4. Now, to draw line D, move the cursor down and click to specify the endpoint of line D.

5. Line E is inclined. Move the cursor to size the line and click to specify the endpoint of line E.

6. The next line you need to draw is line F. Move the cursor vertically downwards and click to specify the endpoint of line F.

7. Now, to draw line G, move the cursor horizontally toward the left and click to specify the endpoint of line G.

8. Move the cursor vertically upwards and click to specify the endpoint of line H.

9. Now, continue drawing the remaining lines that are shown in Figure 1-27. When the sketch is complete, end line creation by pressing the middle mouse button. Notice that the sketched entities are dimensioned automatically as you draw them. These dimensions are weak dimensions and appear in gray color.

Applying the Constraints to the Sketch

Constraints are applied to the sketch to maintain the design intent of the feature and this might sometimes result in less dimensions in the sketch.

1. Choose the **Impose sketcher constraints on the section** button from the **Sketcher Tools** toolbar; the **Constraints** dialog box is displayed.

2. Choose the **Create Equal Lengths, Equal Radii, or Same Curvature constraint** button and select lines F and H. The equal length constraint L_2 is applied to both the lines. The constraint labels such as L_2 or L_3 vary from sketch to sketch.

3. Select lines C and K; the equal length constraint is applied to both the lines.

4. Now, select lines J and N; the equal length constraint is applied to both the lines.

5. Select lines A and B. The equal length constraint is applied to both the lines.

6. Choose the **Make line or two vertices horizontal** button from the **Constraints** dialog box; you are prompted to select a line or two vertices.

7. Select the vertex that is joining lines L and M and the vertex that is joining lines G and H. Both the vertices are aligned horizontally, as shown in Figure 1-27.

8. Select the vertex that is joining lines C and D and the vertex that is joining lines J and K. Both the vertices are aligned horizontally, as shown in Figure 1-27.

Dimensioning the Sketch

Weak dimensions are already applied to the sketch while drawing. You need to dimension only the angle between lines D and E and lines J and I.

1. Choose the **Create defining dimension** button.

2. Select lines D and E using the left mouse button; the selected lines turn red in color. Now, press the middle mouse button to place the dimension close to the vertex where lines D and E join.

3. Similarly, dimension the angle between lines J and I.

   Figure 1-28 shows the sketch after applying dimensions. If your sketch does not have all dimensions shown in this figure, apply them using the **Create defining dimension** button.

Modifying the Dimensions

The dimensions that are applied to the sketch need modification in dimension values.

1. Choose the **Select items** button and then select all dimensions by specifying a window around them.

Creating Sketches in the Sketch Mode-I

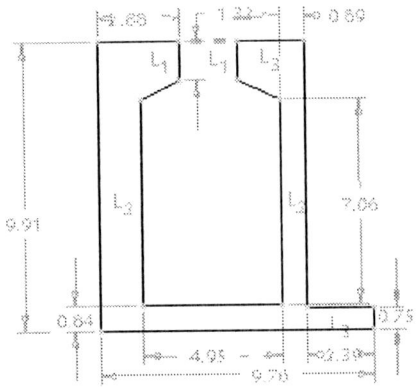

Figure 1-27 Sketch with weak dimensions and constraints

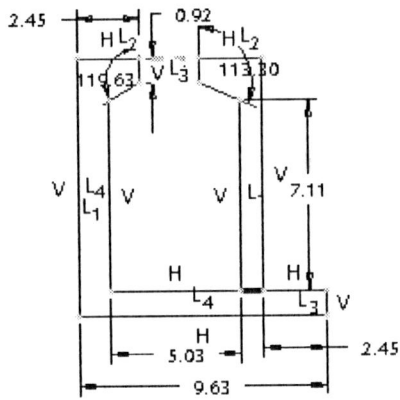

Figure 1-28 Sketch after dimensioning

2. When the dimensions turn red in color, choose the **Modify the values of dimensions, geometry of splines, or text entities** button. The **Modify Dimensions** dialog box is displayed.

3. Clear the **Regenerate** check box and then modify the values of the dimensions. When you clear this check box, the sketch is not regenerated while you modify the dimensions.

   Notice that the dimension you select in the **Modify Dimensions** dialog box is enclosed in a yellow box in the graphics window.

4. When all dimensions are modified, choose the **Regenerate the section and close the dialog** button from the **Modify Dimensions** dialog box. A message **Dimension modifications successfully completed** is displayed in the **Message Area**. The completed sketch is shown in Figure 1-29.

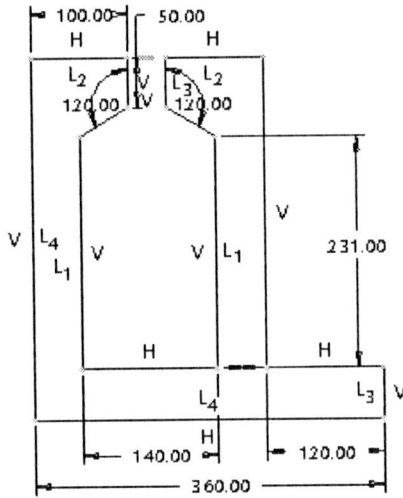

Figure 1-29 Complete sketch with dimensions and constraints

6. Save the sketch as discussed earlier. After saving the sketch, choose **Window > Close** from the menu bar to exit the **Sketch** mode.

Note
You can also modify dimensions individually. But, individual modification of dimensions is recommended only when either there is a minor change in the dimension value or when only one dimension is required to be modified.

Tutorial 3

In this tutorial, you will draw the sketch for the model shown in Figure 1-30. The sketch is shown in Figure 1-31. For your reference, all entities in the sketch are labeled alphabetically.

(Expected time: 30 min)

Figure 1-30 Model for Tutorial 3   *Figure 1-31 Sketch of the model*

The following steps are required to complete this tutorial:

a. Set the working directory and create a new object file.
b. Draw the sketch using the sketcher tools, refer to Figures 1-32 through 1-35.
c. Dimension the sketch and then modify the dimensions of the sketch, refer to Figure 1-36
d. Save the sketch and close the file.

Setting the Working Directory

The working directory was selected in Tutorial 1, and therefore there is no need to select the working directory again. But if a new session of Pro/Engineer is started, then you have to again set the working directory by following the steps given next.

1. Open the Navigator by sliding it out. Click on the plus symbol adjacent to the *ProE-WF-2.0* folder in the Navigator; the contents of the *ProE-WF-2.0* folder are displayed.

2. Now right-click on the *c01* folder to display a shortcut menu. From this shortcut menu, choose the **Set Working Directory** option; the working directory is set to *c01*. Close the Navigator.

Creating Sketches in the Sketch Mode-I   1-35

Starting New Object File

1. Choose the **Create a new object** button from the **File** toolbar. The **New** dialog box is displayed. Select the **Sketch** radio button from the **Type** area of the **New** dialog box. A default name of the sketch appears in the **Name** edit box.

2. Enter *c01tut3* in the **Name** edit box. Choose the **OK** button to enter the sketcher environment of the **Sketch** mode.

Drawing the Circles

1. Choose **Create circle by picking the center and a point on the circle** button from the **Sketcher Tools** toolbar and specify the center of the circle.

2. Move the cursor to size the circle and click to complete the circle.

3. Draw another circle whose center is collinear with the center of the previous circle.

   Figure 1-32 shows the two collinear circles drawn using the **Create circle by picking the center and a point on the circle** button from the **Sketcher Tools** toolbar.

Drawing the Tangent Lines

1. Choose the **Create lines tangent to 2 entities** button from the flyout in the **Sketcher Tools** toolbar. You are prompted to select the start location on the arc or a circle.

2. Select the left circle at the top. A rubber-band line appears whose one end is attached to the circle and the other end is attached to the cursor.

3. Click on the top of the right circle; a tangent that connects the two circles is drawn.

4. Similarly, draw a tangent by selecting the two circles at the bottom.

   Figure 1-33 shows the sketch after drawing the tangent lines.

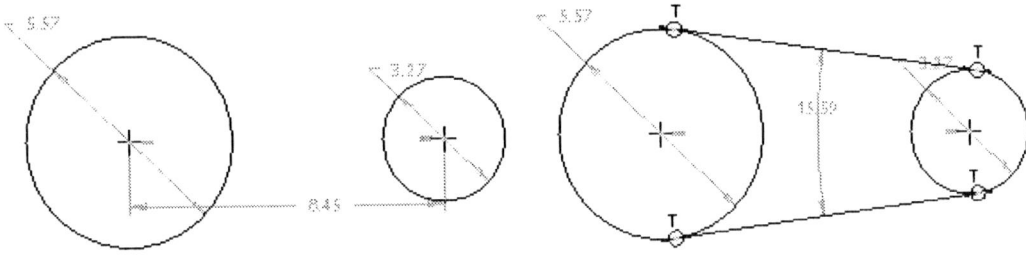

Figure 1-32 The two circles with weak dimensions and constraints

Figure 1-33 Circles joined by lines and the tangent constraint applied to them

Trimming the Circles

As evident from Figure 1-33, the tangents that are drawn intersect the circles at the point where they meet the circle. Therefore, the part of the circle that is not required can be dynamically trimmed.

1. Choose the **Dynamically trim section entities** button from the **Sketcher Tools** toolbar.

2. Select the two circles individually to trim them at the locations shown in Figure 1-34.
   Figure 1-35 shows the two circles after deleting the unwanted portion of the circle.

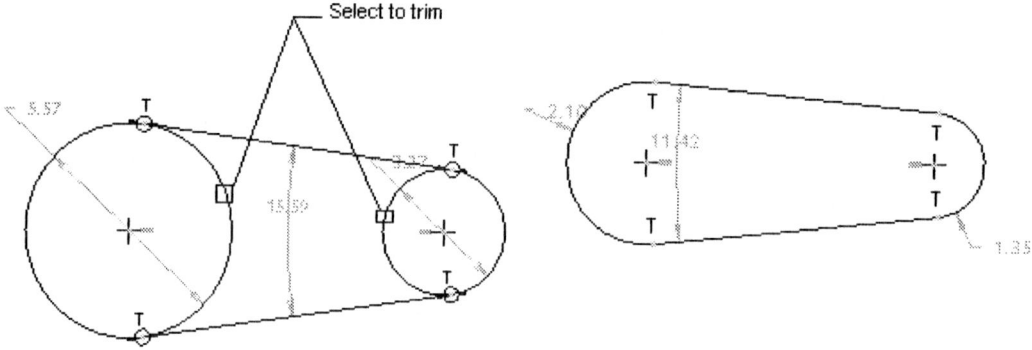

Figure 1-34 Locations to trim *Figure 1-35 Sketch after trimming*

Drawing the Circles

1. Choose the black arrow on the right of the **Create circle by picking the center and a point on the circle** button to display the flyout. From this flyout, choose the **Create concentric circle** button; you are prompted to select an arc.

2. Select arc P and create circle X concentric to the arc. Similarly, select arc Q to create a concentric circle Y.

   Notice that the two arcs are applied radius dimension whereas the circles are applied diameter dimension. This is because by default, the arcs are applied radius dimension and circles are applied diameter dimension.

Dimensioning the Sketch

In order to fully define a sketch, it should be dimensioned.

1. Choose the **Create defining dimension** button.

2. Select the centers of the two circles and place the dimension at the bottom of the sketch.

Modifying the Dimensions

1. Choose the **Select items** button.

Creating Sketches in the Sketch Mode-I

2. Select all dimensions by defining a window.

Note
You can also use CTRL+ALT+A from the keyboard to select all entities and items in the sketch.

3. When all dimensions turn red in color, choose the **Modify the values of dimensions, geometry of splines, or text entities** button; the **Modify Dimensions** dialog box is displayed.

4. Clear the **Regenerate** check box and then modify the values of the dimensions.

   You will notice that the dimension you edit in the **Modify Dimensions** dialog box is enclosed by a yellow box in the graphics window.

5. When all dimensions are modified, choose the **Regenerate the section and close the dialog** button from the **Modify Dimensions** dialog box; a message **Dimension modifications successfully completed** is displayed in the **Message Area**.

   The sketch is completed and is shown in Figure 1-36.

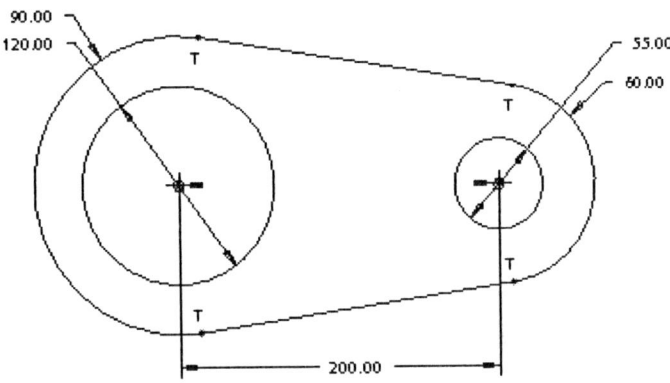

Figure 1-36 The complete sketch with dimensions and constraints

6. Save the sketch as discussed earlier. After saving the sketch, choose **Window > Close** from the menu bar to exit the **Sketch** mode.

Self-Evaluation Test

Answer the following questions and then compare your answers with those given at the end of this chapter.

1. Dimensions and constraints are automatically applied to a sketch when you draw it. (T/F)

2. In Pro/ENGINEER, you can create lines that are tangent to two circles. (T/F)

3. If the **Intent Manager** is on and you draw a line, the cursor snaps to the endpoint of the previous line. (T/F)

4. You can convert a weak constraint to strong by using the shortcut menu that is displayed when you right-click on the weak constraint. (T/F)

5. While drawing a circle, first you need to specify its diameter. (T/F)

6. The _____ menu in the menu bar has the **Modify** option in it.

7. The sketch can be modified by changing its _____.

8. **Intent Manager** is _____ by default when you enter the **Sketch** mode. (on/off)

9. In the **Sketch** mode the tangent constraint is represented by _____ symbol.

10. The **Sketch** mode file is saved with a _____ file extension.

Review Questions

Answer the following questions:

1. What is the need of **Sketch** mode in Pro/ENGINEER?

2. What are the four basic steps to create a sketch?

3. What are the various types of lines you can sketch using the buttons available in the **Sketcher Tools** toolbar?

4. Why is it important to select the working directory before creating a new file?

5. Write all steps involved in creating a sketch that is accepted by Pro/ENGINEER.

6. You can dynamically modify the geometry of a sketch. (T/F)

7. You can use the **Create rectangle** button from the **Sketcher Tools** toolbar to draw a square. (T/F)

8. The _____ button is used to apply constraints manually.

9. You cannot undo a previous operation in the sketcher environment. (T/F)

10. You can also use the options in the menu bar to draw a sketch from the **Sketch** menu. (T/F)

Creating Sketches in the Sketch Mode-I

Exercises

Exercise 1

In this exercise, you will draw the sketch for the model shown in Figure 1-37. The sketch is shown in Figure 1-38.

(Expected time: 30 min)

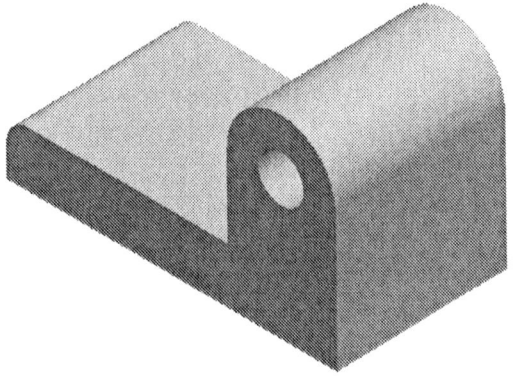

Figure 1-37 Solid model for Exercise 1

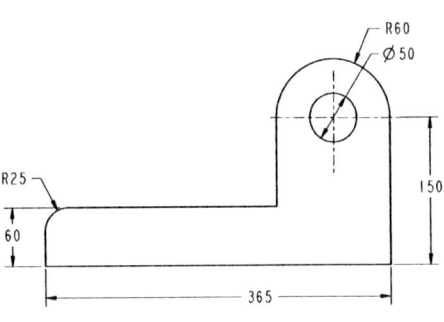

Figure 1-38 Sketch of the model

Exercise 2

In this exercise, you will draw the sketch for the model shown in Figure 1-39. The sketch is shown in Figure 1-40.

(Expected time: 30 min)

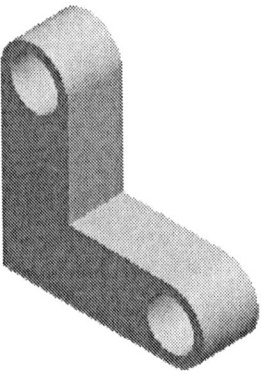

Figure 1-39 Solid model for Exercise 2

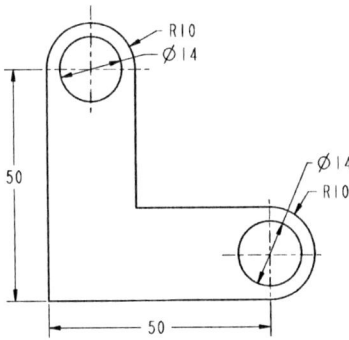

Figure 1-40 Sketch of the model

Exercise 3

In this exercise, you will draw the sketch for the model shown in Figure 1-41. The sketch is shown in Figure 1-42. **(Expected time: 30 min)**

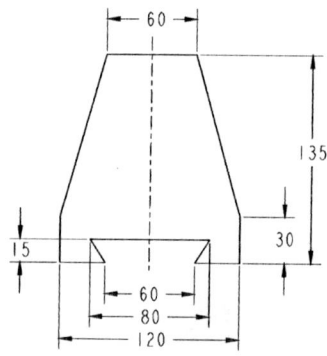

Figure 1-41 Solid model for Exercise 3

Figure 1-42 Sketch of the model

Exercise 4

In this exercise, you will draw the sketch for the model shown in Figure 1-43. The sketch is shown in Figure 1-44. **(Expected time: 30 min)**

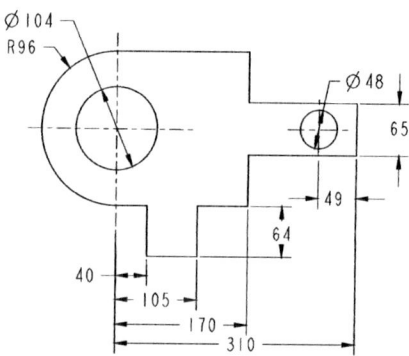

Figure 1-43 Solid model for Exercise 4

Figure 1-44 Sketch of the model

Answers to Self-Evaluation Test

1 - T, 2 - T, 3 - T, 4 - T, 5 - F, 6 - **Edit**, 7 - **dimensions**, 8 - **on**, 9 - T, 10 - **.sec**

Chapter 2

Creating Sketches in the Sketch Mode-II

Learning Objectives

After completing this chapter you will be able to:
- *Use various options to dimension a sketch.*
- *Create fillets.*
- *Place a user-defined coordinate system.*
- *Create, dimension, and modify splines.*
- *Create text.*
- *Scale and rotate entities.*
- *Copy a sketch.*
- *Import 2D drawings.*

DIMENSIONING THE SKETCH

In Chapter 1 you have learned how to dimension a sketch using the **Normal** option or using its equivalent tool, that is, **Create defining dimension** button in the **Sketcher Tools** toolbar. In this chapter, you will learn the **Baseline** option of dimensioning a sketch.

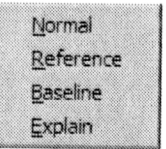

Choose **Sketch > Dimension** from the menu bar; the cascading menu is displayed, as shown in Figure 2-1. The options that are available to dimension a sketch are shown in this figure and are discussed next.

Figure 2-1 Dimensioning options

Dimensioning a Sketch Using the Baseline Option

In Pro/ENGINEER, the **Baseline** option of dimensioning is used to create dimensions in terms of horizontal and vertical location values of an entity with respect to a specified baseline. This type of dimensioning in a drawing can be used for writing a CNC program to manufacture a component.

The **Baseline** option can be used to dimension lines, conics, arcs, and so on. The following steps explain the procedure to create dimensions using the **Baseline** option:

1. Choose **Sketch > Dimension > Baseline** from the menu bar.

2. Select the entity that will act as the baseline (origin or reference). Press the middle mouse button to place the dimension, the dimension **0.00** is displayed where you place the dimension. Note that since the location value of the baseline is taken as the origin, the dimension value of the baseline entity will become 0.00. The dimension values of all other entities dimensioned with reference to the baseline will be measured from this origin.

   Depending upon the entity selected to act as the baseline, the horizontal or the vertical dimension value of the location of the entity will be placed. For example, if you select a vertical line, the value of its location will be placed vertically. Similarly, if you select a horizontal line, the value of its location will be placed horizontally.

   For arcs, circles, and splines there are two options to dimension using the **Baseline** option. When you select a circle center or an arc for baseline dimensioning and press the middle mouse button, the **Dim Orientation** dialog box is displayed, as shown in Figure 2-2. You are prompted to select the dimension orientation; the dimension is placed according to the orientation selected.

   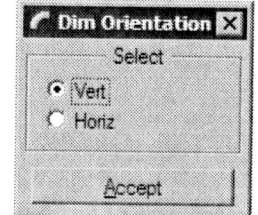

3. Next, choose the **Create defining dimension** button from the **Sketcher Tools** toolbar. Select the baseline dimension that was placed earlier and then select the entity to dimension. Now, press the middle mouse button to place the dimension.

Figure 2-2 The Dim Orientation dialog box

The orientation of the dimension will depending upon the baseline dimension and the entity selected. Figure 2-3 shows a sketch dimensioned using the above-mentioned method. In this figure, the two baselines are dimensioned using the **Baseline** option. Therefore, the dimensions of these lines are displayed as 0.00. The remaining lines are dimensioned by using the **Create defining dimension** button by first selecting the baseline dimension and then the entity to dimension.

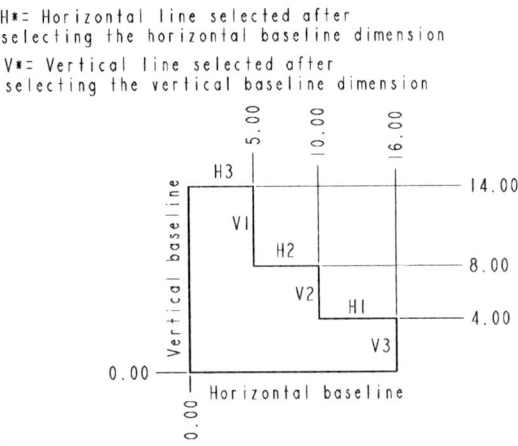

Figure 2-3 Baseline dimensioning of a sketch

Replacing the Dimensions of a Sketch Using the Replace Option

The **Replace** option is used to replace a dimension from a sketch. To use this option you must have a dimensioned sketch. The following steps explain the procedure to dimension a sketch using the **Replace** option:

1. Choose **Edit > Replace** from the menu bar, you will be prompted to select a dimension to be replaced.

2. Select the dimension to be replaced; the selected dimension is erased. Select an entity to dimension and place the dimension at the desired place. The previous dimension is replaced by a new dimension.

CREATING FILLETS

In the sketcher environment, you can create two types of fillets.

1. Circular fillets
2. Elliptical fillets

Creating Circular Fillets

A circular fillet is the arc that is used to join two lines, a line and an arc, or two arcs. The fillet is controlled by its radius or diameter dimension. The resulting fillet will depend on the location where the elements are selected.

Figure 2-4 shows two lines that do not join at the corners and Figure 2-5 shows the circular fillet created between them. The circular fillet that is created is an arc with its endpoints tangent to the two lines. This is evident from the **T** symbol that is automatically applied to the endpoints of the fillet arc.

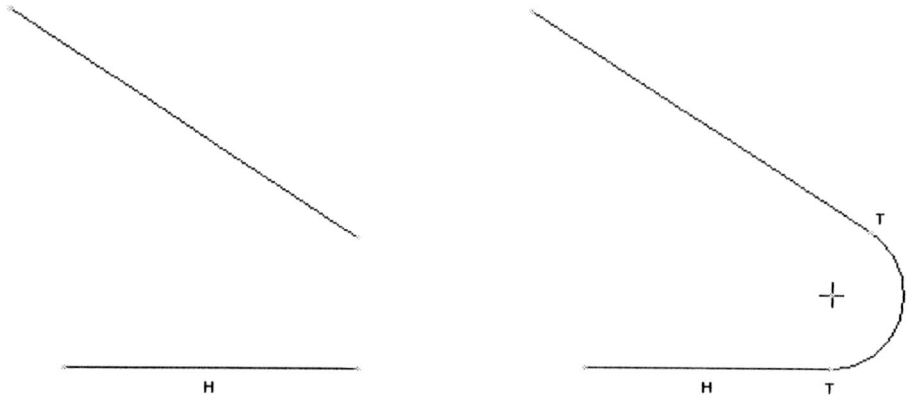

Figure 2-4 Two lines that do not join   *Figure 2-5* Fillet created between the two lines

Figure 2-6 shows two lines that join at the corner and Figure 2-7 shows the circular fillet created that joins the two lines by an arc. The corner is automatically deleted.

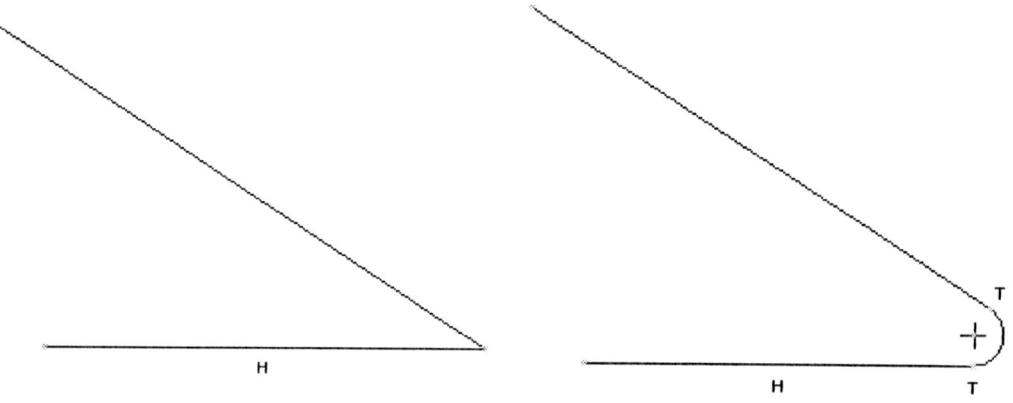

Figure 2-6 Two lines joining at the corner   *Figure 2-7* Filleted corner

Figure 2-8 shows two sets of arcs and Figure 2-9 shows the circular fillet created between the two arcs. The location where you select the arcs to create the fillet is important. The fillet is created tangent to the selection points on the arcs. Here, the endpoints of the arcs are selected to create the fillet.

If the points of selection on the two arcs are away from the endpoints of the arcs, then the fillet is created at the selection points. The portion of the arc that extends beyond the fillet should be manually deleted or trimmed. This case of creating a fillet is explained in Figures 2-10 and 2-11.

Creating Sketches in the Sketch Mode-II

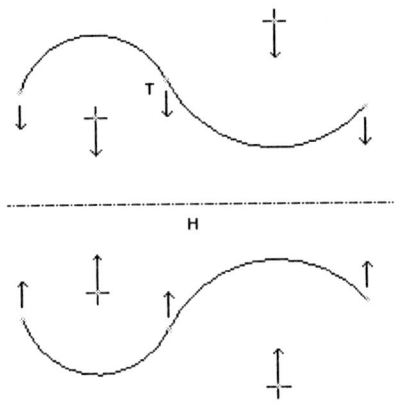

Figure 2-8 Two sets of arcs

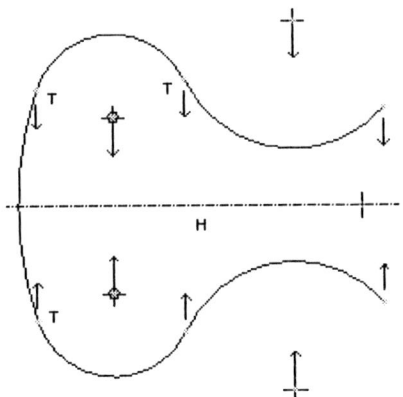

Figure 2-9 Fillet created by selecting the endpoints

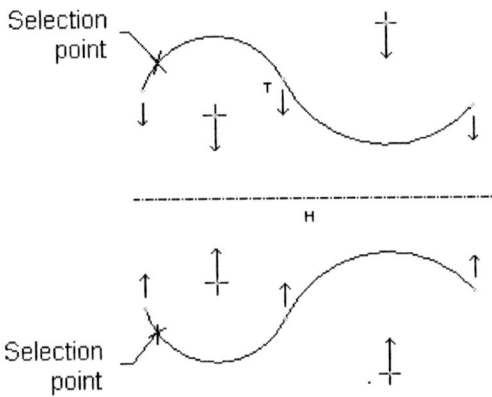

Figure 2-10 Points selected on arcs

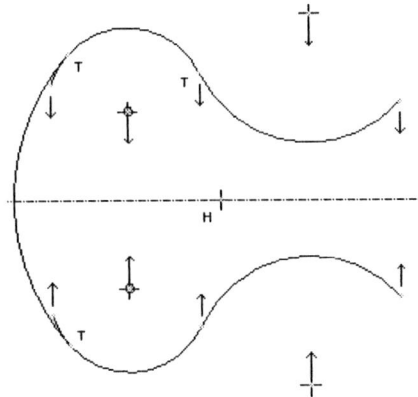

Figure 2-11 Filleted arcs

 The **Create a circular fillet between two entities** button in the **Sketcher Tools** toolbar is used to create circular fillets. The following steps explain the procedure to create a circular fillet:

1. Choose the **Create a circular fillet between two entities** button from the **Sketcher Tools** toolbar; you are prompted to select two entities.

2. Select the first entity for filleting by using the left mouse button; the yellow color of the first entity changes to red. Now, select the second entity. As soon as you select the second entity, if possible, a fillet is drawn between the two selected entities.

3. Repeat step 2 until you have created all fillets.

 Note
You can create a fillet between any two entities except between two parallel lines and between a centerline and an entity.

Creating Elliptical Fillets

An elliptical fillet is the arc in the form of an ellipse that joins two lines, two arcs, or a line and an arc. The geometry of the elliptical fillet depends on the location where you select the entities to create a fillet.

The advantage of elliptical fillets over circular fillets is that elliptical fillets have the geometry of an ellipse. As you know, an ellipse is controlled by dimensions in two directions. Therefore, when an elliptical fillet is dynamically modified, its geometry can be controlled in either the x-direction or the y-direction. Hence, the elliptical fillet when modified, results in more curved geometric shapes than a circular fillet.

Figures 2-12 and 2-13 illustrate the elliptical fillet. Notice, that a strong tangent constraint **T** is automatically applied when you create a fillet.

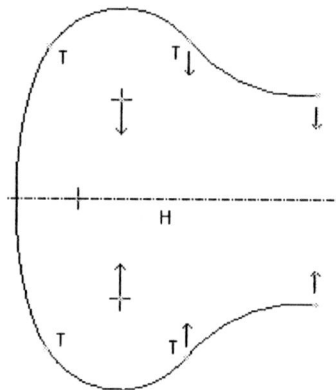
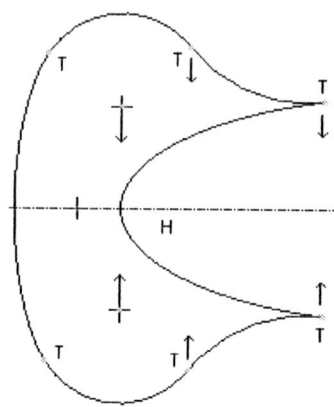

Figure 2-12 Arcs to be filleted   *Figure 2-13 Elliptical fillet created*

 An elliptical fillet is created by choosing the **Create an elliptical fillet between two entities** button in the **Sketcher Tools** toolbar. The following steps explain the procedure to create elliptical fillets:

1. Choose the black arrow on the right of the **Create a circular fillet between two entities** button to display a flyout. Choose the **Create an elliptical fillet between two entities** button from this flyout; you are prompted to select two entities.

2. Select the first entity by clicking; the color of the entity changes to red.

3. Select the second entity. As soon as you select the second entity, the elliptical fillet is created. The shape of the elliptical fillet depends upon the specified points. After the fillet is created, you are again prompted to select two entities for elliptical fillet.

4. Repeat steps 2 and 3 until you have created all fillets.

When you select an elliptical fillet to dimension, the **Ellipse Rad** dialog box is displayed, as shown in Figure 2-14. There are two radio buttons in this dialog box. The **X Radius** radio button when selected dimensions the elliptical fillet radially in the X-direction. The **Y Radius** radio button when selected dimensions the elliptical fillet radially in the Y-direction.

CREATING A REFERENCE COORDINATE SYSTEM

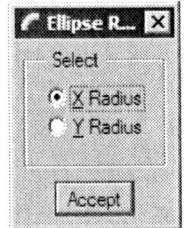

Figure 2-14 Ellipse Radius dialog box

The **Create reference coordinate system** button in the **Sketcher Tools** toolbar is used to create a coordinate system and add it to the created sketch. The coordinate system acts as reference for dimensioning. You can dimension the splines using the coordinate system. Thus, it provides you the flexibility to modify the spline points by specifying different coordinates with respect to the coordinate system.

The user-defined coordinate system is used in blend features to align different sections in a blend. It is also used in the **Assembly** mode and **Manufacturing** mode of Pro/ENGINEER.

The following steps explain the procedure to create a coordinate system:

1. Choose **Create reference coordinate system** button on the flyout that is displayed when you choose the black arrow on the right of the **Create points** button in the **Sketcher Tools** toolbar. You are prompted to select the location for the coordinate system. The coordinate system symbol is attached to the cursor.

2. Place the coordinate system at the desired points on the screen by clicking. The coordinate system will be placed at as many places as you click in the graphics window. You can end coordinate system creation by using the middle mouse button.

Note
If you add a coordinate system to a sketch, it must be dimensioned. Unless the coordinate system is placed at the endpoints of a line, an arc, a spline, or at the center of an arc or a circle. In other words, a coordinate system must be referenced to an entity in a sketch.

WORKING WITH SPLINES

Splines are curved entities that pass through an infinite number of intermediate points. Generally, splines are used to define the outer surface of a model. This is because the splines can provide different shape to curves and the flexibility to modify the surfaces that result from the splines. Splines find application in automobile and aeroplane body designing.

Creating a Spline

To draw a spline, choose the **Create a spline curve** button from the **Sketcher Tools** toolbar.

The following steps explain the procedure to create a spline:

1. Choose the **Create a spline curve** button from the **Sketcher Tools** toolbar; you are prompted to select the location for spline.

2. Use the left mouse button to select the start point for the spline. Similarly, select additional points in the graphics window and a spline will be drawn that passes through the specified points. All points through which the spline passes are called interpolation points.

Note

If you select the start point of the spline as its end point, then the resulting spline will be in form of a continuous loop.

Dimensioning of Splines

When a spline is drawn, the weak dimensions are automatically applied to the spline. A spline can be dimensioned manually in the following ways:

1. Dimensioning the endpoints.
2. Radius of curvature dimensioning.
3. Tangency dimensioning.
4. Coordinate dimensioning.
5. Dimensioning the interpolation points.

Dimensioning the Endpoints

To dimension a spline by selecting the endpoints:

1. Choose the **Create defining dimension** button from the **Sketcher Tools** toolbar.

2. Select the two endpoints of the spline and place the horizontal or vertical dimension by pressing the middle mouse button. Figure 2-15 shows a spline that is dimensioned by selecting the endpoints.

Radius of Curvature Dimensioning

The radius of curvature of a spline can be dimensioned only if its tangency is defined. In other words, radius of curvature of a spline can be dimensioned only if the spline is tangent to an entity. For radius of curvature dimensioning of a spline:

1. Choose the **Create defining dimension** button from the **Sketcher Tools** toolbar.

2. Select the endpoint of the spline where tangency is defined.

3. Press the middle mouse button to place the dimension. Figure 2-16 shows the radius of curvature dimensioning of a spline.

Tangency or Angular Dimensioning

A spline can be dimensioned tangentially. This type of dimensioning is also called angular dimensioning. To angular dimension a spline and a line tangent to it:

Creating Sketches in the Sketch Mode-II

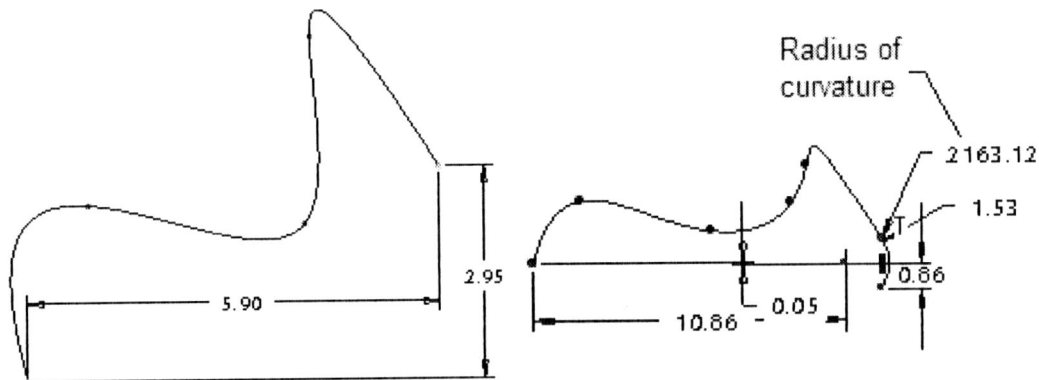

Figure 2-15 Endpoint dimensioning   *Figure 2-16* Radius of curvature

1. Choose the **Create defining dimension** button from the **Sketcher Tools** toolbar.

2. Select the spline by clicking.

3. Select the entity tangent to the spline by clicking.

4. Select the interpolation point of the spline that is to be dimensioned tangentially.

5. Press the middle mouse button to place the dimension.

Coordinate Dimensioning

The spline can be dimensioned with respect to a user-defined coordinate system. Choose the **Create reference coordinate system** button from the flyout in the **Sketcher Tools** toolbar. The coordinate system is attached to the cursor. Place the coordinate system in the graphics window. Now, the spline can be dimensioned with respect to the coordinate system.

Dimensioning the Interpolation Points

A spline can be dimensioned by dimensioning its interpolation points or vertices. This type of dimensioning is used when the designer wants the spline to be standard for all designs. This is because the exact curve can be duplicated if the interpolation points or the vertices of a spline are dimensioned.

 **Tip**: *A dimension can be moved by pressing and holding the left mouse button on the dimension and moving it. The dimension text is replaced by a red colored box. You can drag the dimension to the desired location in the graphics window and release the left mouse button to place the dimension at that point.*

Modifying a Spline

In Pro/ENGINEER Wildfire 2.0, a spline can be modified in the following ways:

1. Moving the interpolation points of the spline.
2. Adding points to a spline.

3. Deleting points of a spline.
4. Creating a control polygon and moving its control points.
5. Modifying the dimensions of the spline.

Moving the Points of the Spline

The position of the interpolation points can be dynamically modified. To modify a spline, select an interpolation point on the spline and drag it to modify the shape of the spline.

You can also use the dashboard to modify a spline. To invoke the dashboard, select the spline and hold the right mouse button down to invoke the shortcut menu. Choose the **Modify** option from the shortcut menu. The **Modify Spline** dashboard is displayed at the bottom of the graphics window with the options and buttons to modify a spline.

The interpolation points of the spline appear in white in the graphics window. Drag the interpolation points to modify the shape of the spline. Choose ✓ from the dashboard to exit it.

Adding Interpolation Points to a Spline

To add interpolation points on a spline, invoke the dashboard. Now, right-click on the spline to invoke the shortcut menu. Choose the **Add Point** option; a point is added to the spline where the spline was selected. The new point appears in white. Note that the **Add Point** option is available only when the dashboard is displayed at the bottom of the graphics window. You cannot increase the length of the spline by adding points before the start point and after the endpoint of the spline.

Deleting Interpolation Points of a Spline

To delete a point or a vertex, invoke the dashboard. Now, select the vertex and hold down the right mouse button to invoke the shortcut menu. Choose the **Delete Point** option from the shortcut menu. The selected point is deleted. You can continue deleting vertices or points from a spline until only two end points are left in the spline.

Creating a Control Polygon and Moving its Control Points

When you draw a spline, it is associated with a control frame. The vertices of its frame are called control points. To create a control polygon, choose the **Modify spline using control points** button from the dashboard. The control polygon is displayed in the graphics window. The control points of this polygon can be moved by dragging to modify the spline shape.

Modifying the Dimensions of the Spline

The shape of the spline is controlled by the position of its interpolation points. Hence by modifying the dimensions, the position of the interpolation points are changed, which results in modification of the shape of the spline.

Tip: *To dynamically modify the shape of the sketch, you need to select an entity of the sketch and drag the mouse to modify the sketch. Remember that if the selected entity is constrained then you cannot modify it. You can modify it only after disabling the constraints.*

WRITING TEXT IN SKETCHER ENVIRONMENT

There are various instances when a designer needs a text command to write text on the model. For example, for creating a label, model number, company name, and so on. In Pro/ENGINEER, you can write this text in the sketcher environment.

In the sketcher environment, the text is written using the **Create text as a part of a section** button from the **Sketcher Tools** toolbar. The following steps explain the procedure to write text in the sketcher environment:

1. Choose the **Create text as a part of a section** button from the **Sketcher Tools** toolbar; you are prompted to select the start point of line to determine the text height and orientation.

2. Specify the start point on the screen by clicking; you are prompted to select the second point of line to determine the text height and orientation.

3. Note that to write the text upright, the second point should be above the start point and in a straight line. If the second point is below the start point, the text will be written upside down. Specify the second point on the screen by clicking; the **Text** dialog box is displayed, as shown in Figure 2-17.

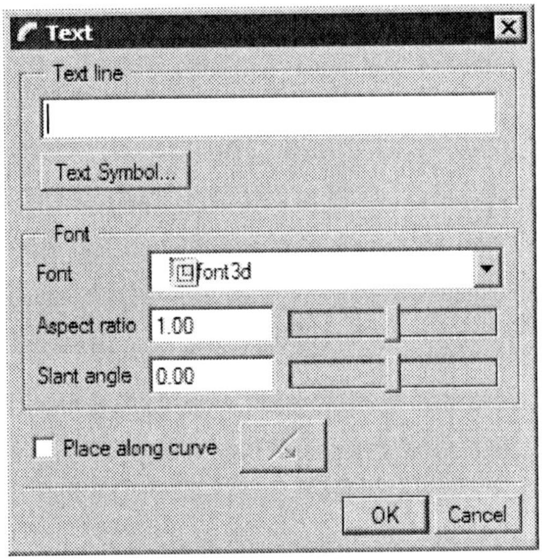

*Figure 2-17 The **Text** dialog*

After specifying the second point, a construction line is drawn having height equal to the distance between the two points. The height and orientation of the text depends on the height and angle of the construction line. If the construction line is drawn at an angle, then the text is written at that angle.

4. In the **Text Line** edit box, enter the text, which can be up to 79 characters. As you enter the text, the text is displayed dynamically in the graphics window. You can choose the

desired font of the text from the **Font** drop-down list. The aspect ratio and the slant angle of the text can be controlled by using the slider bars.

5. Choose the **OK** button in the **Text** dialog box to exit it.

Note

In later chapters of this book, you will learn that to enter the sketcher environment there are other methods also, other than entering through the Sketch mode.

SCALING AND ROTATING ENTITIES

The sketches can be scaled or rotated by using the **Scale and rotate selected entities** button from the **Sketcher Tools** toolbar. To invoke this button, select the sketch and then choose the arrow on the right of the **Mirror selected entities** button. This button is available on the flyout that is displayed. After choosing this tool button, the sketch that is composed of various entities now acts as a single entity. The sketch appears orange in color and is enclosed in a boundary box, as shown in Figure 2-18.

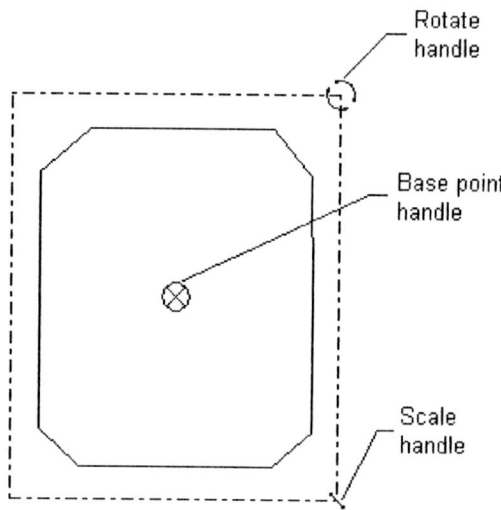

Figure 2-18 Selected entities enclosed in a boundary box with three handles

There are three handles that facilitate in scaling and rotating the selected sketch. The rotate handle is used to dynamically rotate the selected entities. The scale handle is used to dynamically scale the selected entities. The base point handle is used to pick the sketch and place it at any other location in the graphics window. To change the location of any of the three handles, right-click on the handle. The handle symbol is attached to the cursor. Place the symbol at the desired location. The following steps explain the procedure to scale and rotate a sketch:

1. Select the sketch to be rotated and scaled, and then choose the black arrow on the right of the **Mirror selected entities** button from the **Sketcher Tools** toolbar.

Creating Sketches in the Sketch Mode-II

2. From the flyout, choose the **Scale and rotate selected entities** button; the **Scale Rotate** dialog box is displayed, as shown in Figure 2-19. This dialog box contains the **Scale** and **Rotate** edit boxes. The sketch can be scaled and rotated dynamically or by entering a value in these edit boxes.

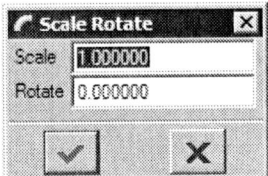

Figure 2-19 The Scale Rotate dialog box

3. To dynamically rotate the sketch, select the rotate handle and then move the cursor; the sketch is rotated as you move the cursor. You can also enter the rotation angle in the **Rotate** edit box.

To scale the sketch, select the scale handle and then move the cursor. As you move the cursor, the sketch is scaled dynamically in the graphics window. You can also enter the scale value in the **Scale** edit box.

4. After the sketch is rotated and scaled, choose the **Accept the changes and close the dialog** button in the **Scale Rotate** dialog box.

COPYING SKETCHED ENTITIES IN SKETCH MODE

The **Make a copy of selected entities** button in the **Sketcher Tools** toolbar is used to copy the entities in the sketcher environment. To invoke this button, select the sketch and then choose the arrow on the right of the **Mirror selected entities** button. Choose the **Make a copy of selected entities** button from the flyout that is displayed. The **Scale Rotate** dialog box is displayed. Using this dialog box, you can simultaneously copy, scale and rotate the selected entities. The following steps explain the procedure to copy the selected entities:

1. Select the entities to be copied and then choose the black arrow on the right of the **Mirror selected entities** button from the **Sketcher Tools** toolbar. From the flyout, choose the **Make a copy of selected entities** button.

   The copy of the selected entity is placed at the top left corner in the graphics window. This copy appears red in color. There are three handles in the copied entities that facilitate in sizing and orienting them. Figure 2-20 shows the copied sketch enclosed in a boundary with the three handles.

   The **Scale Rotate** dialog box is also displayed. Use this dialog box to specify the scale factor and the rotation angle.

2. Select the base point handle; the copied entities are attached to the cursor. Place the entities where needed.

3. Finally, choose the **Accept the changes and close the dialog** button in the **Scale Rotate** dialog box.

IMPORTING 2D DRAWINGS IN THE SKETCH MODE

The two-dimensional (2D) drawings when opened in the sketcher environment can be saved

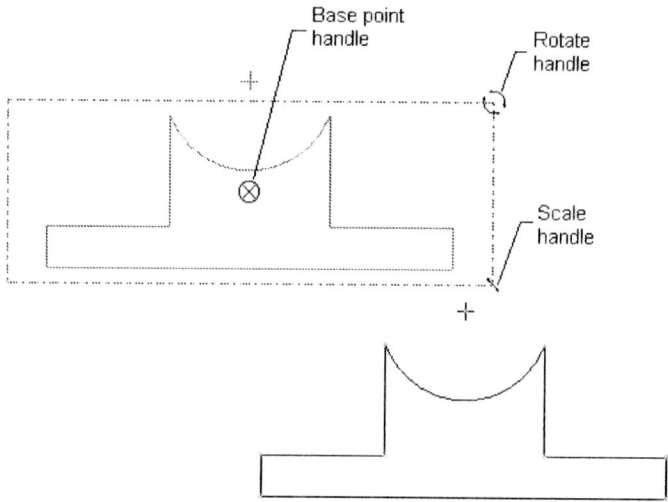

Figure 2-20 Original entities and the copied entities with three handles

in the *.sec* format. The *.sec* file can then be converted to a solid model. The **Data from file** option in the **Sketch** menu in the menu bar is used to import the 2D sketches. This option saves time in drawing the same or similar section again. The file formats from which the data can be imported are shown in Figure 2-21.

Figure 2-21 File formats

When you choose **Sketch** > **Data from file** from the menu bar, the **Open** dialog box is displayed, as shown in Figure 2-22. Use this dialog box to select and open the file.

In the **Open** dialog box, if you select a drawing file that is created in the **Drawing** mode of Pro/ENGINEER Wildfire 2.0, the draft entities in that file are imported. The selected drawing is opened in a sub window. You are prompted to select the entities to copy from the sub window. Select the draft entities and then press the middle mouse button. The selected entities are enclosed in a boundary and the **Scale Rotate** dialog box is displayed. Use this dialog box to set the scale and orientation of the sketch. Note that if the *.drw* file does not consist of draft entities, no data is imported.

In the **Open** dialog box, if you select a *.sec* file that is created in the sketcher environment, the sketch is either displayed in the graphics window enclosed in a boundary or is directly placed in the graphics window. If there are no entities in the current sketcher environment, the selected sketch will be automatically placed in the graphics window. In case, there are some

Creating Sketches in the Sketch Mode-II

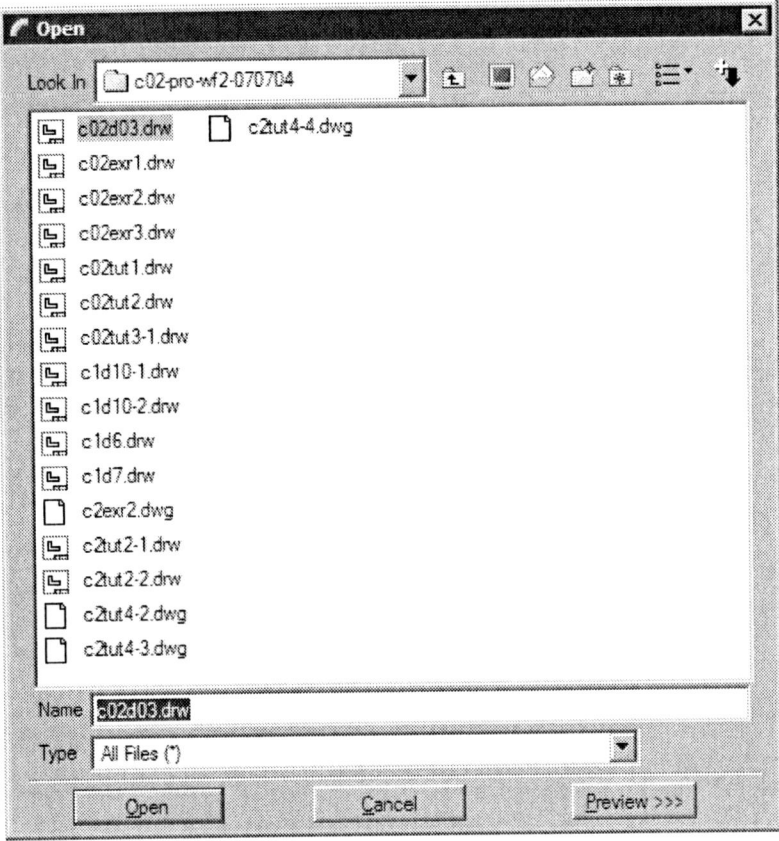

*Figure 2-22 The **Open** dialog box*

entities present in the current sketcher environment, the selected sketch is enclosed in a boundary and the **Scale Rotate** dialog box is displayed. Use this dialog box to set the scale and orientation of the sketch.

The section imported using the **Data from file** option in the current sketch is an independent copy. The imported section is no longer associated with the source section. The units, dimensions, grid parameters, and accuracy are acquired from the current sketch.

 **Tip**: *The display of vertices of the section, the display of dimensions, and the display of constraints can be turned on or off from the **Sketcher** toolbar. This toolbar is in the **Top Toolchest**.*

TUTORIALS

Tutorial 1

In this Tutorial you will import an existing sketch that you had drawn in Tutorial 3 of Chapter 1. After placing the sketch, draw the keyway, as shown in Figure 2-23.

(Expected time: 15 min)

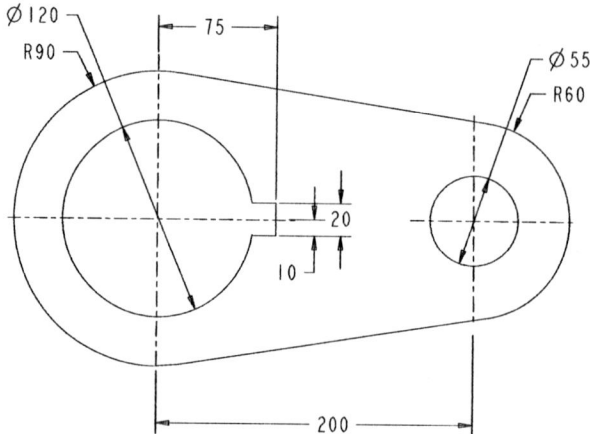

Figure 2-23 Sketch for Tutorial 1

The following steps are required to complete this tutorial:

a. Start Pro/ENGINEER Wildfire 2.0.
b. Set the working directory and create a new object file.
c. Import the section by using the **Data from file** option, refer to Figure 2-24.
d. Draw the keyway and dimension it, refer to Figures 2-25 and 2-26.
e. Modify the dimensions, refer to Figure 2-27.
f. Save the sketch and exit the sketcher environment.

Starting Pro/ENGINEER

1. Start Pro/ENGINEER Wildfire 2.0 by double-clicking on the Pro/ENGINEER icon on the desktop of your computer or by using the **Start** menu.

Setting the Working Directory

When the Pro/ENGINEER session is started, the first task is to set the working directory. As mentioned earlier, working directory is a directory on your system where you can save the work done in the current session of Pro/ENGINEER. You can set any directory existing on your system as the working directory. Since this is the first tutorial of this chapter, you need to create a folder named **c02** in the *C:\ProE-WF-2.0* folder, if it does not exist.

1. Choose the **Set Working Directory** option from the **File** menu. The **Select Working Directory** dialog box is displayed.

2. Browse and select *C:\ProE-WF-2.0*.

3. Choose the **New Directory** button in the **Select Working Directory** dialog box. The **New Directory** dialog box is displayed.

Creating Sketches in the Sketch Mode-II

4. Type **c02** in the **New Directory** edit box. Choose **OK** from the dialog box. You have created a folder named *c02* in *C:\ProE-WF-2.0*.

5. Choose **OK** from the **Select Working Directory** dialog box. You have set the working directory to *C:\ProE-WF-2.0\c02*.

Starting a New Object File

1. Choose the **Create a new object** button from the **File** toolbar; the **New** dialog box is displayed. Select the **Sketch** radio button from the **Type** area of the **New** dialog box. A default name of the sketch appears in the **Name** edit box.

2. Enter *c02tut1* in the **Name** edit box. Choose the **OK** button.

   You are in the sketcher environment of the **Sketch** mode. When you enter the sketcher environment, the Navigator is displayed to the left in the graphics window. Slide in the Navigator by clicking on the sash present on its right edge. Now, the drawing area is increased.

Importing the Section

1. Choose **Sketch > Data from file** from the menu bar; the **Open** dialog box is displayed with the working directory as the current directory.

2. Choose the **Up One Level** button and then open the *c01* directory. Make sure that the **sketch (*.sec)** option is selected in the **Type** drop-down list. Select *c01tut3.sec* and choose the **Open** button from the **Open** dialog box.

   The sketch is placed in the graphics window, as shown in Figure 2-24.

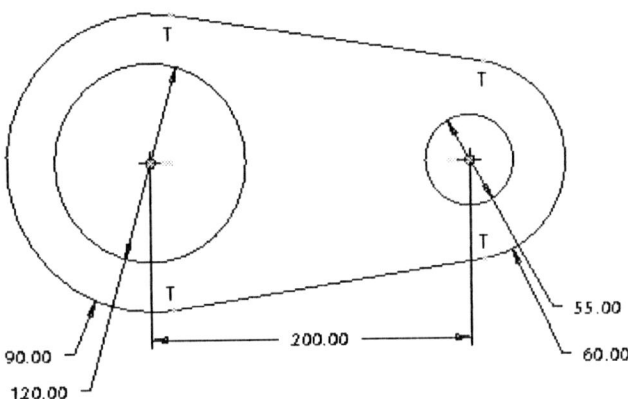

Figure 2-24 Sketch imported and placed in the current file

Drawing the Keyway

To create the keyway, you will sketch the keyway and then the portion of the circle that lies between the horizontal lines will be removed.

1. Choose the **Create 2 points line** button from the **Sketcher Tools** toolbar.

2. Draw the keyway, as shown in Figure 2-25; the weak dimensions and constraints are automatically applied to the sketch of the keyway.

   The horizontal lines of the keyway and the circle intersect at the points where the lines meet the circle. The intersection points are displayed in orange color. The portion of the circle that lies between the two horizontal lines of the keyway needs to be deleted from the circle.

3. Choose the **Zoom In** button from the **View** toolbar in the **Top Toolchest**; the cursor is converted into a magnifying glass symbol.

4. Draw a window around the keyway to zoom in. Now, the display of the keyway is enlarged.

5. Choose the **Dynamically trim section entities** button from the **Sketcher Tools** toolbar.

6. Click to select the part of the circle that lies between the two intersection points. The part of the circle that was between the two horizontal lines is deleted.

7. Choose the **Refit object to fully display it on the screen** button from the **View** toolbar to view the full sketch.

Dimensioning the Keyway

The dimensions that will be applied to the keyway are shown in Figure 2-25.

1. Choose the **Create defining dimensions** button from the **Sketcher Tools** toolbar.

2. Dimension the keyway, as shown in Figure 2-26.

Modifying the Dimensions

The dimensions of the keyway will be modified to the given dimension values.

1. Select the three dimensions of the keyway by using CTRL+left mouse button.

2. Choose the **Modify the values of dimensions, geometry of splines, or text entities** button from the **Sketcher Tools** toolbar. The **Modify Dimensions** dialog box is displayed.

3. Clear the **Regenerate** check box and then modify the values of the dimensions. When you clear the check box, the sketch does not regenerate as you modify the dimensions.

Creating Sketches in the Sketch Mode-II

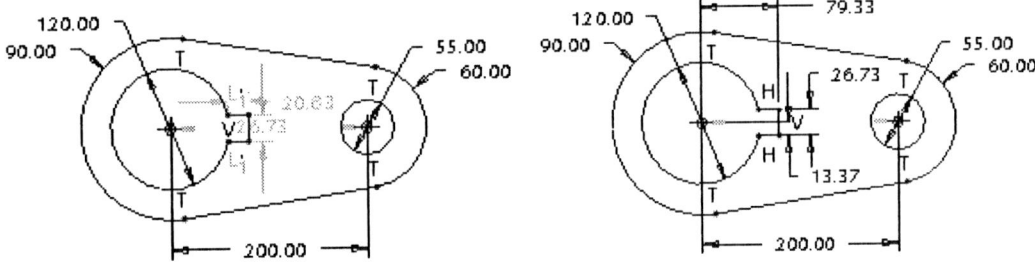

Figure 2-25 Sketch of the keyway with weak dimensions and constraints

Figure 2-26 Sketch after dimensioning the keyway

The dimension that you edit in the **Modify Dimensions** dialog box is enclosed in a yellow box in the sketch.

4. Modify all dimensions. Refer to Figure 2-23 for dimension values.

5. After the dimensions are modified, choose the **Regenerate the section and close the dialog** button from the **Modify Dimensions** dialog box. The message **Dimension modifications successfully completed** is displayed in the **Message Area**.

The sketch after modifying the dimension values of the sketch is shown in Figure 2-27.

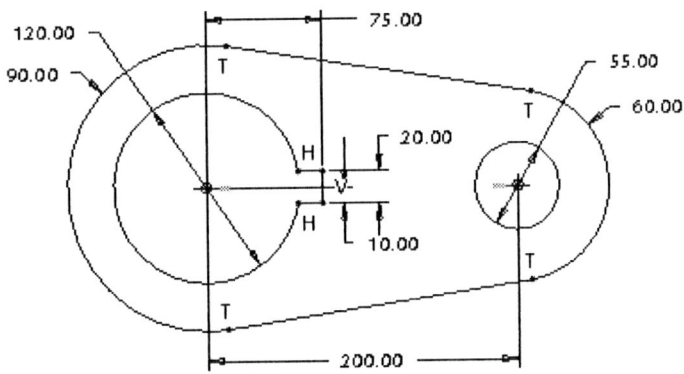

Figure 2-27 Sketch after modifying the dimensions

Saving the Sketch

You have to save the sketch because you may need the sketch later.

1. Choose the **Save the active object** button from the **File** toolbar. The **Save Object** dialog box is displayed with the name of the sketch that you had entered earlier.

2. Choose the **OK** button; the sketch is saved.

3. After saving the sketch, choose **Window > Close** from the menu bar to exit the **Sketch** mode.

Tutorial 2

In this tutorial, you will draw the sketch for the model shown in Figure 2-28. The sketch is shown in Figure 2-29.
(Expected time: 30 min)

Figure 2-28 Model for Tutorial 2

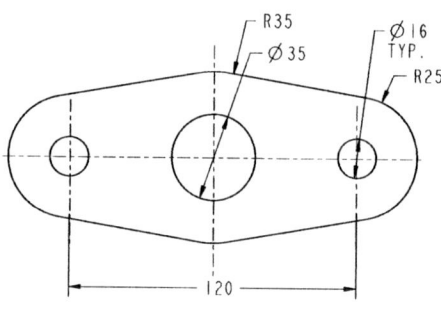

Figure 2-29 Sketch of the model

The following steps are required to complete this tutorial:

a. Set the working directory and create a new object file.
b. Draw the sketch using sketcher tools, refer to Figures 2-30 through 2-33.
c. Apply the required constraints and dimensions to the sketched entities, refer to Figure 2-35.
d. Modify the dimensions of the sketch and then save the sketch, refer to Figure 2-36.
e. Exit the **Sketch** mode.

Setting the Working Directory

The working directory was selected in Tutorial 1, and therefore there is no need to select the working directory again. But if a new session of Pro/Engineer is started, then you have to again set the working directory by following the steps given next.

1. Open the Navigator by clicking the top arrows on the left edge of the Pro/ENGINEER main window; the Navigator slides out.

Creating Sketches in the Sketch Mode-II

2. Click on the plus symbol adjacent to the *ProE-WF-2.0* folder in the Navigator; the contents of the *ProE-WF-2.0* folder are displayed.

3. Now right-click on the *c02* folder to display a shortcut menu. From this shortcut menu, choose the **Set Working Directory** option. The working directory is set to *c02*.

4. Close the Navigator by clicking the sash on the right edge of the Navigator. The Navigator slides in.

Starting a New Object File

1. Start a new object file in the **Sketch** mode. Name the file as *c02tut2*.

Drawing the Sketch

To draw the outer loop, you need to draw three circles and then draw lines tangent to the three circles.

1. Choose the **Create circle by picking the center and a point on the circle** button. Draw the three circles, as shown in Figure 2-30.

2. Choose the black arrow on the right of the **Create 2 points line** button to display a flyout. Choose the **Create lines tangent to 2 entities** button from this flyout.

3. Select the left circle at the top and then select the middle circle at the top. A tangent is drawn from the top of the left circle to the top of the middle circle.

4. Select the right circle at the top and the middle circle at the top. A tangent is drawn from the top of the right circle to the top of the middle circle.

5. Similarly, using the **Create lines tangent to 2 entities** button, draw the other tangents from the bottom of the circles, as shown in Figure 2-31.

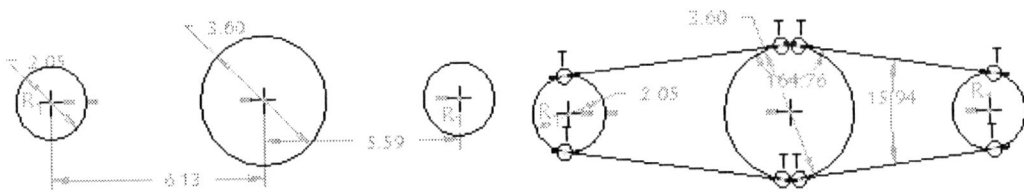

Figure 2-30 Three circles with weak dimensions and constraints

Figure 2-31 Lines joining the three circles with weak dimensions

Trimming the Circles

The inner portions of the circles are not required. Therefore, you need to trim them.

1. Choose the **Dynamically trim section entities** button from the **Sketcher Tools** toolbar.

2. Bring the cursor close to the right portion of the left circle. The right part of the circle turns cyan in color. Click on the right portion of the circle to delete it.

3. Similarly, trim the parts of the middle and the right circle that are not required. The sketch after trimming the circles is shown in Figure 2-32.

Drawing the Circles

1. Choose the black arrow on the right of the **Create circle by picking the center and a point on the circle** button to display the flyout. From this flyout, choose the **Create concentric circle** button. You are prompted to select an arc.

Tip: *While drawing a concentric circle, sometimes the circle snaps to the other circle or arc and it becomes difficult to draw a circle of the size you need. The snapping of the circle to the other circle or arc can be disabled. While drawing the circle when the circle snaps to the other circle, use TAB+right-click to disable the snapping or to disable the equal radii constraint that the system tends to apply.*

2. Click on the left arc. Notice that when you move the mouse, a circle appears. Select a point inside the sketch to complete the circle. Press the middle mouse button.

Note
You may need to zoom in to select the top arc in the next step.

3. Click on the top arc. Notice that when you move the mouse a circle appears. Select a point inside the sketch to complete the circle. Press the middle mouse button to end circle creation.

4. Click on the right arc. Notice that when you move the mouse a circle appears. When R_2 symbol appears on the circle, click to complete the circle. Press the middle mouse button to end circle creation.

The R_2 symbol appears on both the left and the right circle indicating that their radii are same. The constraint symbols like R_1 or R_2 vary from sketch to sketch.

The sketch after drawing the three circles is shown in Figure 2-33.

Applying the Constraints

1. Choose the **Impose sketcher constraints on the section** button from the **Sketcher Tools** toolbar. The **Constraints** dialog box is displayed.

Creating Sketches in the Sketch Mode-II 2-23

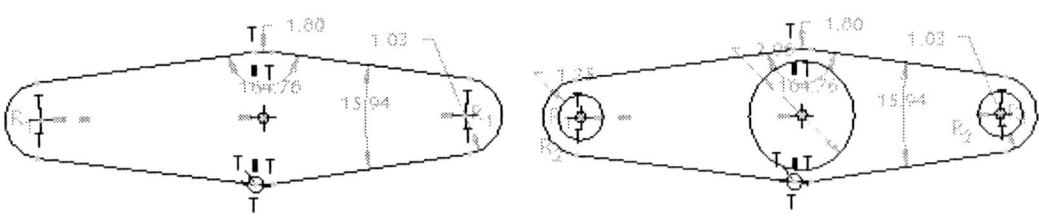

Figure 2-32 Sketch after trimming the circles   *Figure 2-33 Sketch after drawing the three circles*

2. Choose the **Create Equal Lengths, Equal Radii, or Same Curvature constraint** button from the **Constraints** dialog box.

3. Select the left arc and then select the right arc to apply the equal radius constraint.

4. Select the left circle and the right circle to apply the equal radius constraint.

5. Use the CTRL+middle mouse button to zoom in and now select the top arc and the bottom arc to apply the equal radius constraint. If the constraint is already applied, the **Resolve Sketch** dialog box is displayed, as shown in Figure 2-34.

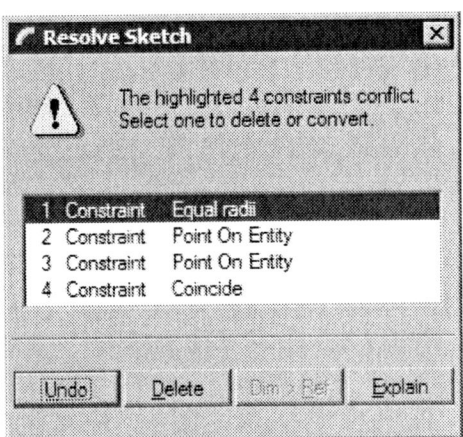

*Figure 2-34 The **Resolve Sketch** dialog box*

6. Select the third constraint in the **Resolve Sketch** dialog box and choose the **Delete** button. The constraint that was conflicting is deleted.

7. Choose the **Make two lines parallel** constraint button from the **Constraints** dialog box.

8. Click to select the tangent line that connects the left circle and the middle circle at the top.

9. Click to select the tangent line that connects the right circle and the middle circle at the bottom.

   It is evident from the parallel constraint symbol that the parallel constraint is applied to the two tangent lines in the sketch.

10. Similarly, apply the parallel constraint to the other set of tangent lines. The sketch after applying the constraints is shown in Figure 2-35.

11. Choose the **Refit object to fully display it on the screen** button from the **View** toolbar to view the full sketch in the graphics window.

Dimensioning the Sketch

Pro/ENGINEER applies the weak dimensions to the sketch automatically. These dimensions are not the dimensions that are needed because these dimensions will not help to machine the model. Therefore, you will dimension the sketch with the dimensions that will be used to machine the model.

1. Choose the **Create defining dimensions** button from the **Sketch Tool** toolbar.

2. Select the center of right and left circles; the centers of the circles turn red in color. Now, using the middle mouse button, place the dimension below the sketch.

   The rest of the weak dimensions are needed and therefore they will be modified.

Modifying the Dimensions

All constraints and dimensions are applied to the sketch and now dimensions will be modified.

1. Select all dimensions using CTRL+ALT+A.

2. Choose the **Modify the values of dimensions, geometry of splines, or text entities** button. The **Modify Dimensions** dialog box is displayed.

3. Clear the **Regenerate** check box and then modify the values of the dimensions. When you clear the check box, the sketch does not regenerate as you modify the dimensions.

   The dimension that you edit in the **Modify Dimensions** dialog box is enclosed in a yellow box in the sketch.

4. Modify all dimensions. Refer to Figure 2-29 for dimension values.

5. After the dimensions are modified, choose the **Regenerate the section and close the dialog** button from the **Modify Dimensions** dialog box. The message **Dimension modifications successfully completed** is displayed in the **Message Area**.

Creating Sketches in the Sketch Mode-II

The sketch after modifying the dimension values is shown in Figure 2-36.

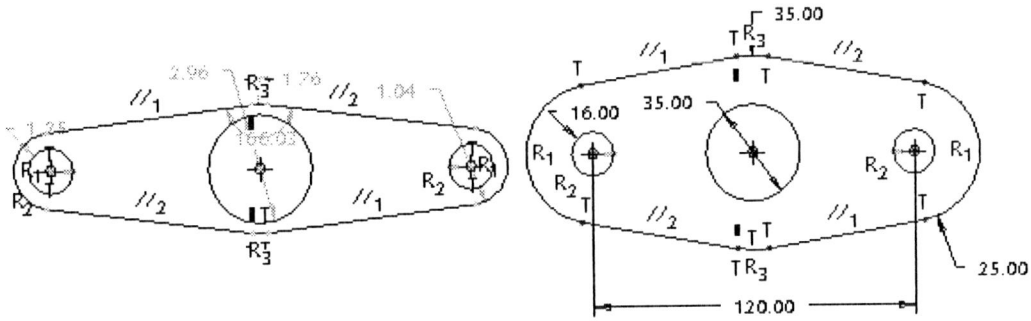

Figure 2-35 Sketch with constraints   *Figure 2-36* Sketch after modifying the dimensions

Saving the Sketch
1. Choose the **Save the active object** button from the **File** toolbar and save the sketch.

Exiting the Sketch Mode
1. After saving the sketch, choose **Window > Close** from the menu bar to exit the **Sketch** mode.

Tutorial 3

In this tutorial, you will draw the sketch for the model shown in Figure 2-37. The sketch is shown in Figure 2-38.
(Expected time: 30 min)

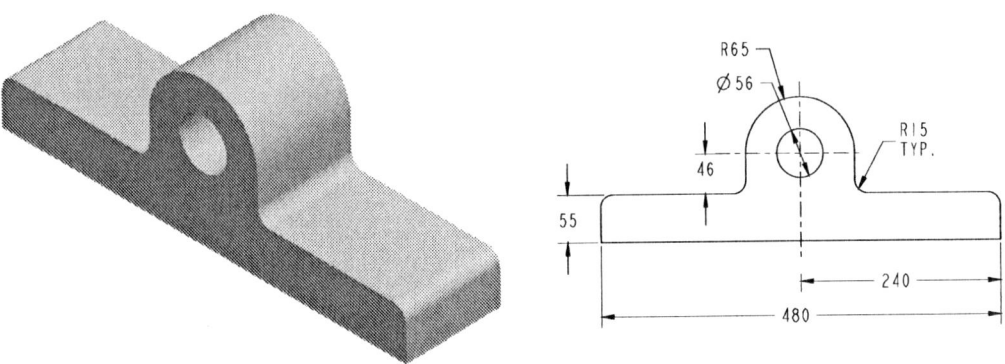

Figure 2-37 Model for Tutorial 3   *Figure 2-38* Sketch of the model

The following steps are required to complete this tutorial:

a. Set the working directory and create a new object file.

b. Draw the sketch using sketcher tools, refer to Figures 2-39 and 2-40.
c. Apply fillets at two corners of the sketch, refer to Figures 2-41 and 2-42.
d. Dimension the sketch, refer to Figure 2-43.
e. Modify dimensions of the sketch, refer to Figure 2-44.
f. Save the sketch and exit the **Sketch** mode.

Setting the Working Directory

The working directory was selected in Tutorial 1, and therefore there is no need to select the working directory again. But if a new session of Pro/Engineer is started, then you have to again set the working directory by following the steps given next.

1. Open the Navigator by clicking the top arrows on the left edge of the Pro/ENGINEER main window; the Navigator slides out.

2. Click on the plus symbol adjacent to the *ProE-WF-2.0* folder in the Navigator. The contents of the *ProE-WF-2.0* folder are displayed.

3. Now right-click on the *c02* folder to display a shortcut menu. From this shortcut menu, choose the **Set Working Directory** option; the working directory is set to *c02*.

4. Close the Navigator by clicking the top arrows on the right edge of the Navigator; the Navigator slides in.

Starting a New Object File

1. Start a new object file in the **Sketch** mode. Name the file as *c02tut3*.

Drawing the Sketch

1. Choose the **Create 2 points line** button from the **Sketcher Tools** toolbar.

2. Draw the lines, as shown in Figure 2-39.

3. Choose the **Create an arc by 3 points or tangent to an entity at its endpoint** button from the **Sketcher Tools** toolbar.

4. Select the endpoint of the left vertical line as the start point of the arc. Complete the arc at the endpoint of the right vertical line.

5. Choose the black arrow on the right of the **Create circle by picking the center and a point on the circle** button to display the flyout. From this flyout, choose the **Create concentric circle** button; you are prompted to select an arc.

6. Click on the arc; a red rubber-band circle appears. Size the circle by moving the cursor and click to complete it.

The sketch after drawing the circle is shown in Figure 2-40.

Creating Sketches in the Sketch Mode-II   2-27

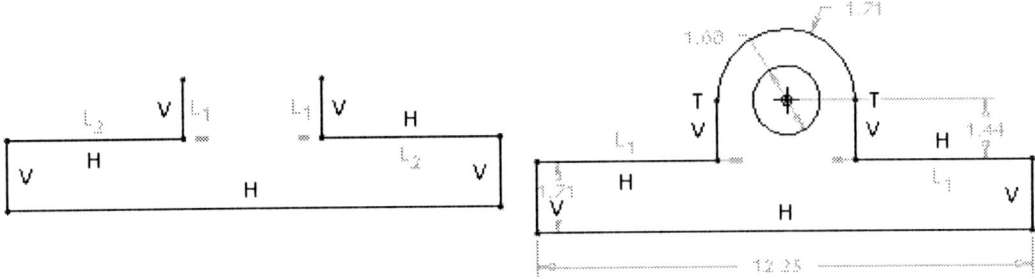

Figure 2-39 Lines in the sketch with the dimensions turned off for clarity

Figure 2-40 Sketch after drawing the arc and the circle

Note
Choose the **Toggle display of dimensions on/off** *button from the* **Sketcher** *toolbar in the* **Top Toolchest** *to turn the dimensions on or off.*

Pro/ENGINEER does not have options like midpoint, endpoint, or center of. However, while drawing a sketch, these options are applied in the form of weak constraints as you sketch. For example, endpoint of any entity snaps the cursor when a new entity is to be drawn. The middle point constraint appears when you approximately bring the cursor near to the middle point of the line to draw a line.

Filleting the Corners

1. Choose the **Create a circular fillet between two entities** button from the **Sketcher Tools** toolbar. You are prompted to select the two entities to be filleted. The corners that you need to fillet are shown in Figure 2-41.

2. Select the two entities one by one using the left mouse button. The corner of the selected lines are filleted. Fillet all required corners shown in Figure 2-41.

   The sketch after filleting is shown in Figure 2-42.

Applying the Constraints

1. Choose the **Impose sketcher constraints on the section** button from the **Sketcher Tools** toolbar. The **Constraints** dialog box is displayed.

2. Choose the **Create Equal Lengths, Equal Radii, or Same Curvature constraint** button from the **Constraints** dialog box.

3. Click to select the fillets and apply the equal constraint to all fillets.

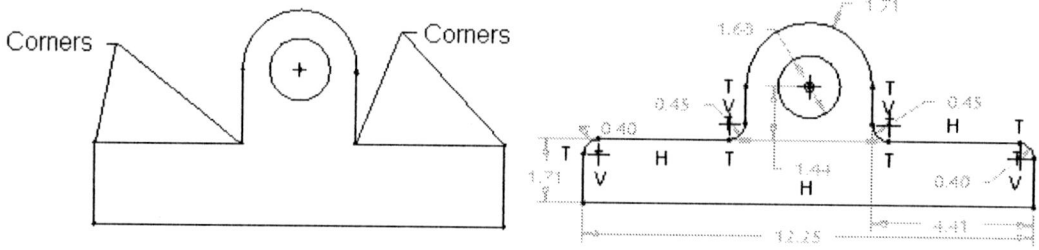

Figure 2-41 Fillet corners   *Figure 2-42* Sketch after filleting

Dimensioning the Sketch

The weak dimensions are applied to the sketch automatically. These are not the required dimensions and therefore you need to manually dimension the sketch.

1. Choose the **Create defining dimensions** button from the **Sketch Tools** toolbar.

2. Dimension the sketch, as shown in Figure 2-43.

Modifying the Dimensions

You need to modify the dimension values that are assigned to the sketch.

1. Select all dimensions using CTRL+ALT+A.

2. Choose the **Modify the values of dimensions, geometry of splines, or text entities** button. The **Modify Dimensions** dialog box is displayed.

3. Clear the **Regenerate** check box and then modify the values of the dimensions. When you clear the check box, the sketch does not regenerate as you modify the dimensions.

    The dimension that you edit in the **Modify Dimensions** dialog box is enclosed in a yellow box in the sketch.

4. Modify all dimensions. Refer to Figure 2-38 for dimension values.

5. After the dimensions are modified, choose the **Regenerate the section and close the dialog** button from the **Modify Dimensions** dialog box.

    The sketch after modifying the dimension values is shown in Figure 2-44.

Creating Sketches in the Sketch Mode-II

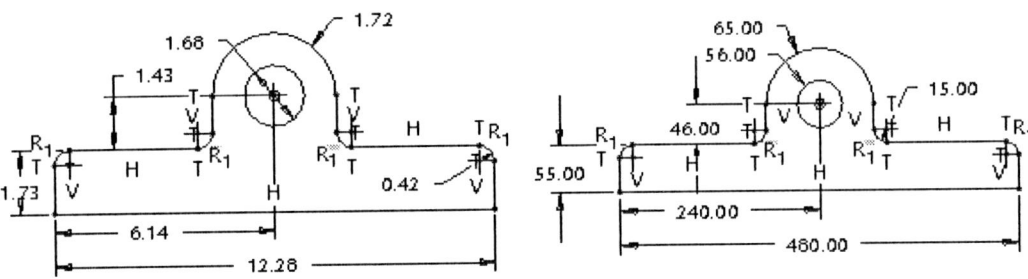

Figure 2-43 *Sketch after dimensioning*   Figure 2-44 *Sketch after modifying the dimensions*

Saving the Sketch
1. Choose the **Save the active object** button from the **File** toolbar and save the sketch.

Exiting the Sketch Mode
1. Choose **Window > Close** from the menu bar to exit the **Sketch** mode.

Self-Evaluation Test

Answer the following questions and then compare your answers with those given at the end of this chapter.

1. You can increase the length of the spline by adding points before the start point and after the endpoint of the spline. (T/F)

2. While copying sketched entities, the **Scale Rotate** dialog box is also displayed. (T/F)

3. When you modify a weak dimension it becomes strong. (T/F)

4. You can dimension the length of a centerline. (T/F)

5. The font of the text written in the sketcher environment is fixed and cannot be modified. (T/F)

6. The _____ of a spline can be dynamically modified.

7. The _____ dialog box is used to modify dimensions.

8. The display of dimensions and constraints can be turned on/off by choosing the _____ and _____ buttons respectively from the **Sketcher** toolbar.

9. The _____ button is used to rotate selected entities.

10. You can delete the entities by selecting them and then using the _____ key on the keyboard.

Review Questions

Answer the following questions:

1. How many handles are displayed in the graphics window while scaling and rotating entities?

    (a) one (b) two
    (c) three (d) four

2. Which of the following mouse buttons is used to place the dimension?

    (a) left (b) middle
    (c) right (d) mouse is not used for dimensioning

3. Which of the following is the default font for the text in the sketcher environment?

    (a) font (b) filled
    (c) font3d (d) isofont

4. Which of the following toolbar is used to toggle the display of dimensions and constraints in the sketcher environment?

    (a) **Sketcher Tools** (b) **Sketcher**
    (c) **File** (d) **Edit**

5. In which type of dimensioning the **Dim Orientation** dialog box is displayed while dimensioning the arcs and circles?

    (a) **Normal** (b) **Perimeter**
    (c) **Baseline** (d) None of the above

6. While placing a section in a new sketch, use the right mouse button to place the section. (T/F)

7. Elliptical fillets can be created in Pro/Engineer . (T/F)

8. While creating text in the sketcher environment, you need to draw a construction line that will define the height of the text. (T/F)

9. You can modify the dimensions dynamically. (T/F)

10. The modification of a spline is possible by moving its interpolation points. (T/F)

Creating Sketches in the Sketch Mode-II

Exercises

Exercise 1

In this exercise, you will draw the sketch for the model shown in Figure 2-45. The sketch is shown in Figure 2-46. **(Expected time: 30 min)**

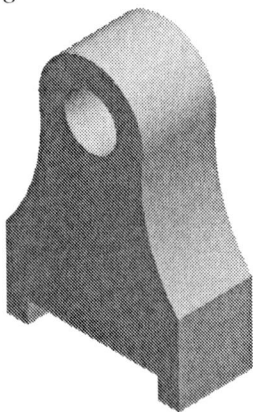
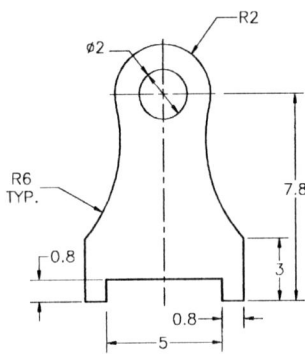

Figure 2-45 Solid model for Exercise 1   *Figure 2-46* Sketch of the model

Exercise 2

In this exercise, you will draw the sketch for the model shown in Figure 2-47. The sketch is shown in Figure 2-48. **(Expected time: 15 min)**

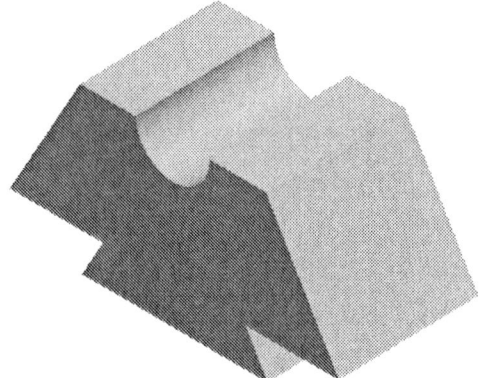
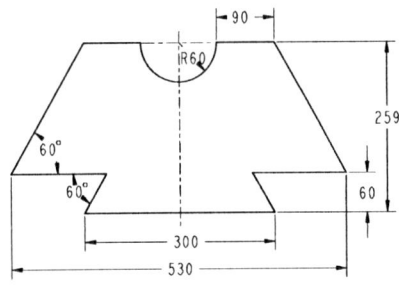

Figure 2-47 Solid model for Exercise 2   *Figure 2-48* Sketch of the model

Exercise 3

In this exercise, you will draw the sketch for the model shown in Figure 2-49. The sketch is shown in Figure 2-50. **(Expected time: 30 min)**

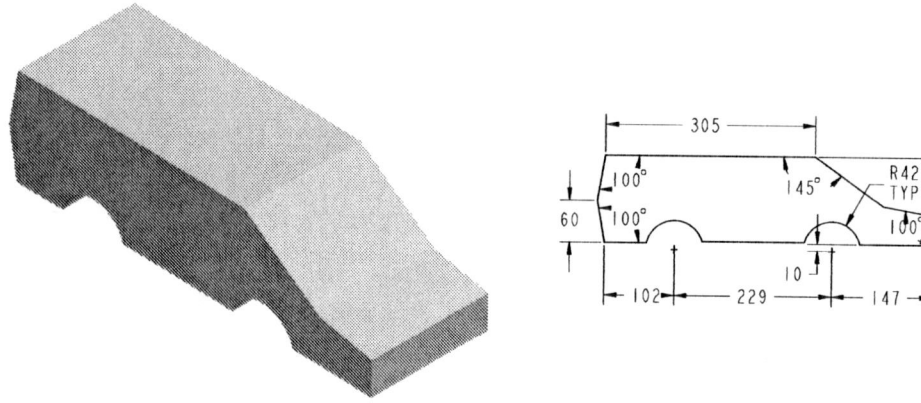

Figure 2-49 Solid model for Exercise 3   *Figure 2-50* Sketch for the model

Answers to the Self-Evaluation Test

1 - F, 2 - T, 3 - T, 4 - F, 5 - F, 6 - shape, 7 - **Modify Dimensions**, 8 - Toggle display of dimensions on/off, Toggle display of constraints on/off, 9 - Scale and rotate selected entities, 10 - DELETE.

Chapter 3

Creating Base Features

Learning Objectives

After completing this chapter you will be able to:
- *Use default datums for the base feature.*
- *Create a solid feature using the Extrude Tool.*
- *Create a thin feature using the Extrude Tool.*
- *Create a solid feature using the Revolve Tool.*
- *Create a thin feature using the Revolve Tool.*
- *Specify depth of extrusion to a solid feature.*
- *Specify angle of revolution to a revolved feature.*
- *Orient the datum planes.*
- *Understand Parent Child relationships.*
- *Understand nesting of sketches.*

CREATING BASE FEATURES

The base feature is the first solid feature created while creating a model in the **Part** mode. The base feature is created using the datum planes. However, they can also be created without using the datum planes. But in that case, you do not have proper control over the orientation of feature and direction of feature creation.

While creating the base feature of a model, the designer should be extra careful in selecting the attributes to create it. This is because if the base feature is created wrong then the features created on it are also created wrong. This results in waste of time and effort. Although, Pro/ENGINEER provides you with the options to redefine a feature, doing that also consumes additional time and effort. To create the base feature, you need to enter the **Part** mode.

Note
It is recommended that you set the working directory before you open a new file.

INVOKING THE PART MODE

To invoke the **Part** mode, choose **New** from the **File** menu or choose the **Create a new object** button from the **File** toolbar. The **New** dialog box is displayed with various modes of Pro/ENGINEER. The **Part** radio button in the **Type** area and the **Solid** radio button in the **Sub-type** area is selected by default in the **New** dialog box. The default name of the part file also appears in the **Name** edit box. You can change the part name as desired and then choose the **OK** button to enter the **Part** mode.

In the **New** dialog box, the **Use default template** check box is selected by default. This means that you have selected to use the default template provided in Pro/ENGINEER. This template has certain parameters that will be assigned to the part file that you will create. For example, the units of this model will be **Inches lbm Second**. The length is in inches, mass in lb, time in seconds, and temperature in fahrenheit. If this check box is not selected, and you choose the **OK** button, the **New File Options** dialog box is displayed, as shown in Figure 3-1.

From the **New File Options** dialog box you can select the required template file. If you want the default system of units to be **mmNs** (millimeter Newton sec), then select the **mmns_part_solid** template and choose the **OK** button from this dialog box.

The **Part** mode is the most commonly used modes of Pro/ENGINEER Wildfire 2.0. This is because solid modeling is done in this mode. It should be noted that a solid model is the start of a product development cycle. Product development cycle refers to the development of a product from scratch to its prototype. If you have created a solid model then it can further be used to generate its drawing views, numerically controlled (NC) machining codes, core and cavity extraction, analysis of the solid model, and so on.

The two-dimensional (2D) sketch drawn in the **Sketch** mode can be converted into a three-dimensional (3D) model in the **Part** mode. The **Part** mode contains the same sketcher environment with similar options to sketch as those available in the **Sketch** mode. There are some sketcher options in the **Part** mode that are not available in the **Sketch** mode because they do not have any use in the **Sketch** mode.

Creating Base Features

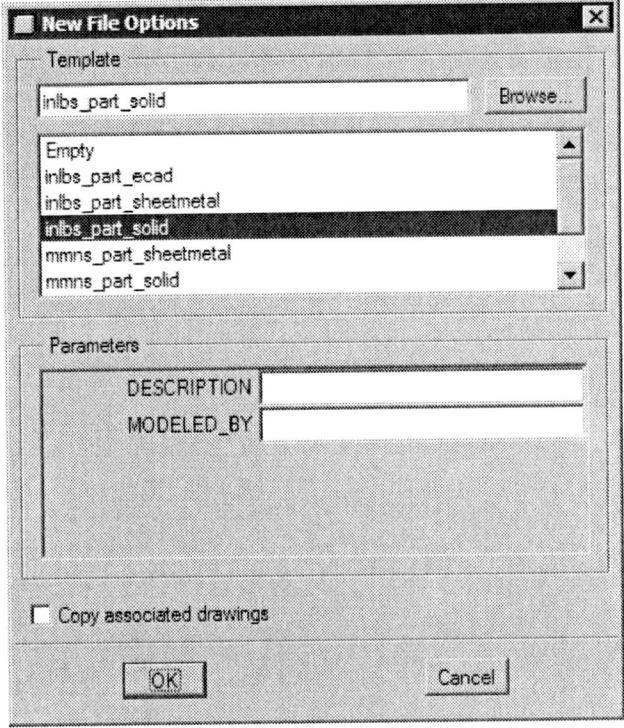

*Figure 3-1 The **New File Options** dialog box*

Figure 3-2 shows you the initial screen appearance in the **Part** mode. The display consists of **Model Tree**, three default datum planes, and different toolbars.

THE DEFAULT DATUM PLANES

Generally, the first feature in the **Part** mode are the three default datum planes. These datum planes are used to create the base feature. Also, these datum planes are used to draw a 2D sketch and then convert it to a 3D model by protrusion. The base feature you create is referenced with the default datum planes.

Note
*Although it is said that the three default datum planes are the first feature in the **Part** mode, in the **Model Tree** the **RIGHT**, **TOP**, and **FRONT** datum planes appear as separate features. If you delete any one of them, only that datum plane is deleted.*

Tip: *It is recommended that you always use the datum planes to create a base feature. This is because the model created using the datum planes can be easily oriented. The uses of datum planes are discussed in Chapter 4.*

These three default datum planes are mutually perpendicular to each other. They are not referenced to each other and are individual features. When a solid model is created, the

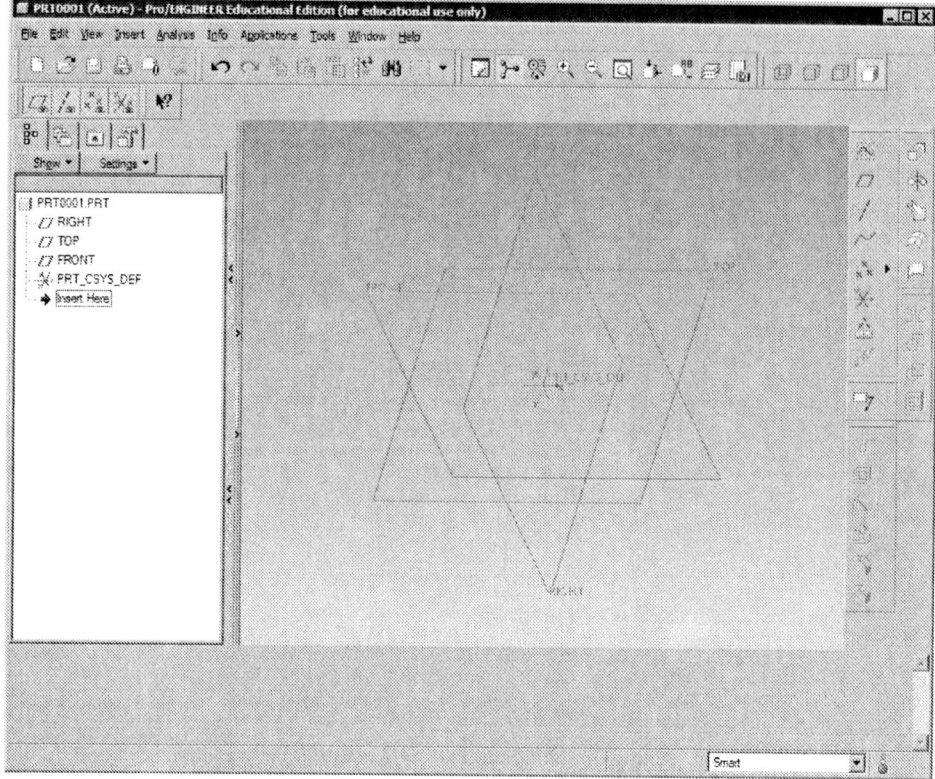

*Figure 3-2 The initial screen appearance in the **Part** mode*

datum planes adjust their size to the size of the model. You can create any number of datum planes as per your requirement. The creation of additional datum planes is discussed in Chapter 4.

In the releases prior to Release $2000i^2$ of Pro/ENGINEER, you had to create the default datum planes. But in recent releases, a default template is provided, with the three default datum planes. However, if you do not need the default datum planes, then at the time of creating a new file, you need to clear the **Use default template** check box in the **New** dialog box and select the **Empty** template.

Note
In this chapter, you will learn to use the three default datum planes to create the base features. It is important to remember the feature-based nature of Pro/ENGINEER while you are working on this chapter. The feature-based nature of Pro/ENGINEER has been discussed in the Introduction.

It is important to create a model in the correct orientation. Its correct orientation is the one in which it will be later used either in assembly or other modes. For example, the Casting component of the Plummer block is modeled in such a way such that it always stands vertical.

CREATING A PROTRUSION

Protrusion is defined as the process of adding material defined by a sketched section. In Pro/ENGINEER, there are various options of adding material such as **Extrude**, **Revolve**, **Sweep**, and so on. These options can be selected from the **Insert** menu in the menu bar. Figure 3-3 shows the options available in the **Insert** menu.

The **Base Features** toolbar shown in Figure 3-4 can also be used to invoke the protrusion options. This toolbar is available in the **Right Toolchest** when you are in the **Part** mode.

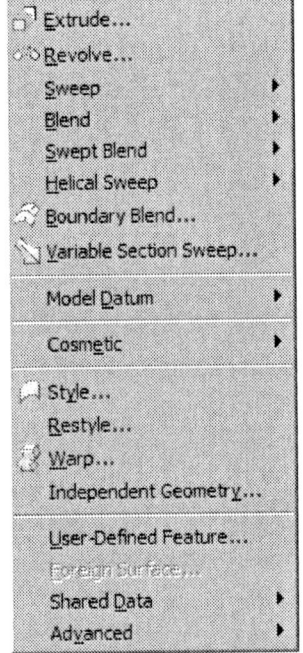

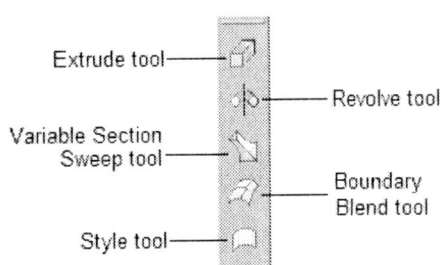

Figure 3-3 Options in partial view of the **Insert** menu

Figure 3-4 The **Base Features** toolbar

Extruding a Sketch

 The **Extrude Tool** button in the **Base Features** toolbar adds material defined by a sketch drawn on a sketching plane. The material is added in the direction of feature creation.

The procedure to create a base feature using the **Extrude Tool** button is explained next.

1. When you choose the **Extrude Tool** button from the **Base Features** toolbar or choose **Insert > Extrude** from the menu bar, a dashboard is displayed below the graphics window, as shown in Figure 3-5. This dashboard has all options to define the extrude feature.

2. Select the **Placement** tab to display the slide-up panel. Choose the **Define** button from the slide-up panel to display the **Sketch** dialog box, as shown in Figure 3-6. You are

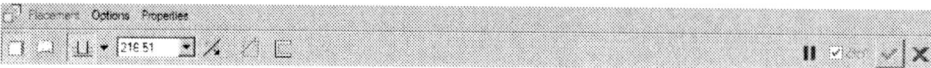

*Figure 3-5 The **Extrude** dashboard*

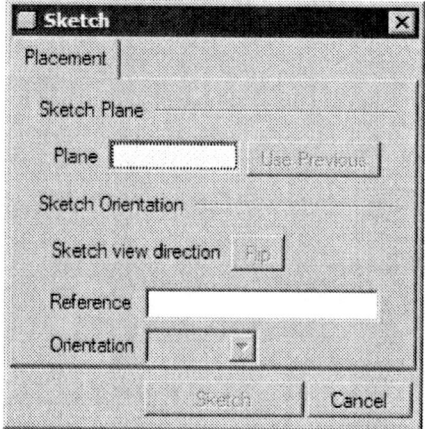

*Figure 3-6 The **Sketch** dialog box*

prompted to select a sketching plane. Select the datum plane named **FRONT** from the graphics window.

In the **Sketch** dialog box, the name of the plane that you have selected appears in the **Plane** collector. At the same time, the reference plane is also selected automatically. The name of reference datum plane appears in the **Reference** collector of the **Sketch** dialog box. Pro/ENGINEER also sets the orientation of the reference plane automatically.

3. After selecting the sketching plane and the reference plane, choose the **Sketch** button in the **Sketch** dialog box.

4. Now, you have entered the sketcher environment. You will notice that the **References** dialog box is displayed on the top right corner of the screen. The status displayed under the **Reference status** area is **Fully Placed**. This suggests that the references are selected by default. The **FRONT** datum plane is the sketching plane and is oriented parallel to the screen.

5. Close the **Model Tree** by clicking the sash on its right edge. This increases the drawing space in the graphics window and its appearance is similar to that shown in Figure 3-7.

6. Close the **References** dialog box by choosing the **Close** button. Now, choose the **Create 2 point lines** button to draw the right half of a I-section and then draw a vertical center line aligned with the **RIGHT** datum plane. Select all lines in the right half of the I-section and choose the **Mirror selected entities** button. You will be prompted to select a center line. Select the center line to mirror the right half and to create the I-section, as

Creating Base Features   3-7

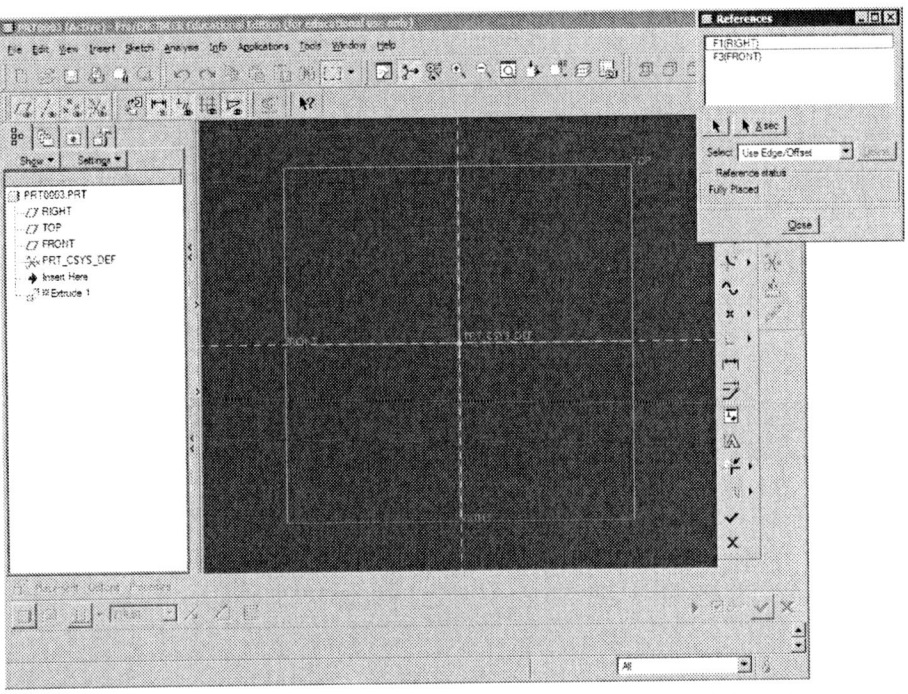

Figure 3-7 *The graphics window after entering the sketcher environment and the **References** dialog box*

shown in Figure 3-8. Assume the dimensions. After the I-section is sketched, choose
the **Continue with the current section** button from the **Sketcher Tools** toolbar.

7. After doing so, the dashboard is enabled and displayed below the graphics window. The model is created by assuming some default attributes and it is displayed in yellow color. The attributes that a model has are available on the dashboard and are discussed later in this chapter.

8. On the dashboard, there is a drop-down list containing some default value. This value is the depth of extrusion of the sketch you have created. Enter an appropriate value in this drop-down list and press ENTER. You can also hold and drag the handle, displayed on the model, to dynamically specify the depth of extrusion. This drag handle will be visible only after the model is rotated slightly using the middle mouse button.

9. From the dashboard, choose the green button. The required 3D model is displayed in the graphics window. However, it appears as a 2D entity. You need to change the display such that you can also view the depth of the model. Change the display of the part from the **View** toolbar by choosing the **Saved view list** button and selecting the **Default Orientation** option from the drop-down list that appears. Figure 3-8 shows the I-section after extruding it to a certain depth.

Figures 3-9 and 3-10 show some models created using the **Extrude Tool** button.

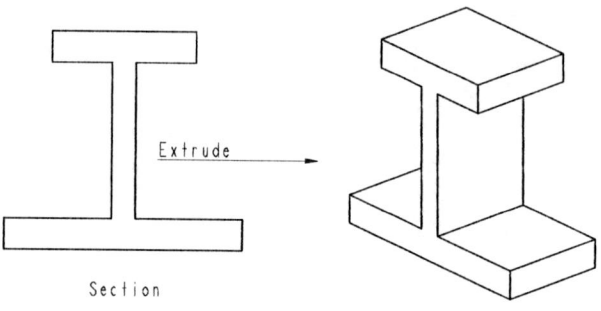

Figure 3-8 The I-section extruded to a certain depth

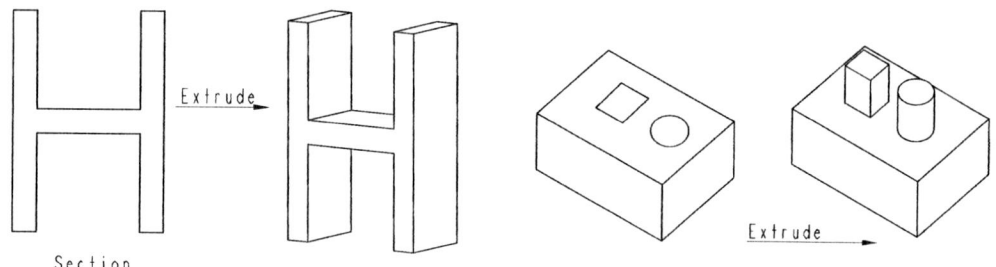

Figure 3-9 Model created using the **Extrude Tool** button

Figure 3-10 Model created using the **Extrude Tool** button

The above steps explain how to construct a 3D model using the **Extrude Tool** button. The **Extrude** dashboard, the **Sketch** dialog box, and the **Reference** dialog box that you came across while creating the 3D model from an I-section are discussed next.

The Extrude Dashboard

The options and buttons in the **Extrude** dashboard shown in Figure 3-11 are used to extrude the sketch and to specify certain attributes related to the model. These attributes can be assigned to the model before or after the sketch of the model is drawn. The tabs and the tool buttons that are available in the dashboard are discussed next.

Figure 3-11 The **Extrude** dashboard

Placement Tab

When you choose the **Placement** tab, the slide-up panel is displayed, as shown in Figure 3-12. The collector in this slide-up panel displays **Select 1 item** because the section has not been defined yet. You can select a sketch that is drawn using a datum curve or you can choose the **Define** button to draw a sketch. When you choose the

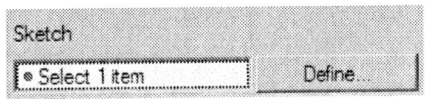

*Figure 3-12 The **Placement** tab slide-up panel*

Define button from this slide-up panel, the **Sketch** dialog box is displayed.

Options Tab

When you choose the **Options** tab, the slide-up panel is displayed, as shown in Figure 3-13. This slide-up panel is used to specify whether you want the sketch to extrude to one side of the sketching plane or to both sides of the sketching plane. It has **Side 1** and **Side 2** drop-down lists. The options in the **Side 1** drop-down list are explained next.

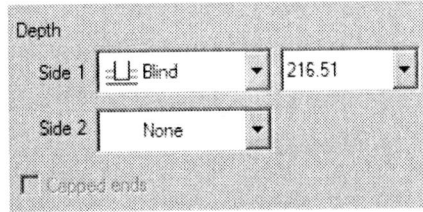

*Figure 3-13 The **Options** tab slide-up panel*

Blind

The **Blind** option is one of the most commonly used options to define the extrusion depth of the sketch by specifying a depth value. When you select this option from the **Side 1** drop-down list, a default value appears in the dimension box that is adjacent to the **Side 1** drop-down list. You can enter a value in this dimension box and press ENTER. The depth of extrusion created by this value can be modified if needed.

Symmetric

If you choose the **Symmetric** option from the **Side 1** drop-down list, the material is added equally in both the directions of the sketching plane. When you select this option, the **Side 2** drop-down list becomes inactive. This means that you cannot choose any option from the **Side 2** drop-down list.

To Selected

The **To Selected** option allows you to select a point, surface, plane, or a curve up to which the section is to be extruded.

The options in the **Side 2** drop-down list are similar to the options available in the **Side 1** drop-down list. The **None** option in this drop-down list allows you to extrude the sketch in the first direction only. The **Blind** and **To Selected** options give the depth of extrusion to the sketch in the second direction.

Properties Tab

When you choose the **Properties** tab, the slide-up panel is displayed. This slide-up panel displays the feature identity in the **Name** collector. The **i** button in the slide-up panel when selected opens the browser and all information about the feature you are creating is displayed in it. The browser is discussed in the Introduction chapter.

Tip: *The features created using the **To Selected** option do not have a dimension associated with them, and hence, they cannot be modified by changing the dimension value. However, changing the terminating surface changes the depth of the feature. You will understand this better when you learn the modification of an existing feature.*

*The extrusion depth given using the **Symmetric** option does not appear when you generate the dimensions in the drawing views of the model. The drawing views of the model are generated in the Drawing mode of Pro/ENGINEER and are discussed in Chapter 10 of this book..*

Extrude as solid Button

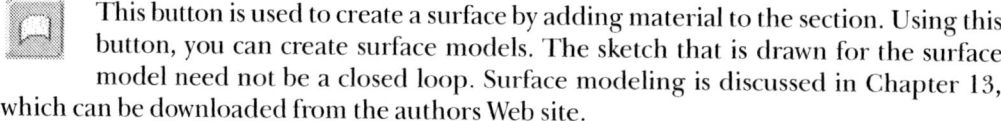
This button is chosen by default and is used to create a solid by adding material to the section. When you select a sketch to extrude using this button, the sketch should be a single closed loop.

Extrude as surface Button

This button is used to create a surface by adding material to the section. Using this button, you can create surface models. The sketch that is drawn for the surface model need not be a closed loop. Surface modeling is discussed in Chapter 13, which can be downloaded from the authors Web site.

Note
*The tabs and the tool buttons that are available on the **Extrude** dashboard can either be used before drawing the sketch or after drawing it.*

Change depth direction of extrude to other side of sketch Button

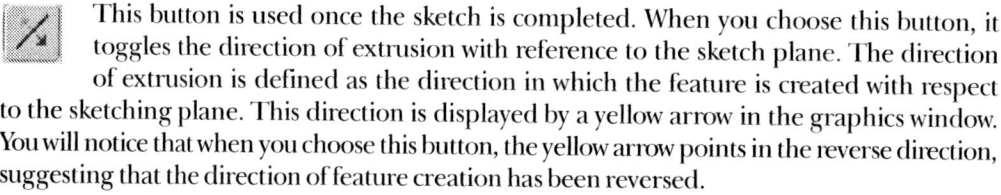
This button is used once the sketch is completed. When you choose this button, it toggles the direction of extrusion with reference to the sketch plane. The direction of extrusion is defined as the direction in which the feature is created with respect to the sketching plane. This direction is displayed by a yellow arrow in the graphics window. You will notice that when you choose this button, the yellow arrow points in the reverse direction, suggesting that the direction of feature creation has been reversed.

Note
The direction of the yellow arrow depends on the type of extrude being created. If material has to be added to a feature then the arrow by default points in the direction toward the user, that is, out of the screen. But, if a cut feature is being created in which material has to be removed from a feature, then, by default the arrow points into the screen.

If you spin the model using the middle mouse button then the direction of the arrow can be easily recognized.

Remove Material Button

The **Remove Material** button is available on the dashboard only after the base feature is created. This is because this button is used to remove material from an existing feature. Therefore, when you create a base feature this button is not available. The use of this button is explained in later chapters.

Thicken Sketch Button

The **Thicken Sketch** button is used to create thin parts with specified thickness. When you choose this button, the **Change direction of extrude between one side, other side, or both sides of sketch** button and a dimension box appear on the right of this button.

The **Change direction of extrude between one side, other side, or both sides of sketch** button is used to set the direction with respect to the sketch where the thickness should be added. This button appears on the dashboard only when the sketch is completed and you exit the sketcher environment. This button serves three functions. It is used to set the thickness direction to either side of the sketch and even to both sides of the sketch symmetrically. When the sketch is completed, the preview of the model is displayed in the graphics window. By default, the thickness to the sketch is applied on one side. Now, when you choose the **Change direction of extrude between one side, other side, or both sides of sketch** button, the thickness is applied to the other side of the sketch. When you choose this button for the second time, the thickness is applied symmetrically to both the sides of the sketch.

The dimension box that appears when you choose the **Thicken Sketch** button is used to enter the thickness value of the feature to be created. This edit box is available only when you exit the sketcher environment.

Figure 3-14 shows the arrow pointing away from the side of the section wall where the material will be added. If you want to change the direction of the arrow, choose the **Change direction of extrude between one side, other side, or both sides of sketch** button. Figure 3-15 shows the thickness of the material added to one side of the wall pointed by the arrow.

You can choose the **Change direction of extrude between one side, other side, or both sides of sketch** button such that the material is added symmetrically to both the sides of the section wall, as shown in Figure 3-16. Figure 3-17 shows the model created using the **Thicken Sketch** button.

Pause tool Button

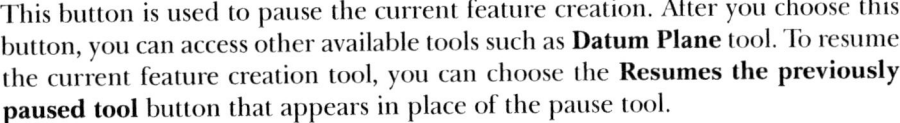
This button is used to pause the current feature creation. After you choose this button, you can access other available tools such as **Datum Plane** tool. To resume the current feature creation tool, you can choose the **Resumes the previously paused tool** button that appears in place of the pause tool.

Geometry preview/Feature preview Button

This button has two functions. When the check box on this button is selected, the temporary preview of the created geometry is visible along with its dimensions. When you choose the **Feature Preview** button, the system allows you to preview the feature.

Note that the preview of the feature appears to be a solid model where as in the temporary preview the feature appears yellow and translucent.

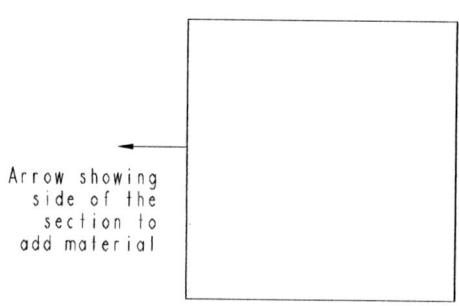

Figure 3-14 The sketch drawn to thicken   *Figure 3-15* Adding thickness to one side of the sketch

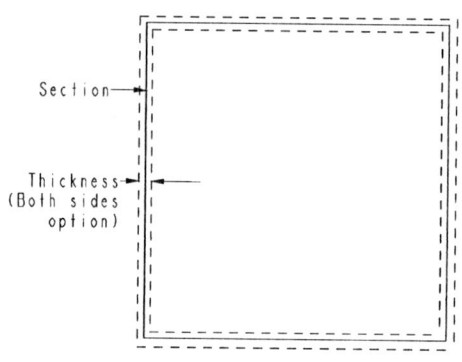

Figure 3-16 Adding thickness to both the sides of the sketch

Figure 3-17 Model created using the **Thicken Sketch** button

 Note

*By default, the check box in the **Geometry Preview/Feature Preview** button remains selected. If this check box is cleared then the temporary preview of the feature will not be displayed.*

Build feature Button

 This button is used to confirm the feature creation and exit the current feature creation tool.

Close Button

 This button is used to abort the feature creation. When you choose this button, the dashboard is closed.

The Sketch Dialog Box

The **Sketch** dialog box is displayed when you choose the **Define** button from the **Placement** slide-up panel. The **Sketch** dialog box is shown in Figure 3-18. This dialog box is used to select the sketching plane and to set its orientation. As soon as you select the sketching plane, the reference plane and its orientation are selected automatically.

Creating Base Features

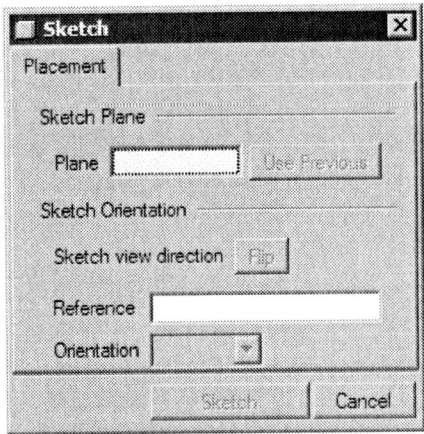

*Figure 3-18 The **Sketch** dialog box*

If you want to change the sketching plane that is already selected, hold down the right mouse button in the graphics window to display a shortcut menu shown in Figure 3-19. There are three options in this shortcut menu. In the figure, the **View Orientation** option is selected. If you choose the **Clear** option, the reference plane and its orientation, which were selected automatically, are cleared and now you can manually select them. Similarly, to change the sketching plane, select the **Placement** option in the shortcut menu. Again invoke the shortcut menu and choose the **Clear** option.

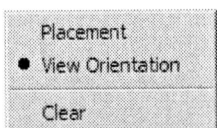

*Figure 3-19 Shortcut menu invoked when the **Sketch** dialog box is displayed*

Tip: *You can also remove the selected item using the **Sketch** dialog box. To remove a selected item in the **Sketch** dialog box, select the item in the dialog box and right-click to invoke a shortcut menu. From this shortcut menu, choose **Remove**.*

However, after you select the sketching plane, no matter which reference plane is selected automatically, you can select the reference plane manually. The reference you select manually replaces the old reference. If the background of **Plane** collector or any other collector is displayed yellow, it indicates that you can select its reference from the model. This point should be remembered because this is true with other dialog boxes available in Pro/ENGINEER and this you will learn later in this book.

The **Flip** button in the **Sketch** dialog box can be used after you select the sketching plane and its orientation. This button is used to set the direction of viewing the sketching plane. When you choose this button, the yellow arrow that is displayed on the sketching plane changes its direction.

In the **Sketch** dialog box, the options in the **Orientation** drop-down list are used to specify the orientation of the horizontal or vertical reference for the sketching plane. Generally, the view is normal to the sketching plane you have selected. In order to orient the sketching plane normal to the viewing direction, you have to specify a plane or a planar surface that is perpendicular to the sketching plane. For example, if you select the **RIGHT** datum plane as the reference plane and then select the **Top** option from the drop-down list, then the **RIGHT** datum plane will be placed at the top while sketching. The options in this drop-down list are common to other feature creation tools and are available whenever you need to draw a sketch.

Tip: *Remember that if you are creating a protrusion (a material addition process) then no matter what the direction of viewing you choose, the protrusion takes place in the direction toward the user, that is, out of the screen.*

Before you proceed further, you need to understand the three default datum planes that are displayed when you open a new part file. You can view the feature that you have created from different directions using the **Saved view list** drop-down list shown in Figure 3-20. This drop-down list is available in the **View** toolbar present in the **Top Toolchest**. In this drop-down list there are some standard preset views that are provided by Pro/ENGINEER. If you see the feature from the right then your viewing direction is normal to the **RIGHT** datum plane. If you see the feature from the top then your viewing direction is normal to the **TOP** datum plane. Similarly, if you see the feature from the front then your viewing direction is normal to the **FRONT** datum plane.

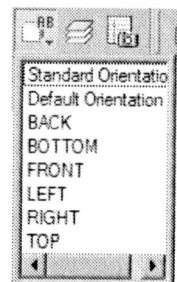
Figure 3-20 The Saved view list drop-down list

The options in this drop-down list are used to set the orientation of the model in the graphics window.

The options in the **Orientation** drop-down list are very important and you need to select the reference plane very carefully, especially for the base feature. The importance of selecting the reference plane is explained using Figure 3-21. This figure shows two cases where the sketching plane for both the solid models was the same, the same sketch was extruded, and the same option **Top** was chosen from the **Orientation** drop-down list in the **Sketch** dialog box but different reference planes were selected. The same sketch was used to extrude both the models and the top curve in the sketch was on the top while sketching. Note, the difference in orientation of the resulting models in their default trimetric orientations. The model can be oriented in its default orientation using CTRL+D or by selecting the **Default Orientation** option from the **Saved view list** drop-down list.

Once a plane has been selected for sketching and the reference plane is oriented, you can enter the sketcher environment. To enter the sketcher environment, choose the **Sketch** button in the **Sketch** dialog box. Now, you enter the sketcher environment. The **References** dialog box is displayed on the right of the main window.

Tip: *The **Orientation** dialog box that is displayed when you choose the **Reorient view** button from the **View** toolbar is used to orient the model in the graphics window. This dialog box provides some advanced options to orient the model and also allows you to save the orientation of the model. Using the options in this dialog box, you can dynamically orient a model, orient by selecting references, and orient by using other options.*

Creating Base Features 3-15

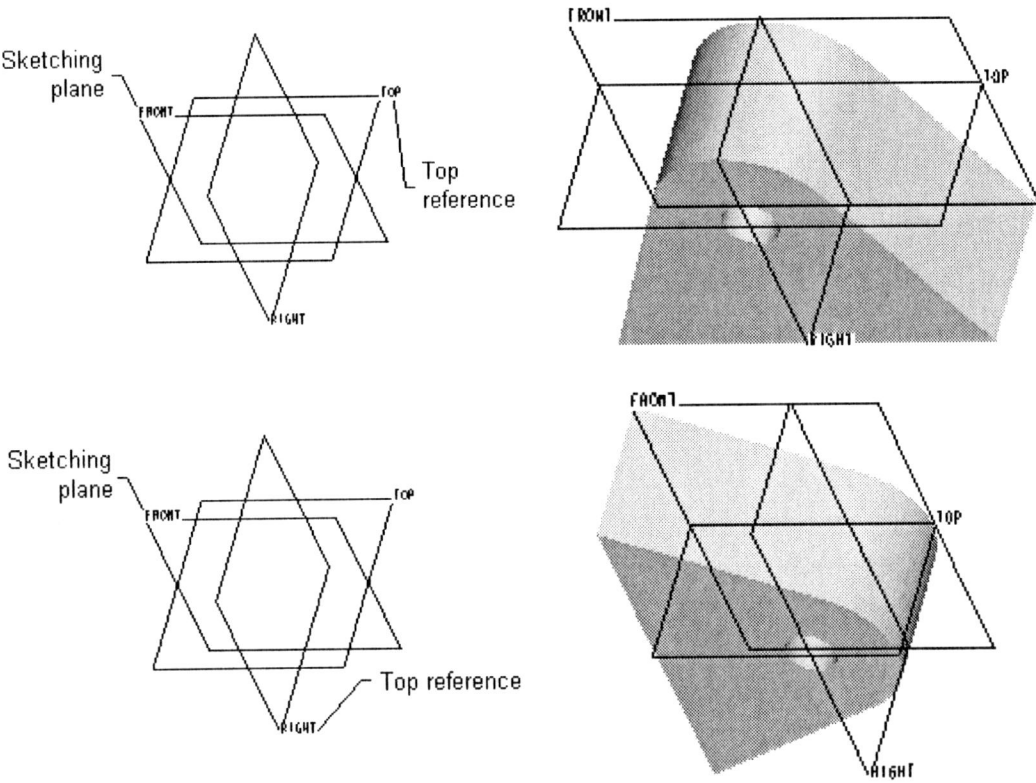

Figure 3-21 *Importance of selecting reference planes*

References Dialog Box

When you enter the sketcher environment, the **References** dialog box is displayed, as shown in Figure 3-22. This dialog box is used to specify the references for dimensioning the sketch. Pro/ENGINEER by default selects the references, but you can change the references by deleting the default references. Sometimes when Pro/ENGINEER is not able to determine the references, you need to select the references manually. You can select datum planes, edges, axes, datum curves, and so on as references. Generally, two references are required for a sketch. But if you select a vertex or a datum axis, then they alone are enough to determine the references.

When you draw a sketch, the weak dimensions are applied to the sketch with references selected in the **References** dialog box. For example, if you select the two datum planes as references, then the sketch is by default dimensioned with the two datum planes. If you force a dimension and dimension the sketch with an axis then the axis is also displayed in the **References** dialog box. The **References** dialog box in the sketcher environment can be invoked by choosing **Sketch > References** from the menu bar.

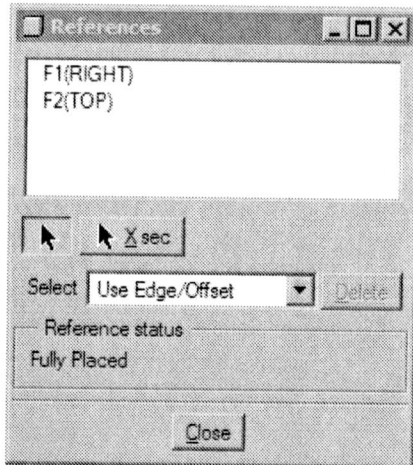

*Figure 3-22 The **References** dialog box*

Revolving a Sketch

The **Revolve Tool** button allows you to revolve the sketched section through the specified angle about a center line. By revolving a sketch, you can add material (protrusion) and you can even remove material (cut). Here, you will learn to use the **Revolve Tool** to add material. The revolved feature can be revolved on one side of the sketching plane or on both the sides of the sketching plane. This and the other attributes can be specified to the sketch by using the **Revolve** dashboard. The **Revolve** dashboard is shown in Figure 3-23 and it is displayed when you choose the **Revolve Tool** button from the **Base Features** toolbar in the **Right Toolchest**.

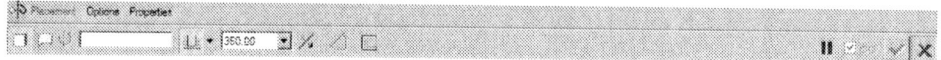

*Figure 3-23 The **Revolve** dashboard*

Some of the points to be kept in mind while creating a revolve feature are given next.

1. If the revolve feature is a base feature, then the section drawn should be a closed section for revolving the sketch as a solid.

2. The sketch of the revolved feature will not be completed until you have drawn a center line.

3. The section drawn should be on one side of the center line.

4. If there are more than one center lines in the sketch, then Pro/ENGINEER uses the center line that is drawn first and considers it the axis of rotation.

Revolving a Sketch as a Solid

The **Revolve as solid** button in the **Revolve** dashboard is used to revolve the sketch as a solid. To revolve a sketch as a solid, the sketch should be a closed loop. Figures 3-24 and 3-25 explain the use of **Revolve as solid** tool to revolve a sketch.

Creating Base Features

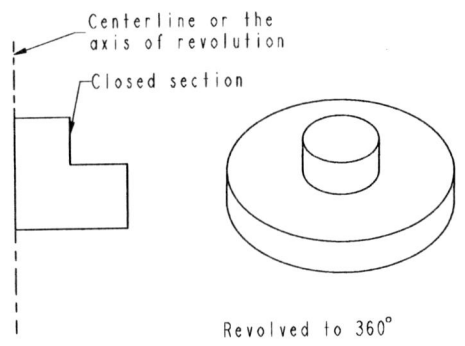
Figure 3-24 Revolving a sketch as a solid

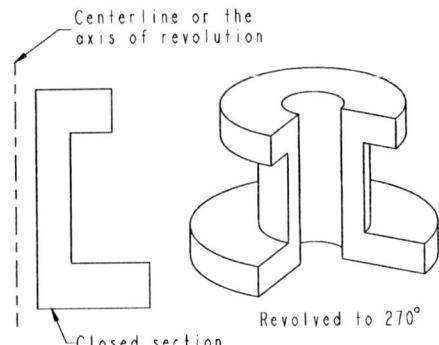
Figure 3-25 Revolving a sketch as a solid

Revolving a Sketch with Thickness

The **Thicken Sketch** button is used in combination with the **Revolve as solid** button to create revolved features having a certain thickness. Unlike in revolving the sketch as a solid, when you revolve a sketch with thickness, the section need not be a closed loop. When you exit the sketcher environment after the section is created, you can specify the side of the section where the material will be added. After specifying the side for the material addition, you can enter the thickness value in the edit box that appear on the dashboard.

Revolving the sketch with thickness is explained in Figures 3-26 and 3-27.

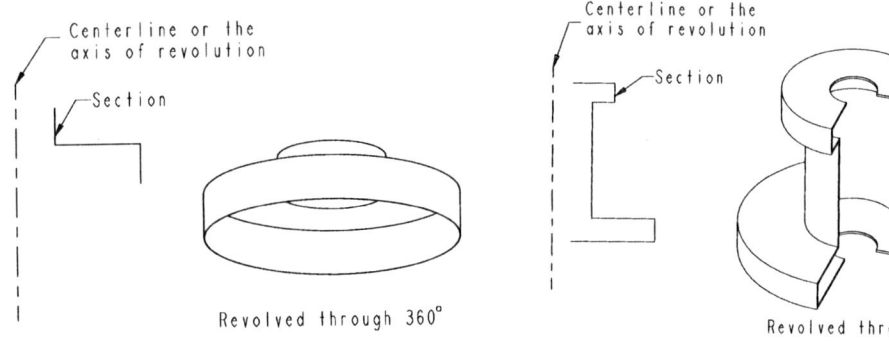

Figure 3-26 Revolving a sketch with thickness

Figure 3-27 Revolving a sketch with thickness

UNDERSTANDING THE ORIENTATION OF DATUM PLANES

Consider a case where you need to revolve a sketch through an angle of 270-degree. The trimetric view of the solid model is shown in Figure 3-28. Now, when you start to create the solid model, the first step is to decide the datum plane on which you need to create the sketch.

Figure 3-28 Default trimetric view of a shaded model

As evident from the model, the axis of revolution of the model is perpendicular to the **TOP** datum plane. Therefore, the **TOP** datum plane will be selected to be at the top while drawing the sketch. Now, only two options are left for selecting the sketching plane, the **FRONT** and the **RIGHT** datum planes. You can draw the section of the revolve model on any of the two planes. This is because when you view the model, you notice that the cross-section of the model is parallel to the **RIGHT** datum plane as well as to the **FRONT** datum plane.

To draw the cross-section sketch on the two datum planes to achieve the desired orientation of the model is discussed next.

Case-1
Drawing the Sketch on the RIGHT Datum Plane

1. When the **Sketch** dialog box is displayed, select the **RIGHT** datum plane as the sketching plane.

   The selected datum plane gets highlighted and the TOP datum plane is automatically selected as the reference plane. The yellow colored arrow also appears on the sketching plane. This arrow indicates the direction of feature creation. You can change the direction of arrow but at this stage accept its default direction.

2. From the **Orientation** drop-down list, select the **Top** option.

   Now, when you draw the sketch, the **RIGHT** datum plane will be parallel to the screen and the **TOP** datum plane will be at the top.

3. Choose the **Sketch** button from the **Sketch** dialog box to enter the sketcher environment.

4. After closing the **References** dialog box, draw the sketch of the cross-section, dimension it, and then modify the dimensions, as shown in Figure 3-29.

5. Exit the sketcher environment by choosing the **Continue with the current section** button.

Creating Base Features 3-19

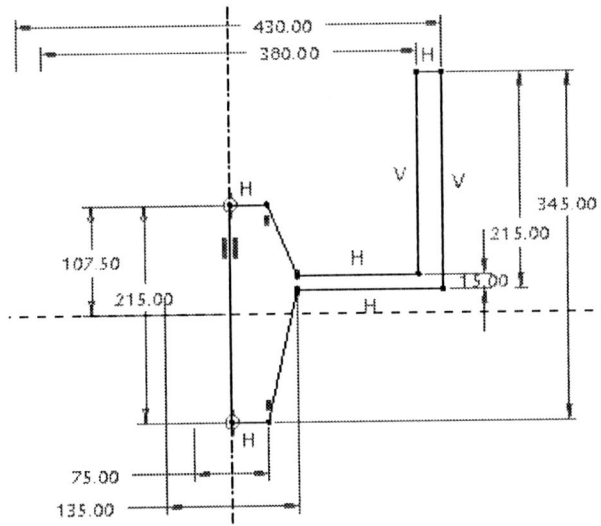

Figure 3-29 Sketch with dimensions and constraints

6. The model appears translucent and in yellow color. The dashboard under the graphics window again becomes active.

7. In the drop-down list that is present on the dashboard, select a value of **270**.

8. From the **Saved view list** button, choose the **Default Orientation** option. The model orients in its default orientation, that is, trimetric view, as shown in Figure 3-30.

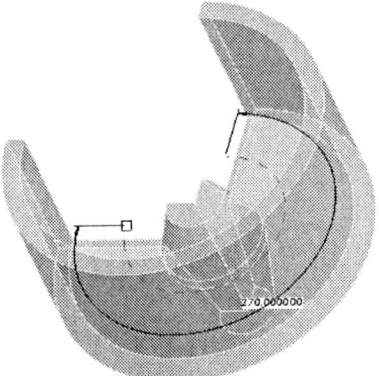

Figure 3-30 Default trimetric view of the model

But the view that is shown in Figure 3-30 is not the view that is needed. The model is not oriented correctly. This means that the direction of feature creation that was selected was not correct in order to get the desired orientation.

Note
The model is created but its orientation is not the desired one. This means that at the time of selecting the sketching plane, when the yellow arrow appeared, it was pointing in the wrong direction.

Tip: *You might have difficulty in visualizing the cross-section of a revolved model that has an angle of revolution of 360-degree. This is because the cross-section of the model is not visible. Therefore, you cannot decide the sketch to be drawn for the cross-section of the model.*

Whenever you come across a revolve model, just imagine that if you cut the model along its axis of revolution and remove one quarter of the revolved feature then the section obtained in the one quarter of the revolved feature is the section of the revolved model. Therefore, you need to draw this section in the sketcher environment to create the desired model.

Figure 3-31 shows you the sketch that you have drawn on the **RIGHT** datum plane. The sketch is always revolved in the clockwise direction. Remember that the arrow shows the direction in which you will be viewing the sketching plane while in the sketcher environment. Considering the mentioned facts, to orient the sketching plane correctly, the arrow direction should have been flipped by using the **Flip** button in the **Sketch** dialog box. Now, since you have not achieved the desired orientation of the model, exit this feature creation tool, again invoke the **Revolve Tool** button, and this time after you select the sketching plane, choose the **Flip** button to reverse the direction of viewing the sketch. Remember that the case discussed here is when you select the **RIGHT** datum plane as the sketching plane.

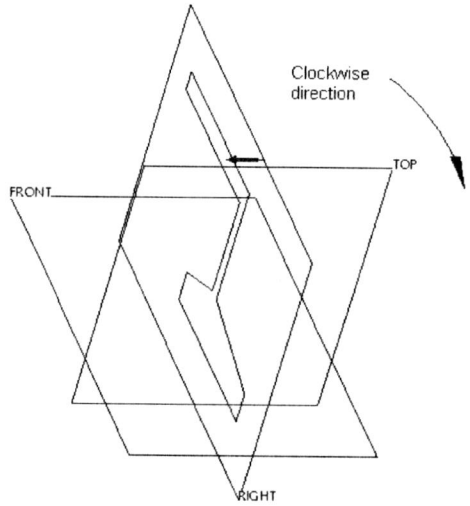

*Figure 3-31 Sketch drawn on the **RIGHT** datum plane*

Figure 3-32 shows the sketch on the sketching plane and the arrow direction reversed. In this case, the sketch when rotated in clockwise direction results in the desired orientation of the model, as shown in Figure 3-33.

Creating Base Features

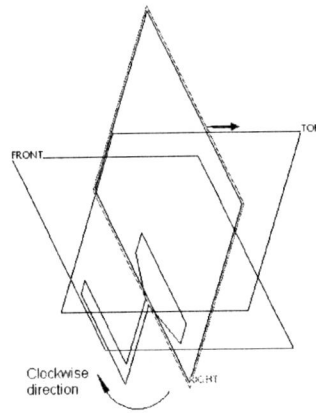

Figure 3-32 Sketch drawn on the **RIGHT** datum plane

Figure 3-33 Desired trimetric view of the model

 **Tip**: By default, when Pro/ENGINEER creates a revolved model, the material is added in the clockwise direction. You can revolve the model in counterclockwise direction by using the **Change angle direction of revolve to other side of sketch** button from the Revolve dashboard. Also, to revolve the model in the anti-clockwise direction you can enter a negative value for the angle of revolution.

Case 2
Drawing the Sketch on the FRONT Datum Plane

In this case, you will select the **FRONT** datum plane as the sketching plane and the **TOP** datum plane will be selected as reference plane. It is advisable to decide the direction of the arrow at this stage so that you create the model shown in Figure 3-28 in the required orientation. Figures 3-34 shows the sketch drawn on the **FRONT** datum plane and the default direction of viewing the sketching plane. Remember that the arrow points in the direction along which you will view the plane in the sketcher environment. Figure 3-35 shows the default trimetric view of the solid model created in this case.

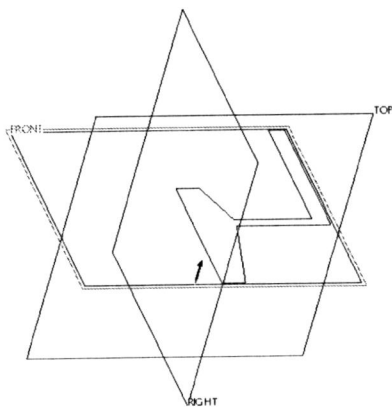

Figure 3-34 Sketch drawn on the **FRONT** datum plane

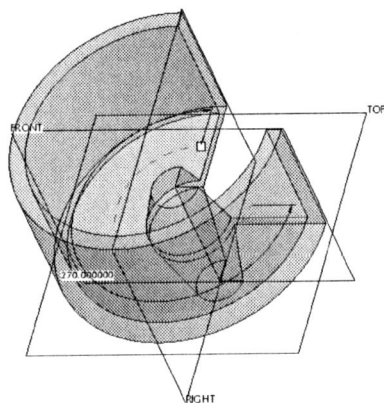

Figure 3-35 Default trimetric view of the model

This is not the desired default orientation of the model. You can still get the desired orientation of the model by choosing the **Change angle direction of revolve to other side of the sketch** button from the **Revolve** dashboard, but, you should understand the default direction of the yellow arrow on the sketching plane. The desired orientation of the model is shown in Figure 3-36.

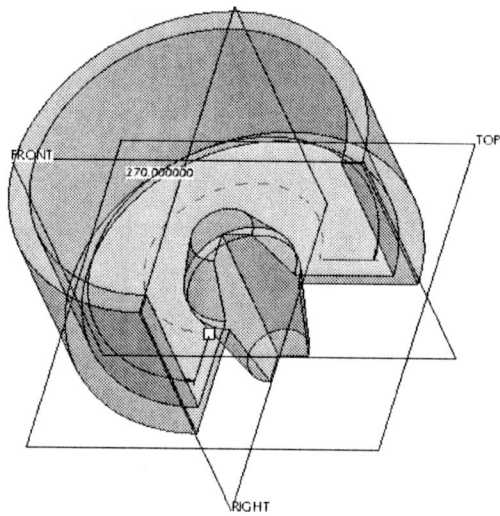

Figure 3-36 Desired trimetric view of the model

Figure 3-37 shows the sketch drawn on the **FRONT** datum plane and the arrow is showing the direction of viewing the sketching plane. The direction of the yellow arrow is reversed by choosing the **Flip** button in the **Sketch** dialog box. Figure 3-38 shows the default trimetric view of the model in this case.

In Figure 3-37, the sketch is revolved in clockwise direction to create the model shown in Figure 3-38.

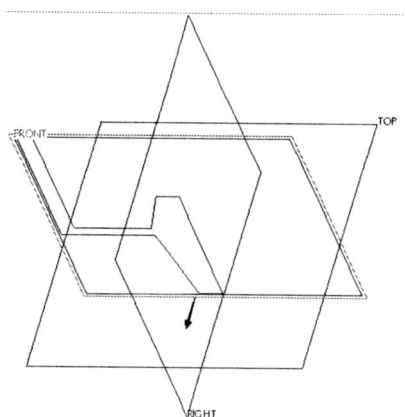

Figure 3-37 Sketch drawn on the **FRONT** datum plane

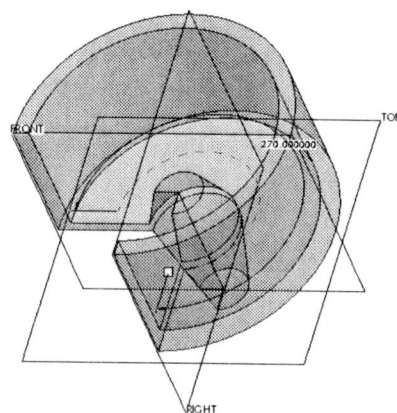

Figure 3-38 Default trimetric view of the model

Creating Base Features

From the two cases discussed, the conclusion is that the arrow that is displayed when you select the sketching plane is of great importance in orienting the model. Also, the direction of feature creation can be changed once it is created, using the **Change angle direction of revolve to other side of the sketch** button from the **Revolve** dashboard.

PARENT-CHILD RELATIONSHIP

Every model created in Pro/ENGINEER is composed of features that in some way or the other are related to other features in the model. The feature that occurs first in the **Model Tree** is called the parent feature, and any feature(s) that is related to this feature is called the child feature(s).

There are two types of relationships that can exist between two features, Implicit relationship and Explicit relationship.

Implicit Relationship

This type of relationship exists when the two features are related through equations. These equations are formed using relations. Relations are explained in Chapter 8.

Explicit Relationship

The explicit relationship is developed when a feature is used as a reference to create another feature. For example, one of the planar surfaces of a feature is used to create another feature, or an edge of a feature is used to dimension the other feature. In this case, the first feature is called the parent feature and the second feature is called the child feature of the first feature. Another example of this type of relationship is, when a hole is referenced to the edges of the surface it is placed on, or to the edges of some other surface. In this case, the hole is the child feature of the feature it is referenced to.
Remember that if the parent feature is modified then the child feature is also modified. Similarly, if the parent feature is deleted then the child feature is also deleted. For example, if you have used the three default datum planes to create the base feature and you delete any one of the datum planes, the base feature is also deleted. In Chapter 5, you will learn how to break the parent-child relationship.

> **Tip**: *If you want to check the parent-child relationship of features in a model, choose* **Info > Parent/Child** *from the menu bar. You are prompted to select a feature. After you select the feature, the* **Reference Information Window** *is displayed. The child features of the selected feature are displayed in the* **Children of Current Feature** *area and all parent features are displayed in the* **Parents of Current Feature** *area.*

NESTING OF SKETCHES

If more than one closed loops are drawn one inside the other in the sketch of a single feature, it is called nesting of sketches. These sketches are drawn in the sketcher environment. Figure 3-39 shows a sketch in which the two circles are created on the base profile.

Advantages of Nesting the Sketches

1. One of the advantages of nesting of sketches is that the number of features used to create a model is reduced. In Figure 3-40, the model has two features. The base feature is the base plate and the second feature are the holes. But, when you nest the two sketches to create the model, you are using only one feature to create the model.

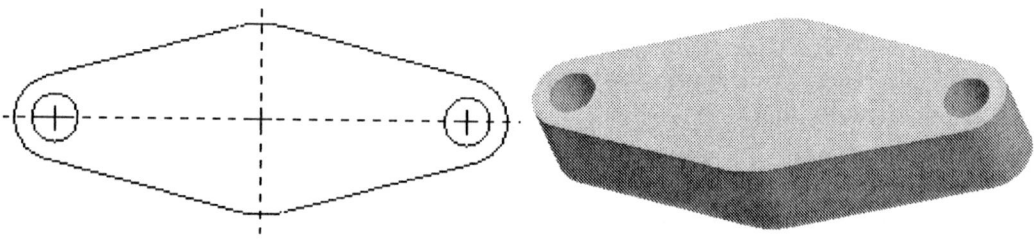

Figure 3-39 Nested sketch   *Figure 3-40* Solid model of the sketch

2. There is no parent-child relationship that exists in a nested sketch.

3. The depth of the hole is equal to the depth of extrusion of the base feature. This is because the two circles and the base profile are part of the same sketch. This depends on the designer whether he wants to use the nested sketch to create a part.

Disadvantages of Nesting the Sketches

1. In nesting, since the two features on the model are combined into one feature, therefore there is no flexibility in editing the features of a model.

2. If at a later stage, the designer needs to convert the circular holes into elliptical holes with depth as half the depth of extrusion of the base feature, it consumes a lot of time to edit the model.

After understanding the advantages and disadvantages of nesting of sketches, it is recommended to divide a model into separate features. Draw all features as individual features so that the created model is flexible. However, it depends on the need of the designer and the need of the model how the model is approached for creating.

TUTORIALS

Tutorial 1

In this tutorial you will create the model shown in Figure 3-41. The dimensions for the model are shown in Figure 3-42. **(Estimated time: 30 min)**

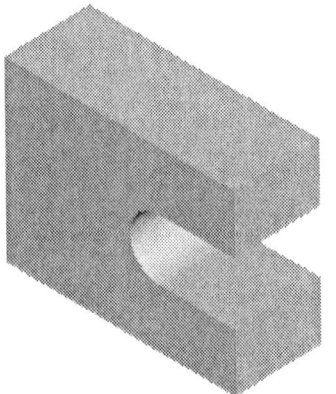
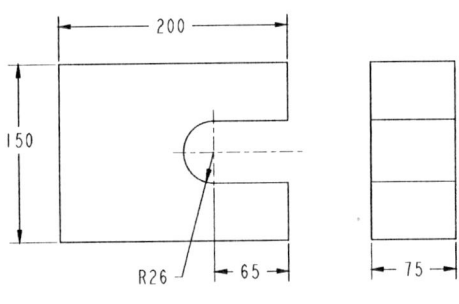

Figure 3-41 Isometric view of the model

Figure 3-42 Front view and the right-side view of the model with dimensions

The following steps are required to complete this tutorial:

a. Set the working directory and create a new object file in the **Part** mode.
b. First examine the model and then determine the type of protrusion for the model.
c. Select the sketching plane for the model and orient it parallel to the screen.
d. Draw the sketch using the sketching tools and apply constraints and dimensions, refer to Figures 3-43 through 3-47
e. Exit the sketcher environment and define the model attributes, refer to Figures 3-48 and 3-49.

Setting the Working Directory

When Pro/ENGINEER session is started, the first task is to set the working directory. A working directory is a directory on your system where you can save the work done in the current session of Pro/ENGINEER. You can set any directory existing on your system as the working directory. Since this is the first tutorial of this chapter, you need to create a folder named *c03*, if it does not exist.

1. Choose **File > Set Working Directory** from the menu bar. The **Select Working Directory** dialog box is displayed.

2. Browse and select *C:\ProE-WF-2.0*. It is assumed that the *ProE-WF-2.0* folder exists.

3. Choose the **New Directory** button in the **Select Working Directory** dialog box. The **New Directory** dialog box is displayed.

4. Type **c03** in the **New Directory** edit box. Choose **OK** from the dialog box. You have created a folder named *c03* in *C:\ProE-WF-2.0*.

5. Choose **OK** from the **Select Working Directory** dialog box. You have set the working directory to *C:\ProE-WF-2.0\c03*.

Starting a New Object File

Solid models are created in the **Part** mode of Pro/ENGINEER. The file extension for the files created in this mode is *.prt*.

1. Choose the **Create a new object** button from the **File** toolbar. The **New** dialog box is displayed. The **Part** radio button is selected by default in the **Type** area and the **Solid** radio button is selected by default in the **Sub-type** area of the **New** dialog box.

Note
*Before choosing the **OK** button from the **New** dialog box, make sure that the **Use default template** check box is selected. If this check box is cleared then you will need to select the template that you will use to create the model.*

2. Enter the file name as **c03tut1** in the **Name** edit box and choose the **OK** button.

The three default datum planes are displayed in the graphics window. The **Model Tree** also appears on the left of the graphics window in the Navigator.

3. Close the **Model Tree** by clicking on the sash present on the right edge of the Navigator. Now, the drawing area is increased.

Selecting the Protrusion Option

The given solid model is created by extruding the sketch to a distance of 75. Therefore, the sketch will be extruded as a solid to create the model. There are two methods to invoke the **Extrude** option. The first method is to use the menu bar present on the top of the screen and the second method is to use the **Extrude Tool** button present on the **Base Features** toolbar.

1. Choose **Insert > Extrude** from the menu bar or choose the **Extrude Tool** button present on the **Base Features** toolbar. The **Extrude** dashboard is displayed below the graphics window.

All attributes that are needed to create the model will be defined after the sketch is drawn.

Selecting the Sketching Plane

To create the sketch for the extruded feature, you first need to select the sketching plane for the model. The **FRONT** datum plane will be selected as the sketching plane. The sketching plane is selected such that the direction of extrusion of the solid model is perpendicular to it. From the isometric view of the model shown in Figure 3-41, it is evident that the direction of extrusion of the model is perpendicular to the **FRONT** datum plane.

Creating Base Features

1. Choose the **Placement** tab to display the slide-up panel. Choose the **Define** button to display the **Sketch** dialog box.

2. Select the **FRONT** datum plane as the sketching plane. As you select the sketching plane, the reference plane and its orientation are set automatically. The reference plane is selected in order to orient the sketching plane.

   The yellow arrow that appears on the sketching plane indicates the direction of viewing the sketch.

   In the **Sketch** dialog box, the **Reference** box displays **RIGHT:F1(DATUM PLANE)**. This indicates that the **RIGHT** datum plane is selected as the reference plane. In the **Orientation** drop-down list, the **Right** option is selected by default. This means that while drawing the sketch the **RIGHT** datum plane will be on the right.

   The **RIGHT** datum plane will be perpendicular to the sketching plane and the sketching plane will be parallel to the screen.

3. Choose the **Sketch** button in the **Sketch** dialog box.

Specifying References

Now, you have entered the sketcher environment and the **References** dialog box is displayed on the top right corner of the screen. You will be prompted to select a perpendicular surface, an edge, or a vertex relative to which the section will be dimensioned or constrained. Pro/ENGINEER does not use any coordinate system, and therefore it becomes necessary for the user to locate the sketch with some reference. The reference can consist of part surfaces, datums, edges, or axes.

The status displayed in the **Reference status** area is **Fully Placed**. This indicates that the references required for the sketch are automatically defined.

1. Choose the **Close** button from the **References** dialog box to exit it.

   The **FRONT** datum plane is the sketching plane and is parallel to the graphics window. This can be verified by performing the next step.

2. Spin the datum planes by holding the middle mouse button down and dragging the cursor. Now, from this view of the datum planes, it is clear that the sketching plane was parallel to the graphics window and the other two datum planes were perpendicular to the sketching plane.

   The above step is just to understand the orientation of the three default datum planes when you enter the sketcher environment.

3. Choose the **Orient the sketching plane parallel to the screen** button from the **Sketcher** toolbar in the **Top Toolchest**. The sketching plane and the other two perpendicular datum planes are reoriented on the screen.

Drawing the Sketch

You need to draw the sketch of the solid model that will later be extruded to create the 3D model.

1. Choose the **Create rectangle** button from the **Sketcher Tools** toolbar.

2. Draw a rectangle by defining its lower left corner and the upper right corner. The rectangle is created and weak dimensions are applied to it.

   You will notice that strong vertical and horizontal constraints are applied to the lines composing the rectangle. This is because drawing a rectangle is itself a constraint to the lines composing the rectangle.

3. Choose the **Select items** button from the **Sketcher Tools** toolbar.

4. Select the right vertical line; the line turns red in color. Press DELETE to delete it. The sketch after deleting the vertical line is shown in Figure 3-43.

5. Now, draw the lines and the arc. Some weak dimensions and constraints are applied to the sketch, as shown in Figure 3-44.

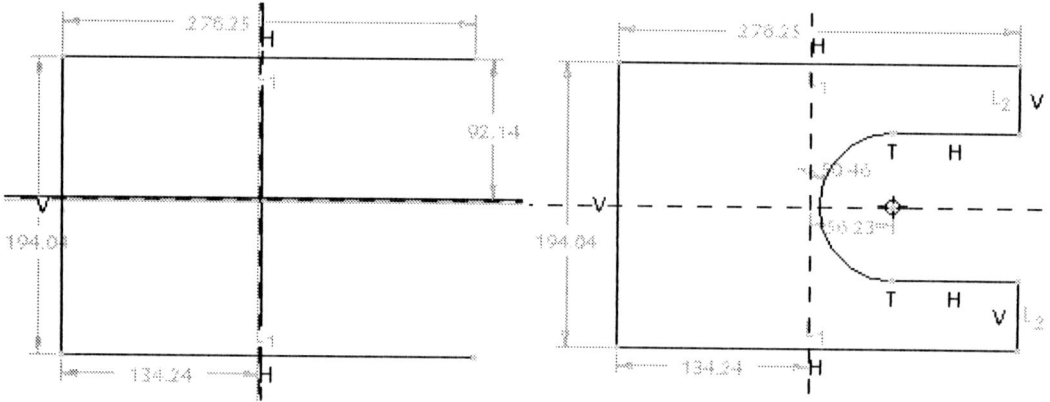

Figure 3-43 Outer loop of the sketch with weak dimensions

Figure 3-44 Sketch before modifying the weak dimensions with all datums turned off

 Note

As evident from Figure 3-44, strong constraints are also applied to the sketch while drawing it. However, if these strong constraints are not applied when you draw the sketch, you need to apply these constraints manually.

*The center of the arc and the **TOP** datum plane are aligned by default. In case, they are not aligned, you need to align them. Choose the **Create same points, points on entity or collinear constraint** button from the **Constraints** dialog box, which is displayed after choosing the **Impose sketcher constraint on the section** button form the **Sketcher Tools** toolbar. Select the center of the arc and then select the **TOP** datum plane. Now, the center and the datum plane are aligned.*

Creating Base Features

Applying Constraints to the Sketch

You need to apply equal length constraints to the sketch in order to maintain the design intent of the model.

1. Choose **Impose sketcher constraints on the section** button from the **Sketcher Tools** toolbar. The **Constraints** dialog box is displayed.

2. Choose the **Create Equal Lengths, Equal Radii, or Same Curvature constraint** button and select the two vertical lines on the right of the sketch. The weak constraint L_2 was applied when the two vertical lines were drawn. Now, the equal length constraint L_2 is applied to both the lines. The constraint labels like L_2 or L_3 vary from sketch to sketch.

3. Select the two horizontal lines that are connected to the arc to apply the equal length constraint. The equal length constraint L_2 is applied to both the lines and the constraint symbol L_2 on the two right vertical lines is changed to L_1.

The sketch after applying the equal length constraints is shown in Figure 3-45.

Dimensioning the Sketch

Although some weak dimensions are applied to the sketch, you still need to add a dimension to the sketch.

1. Choose the **Create defining dimension** button from the **Sketcher Tools** toolbar.

2. Select the center of the arc and the upper right vertical line and place the dimension, as shown in Figure 3-46.

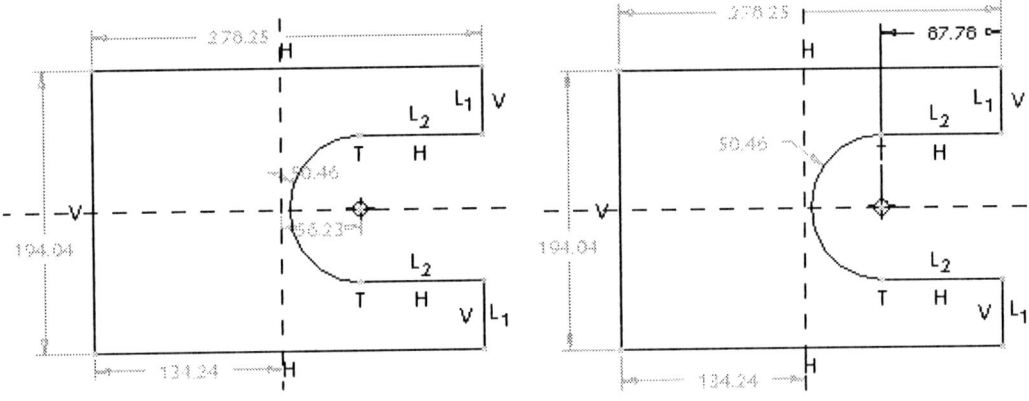

Figure 3-45 Equal length constraints applied to the sketch

Figure 3-46 Dimension added to the sketch

You need to dimension only these entities because the rest of the weak dimensions are useful dimensions and will be modified directly.

Note

If the dimensions in your sketch are different from those shown in Figure 3-46, add the missing dimensions and delete the dimensions that are not required.

Modifying the Dimensions

You need to modify the dimension values of the sketch. You will notice that the default dimensions shown in Figure 3-46 also include the length and width of the rectangle and the distance of the sketched section from the selected references. It is recommended that you draw the base feature symmetrical with the other two datum planes. Hence, the distance from the **RIGHT** datum plane is 100 (200 divided by 2 is equal to 100).

1. Select the sketch and dimensions using CTRL+ALT+A.

2. Choose the **Modify the values of dimensions, geometry of splines, or text entities** button from the **Sketcher Tools** toolbar. The **Modify Dimensions** dialog box is displayed.

   All dimensions in the sketch are displayed in this dialog box and each dimension has a separate thumbwheel and an edit box. You can use the thumbwheel or the edit box to modify the dimensions. It is recommended to use the edit boxes to modify the dimensions if the desired value is not obtained using the thumbwheel.

3. Clear the **Regenerate** check box. If you clear this check box, then any modification in a dimension value does not update the sketch during the modification. The dimensions will be modified after you exit the **Modify Dimensions** dialog box. It is recommended to clear the **Regenerate** check box when more than one dimension has to be modified.

4. Modify all dimensions one by one, as shown in Figure 3-47. You will notice that the dimension you select in the **Modify Dimensions** dialog box is enclosed in a yellow box in the graphics window.

5. After modifying the dimensions, choose the **Regenerate the section and close the dialog** button from the **Modify Dimensions** dialog box. A message **Dimension modifications successfully completed** is displayed in the **Message Area**.

6. Choose the **Continue with the current section** button to exit the sketcher environment.

Specifying the Model Attributes

Next, you need to specify the attributes to create the model.

1. Choose the **Saved view list** button from the **View** toolbar in the **Top Toolchest**. Choose the **Default Orientation** option from the drop-down list.

   The default trimetric view is displayed. This display gives you a better view of the sketch in the 3D space. The model is displayed in yellow color and is translucent. The yellow colored

Creating Base Features 3-31

arrow is also displayed on the model, indicating the direction of extrusion. The model may not fit fully in the graphics window.

2. Press CTRL+middle mouse button and drag the mouse upwards in the graphics window. Notice a red rubber-band line attached to the cursor. The length of the rubber-band line gives an idea of the extent you need to zoom the model. After the model is in the graphics window, release the middle mouse button and the CTRL key.

Note
*Alternatively, to fit the model in the graphics window, choose the **Refit object to fully display on the screen** button.*

The model appears, as shown in Figure 3-48. All attributes that are selected by default in the **Extrude** dashboard will be used to create the model. You need to change only the depth of extrusion.

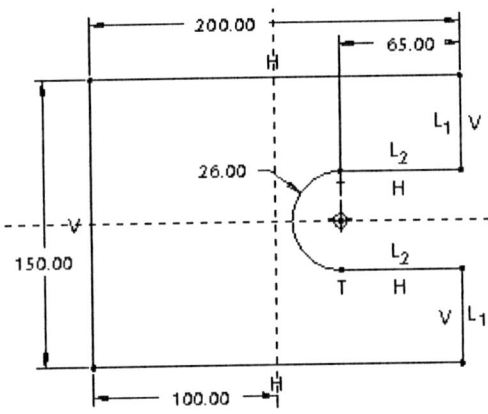
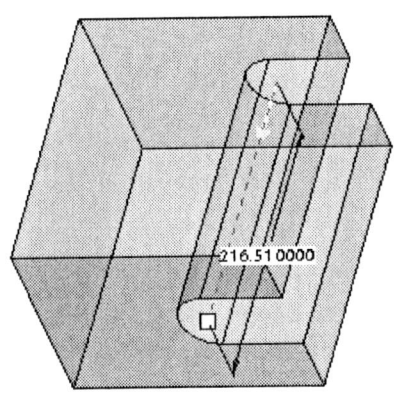

Figure 3-47 Sketch after modifying the dimensions

Figure 3-48 Arrow showing the direction of feature creation

3. In the edit box present on the **Extrude** dashboard, enter the depth of extrusion equal to **75** and press ENTER. The model in the graphics window is displayed with the specified depth of extrusion.

4. Choose the **Build feature** button in the **Extrude** dashboard.

5. Again choose the **Default Orientation** option from the **Saved view list** drop-down list. The model appears, as shown in Figure 3-49.

Note
*In Figure 3-49, the display of datum planes and coordinate system are turned off by choosing the **Datum planes on/off** and **Coordinate systems on/off** buttons respectively.*

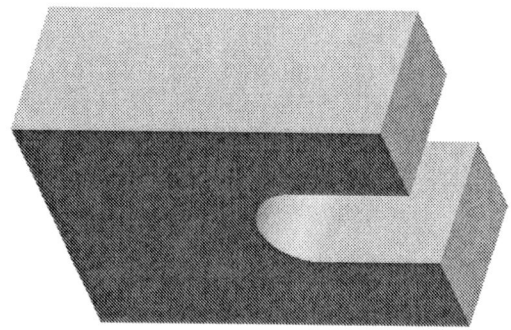

Figure 3-49 Default trimetric view of the model

Saving the Model

1. Choose the **Save** option from the **File** menu or choose the **Save the active object** button from the **File** toolbar. The **Save Object** dialog box is displayed with the name of the object file that you had specified earlier.

2. Choose the **OK** button from **Save Object** dialog box to save the file.

Closing the Current Window

The given model is created and is also saved. Now, you can close the current window.

1. Choose **File > Close Window** or choose the **Window > Close** from the menu bar.

Tutorial 2

In this tutorial you will create the model shown in Figure 3-50. The dimensions are shown in Figure 3-51.   **(Expected time: 30 min)**

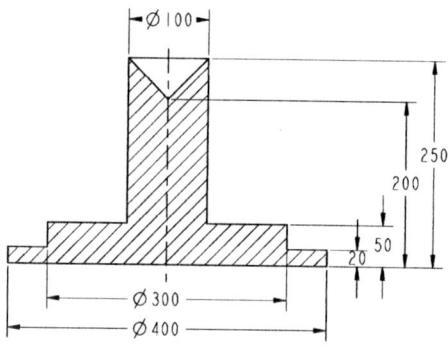

Figure 3-50 Isometric view of the solid model   *Figure 3-51 Front section view of the solid model*

Creating Base Features

3-33

The following steps are required to complete this tutorial:

a. Create a new object file in the **Part** mode.
b. First examine the model and then determine the type of protrusion for the model.
c. Select the sketching plane for the model.
d. Draw the sketch for the revolve feature and a center line to revolve it using the sketching tools and apply dimensions, refer to Figures 3-52 through 3-54.
e. Exit the sketcher environment and define the model attributes, refer to Figures 3-55 and 3-56.

Setting the Working Directory

The working directory was selected in Tutorial 1, and therefore there is no need to select the working directory again. But if a new session of Pro/Engineer is started, then you have to again set the working directory by following the steps given next.

1. Open the **Navigator** by clicking the top arrows on the left edge of the Pro/ENGINEER main window. The Navigator slides out.

2. Choose the **Folder Browser** button in the Navigator to view the folders.

3. Click on the plus symbol adjacent to the *ProE-WF-2.0* folder in the navigator. The contents of the *ProE-WF-2.0* folder are displayed.

4. Now right-click on the *c03* folder to display a shortcut menu. From this shortcut menu, choose the **Make Working Directory** option. The working directory is set to *c03*.

5. Close the Navigator by clicking the sash on its right edge.

Starting a New Object File

1. Open a new object file in the **Part** mode. Name the file as **c03tut2**.

   The three default datum planes are displayed in the graphics window. However, if the default datum planes were turned off in the previous tutorial, then they will not appear in the graphics window.

2. Turn on the display of datum planes by selecting the **Datum planes on/off** button.

Selecting the Protrusion Option

The given solid model is created by revolving the sketch through an angle of 360-degree about an axis. Therefore, the **Revolve Tool** button will be used to create the model. There are two methods to invoke the **Revolve** option. The first method is to use the menu bar present on the top of the screen and the second method is to use the **Base Features** toolbar present on the **Right Toolchest**.

1. Choose **Insert > Revolve** from the menu bar or choose the **Revolve Tool** button from the **Base Features** toolbar. The **Revolve** dashboard is displayed below the graphics window.

Some of the attributes like rotation angle and the direction of rotation of the sketch will be defined after the sketch is created.

Selecting the Sketching Plane

To create the sketch of the model, you first need to select the sketching plane for the model. Note that the axis of revolution of the revolved feature is normal to the **TOP** datum plane in the model. Therefore, any of the other two datum planes other than the **TOP** datum plane can be selected as the sketching plane. Here you will select the **FRONT** datum plane as the sketching plane.

1. Choose the **Placement** tab to display the slide-up panel. Choose the **Define** button. The **Sketch** dialog box is displayed.

2. Select the **FRONT** datum plane as the sketching plane. As you select the sketching plane, the reference plane and its orientation are set automatically. The reference plane is selected in order to orient the sketching plane.

   In the **Sketch** dialog box, the **Reference** collector displays **RIGHT:F1(DATUM PLANE)**. This indicates that the **RIGHT** datum plane is selected as the reference plane. In the **Orientation** drop-down list, the **Right** option is selected by default. This means that while drawing the sketch the **RIGHT** datum plane will be on the right.

   The **RIGHT** datum plane will be perpendicular to the sketching plane and the sketching plane will be parallel to the screen.

3. Choose the **Sketch** button in the **Sketch** dialog box.

Specifying References

Now, you have entered the sketcher environment and the **References** dialog box is displayed on the top right corner of the screen. You will be prompted to select a perpendicular surface, an edge, or a vertex relative to which the section will be dimensioned or constrained. Pro/ENGINEER does not use any coordinate system, and therefore it becomes necessary for the user to locate the sketch with some reference. The reference can consist of part surfaces, datums, edges, or axes.

The status displayed in the **Reference status** area is **Fully Placed**. This indicates that the references required for the sketch are automatically defined.

1. Choose the **Close** button from the **References** dialog box to exit it.

Drawing the Sketch

You need to draw the sketch of the revolved feature. The sketch that will be drawn is the cross-section of the revolved feature, which will be revolved about the center line.

1. Choose the **Create 2 point lines** button from the **Sketcher Tools** toolbar. Draw the sketch, as shown in Figure 3-52. The sketch should be a closed loop and the bottom horizontal line should be aligned to the **TOP** datum plane.

Creating Base Features

As you draw the sketch, weak dimensions and strong constraints are applied to the sketch.

2. Choose the black arrow on the right of the **Create 2 point lines** button to display the flyout. From this flyout, choose the **Create 2 point centerlines** button. Draw the center line for the axis of revolution. The center line should be drawn such that it is aligned with the **RIGHT** datum plane, see Figure 3-52.

Dimensioning the Sketch

The weak dimensions are automatically applied to the sketch. Since the model is a revolved feature, therefore, you need to manually apply linear diameter dimensions to the sketch. The linear diameter dimensions are applied using the center line that was drawn in the sketch.

 **Tip**: *Linear diameter dimensioning is necessary for all revolved features. This is because mostly all revolved models are machined on a lathe. Hence, while machining a revolved model it is necessary that the operator of the machine has a drawing of the model that is diametrically dimensioned.*

1. Choose the **Create defining dimension** button from the **Sketcher Tools** toolbar.

2. Select the center line, the first right vertical line, and then again the center line.

3. Now, use the middle mouse button to place the dimension on top of the sketch. The diameter dimension is placed.

   The method for diametrically dimensioning a sketch has been discussed in Chapter 1.

4. Select the center line, the second right vertical line, and then again the center line.

5. Place the dimension below the sketch.

6. Select the center line, the third right vertical line, and then again the center line. Now, place the dimension below the previous dimension.

   Dimension the remaining entities in the sketch, as shown in Figure 3-53.

Modifying the Dimensions

When you dimension a sketch, default dimension values are applied to the sketch. You need to modify the dimension values of the dimensions.

1. Select the sketch and dimensions using CTRL+ALT+A.

2. Choose the **Modify the values of dimensions, geometry of splines, or text entities** button from the **Sketcher Tools** toolbar. The **Modify Dimensions** dialog box is displayed.

3. Clear the **Regenerate** check box and then modify the values of the dimensions, as shown in Figure 3-54. If you clear this check box, then any modification in

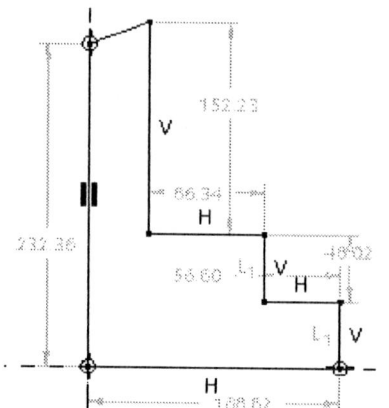

Figure 3-52 Sketch with weak dimensions   *Figure 3-53 Sketch after dimensioning*

a dimension value does not update the sketch. It is recommended that you clear the **Regenerate** check box when more than one dimension has to be modified.

You will notice that the dimension you select in the **Modify Dimensions** dialog box is enclosed in a yellow box in the graphics window.

4. After modifying all dimensions, choose the **Regenerate the section and close the dialog** button from the **Modify Dimensions** dialog box. The message **Dimension modifications successfully completed** is displayed in the **Message Area**.

5. Choose the **Continue with the current section** button to exit the sketcher environment.

Specifying the Model Attributes

When you exit the sketcher environment, the **Revolve** dashboard below the graphics window is enabled again. Using this dashboard, you will specify the angle of revolution for the revolved feature.

1. Choose the **Saved view list** button from the **View** toolbar. Choose the **Default Orientation** option from the drop-down list.

Note

If the model is not fully displayed in the graphics window, you may need to zoom out.

The feature will orient in its default orientation, as shown in Figure 3-55. This display gives you a better view of the sketch in the 3D space. Model appears in yellow color and is translucent. The drag handle is also available on the model which can be used to modify the angle of revolution dynamically.

All attributes that are needed to create the solid model are selected by default in the **Revolve** dashboard.

Creating Base Features

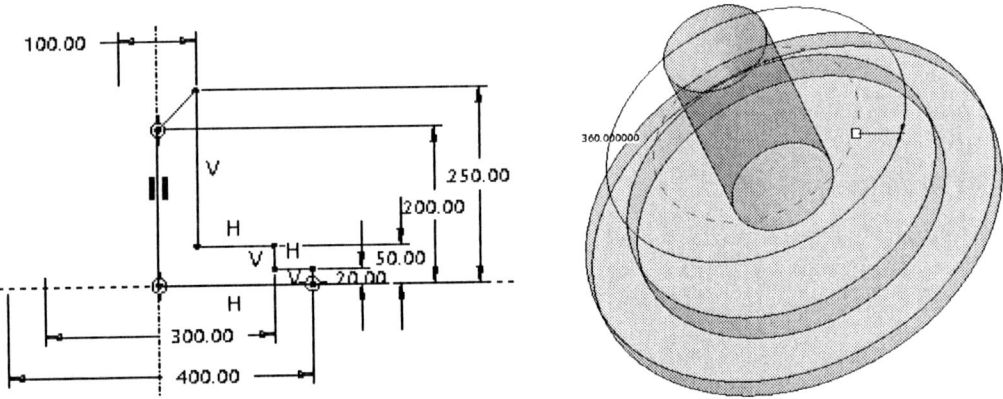

Figure 3-54 Sketch after modifying the dimensions

Figure 3-55 The preview of the model in default trimetric view

2. Choose the **Build feature** button in the **Extrude** dashboard. The model appears, as shown in Figure 3-56.

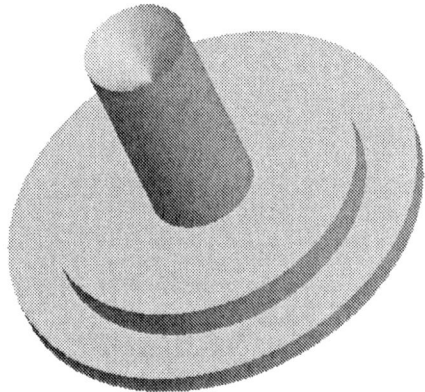

Figure 3-56 Default trimetric view of the model

Saving the Model

1. Choose the **Save** option from the **File** menu or choose the **Save the active object** button from the **File** toolbar. The **Save Object** dialog box is displayed with the name of the object file that you had specified earlier.

2. Choose the **OK** button.

Closing the Current Window

The given model is completed and is also saved. Now you can close the current window.

1. Choose **File > Close Window** from the menu bar.

Note

*If you need to view the **Model Tree** of the model you have created then you need to open it. Click on the sash present on the left edge of the graphics window. The **Model Tree** slides out. To again close it, click on the sash again. The **Model Tree** slides in.*

Tutorial 3

In this tutorial you will create the model shown in Figure 3-57. Figure 3-58 shows the dimensions.

(Expected time: 30 min)

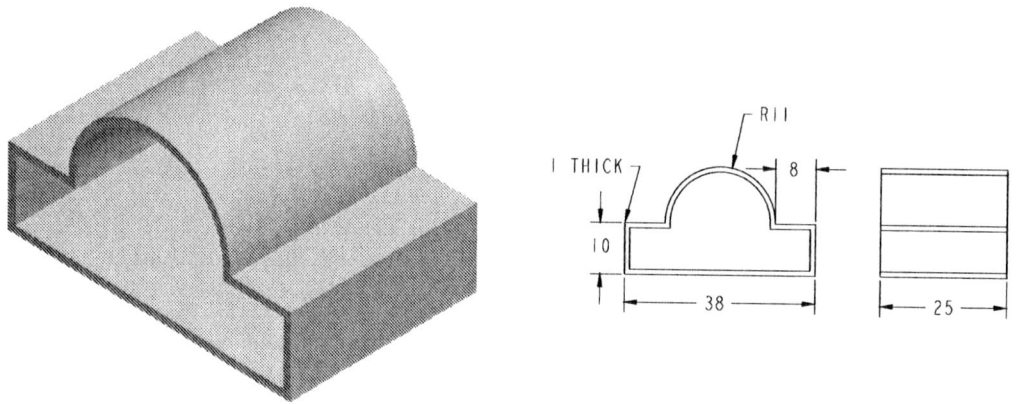

Figure 3-57 Isometric view of the model   *Figure 3-58 Front view and the right-side view*

The following steps are required to complete this tutorial:

a. Create a new object file in the **Part** mode.
b. First examine the model and then determine the type of protrusion for the model.
c. Select the sketching plane for the model.
d. Draw the sketch using the sketching tools and apply dimensions, refer to Figures 3-59 through 3-61.
e. Specify the attributes to create the model and save the model, refer to Figures 3-63 and 3-64.

The working directory was selected in Tutorial 1, and therefore there is no need to select the working directory again. But if a new session of Pro/Engineer is started, then set the working directory using the **Navigator**.

Starting a New Object File

1. Open a new object file in the **Part** mode. Name the file as **c03tut3**.

2. If the default datum planes were not turned off in the previous tutorial, they will appear in the graphics window. If the datum planes are not displayed, turn them on using the **Datum planes on/off** button.

Creating Base Features

Selecting the Protrusion Option

The given thin solid model is created by extruding the sketch to a distance of 25. Therefore, the **Thicken Sketch** tool button on the **Extrude** dashboard will be used to create the model.

1. Choose **Insert > Extrude** from the menu bar or choose the **Extrude Tool** button from the **Base Features** toolbar. The **Extrude** dashboard is displayed below the graphics window.

   Most of the attributes that are needed to create the model will be defined after the sketch is drawn.

Selecting the Sketching Plane

From the isometric view of the model shown in Figure 3-57, it is evident that the direction of extrusion of the solid model is perpendicular to the **FRONT** datum plane. Therefore, the **FRONT** datum plane will be selected as the sketching plane.

1. Choose the **Placement** tab to display the slide-up panel. Choose the **Define** button. The **Sketch** dialog box is displayed.

2. Select the **FRONT** datum plane as the sketching plane. As you select the sketching plane, the reference plane and its orientation are set automatically.

   In the **Sketch** dialog box, the **Reference** collector displays **RIGHT:F1(DATUM PLANE)**. This indicates that the **RIGHT** datum plane is selected as the reference plane. In the **Orientation** drop-down list, the **Right** option is selected by default. This means that while drawing the sketch the **RIGHT** datum plane will be on the right.

   The **RIGHT** datum plane will be perpendicular to the sketching plane and the sketching plane will be parallel to the screen.

3. Choose the **Sketch** button in the **Sketch** dialog box.

Specifying References

Now, you have entered the sketcher environment and the **References** dialog box is displayed on the top right corner of the screen. You will be prompted to select a perpendicular surface, an edge, or a vertex relative to which the section will be dimensioned or constrained.

The status displayed in the **Reference status** area is **Fully Placed**.

1. Choose the **Close** button from the **References** dialog box to exit it.

Drawing the Sketch

Using the sketcher tools, you need to draw the sketch for the thin extruded model.

1. Draw the arc shown in Figure 3-59 using the **Create an arc by 3 points or tangent to an entity at its endpoint** button from the **Sketcher Tools** toolbar. Complete the sketch, as shown in Figure 3-59.

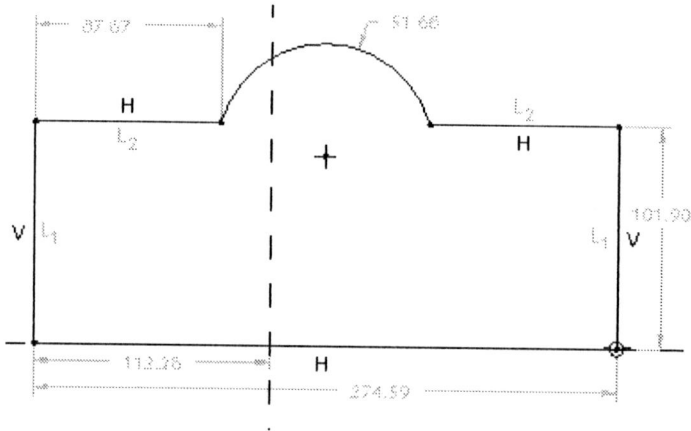

Figure 3-59 Sketch for the thin extruded model

While drawing the sketch, weak dimensions are applied to the sketch. These dimensions appear gray in color.

Applying Constraints to the Sketch

Some weak constraints are applied to the sketch while drawing, but you need to apply the constraints using the **Constraints** dialog box.

1. Choose the **Impose sketcher constraints on the section** button from the **Sketcher Tools** toolbar. The **Constraints** dialog box is displayed.

2. Choose the **Create same points, points on entity or collinear constraint** button from the **Constraint** dialog box.

3. Select the center of the arc and then select the **RIGHT** datum plane. The center of the arc is aligned with the **RIGHT** datum plane.

4. Select the bottom horizontal line and then select the **TOP** datum plane. Now, the line is aligned with the plane. If the two horizontal lines on the top are aligned with the **TOP** datum plane, skip this point.

5. Choose the **Create Equal Lengths, Equal Radii, or Same Curvature constraint** button and select the two horizontal lines on the top to apply the equal length constraint.

6. Choose the **Make two entities perpendicular** button and select the left horizontal line and the arc. The perpendicular constraint symbol is applied. Similarly, make the right horizontal line and the arc perpendicular. If these constraints are already applied, the **Resolve Sketch** dialog box will be displayed. Choose **Undo** from this dialog box.

Creating Base Features 3-41

The sketch after applying the constraints is shown in Figure 3-60.

Note
When you apply constraints, some of the weak dimensions are automatically deleted. Also, some constraints are applied when you draw the sketch, and therefore you do not need to apply those constraints again.

Modifying Dimensions

You need to modify the dimension values of the weak dimensions.

1. Select the sketch and dimensions using CTRL+ALT+A.

2. Choose the **Modify the values of dimensions, geometry of splines, or text entities** button from the **Sketcher Tools** toolbar. The **Modify Dimensions** dialog box is displayed.

3. Clear the **Regenerate** check box and then modify the values of the dimensions, as shown in Figure 3-61. When you clear the **Regenerate** check box, any modification in a dimension value does not update the sketch. As mentioned earlier, it is recommended that you clear the **Regenerate** check box when more than one dimension has to be modified.

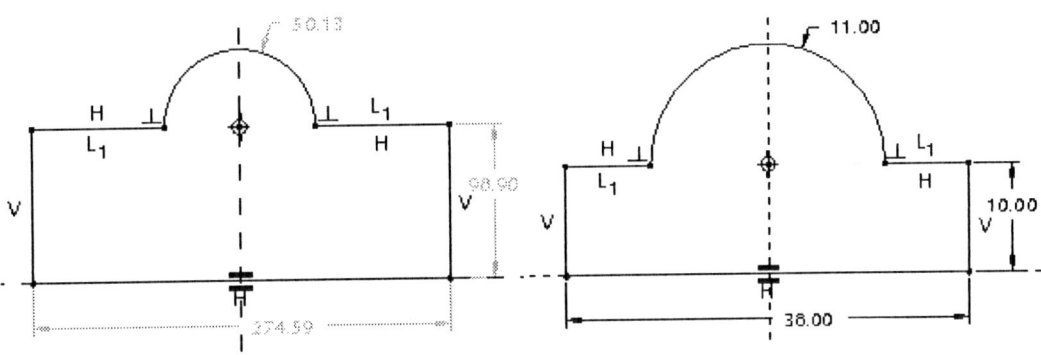

Figure 3-60 Sketch after applying constraints   *Figure 3-61 Figure after modifying the dimensions*

You will notice that the dimension you select in the **Modify Dimensions** dialog box is enclosed in a yellow box in the graphics window.

4. After modifying all dimensions, choose the **Regenerate the section and close the dialog** button from the **Modify Dimensions** dialog box. A message **Dimension modifications successfully completed** is displayed in the **Message Area**.

5. Choose the **Continue with the current section** button to exit the sketcher environment.

Specifying the Model Attributes

The attributes that will create the model needs to be selected from the **Extrude** dashboard.

1. Choose the **Thicken Sketch** button from the **Extrude** dashboard. The model changes to a thin model of certain thickness, as shown in Figure 3-62.

2. Choose the **Saved view list** button from the **View** toolbar. Choose the **Default Orientation** option from the drop-down list.

   The default trimetric view of the model is displayed. The model appears yellow in color and is translucent. If the model does not fully fit on the screen then you need to zoom out. This display gives you a better view of the model in the 3D space.

   On the **Extrude** dashboard, there are two drop-down lists. The drop-down list that is present on the left is used to enter the depth of extrusion. The drop-down list that is present on the right appears only when the **Thicken Sketch** button is selected. This drop-down list is used to specify the thickness value of the thin model.

3. Enter a value of **25** in the left edit box and press ENTER.

4. Enter a value of **1** in the right edit box and press ENTER. The model appears, as shown in Figure 3-63. The model is completed and now you need to exit the **Extrude** dashboard.

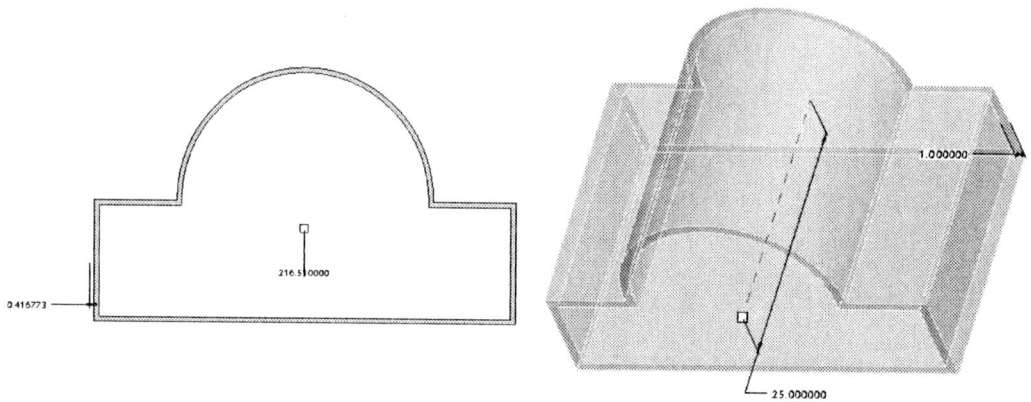

Figure 3-62 Arrow showing the direction of material addition with respect to the boundary of the section

Figure 3-63 The default trimetric view of the model

5. Choose the **Build feature** button from the **Extrude** dashboard. The model will be displayed, as shown in Figure 3-64.

Saving the Model

1. Choose the **Save** option from the **File** menu or choose the **Save the active object** button from the **Top Toolchest**. The **Save Object** dialog box is displayed with the name of the object file that you had specified earlier.

2. Choose the **OK** button.

Creating Base Features

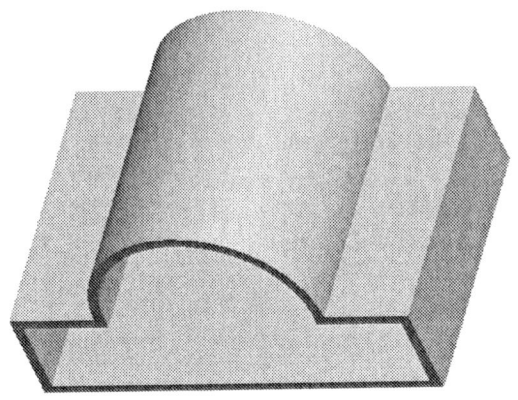

Figure 3-64 Default trimetric view of the solid model

Closing the Current Window

The given model is completed and is also saved. Now you can close the current window.

1. Choose **File > Close Window** from the menu bar.

Tutorial 4

In this tutorial you will create the model shown in Figure 3-65. Figure 3-66 shows the dimensions of the model. **(Expected time: 30 min)**

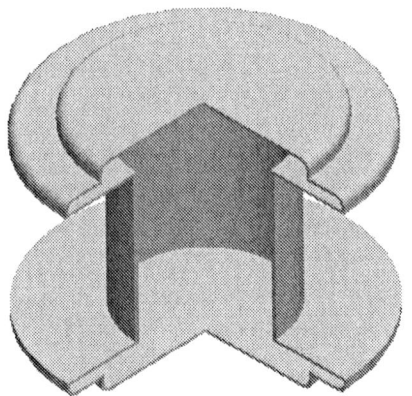
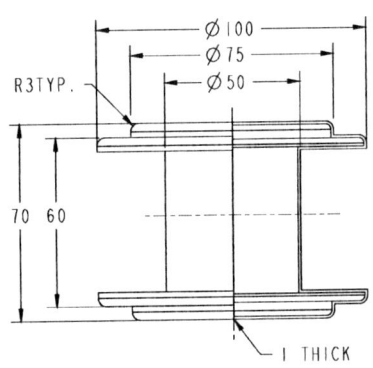

Figure 3-65 Isometric view of the solid model *Figure 3-66* Front view of the solid model

The following steps are required to complete this tutorial:

a. Create a new object file in the **Part** mode.
b. First examine the model and then determine the type of protrusion for the model.
c. Select the sketching plane for the model.
d. Draw the sketch using sketching tools, apply dimensions, and modify dimension values, refer to Figure 3-68 through 3-70.

e. Specify the attributes of the model and then save it, refer to Figures 3-71 and 3-72.

The working directory was selected in Tutorial 1, and therefore there is no need to select the working directory again. But if a new session of Pro/Engineer is started, then set the working directory using the **Navigator**.

Starting a New Object File

1. Start a new object file in the **Part** mode. Name the file as **c03tut4**.

2. If the default datum planes were not turned off in the previous tutorial, they will appear in the graphics window. If the datum planes are not displayed, turn them on using the **Datum planes on/off** button.

Selecting the Protrusion Option

The given revolved thin model is created by revolving the sketch through a given angle. Therefore, the **Thicken Sketch** button in the **Revolve** dashboard will be used to create the model. There are two methods to invoke the **Revolve** option. The first method is to use the menu bar present on the top of the screen and the second method is to use the **Base Features** toolbar.

1. Choose **Insert > Revolve** from the menu bar or choose the **Revolve Tool** button from the **Base Features** toolbar. The **Revolve** dashboard appears below the graphics window.

2. Choose the **Thicken Sketch** button from the dashboard. The **Revolve as solid** button is selected by default.

In this tutorial, the **Thicken Sketch** button is selected before drawing the sketch because the sketch that will be drawn will be an open section. Therefore, with the default attributes selected in the **Revolve** dashboard, it is not possible to draw an open sketch that can be accepted by Pro/ENGINEER.

3. Choose the **Placement** tab to display the slide-up panel. Choose the **Define** button. The **Sketch** dialog box is displayed.

Selecting the Sketch Plane

Note that in Figure 3-65, the imaginary axis of revolution of the revolved feature is normal to the **TOP** datum plane. Therefore, the **TOP** datum plane will be selected to be at the top while drawing the sketch. Now, you need to decide the sketching plane from the datum planes **RIGHT** and **FRONT**. Any of the two datum planes can be selected as the sketching plane. The **RIGHT** datum plane will be selected as the sketching plane.

1. Select the **RIGHT** datum plane as the sketching plane. As you select the sketching plane, the reference plane and its orientation are set automatically.

In the **Sketch** dialog box, the **Reference** box displays **TOP:F2(DATUM PLANE)**. This

Creating Base Features

3-45

indicates that the **TOP** datum plane is selected as the reference plane. But you need to change the orientation of the plane.

2. From the **Orientation** drop-down list, choose the **Top** option.

The **TOP** datum plane will be perpendicular to the sketching plane and the sketching plane will be parallel to the screen.

3. Change the direction of the yellow arrow that appears on the sketching plane by choosing the **Flip** button. Now, the direction of viewing the sketching plane is reversed.

4. Choose the **Sketch** button in the **Sketch** dialog box.

Specifying References

Now, you have entered the sketcher environment and the **References** dialog box is displayed on the top right corner of the screen. You will be prompted to select a perpendicular surface, an edge, or a vertex relative to which the section will be dimensioned or constrained. The status displayed in the **Reference status** area is **Fully Placed**.

1. Choose the **Close** button from the **References** dialog box to exit it.

Drawing the Sketch

You need to draw the sketch of the thin extruded model. The sketch can be a closed or an open loop. Here, you will draw an open sketch.

1. Draw a center line and then draw the sketch, as shown in Figure 3-67. The center line is the axis of revolution. To fillet the corners, choose the **Create a circular fillet between two entities** button from the **Right Toolchest**.

While drawing the sketch, weak constraints are applied to the sketch.

Applying Constraints to the Sketch

Weak constraints are applied to the sketch while drawing but you need to apply the constraints using the **Constraints** dialog box.

1. Choose the **Impose sketcher constraints on the section** button from the **Sketcher Tools** toolbar. The **Constraints** dialog box is displayed.

2. Choose the **Create Equal Lengths, Equal Radii, or Same Curvature constraint** button from the **Constraints** dialog box.

3. One by one select the two horizontal lines with the constraint symbol L_1 shown in Figure 3-68 to apply the equal length constraint.

4. One by one select the two horizontal lines with the constraint symbol L_2 shown in Figure 3-68 to apply the equal length constraint.

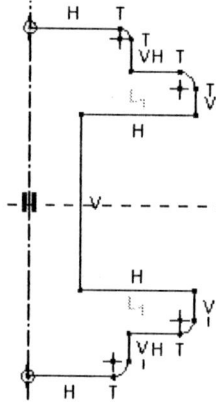 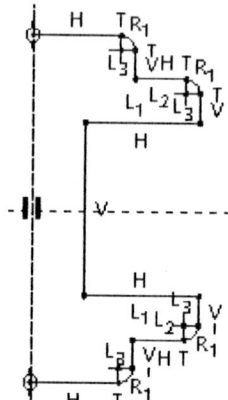

Figure 3-67 Sketch for the revolved feature with the weak dimensions turned off for clarity

Figure 3-68 Sketch after applying constraints with the weak dimensions turned off for clarity

5. Select the vertical lines with the constraint symbol **L₃** shown in Figure 3-68 to apply the equal length constraint.

6. Select all fillets to apply equal radii constraint.

Applying Dimensions to the Sketch

The weak dimensions are automatically applied to the sketch. Remember that since the model is a revolved feature, therefore, you need to apply linear diameter dimensions to the sketch manually. The linear diameter dimensions are applied using the center line that was drawn in the sketch.

1. Choose the **Create defining dimensions** button from the **Sketcher Tools** toolbar.

2. Dimension the sketch, as shown in Figure 3-69.

Modifying the Dimensions

You need to modify the dimension values of the weak dimensions.

1. Select the sketch and dimensions using CTRL+ALT+A.

2. Choose the **Modify the values of dimensions, geometry of splines, or text entities** button from the **Sketcher Tools** toolbar. The **Modify Dimensions** dialog box is displayed.

3. Clear the **Regenerate** check box and then modify the values of the dimensions, as shown in Figure 3-70.

You will notice that the dimension you select in the **Modify Dimensions** dialog box is enclosed in a yellow box in the graphics window.

Creating Base Features

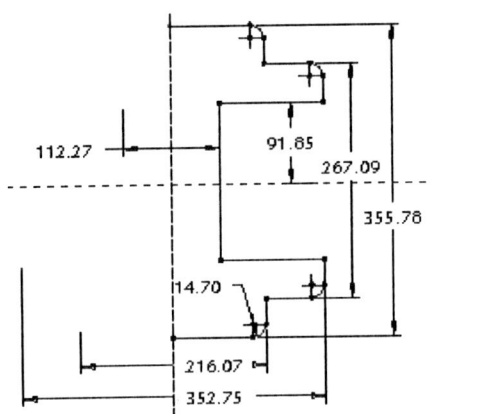

Figure 3-69 Sketch after dimensioning with constraints turned off for clarity

Figure 3-70 Sketch after modifying the dimensions with constraints turned off for clarity

4. After modifying all dimensions, choose the **Regenerate the section and close the dialog** button from the **Modify Dimensions** dialog box. The message **Dimension modifications successfully completed** is displayed in the **Message Area**.

5. Choose the **Continue with the current section** button to exit the sketcher environment.

Specifying the Model Attributes

After completing the sketch, you need to specify the side to which thickness of material will be applied. The thickness can be applied outside the section, inside the section, or symmetrically to both the sides of the section boundary. Here, you will apply the thickness inside the section.

When you exit the sketcher environment, the model assumes some default attributes and therefore it does not display in the graphics window.

1. In the drop-down list that is present at the right on the **Revolve** dashboard, enter the thickness value as **1** and press ENTER.

2. In the edit box that is present at the left on the **Revolve** dashboard, the default value of angle of revolution is **360**. Enter a value of **270** and press ENTER.

3. Choose the **Saved view list** button from the **View** toolbar. Choose the **Default Orientation** option from the drop-down list.

The trimetric view of the model after specifying all attributes is shown in Figure 3-71.

4. Choose the **Build feature** button from the **Revolve** dashboard. The default trimetric view of the model is shown in Figure 3-72.

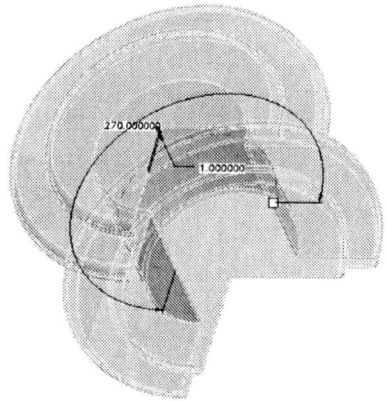

Figure 3-71 Model after specifying attributes   *Figure 3-72* Default trimetric view of the model

Saving the Model

1. Choose the **Save** option from the **File** menu or choose the **Save the active object** button from the **Top Toolchest**. The **Save Object** dialog box is displayed with the name of the object file that you had specified earlier.

2. Press ENTER to save the file.

Closing the Current Window

The given model is completed and is also saved. Now you can close the current window.

1. Choose **File > Close Window** from the menu bar.

Self-Evaluation Test

Answer the following questions and then compare your answers with those given at the end of this chapter.

1. All features created in the **Part** mode are called the base features. (T/F)

2. You can extrude a sketch to both the sides of the sketching plane symmetrically. (T/F)

3. When the sketcher environment is invoked, the references required for creating a sketch are automatically selected. (T/F)

4. The **Sketch** dialog box is used to select the sketching plane. (T/F)

5. After selecting the sketching plane, the yellow arrow displayed on it shows the direction in which you will view the sketching plane. (T/F)

6. The _____ button is selected by default when the **Extrude** dashboard is displayed.

Creating Base Features

7. The _____ button in the **Revolve** dashboard is used to create a thin revolve model.

8. The **Orientation** dialog box is displayed when you choose the _____ button.

9. After you exit the sketcher environment, the _____ appears on the model to dynamically modify the extrusion depth or the angle of revolution.

10. The **Revolve Tool** button revolves the sketched section about a _____ through the specified angle.

Review Questions

Answer the following questions:

1. By default a sketch is revolved about a center line in the _____ direction.

   (a) Clockwise (b) Counterclockwise
   (c) Right (d) None of the above

2. Which of the following menus in the menu bar contains the **Extrude** option?

   (a) **Edit** (b) **View**
   (c) **Insert** (d) **File**

3. How many datum planes are available when you enter the **Part** mode?

   (a) 4 (b) 3
   (c) 2 (d) None

4. Which of the following toolbar is used to turn off the display of datum planes?

   (a) **View** (b) **Sketcher**
   (c) **Model Display** (d) **Datum Display**

5. Which of the following combination is used to change the orientation of the model to the default orientation?

   (a) CTRL+D (b) CTRL+left mouse button
   (c) CTRL+right mouse button (d) None of the above

6. The features created using the Blind option do not have a dimension associated with them, and hence they cannot be modified by changing the dimension value. (T/F)

7. It is recommended not to use the default datum planes to create the base feature. (T/F)

8. The section drawn for revolving as a solid should be a closed loop. (T/F)

9. The revolved section should have a center line. (T/F)

10. A revolved section can be drawn on both the sides of the center line. (T/F)

Exercises

Exercise 1

Create the model shown in Figure 3-73. The dimensions of the model are shown in Figure 3-74.

(Expected time: 20 min)

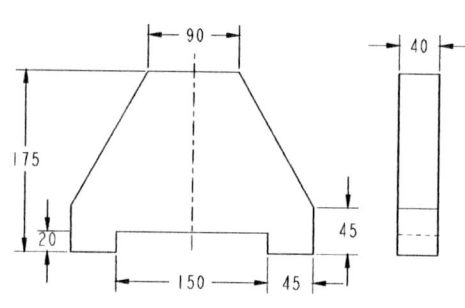

Figure 3-73 Isometric view of the model

Figure 3-74 Front and right-side views of the model

Exercise 2

Create the model shown in Figure 3-75. The dimensions of the model are shown in Figure 3-76.

(Expected time: 30 min)

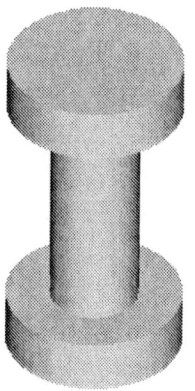

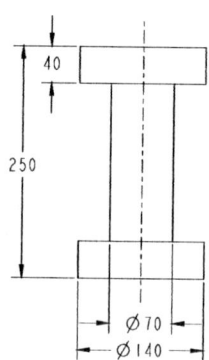

Figure 3-75 Isometric view of the model

Figure 3-76 Front view of the model

Creating Base Features

Exercise 3

Create the model shown in Figure 3-77. The dimensions of the model are shown in Figure 3-78.

(Expected time: 30 min)

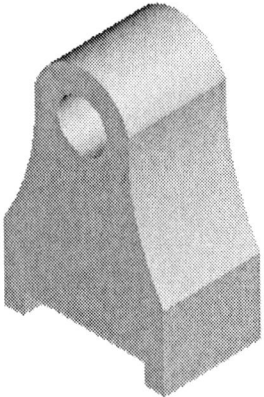

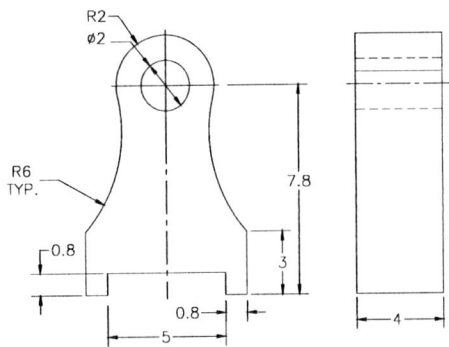

Figure 3-77 Isometric view of the model

Figure 3-78 Front view of the model

Exercise 4

Create the model shown in Figure 3-79. The dimensions of the model are shown in Figure 3-80.

(Expected time: 30 min)

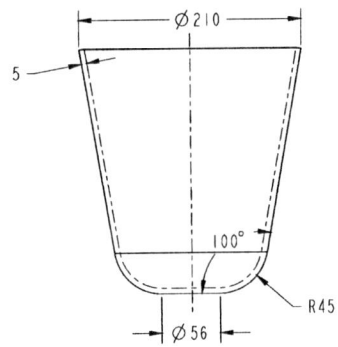

Figure 3-79 Isometric view of the model

Figure 3-80 Front view of the model

Answers to the Self-Evaluation Test
1 - F, **2** - T, **3** - T, **4** - T, **5** - T, **6** - **Extrude as solid**, **7** - **Thicken Sketch**, **8** - **Reorient view**, **9** - handles, **10** - center line.

Chapter 4

Datums

Learning Objectives

After completing this chapter you will be able to:
- *Understand the three default datum planes.*
- *Understand selection methods in Pro/ENGINEER Wildfire 2.0.*
- *Create the datum planes using different constraints available.*
- *Create datum planes on-the-fly.*
- *Create datum axes using the different constraints available.*
- *Create the datum points.*
- *Create extrude and revolve cuts.*

DATUMS

Datums are imaginary features with no mass or volume and are available to help you in creating models. They act as reference for sketching a feature, orienting the model, assembling components, and so on. Remember that the datums play a very important role in creating complex models in Pro/ENGINEER and therefore you must have a good understanding of datums. Datums are considered to be features but not model geometry. In Pro/ENGINEER, datums exist as datum plane, datum curve, datum point, datum coordinate system, datum graph, and so on.

Default Datum Planes

When you enter the **Part** mode or the **Assembly** mode, the three datum planes are by default displayed in the graphics window. These datum planes are known as the default datum planes and they are mutually perpendicular to each other. The only difference between the default datum planes of the **Part** mode and those of the **Assembly** mode lies in the names of the datum planes.

The default datum planes in the **Part** mode are named **FRONT**, **TOP**, and **RIGHT**. In case of the **Assembly** mode, the default datum planes are named **ASM_FRONT**, **ASM_TOP**, and **ASM_RIGHT**. However, the names of the default datum planes can be changed if required.

To change the names, choose **Edit > Setup**. The **PART SETUP** menu is displayed, as shown in Figure 4-1. Choose the **Name** option from this menu. You are prompted to select a feature to change the name. Select the datum plane you want to rename. When the **Message Input Window** appears, enter the desired name in it and choose the **Accept value** button.

The **PART SETUP** menu can also be used to set the material, units, and other parameters related to the model.

Figure 4-1 PART SETUP menu

Note

*The default units of a part are **Inch lbm Second**. The length is measured in inches, mass in lbm, time in seconds, and temperature in Fahrenheit. There are other system of units that are available and you can change the units as desired.*

NEED FOR DATUMS IN MODELING

Generally, most of the engineering components or designs consist of more than one feature. First the base feature of the model is created and then other features of the model are added. Since all features of a model cannot be drawn on a single plane, therefore, to draw rest of the features sometimes additional planes have to be created or selected. Also, most of the times, the three default datum planes are not enough to create a complex model having many features. For example, Figure 4-2 shows a simple model that consists of two features that require two different planes.

Datums

Tip: *Whenever you come across any solid model, first try to visualize the number of features in that model and then decide which feature is to be considered as the base feature.*

In Figure 4-2 any one of the two features that are defined on two different planes can be considered as the base feature. However, in this discussion, the feature that is selected as the base feature, is shown in Figure 4-3. After creating the base feature, the next feature will be created. For the next feature, a sketching plane has to be defined. Therefore, an additional plane has been created on which you can draw the sketch for the second feature.

Figure 4-2 Model having two extruded features   *Figure 4-3 Base feature of the model*

As shown in Figure 4-4, the plane that is used to create the base feature is highlighted by a mesh. To create the second feature, a new plane is created which is shown in Figure 4-5. The sketch of the second feature is drawn on this plane.

Note
Throughout the book, at some instances the datum planes are shown by a mesh plane as evident from Figure 4-4. This view of the datum plane is only for explanation. In Pro/ENGINEER, when you create a datum plane, they do not appear in the form of mesh.

To create various types of datums, you need to select the references on the model. Based on the selection you make on the model, Pro/ENGINEER applies constraints to create the datum.

Selection Method in Pro/ENGINEER Wildfire 2.0

The selection method in Pro/ENGINEER Wildfire 2.0 is divided into two types:

1. Selection
2. Collection

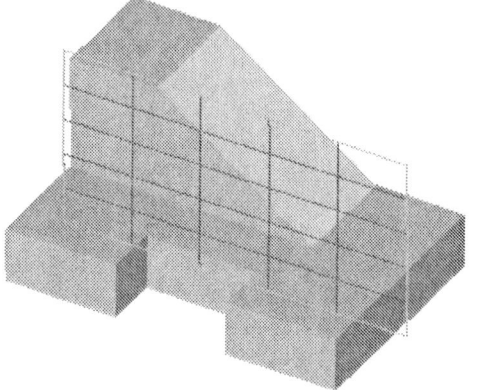

Figure 4-4 Plane selected for the base feature   *Figure 4-5 Plane selected for the second feature*

Selection

In Pro/ENGINEER Wildfire 2.0, selection refers to the action in which you first select the entities like edge, plane, face, axis, coordinate system, and so on, and then invoke a feature creation tool. This property of a design package is also known as Object Action. To make your selection process easier, filters are available in Pro/ENGINEER Wildfire 2.0. Filters are the options that are available in the **Filter** drop-down list located at the lower right corner of the main window in the **Status** bar. The options in the **Filter** drop-down list change with the mode that is active. The options in this drop-down list when you are in the **Part** mode are shown in Figure 4-6.

As you select the entities on the model, the number of selected entities appear to the left of the **Filter** drop-down list in the **Status** bar. When you double-click on the number, the **Selected Items** dialog box is displayed, as shown in Figure 4-7. All selections that you made on the model are available in this dialog box. You can use the dialog box to remove the selected entities. These selections appear highlighted in red color on the model. To remove a selection on the model, press the CTRL key and then select the entity in red.

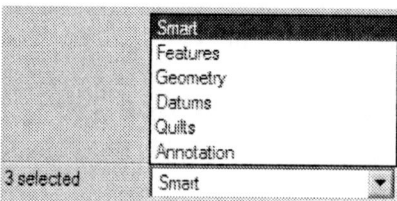

*Figure 4-6 The **Filter** drop-down list*   *Figure 4-7 The **Selected Items** dialog box*

By default, in the **Filter** dorp-down list the **Smart** filter is selected. The second filter in the **Filter** drop-down list is **Features**. This filter is used to select the features on the model. To redefine a feature, select this option from the drop-down list. The third filter is **Geometry**. This filter

Datums

allows you to select only the edges, vertices, and surfaces. The fourth filter is **Datums**. This filter allows you to select the datum features. The fifth filter is **Quilts**. This filter allows you to select surfaces. The sixth filter is **Annotation**. This filter is used to select the notes from the graphics window.

Collection
In Pro/ENGINEER Wildfire 2.0, collection refers to the action when you have invoked a feature creation tool and then select the entities to collect the references in order to create that feature. To make your selection process easy, filters can be used. The filters available in the **Filter** drop-down list depend on the feature creation tool that you have invoked.

DATUM OPTIONS
After discussing the default datum planes, which are the first feature in the **Part** mode, you must know the various other features created using the datum options. Datums are also considered as features having no geometry. Figure 4-8 shows the **Datum** toolbar and Figure 4-9 shows the method of invoking various types of datum features from the menu bar.

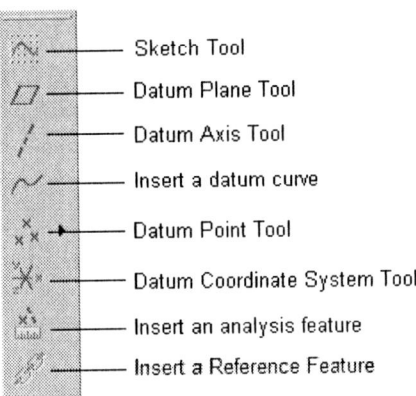

Figure 4-8 The **Datum** *toolbar*

Datum Planes

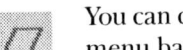

You can create datum planes, other than the three default datum planes, by using the menu bar or the **Datum** toolbar. The datum planes can be created at anytime when required. The display of the datum planes can be turned on or off by using the **Datum planes on/off** button from the **Datum Display** toolbar. Before discussing the procedure to create datum planes by using the avaialble options, it is important to understand the use of datum planes. Some of the uses of datum planes in Pro/ENGINEER Wildfire 2.0 are listed below.

1. Datum planes are used as sketching planes to create sketches for the features of a model.

2. Datum planes are used as reference planes for sketching.

3. Datum planes are used as references for placing holes and for assembly.

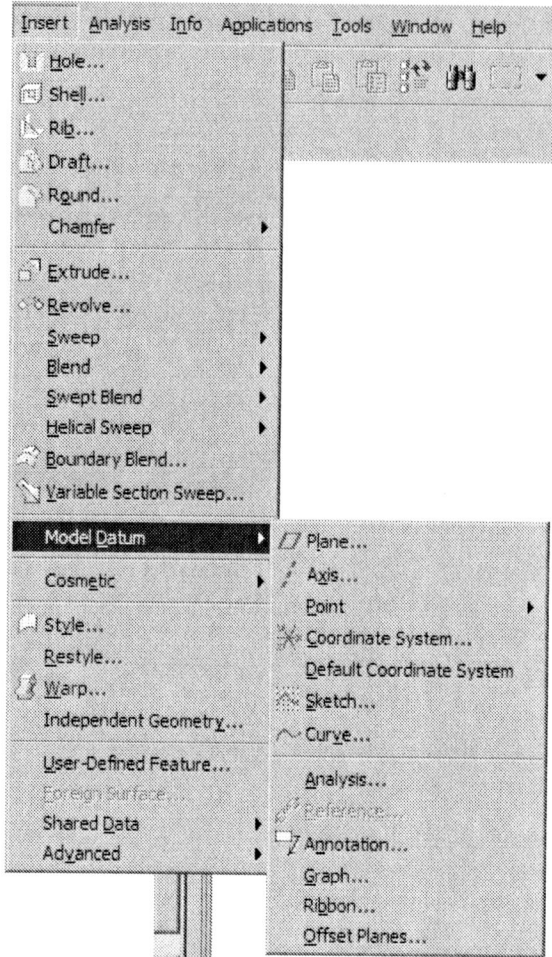

*Figure 4-9 Invoking the datum options from the **Insert** menu in the menu bar*

4. Datum planes are used as a reference for mirroring features, copying features, creating a cross-section, and for orientation of references.

Pro/ENGINEER provides you with various constraints to create additional datum planes. Additional datum planes are created with the help of constraints and the filters available in the **Filter** drop-down list. Datums can also be created while you are in the sketcher environment. When you choose **Insert > Model Datum > Plane** from the menu bar or **Datum Plane Tool** button from the **Datum** toolbar, the **DATUM PLANE** dialog box is displayed, as shown in Figure 4-10. Note that in this dialog box, some references have already been selected. Separate buttons for creating datum axis, datum curve, datum point, datum coordinate system, are available in the **Datum** toolbar.

Tip: *Generally, for the base feature creation, the three default datum planes are used and as the part becomes complex or in other words as the number of features increase, the need for additional datum planes arises.*

Datums

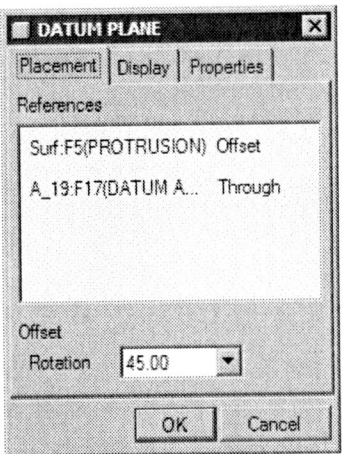

Figure 4-10 The **DATUM PLANE**
dialog box with **Placement** *tab selected*

Figure 4-10 shows various options available in the **DATUM PLANE** dialog box to create datum planes. The options available in different tabs of the **DATUM PLANE** dialog box are discussed next.

Placement Tab

The **Placement** tab in the **DATUM PLANE** dialog box is selected by default. Under this tab, the **References** area displays the references that are selected to create the datum plane. The constraints are displayed on the right of the references. These constraints are applied automatically based on the reference you select. To change a constraint, select the constraint in the dialog box; you will notice that a drop-down list appears in its place. From this drop-down list, select a different constraint. This drop-down list is available only for those constraints in the dialog box, which can be substituted by another constraint. The constraints that are used to create a datum plane are discussed later in the chapter.

The **Offset** area is available only when the **Offset** constraint is used to create a datum plane. The **Rotation** edit box is used to enter the rotation angle of the new datum plane. The angle can also be set dynamically on the model by using the drag handle displayed on the model.

Display Tab

The **Display** tab of the **DATUM PLANE** dialog box is shown in Figure 4-11. The **Flip** button in this dialog box is used to change the normal direction of the datum plane being created. Every datum plane has two sides, one is colored brown and the other is black. These colors are visible when the plane is rotated such that its back side comes into view. By default the arrow on the plane points toward the brown side. The arrow direction and hence the direction of the plane can be changed by using the **Flip** button.

The **Adjust Outline** check box when selected, allows you to set the size for the datum plane.

Properties Tab

The **Properties** tab of the **DATUM PLANE** dialog box is shown in Figure 4-12. This tab when selected, allows you to name the datum plane you are creating. By default, the first datum plane you create is named DTM1 and then every datum plane created is successively numbered.

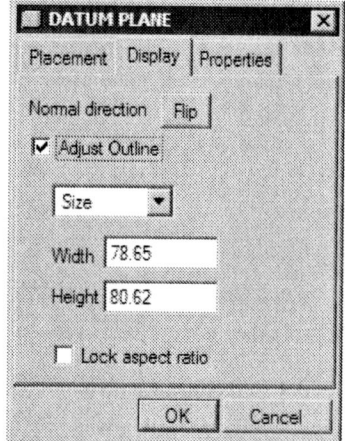 

Figure 4-11 The **DATUM PLANE** dialog box with **Display** tab selected

Figure 4-12 The **DATUM PLANE** dialog box with **Properties** tab selected

Creating Datum Planes

When you create a datum plane, the constraints are applied automatically based on the selection you make on the model. These constraints appear in the **DATUM PLANE** dialog box. Sometimes, while applying constraints to define a datum plane, a single constraint is enough to define the datum plane and sometimes you need more than one constraint to do the same. When only one constraint is used to create a datum plane, it is called stand-alone constraint. The stand-alone constraints are sufficient by themselves to constrain a datum plane definition. The constraints that are used to create a datum plane are discussed next.

Through Constraint

The **Through** constraint is used to create a datum plane through any specified axis, edge, curve, point/vertex, plane, cylinder, or coordinate system. This constraint can be used in combination with other constraints that are discussed next.

However, the **Through** constraint can also be used as a stand-alone constraint when you select a plane. Figure 4-13 shows the datum plane constraint combinations using the **Through** constraint. Datum planes can be created using any of the combinations shown in the figure. The possible combinations of datum plane creation are referred to as **Yes** and the combinations that are not possible are referred to as **No** in the figure.

While reading the table shown in Figure 4-13, first preference is given to the text written in the first column and then the text in the first row should be read. For example, if you want to make a datum plane that is passing through a cylinder and normal to a plane then look for

Datums

DATUM PLANE CONSTRAINT COMBINATIONS (USING THROUGH CONSTRAINT)		Through			Normal	Parallel	Angle	Tangent	Standalone Constraints
		Axis/Edge/Curve	Point/Vertex	Cylinder	Plane	Plane	Plane	Cylinder	
Through	Axis/Edge/Curve	Yes	Yes	Yes	Yes	Yes	Yes	Yes	No
	Point/Vertex	Yes	Yes	Yes	Yes	Yes	No	Yes	No
	Plane	No	No	No	No	No	No	No	Yes
	Cylinder	Yes	Yes	Yes	Yes	Yes	Yes	Yes	No

Figure 4-13 Datum plane constraint combinations using the **Through** constraint

Through in the first column and then for **Cylinder** in the second column. Now, look for **Normal** in the first row and for **Plane** in the second row. After finding both the combinations trace them in the respective column and row till they intersect. You will find **Yes**. This suggests that the creation of a datum plane that passes through a cylinder and is normal to a plane is possible. While reading the table shown in Figure 4-13, remember that the constraints that are not stand-alone have to be applied in pairs.

The options in the drop-down list shown in Figure 4-14 are used to make selection on the model. This drop-down list is located at the bottom right corner of the main window. The options in this drop-down list change depending on the operation being performed. For example, when you are creating a datum plane, the options shown in Figure 4-14 are available. Once you exit the datum plane creation, the options in this drop-down list change.

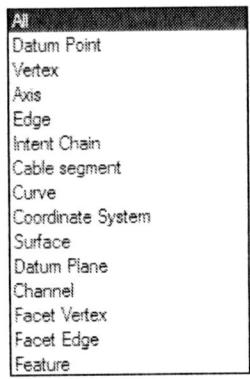

Figure 4-14 **Filter** drop-down list with options to make selection

Figure 4-15 shows that the cylindrical surface and the default datum plane used to create a datum plane at an angle to the selected default datum plane and passing through the center of the cylindrical surface, as shown in Figure 4-16. When you select the cylindrical surface, it is highlighted in red, indicating that it is selected. To make the second selection, select the **Plane** filter from the **Filter** drop-down list. Now, press the CTRL button and then select the datum plane. This method of selecting by pressing the CTRL button is known as collection. Click on the constraint displayed against the selected plane in the **DATUM PLANE** dialog box and select the **Offset** option from the drop-down list. Enter the angle value in the **Rotation** dimension box. Choose the **OK** button to create the datum plane.

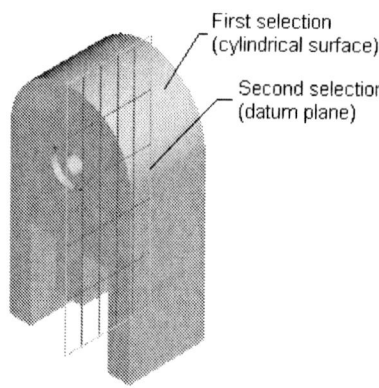
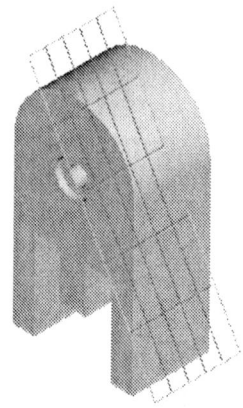

Figure 4-15 Selecting a cylindrical surface and a default datum plane to create a datum plane

Figure 4-16 Resulting datum plane passing through the center of a cylinder and at an angle

Note
The first reference to constrain a datum plane is selected using the left mouse button. The second reference is selected using the CTRL+left mouse button.

*Make sure you change the option in the **Filter** drop-down list when you invoke any other tool.*

Normal Constraint

The **Normal** constraint is used to create a datum plane normal to any specified axis, edge, curve, or plane. This constraint is used in combination with other constraints. The **Normal** constraint cannot be used as a stand-alone constraint. Figure 4-17 shows the datum plane constraint combinations using the **Normal** constraint.

DATUM PLANE CONSTRAINT COMBINATIONS (USING NORMAL CONSTRAINT)		Through			Normal	Parallel	Angle	Tangent	Standalone Constraints
		Axis/Edge/ Curve	Point/ Vertex	Cylinder	Plane	Plane	Plane	Cylinder	
Normal	Axis/Edge/ Curve	Yes	Yes	Yes	No	No	No	No	No
	Plane	Yes	Yes	Yes	Yes	Yes	No	Yes	No

Figure 4-17 Datum plane constraint combinations using the **Normal** constraint

The possible combinations of datum plane creation are referred to as **Yes** and the combinations that are not possible are referred to as **No**.

Datums

Figure 4-18 shows a planar face and a cylindrical face of a solid model. The planar face is selected as the normal surface and the cylindrical face is selected to be tangent to the datum plane. The following steps explain the procedure to create this datum plane:

1. Invoke the **DATUM PLANE** dialog box. Select the first reference shown in Figure 4-18. The **Offset** constraint is automatically applied and is displayed in the **References** collector.

2. Left-click on the **Offset** constraint in the collector; a drop-down list appears. Select the **Normal** constraint from this drop-down list.

3. Now, use CTRL+left mouse button to select the second reference shown in Figure 4-18. The **Through** constraint is displayed in the **References** collector.

4. Change this constraint to **Tangent** from the drop-down list that appears when you click on the constraint in the **References** collector. The datum plane that is created is shown in Figure 4-19.

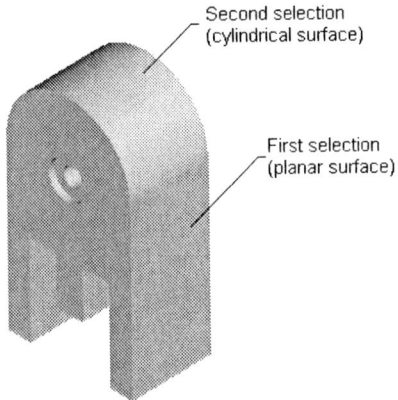 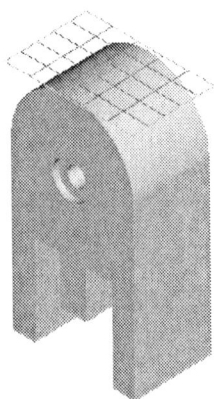

Figure 4-18 Selecting a planar surface and a cylindrical surface to create a datum plane

Figure 4-19 Resulting datum plane

5. Choose the **OK** button from the **DATUM PLANE** dialog box.

Parallel Constraint

The **Parallel** constraint is used to create a datum plane parallel to any specified datum plane or planar face. This option is used in combination with other constraints. The **Parallel** constraint cannot be used as a stand-alone constraint. Figure 4-20 shows various datum plane constraint combinations using the **Parallel** constraint. The possible combinations of datum plane creation are referred to as **Yes** and the combinations that are not possible are referred to as **No** in the figure.

Figure 4-21 shows the selection of a default datum plane and an axis to create a datum plane. The resulting datum plane is parallel to the selected datum plane and passes through the axis, as shown in Figure 4-22. The following steps explain the procedure to create this datum plane:

DATUM PLANE CONSTRAINT COMBINATIONS (USING PARALLEL CONSTRAINT)		Through			Normal	Parallel	Angle	Tangent	Standalone Constraints
		Axis/Edge/ Curve	Point/ Vertex	Cylinder	Plane	Plane	Plane	Cylinder	
Parallel	Plane	Yes	Yes	Yes	No	No	No	Yes	No

Figure 4-20 *Datum plane constraint combinations using the **Parallel** constraint*

1. Invoke the **DATUM PLANE** dialog box. Select the first reference shown in Figure 4-21. The **Offset** constraint is automatically applied and is displayed in the **References** collector.

2. Left-click on the **Offset** constraint in the collector; a drop-down list appears. Select the **Parallel** constraint from this drop-down list.

3. Now, use CTRL+left mouse button to select the second reference shown in Figure 4-21. The **Through** constraint is displayed in the **References** collector.

4. Choose the **OK** button from the **DATUM PLANE** dialog box. The datum plane that is created is shown in Figure 4-22.

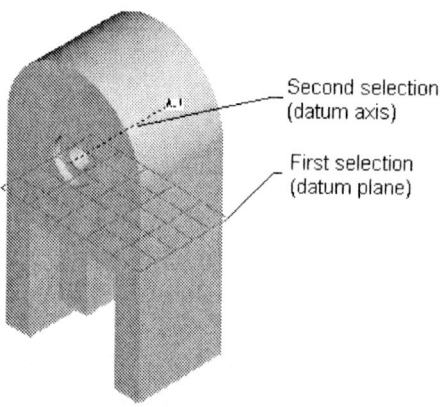
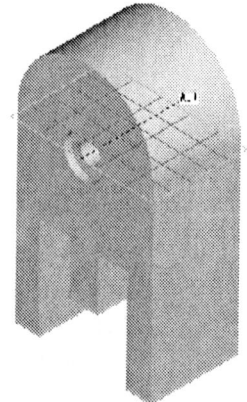

Figure 4-21 *Selecting a datum plane and an axis to create a datum plane*

Figure 4-22 *Resulting datum plane*

Offset Constraint

The **Offset** constraint is used to create a datum plane at an offset distance to any specified plane or coordinate system. This option is used in combination with other constraints. However, the **Offset** constraint can be used as a stand-alone constraint when you select a plane to offset from.

Datums

DATUM PLANE CONSTRAINT COMBINATIONS (USING OFFSET CONSTRAINT)		Through			Normal	Parallel	Angle	Tangent	Standalone Constraints
		Axis/Edge/ Curve	Point/ Vertex	Cylinder	Plane	Plane	Plane	Cylinder	
Offset	Plane	Yes	No	Yes	No	No	No	No	Yes
	Coord System	No	No	No	No	No	No	No	Yes

*Figure 4-23 Datum plane constraint combinations using the **Offset** constraint*

When the **Offset** constraint is applied, the **Translation** dimension box appears. This dimension box list is used to specify an offset distance. In the case of angular planes the **Translation** dimension box changes to the **Rotation** dimension box in which you need to specify the angle. An arrow appears on the model that shows the positive direction of the offset distance or angle.

Figure 4-23 shows various datum plane constraint combinations using the **Offset** constraint. The possible combinations of datum plane creation are referred to as **Yes** and the combinations that are not possible are referred to as **No** in the figure.

Figure 4-24 shows the selection of a default datum plane and a vertex to define an offset datum plane. The resulting datum plane is at an offset to the selected datum plane and passes through the vertex, as shown in Figure 4-25. The following steps explain the procedure to create this datum plane:

1. Invoke the **DATUM PLANE** dialog box. Select the first reference shown in Figure 4-24. The **Offset** constraint is automatically applied and is displayed in the **References** collector. The **Translation** dimension box appears.

2. Now, use CTRL+left mouse button to select the second reference, which is a vertex shown in Figure 4-24. The **Through** constraint is displayed in the **References** collector.

3. Choose the **OK** button from the **DATUM PLANE** dialog box. The datum plane that is created is shown Figure 4-25.

Note
*The **OK** button in the **DATUM PLANE** dialog box is enabled only when the datum plane you are creating is fully constrained.*

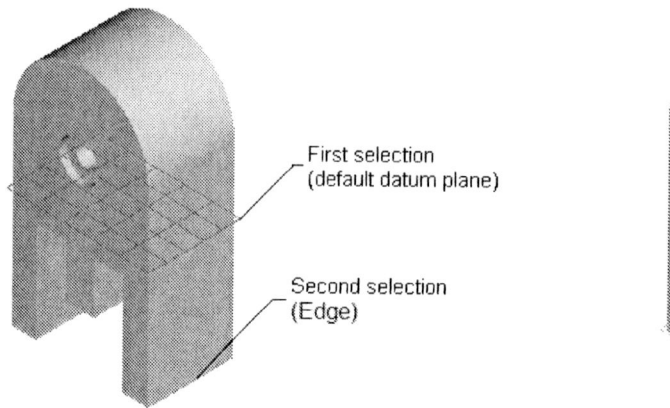 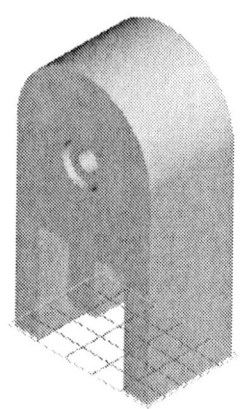

Figure 4-24 Selecting a datum plane and an axis to create a datum plane

Figure 4-25 Resulting datum plane

Tangent Constraint

The **Tangent** constraint creates datum planes tangent to cylindrical features. This constraint is also used with other constraints to create various types of datum planes. Figure 4-26 shows the datum plane constraint combinations using the **Tangent** constraint. The possible combinations of datum plane creation are referred to as **Yes** and the combinations that are not possible are referred to as **No** in the figure.

DATUM PLANE CONSTRAINT COMBINATIONS (USING TANGENT CONSTRAINT)		Through			Normal	Parallel	Angle	Tangent	Standalone Constraints
		Axis/Edge/ Curve	Point/ Vertex	Cylinder	Plane	Plane	Plane	Cylinder	
Tangent	Cylinder	Yes	No	No	Yes	Yes	No	No	No

Figure 4-26 Datum plane constraint combinations using the **Tangent** constraint

Note

To remove a selected reference, select the reference in the dialog box and press DELETE.

Datum Planes Created On-The-Fly

The term **On-the-Fly** refers to the creation of a datum plane when the system prompts you to select or create a plane. At this step, if you choose the **Datum Plane Tool** button from the **Datum** toolbar, the datum plane created will be called datum plane on-the-fly. When you create a datum plane on-the-fly, the datum plane is neither visible in the graphics window nor is displayed under the default display of the **Model Tree** once the feature is completed.

Datums

You can try making a datum plane on-the-fly by following these steps:

1. Choose the **Extrude Tool** button; the **Extrude** dashboard is displayed.

2. In the **Extrude** dashboard choose the **Placement** tab and then click on the **Define** button to invoke the **Sketch** dialog box.

3. Now, you need to select a sketching plane. You can select an existing datum plane, a face as the sketch plane, or you can create a datum plane on-the-fly.

4. Choose the **Datum Plane Tool** button from the **Datum** toolbar. Select the references and constraints to create the datum plane. This datum plane when created is automatically selected as the sketching plane.

When you need to select the references to orient the sketching plane, at this step also you can create a datum plane on-the-fly. Once, the datum plane on-the-fly is created, it cannot be referenced by other features.

The datum plane created on-the-fly does not appear in the **Model Tree** under the default display. But it can be seen in the **Model Tree** when you click the plus sign (+) that appears to the left of the feature you have created. The datum plane and the feature that you have created are grouped and appears as a grouped feature in the **Model Tree**. This means that the datum on-the-fly that was created during feature creation belongs to the feature that is created.

Note
Unlike the previous releases of Pro/ENGINEER, now you can create a rotational pattern without creating a datum on-the-fly.

Datum Axes

Datum axis is an imaginary axis that is created in Pro/ENGINEER to help you in creating a model. Datum axes can be created manually. They are also created automatically when any cylindrical feature is created. The display of a datum axis can be turned on or off by using the **Datum axes on/off** button from the **Datum Display** toolbar. The uses of datum axes are discussed next:

1. Datum axes act as reference for feature creation.

2. They are used in creating a datum plane along with different constraint combinations.

3. They are used in placing features co-axially.

4. They are also used to create rotational patterns. You will learn to create patterns in Chapter 6.

Note
*You can also create a datum axis perpendicular to the sketch plane in the sketcher environment. To create a datum axis in the sketcher environment, choose **Sketch** > **Axis Point** from the menu bar. Then select a point in the graphics window through which the axis will pass.*

Datum axes are named by default in Pro/ENGINEER. The default name of a datum axis is **A_(Number)**, where **Number** represents the number of datum axis. However, the default name of the datum axes can be changed in the same way as that of the datum planes.

When you choose **Insert** > **Model Datum** > **Axis** from the menu bar or the **Datum Axis Tool** button from the **Datum** toolbar, the **DATUM AXIS** dialog box appears, as shown in Figure 4-27. The options avaialble in this dialog box are discussed next.

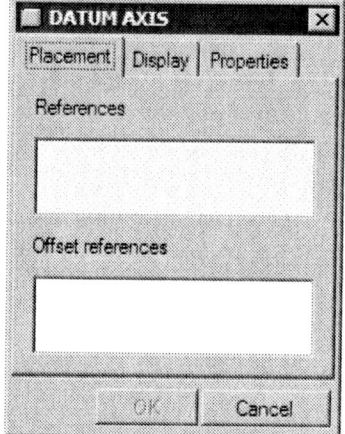

*Figure 4-27 The DATUM AXIS dialog box with the **Placement** tab selected*

Placement Tab

When you invoke the **DATUM AXIS** dialog box, the **Placement** tab is selected by default. Under this tab, the **References** collector allows you to select the references that will create the datum axis. The constraints are displayed on the right of the references. These constraints are applied automatically based on the reference you select. The constraints that are used to create a datum axis are discussed later in this chapter.

The **Offset References** collector is available only when the **Normal** constraint is used in the **References** collector to define a datum axis. To define the datum axis, you need to select two references. These references can be an edge, datum plane, face, or axis. The references can be specified dynamically on the model by dragging the handles displayed on the datum axis. The handles on the datum axis are displayed only when you select a reference. You can also select the references using the **DATUM AXIS** dialog box.

There are three drag handles that are available on the model, as shown in Figure 4-28.

Datums

These handles appear like a white square. The middle handle is used to move the position of the axis you are creating. The other two handles are used to specify the references for dimensioning.

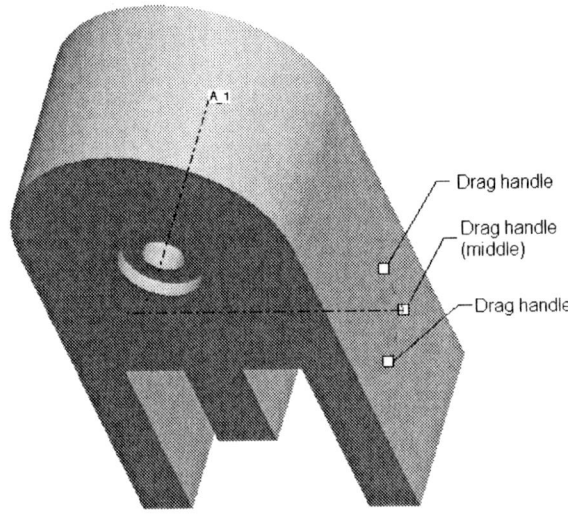

Figure 4-28 Solid model with the datum axis and three drag handles

Display Tab

The options in this tab are used to control the length of the axis. The **DATUM AXIS** dialog box with the **Display** tab is shown in Figure 4-29. Select the **Adjust Outline** check box to activate the **Length** edit box. You can enter a value in this edit box or drag the handles displayed at both ends of the datum axis to modify its length.

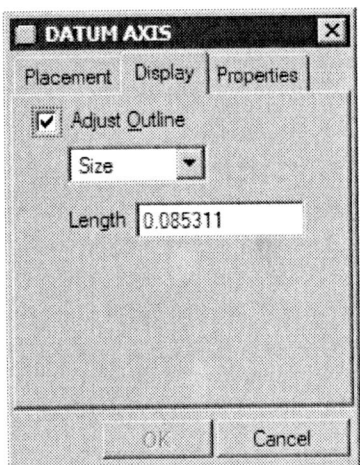

*Figure 4-29 The **DATUM AXIS**
dialog box with the **Display** tab selected*

Properties Tab

The **Properties** tab of the **DATUM AXIS** dialog box is shown in Figure 4-30. This tab, when selected, allows you to name the datum axis you are creating. By default, the first datum axis you create is named A_1 and then additional datum axis created are successively named.

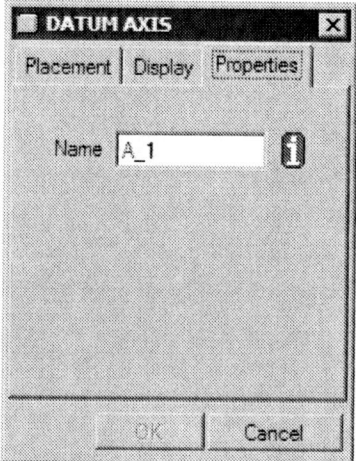

*Figure 4-30   The **DATUM AXIS**
dialog box with the **Properties** tab selected*

As mentioned earlier, a datum axis is created using constraints which are applied automatically, and filters available in the **Filter** drop-down list located in the **Status** bar. While creating a datum axis, the filters that are available are shown in Figure 4-31. The constraints that are used to create a datum axis are discussed next.

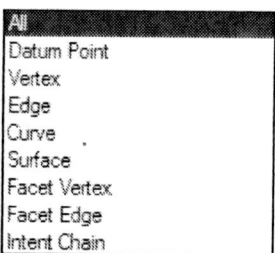

*Figure 4-31 Filters available in
the **Filters** drop-down list*

Datum Axis Passing Through an Edge

The **Through** constraint is used to create a datum axis passing through any selected edge. This constraint is displayed in the dialog box when you select an edge, as shown in Figure 4-32. In Figure 4-32, **A_1** is the datum axis created using this option.

Note

While creating a datum axis, you may need to create datum points in order to create the datum axis passing through them. Therefore, in the various cases that are discussed next you may need to create datum points. Datum points are discussed later in this chapter.

Datum Axis Normal to a Plane

The **Normal** constraint is used to create a datum axis normal to a selected face or datum plane. When you select a face or a datum plane, the **Normal** constraint is displayed in the dialog box and you are prompted to select two references to place the axis. Notice that three drag handles appear on the model. Use the middle handle to change the location of the axis on the selected face. After you specify its placement location, you need to select two edges, axes, datums, or faces to specify the linear dimension for the placement of the datum axis. Click in the **Offset References** collector. The collector turns yellow in color. When you select the first edge for the placement dimensions of the axis, the offset value is displayed in the **Offset References** collector. You can accept the default dimension or click on it to change its value. Similarly, select the second edge for dimensioning by pressing CTRL+left mouse button and enter the dimension value. In Figure 4-33, the preview of the datum axis with drag handles is shown. It should be noted that the references for dimensioning can also be selected dynamically using these drag handles.

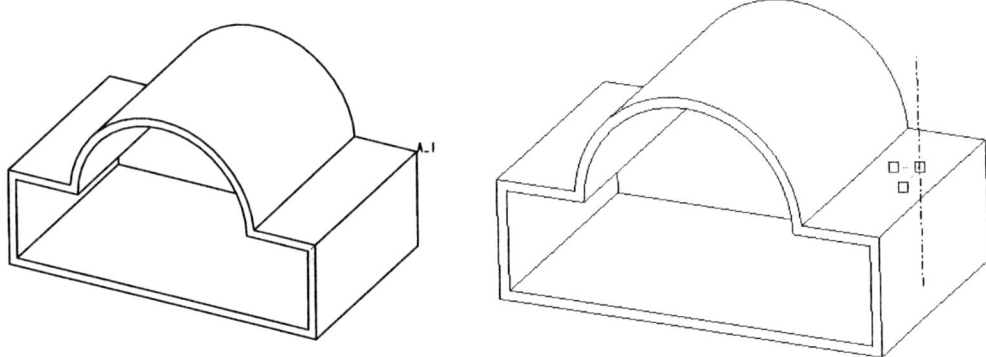

Figure 4-32 Datum axis created along the edge

Figure 4-33 Datum axis created normal to the plane

Note

*When you select a reference for creating the datum axis and you are prompted to select the placement references, you need to click in the **Offset References** collector to make it yellow in color. The yellow color of this area in the **DATUM AXIS** dialog box indicates that now you can select the references to place the axis. Until the **Offset References** collector is made yellow in color, you will not be able to select the placement references.*

Datum Axis Passing Through a Datum Point and Normal to a Plane

The **Normal** constraint creates a datum axis passing through a datum point or vertex and normal to any face or datum plane. When you select a face or a datum plane to which the datum axis will be normal, you are prompted to select two references to place the axis. Use CTRL+left mouse button to select the datum point or vertex to create an axis passing through it. Choose the **OK** button to exit the **DATUM AXIS** dialog box. In Figure 4-34, the preview of the datum axis and the datum point is shown.

Datum Axis Passing Through the Center of a Round Surface

The **Through** constraint is used to create a datum axis passing through the center of a cylindrical or round surface. To create this datum axis, select the round surface through which you need to pass the datum axis. The axis is automatically created and it passes through the center. In Figure 4-35, the preview of the datum axis is shown. The selected cylindrical surface is highlighted in this figure.

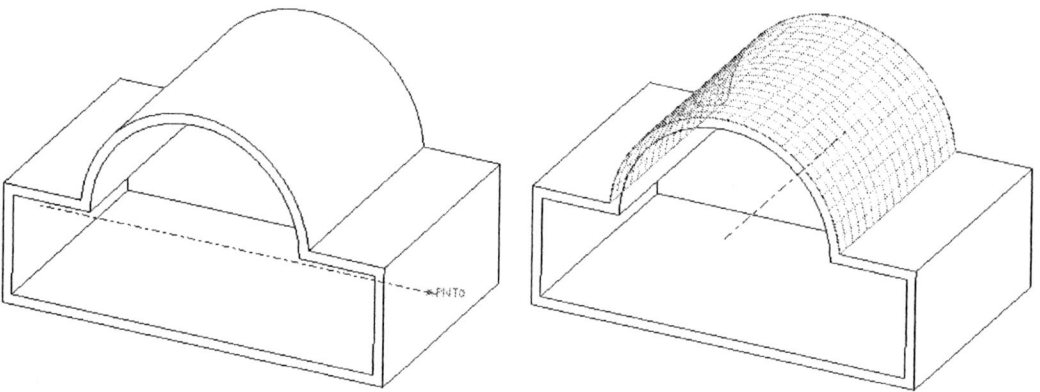

Figure 4-34 Datum axis passing through the datum point and normal to the plane

Figure 4-35 Datum axis passing through the center of a cylinder

Datum Axis Passing Through the Edge Formed by Two Planes

The **Through** constraint is used to create a datum axis passing through the intersection edge of two planar faces or planes. When you select the face, the **Normal** constraint is applied automatically. Click on it to open the drop-down list and select the **Through** constraint. Now, select the second face using the CTRL+left mouse button. A datum axis is created passing through the edge formed by the two faces or planes. In Figure 4-36, the preview of the datum axis is shown. The two faces that were selected are also highlighted in this figure.

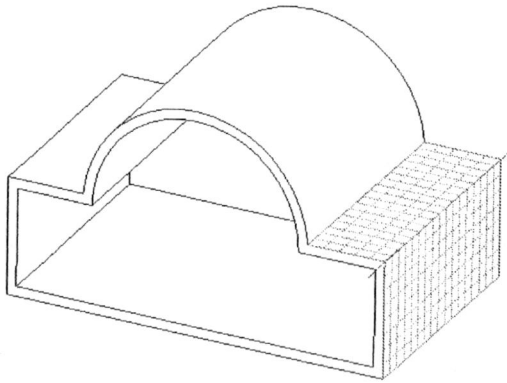

Figure 4-36 Datum axis created on the edge where the two selected planes meet

Datums

Datum Axis Passing Through Two Datum Points or Vertices

The **Through** constraint is used to create a datum axis between two datum points or vertices. To create this datum axis, select the first vertex and then using CTRL+left mouse button, select the second vertex. The datum axis is created along the two selected datum points or vertices. In Figure 4-37, the preview of the datum axis is shown with the two vertices highlighted.

Datum Axis Tangent to a Curve and Passing Through its Vertex

The **Tangent** and **Through** constraints create a datum axis tangent to a curve and passing through one of its vertex. Select the edge of the cylindrical surface as the first selection. Make sure you do not select the cylindrical surface. After you select a curve or an edge, use CTRL+left mouse button to select one vertex of the edge. The datum axis is created tangent to the curve and passing through its selected vertex. In Figure 4-38, the preview of the datum axis that is created using the **Tangent** and **Through** constraints is shown. The curved edge is also highlighted in this figure.

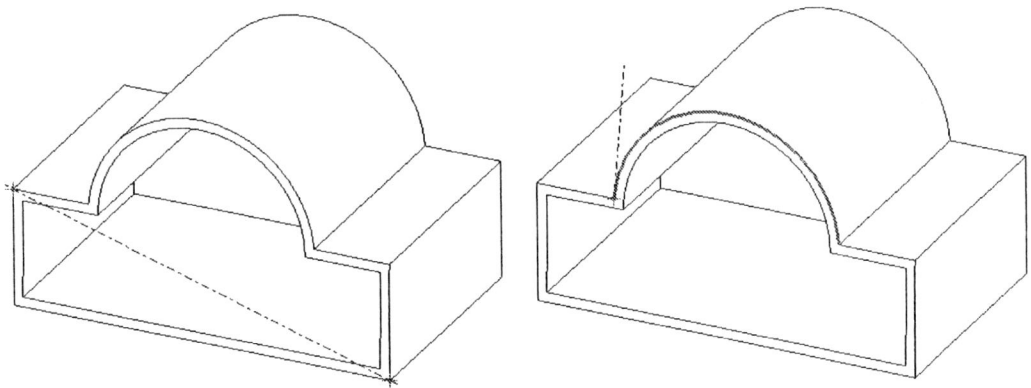

Figure 4-37 Datum axis created between the two selected vertices

Figure 4-38 Datum axis created tangent to the selected curve

Datum Points

Datum points are imaginary points created in Pro/ENGINEER to aid in creating models, drawings, analyzing models, and so on. The uses of datum points are as follows.

1. To create datum planes and axes.

2. To associate note in the drawings and attach datum targets.

3. To create coordinate system.

4. To specify point loads for mesh generation.

5. To create pipe features.

The default name associated with a datum point in Pro/ENGINEER is **PTN(Number)** where **Number** indicates the number of datum points created in a particular model. However, you can change the default name associated with the datum points.

When you choose **Insert > Model Datum > Point** from the menu bar or the **Datum Point Tool** button from the **Datum** toolbar, the **DATUM POINT** dialog box appears with various options to create datum points, as shown in Figure 4-39. This dialog box can be used to create more than one datum point. The options in the **DATUM POINT** dialog box are explained next.

Placement Tab

The **Placement** tab in the **DATUM POINT** dialog box has the options that change with the reference selected. However, the **References** display box shown in Figure 4-39 is always present. This area is used to select the references which aid in placing the datum point. The constraints available in the drop-down list in the **References** collector depends on the references you select from the model. This means that the references narrow down the type of constraints you can apply on the datum point that you are creating.

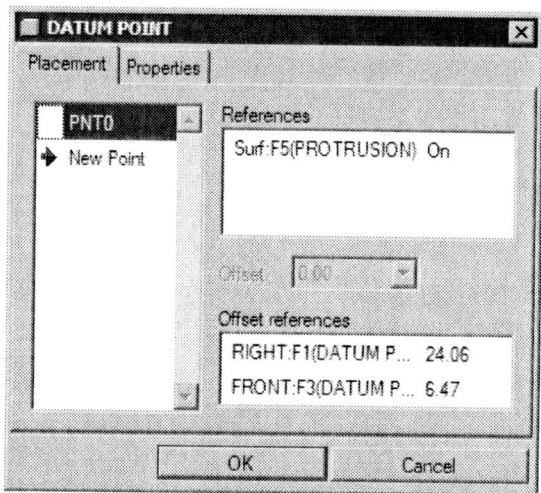

Figure 4-39 The **DATUM POINT** dialog box with the **Placement** tab selected

 **Tip**: *While creating a datum feature, the references selected from the model help you to choose from the available constraints that you can apply to constrain the datum feature. References reduce the range of available constraints, thus reducing the time taken in selecting the appropriate constraint. For example, the lesser the number of options available for a person, the lesser the time he needs to decide among them.*

Various cases of creating datum points are discussed in the explanation that is discussed next.

Datum Point on a Face or a Datum Plane

The **On** constraint is used to create datum points on a face or a datum plane. When you select a face or a datum plane to place the datum point, a yellow colored point is displayed at the selected location on the surface with the three drag handles, as shown in Figure 4-40.

Datums

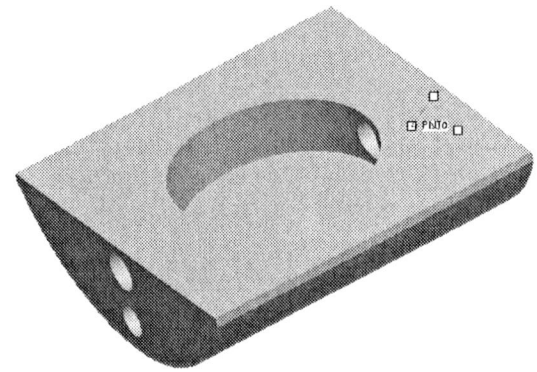

Figure 4-40 Datum point on a face and the three drag handles

Next, click in the **Offset References** collector to make it yellow in color. You are prompted to select two references to specify the linear dimensions for the placement of the datum point. Note that the second selection to select the reference should be made by using CTRL+left mouse button. After you select the two planes or edges for the placement of dimension of the point, a default value is displayed in the **Offset References** collector. You can accept the default value or change it to the required value. After the datum point is located at the desired position on the face or datum plane, choose the **OK** button to exit the dialog box.

Datum Point Offset to a Face or a Datum Plane

The **Offset** constraint creates datum points at an offset distance from a specified face or a datum plane in a specified direction. Select a face or a datum plane from where the offset distance for the placement of the datum point will be measured. The **On** constraint is applied automatically. From the drop-down list in the **References** collector, change the constraint to **Offset**. The model and the **DATUM POINT** dialog box appear, as shown in Figures 4-41 and 4-42.

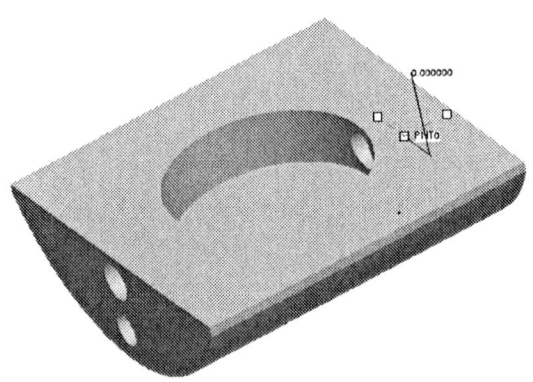

Figure 4-41 Datum point on the top face with offset **0** and the three drag handles

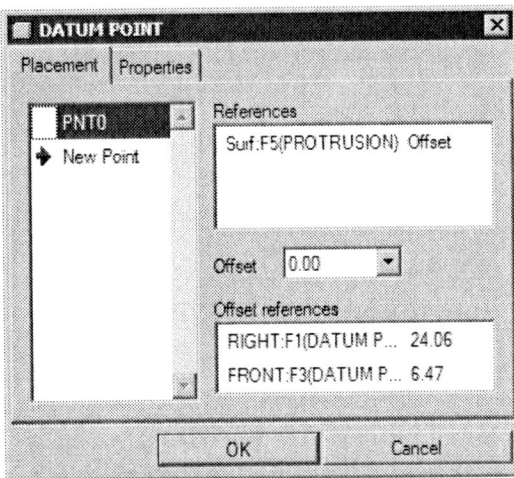

Figure 4-42 The **DATUM POINT** dialog box with **Offset** constraint selected

In the **Offset** edit box under the **References** collector, enter the offset value. You will be prompted to select the planes or the edges for dimensioning the point. Select two planes or edges for dimensioning and enter the distances from the highlighted references.

You can also specify the offset distance dynamically by using the middle drag handle. The two references for dimensioning the datum point can be specified by using the drag handles or selected directly on the model.

Note
*When you use the **Offset** constraint, the middle drag handle is used to specify the offset distance and when you use the **On** constraint, it is used to specify the location of the datum point on the selected face.*

Datum Point at the Intersection of Three Surfaces

The **On** constraint is used to create a datum point at the intersection of three surfaces. To create this datum point, select the three surfaces and the datum point is created, as shown in Figure 4-43. The datum point is created on the vertex that is common to the three surfaces. Remember that the first reference is selected using the left mouse button. The second and third references are selected using CTRL+left mouse button. After selecting the three surfaces, the **DATUM POINT** dialog box appears, as shown in Figure 4-44.

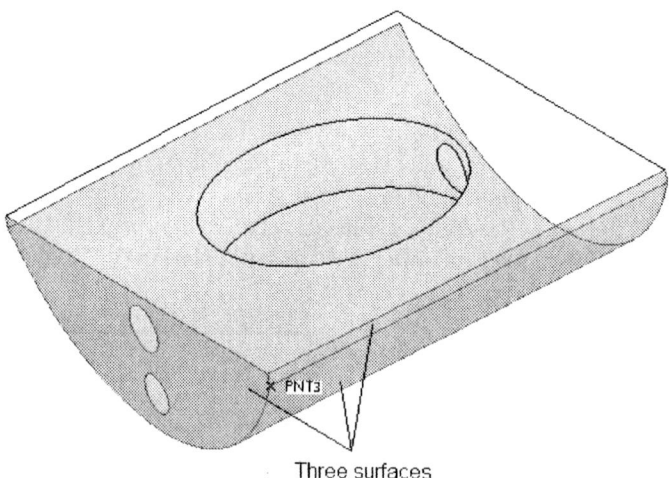

Figure 4-43 The three highlighted surfaces and the datum point at the intersection of the three surfaces

It might so happen that with the current selection of references, more than one intersection point exist. In such a case, use the Next Intersection button in the DATUM POINT dialog box to select the other intersection points to create the datum point on them.

Datum Point on a Vertex

The **On** constraint is used to create a datum point on the vertex of a face, an edge, or a datum

Datums

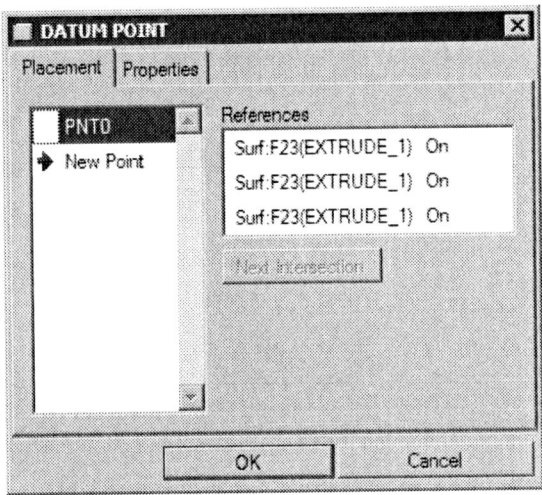

*Figure 4-44 The **DATUM POINT** dialog box*

curve. Invoke the **DATUM POINT** dialog box and then select a vertex to create the datum point. Choose the **OK** button to exit the dialog box.

Note
You can create more than one datum point by simply selecting the desired vertex.

Datum Point at the Center of a Curved Edge

The **Center** constraint creates a datum point at the center of an arc or a curved edge. Invoke the **DATUM POINT** dialog box and select the curve edge. The **On** constraint is applied by default. To modify the constraint, select the constraint in the **References** collector; the drop-down list appears to the right of the constraint. Choose the **Center** constraint from it to create the datum point at the center of the curved edge.

Datum Point on an Edge or a Curve

The **On** constraint is used to create a datum point on an edge or a curve. Invoke the **DATUM POINT** dialog box, and select an edge or a curve. The datum point is placed on the selected edge, as shown in Figure 4-45 and the **DATUM POINT** dialog box appears, as shown in Figure 4-46.

In the **DATUM POINT** dialog box, after the point is placed, the dimension type can be defined. The **Offset** dimension box displays the offset distance of the datum point from one end of the edge. The offset distance is measured as a ratio or as a real value from the end of the edge, which can be selected from the drop-down list, as shown in Figure 4-46.

In the **Offset References** area, the **Next End** button is used to flip the end of the edge from where the offset distance is measured. This button is available only when the **End of curve** radio button is selected. If you select the **Reference** radio button, you need to select a reference from which the datum point will be dimensioned.

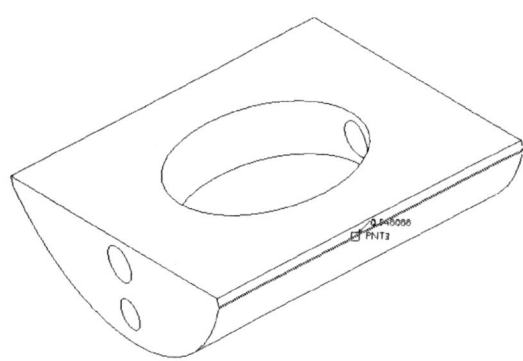

Figure 4-45 *The highlighted edge and the datum point at some offset distance from one end of edge*

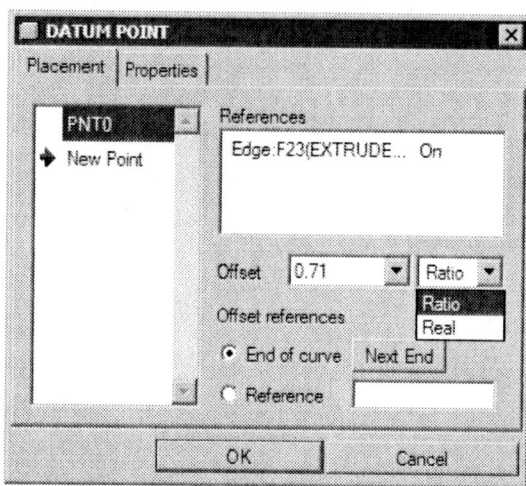

Figure 4-46 *The **DATUM POINT** dialog box*

Sketched Datum Point Tool Button

 When you choose the black arrow on the right of the **Datum Point Tool** button, a flyout is displayed. Choose the **Sketched Datum Point Tool** button from this flyout; the **Sketch Datum Point** dialog box is displayed. The function of this dialog box is same as the **Sketch** dialog box. Using the **Sketch Datum Point** dialog box, select the sketching plane and its orientation. When you choose the **Sketch** button, the system takes you to the sketcher environment. In the sketcher environment you can use the construction tools and the **Create Points** button to sketch the datum points. The datum points are dimensioned with the references. You can also modify these dimensions. Remember that you can exit the sketcher environment only if at least one point is drawn.

To Create an Array of Datum Points from a Coordinate System

The **Offset Coordinate System Datum Point Tool** button is used to create an array of datum points at an offset distance from a coordinate system. You can change the array of the points by redefining the array. This button is available on the flyout. Note that the Datum Coordinate system must be defined before you create an array of datum points.

When you choose this button, the **Offset CSys Datum Point** dialog box is displayed, as shown in Figure 4-47 and you are prompted to select a coordinate system. After selecting a coordinate system, you can enter the values of the coordinates for the datum points.

To add a point, click under the **Name** column in the list box; the first row gets activated. Click under the **XAxis** column; and edit box appears in which you can enter an offset value. Similarly, enter the offset values for **Y** and **Zaxis**. To add another point click in the next row. From the **Type** drop-down list, select the type of coordinate system: Cartesian, Cylindrical, or Spherical. In the Figure 4-47, the dialog box is shown with the coordinates of three datum points are entered.

Datums

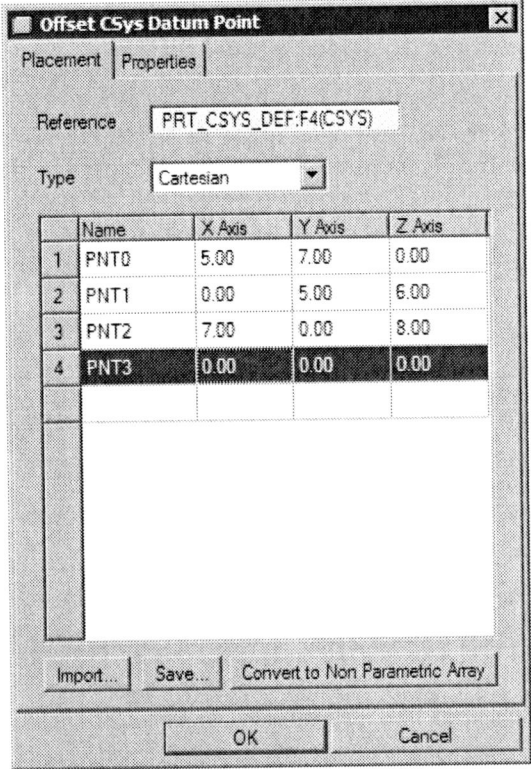

*Figure 4-47 The **Offset CSys Datum Point** dialog box*

To Create a Datum Point by Clicking

When you choose the **Field Datum Point Tool** button from the flyout, the **Field Datum Point** dialog box is displayed, as shown in Figure 4-48. Using this dialog box, you can create a datum point anywhere on the surface of the model. You just need to specify the location using the left mouse button to place a datum point.

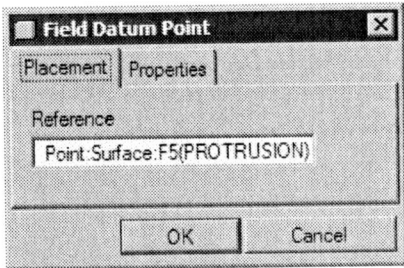

*Figure 4-48 The **Field Datum Point** dialog box*

CREATING CUTS

Cut is a material removal process and this option is available only when at least one base feature exists in the graphics window. The **Remove Material** button available on the dashboard of the feature creation tools is used to create the cuts. In this chapter, you will learn the extrude cut and the revolve cut. The procedure to create a cut on an existing feature is similar to that of adding material or protrusion.

Removing Material by Extruding a Sketch

The **Extrude Cut** is an extruded feature that is created by removing material from an existing feature. The material that is removed is defined by a sketch.

After drawing the sketch for the cut feature, you can specify the direction of material removal with respect to the sketch. For example, the yellow arrow in Figure 4-49 shows the direction of material removal. If the direction shown by the arrow is accepted then the cut feature will be created, as shown in Figure 4-50.

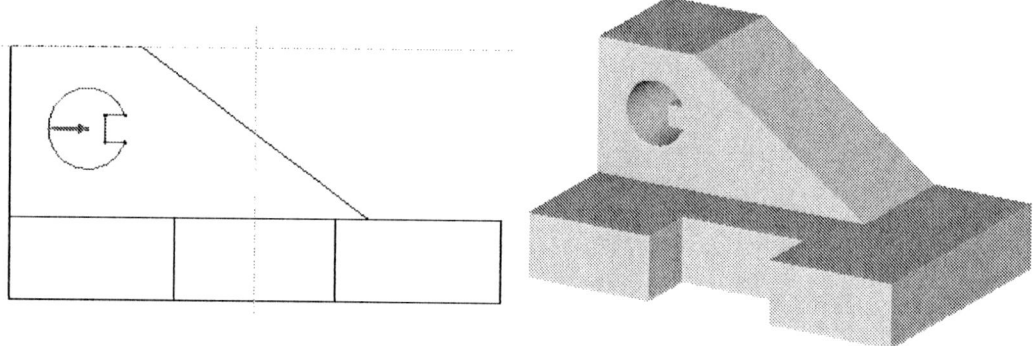

Figure 4-49 Sketch for the extrude cut and arrow showing the direction of material removal

Figure 4-50 Cut feature created on the selected plane

However, if you choose the **Change material direction of extrude to other side of sketch** button from the **Extrude** dashboard, the arrow points in the direction shown in Figure 4-51. All material on the plane selected for sketching will be removed leaving the extruded cut feature, as shown in Figure 4-52.

Note
*A straight hole can also be created by drawing its cross-section, that is a circle, and then creating an extrude cut. But, Pro/ENGINEER provides predefined placement for a hole feature, which can be more desirable than dimensioning the cross-section of a cut feature. Straight holes do not require a sketch if you use the **Hole** dashboard. The **Hole** dashboard is discussed in Chapter 5.*

Datums

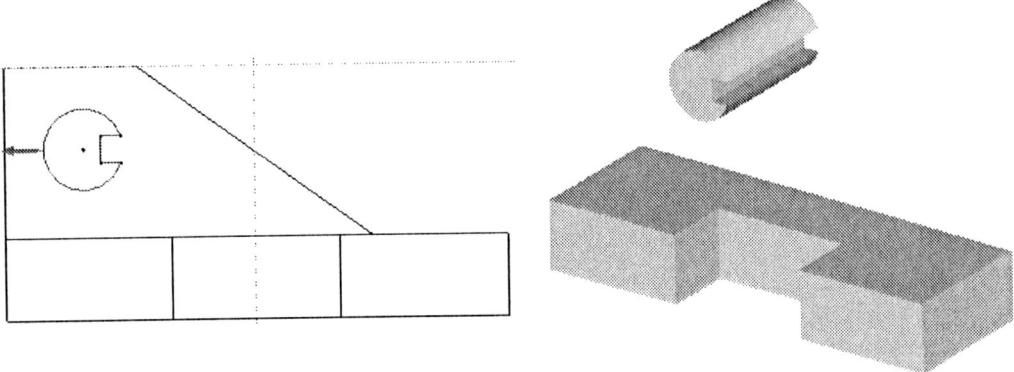

Figure 4-51 Arrow showing the direction of material removal

Figure 4-52 Cut feature created in the direction shown in the adjacent figure

 **Tip**: *In the model shown in Figure 4-49, the sketching plane selected for creation of the extruded cut is not a datum plane but the front face of the base feature. You can also create a datum plane on the surface of an existing feature and select it as the sketching plane. But it is not recommended to create a datum plane if a planar surface of the feature can be used as a sketching plane.*

Removing Material by Revolving a Sketch

The **Revolve Cut** is a revolved feature that is created by removing material from an existing feature. The material that is removed is defined by the sketch you draw. Remember that the center line is necessary in a revolve features. Figure 4-53 shows the section drawn that is to be revolved. The front surface of the second extruded feature is selected as the sketching plane. Figure 4-54 shows the revolve cut created on the selected surface.

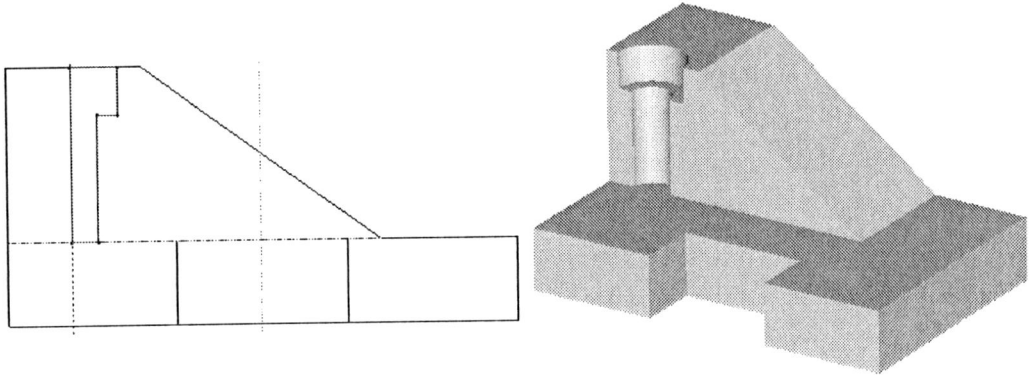

Figure 4-53 The section for revolve cut

Figure 4-54 Revolve cut created

 Note
*The **Sweep Cut** is explained in Chapter 7.*

TUTORIALS

Tutorial 1

In this tutorial you will create the model shown in Figure 4-55. The front and the right-side view of the solid model are shown in Figure 4-56. **(Expected time: 30 min)**

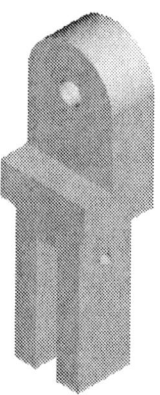

Figure 4-55 Model for Tutorial 1

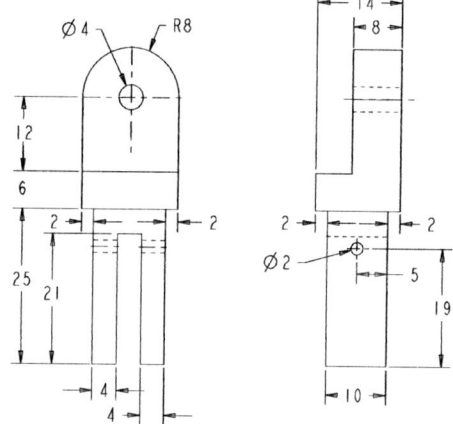

Figure 4-56 Front and side views of the model

The following steps are required to complete this tutorial:

a. Examine the model and determine the number of features in it, refer to Figure 4-55.
b. Create the base feature, refer to Figures 4-57 and 4-58.
c. Create the second extrude feature that is at the bottom of the base feature, refer to Figures 4-59 through 4-62.
d. Create the third feature on an offseted plane, refer to Figures 4-63 through 4-68.
e. Create the circular cut feature, refer to Figures 4-69 through 4-72.

Datums

After understanding the procedure for creating the model, you are now ready to create it. When Pro/ENGINEER session is started, the first task is to set the working directory. Since this is the first tutorial of this chapter, you need to create a folder named *c04*, if it does not exist. Choose the **New Directory** button in the **Select Working Directory** dialog box and create a folder named *c04* in *C:\ProE-WF-2.0*.

Starting a New Object File

1. Open a new part file and name it as **c04tut1**. The three default datum planes are displayed in the graphics window. The **Model Tree** also appears on the left in the Navigator. Close the **Model Tree** by clicking on the sash on its right edge.

Selecting the Sketching Plane for the Base Feature

To create the sketch for the base feature, you need to first select the sketching plane. In this model, you need to draw the base feature on the **FRONT** datum plane because from the isometric view of this model, it is evident that the direction of extrusion for this feature is perpendicular to the **FRONT** datum plane.

Note
The model can be created by selecting any plane as the sketching plane for the base feature. But when the base feature is created, the orientation of the base feature will not be proper. Hence, the final model will be oriented wrongly. You will have to be careful while defining the sketching plane for the base feature. The desired orientation of the model is shown in Figure 4-55.

1. Choose the **Extrude Tool** button from the **Base Features** toolbar. The **Extrude** dashboard is displayed below the graphics window.

2. Choose the **Placement** tab and from the slide-up panel choose the **Define** button. The **Sketch** dialog box is displayed.

3. Select the **FRONT** datum plane as the sketching plane.

   A yellow arrow is displayed on the **FRONT** datum plane, pointing in the direction of viewing the sketch.

4. Select the **TOP** datum plane from the graphics window and then select the **Top** option from the **Orientation** drop-down list.

   The **TOP** datum plane is selected in order to orient the sketching plane.

5. Choose the **Sketch** button from the **Sketch** dialog box. Now you enter the sketcher environment.

Specifying References

In the sketcher environment, the **References** dialog box is displayed at the top right corner of the screen. The status displayed in the **Reference status** area is **Fully Placed**. Close the **References** dialog box by choosing the **Close** button from the dialog box.

Creating and Dimensioning the Sketch for the Base Feature

The base feature can be created by drawing the sketch and then extruding it to the given distance.

1. Draw the sketch using various sketcher tools and add the required constraints and dimensions shown in Figure 4-57. When you initially draw the sketch, it is dimensioned automatically and some weak dimensions are assigned to it.

   Tip: *It is recommended that you use the **Modify the values of dimensions, geometry of splines, or text entities** button to modify the weak dimensions. In the **Modify Dimensions** dialog box that appears, clear the **Regenerate** check box and then modify the dimensions using the thumbwheel or the dimension edit box. This way the sketch will not regenerate as you edit dimensions.*

2. Modify the dimension values to those shown in Figure 4-57.

3. After the sketch is completed, choose the **Continue with the current section** button. Now, you are out of the sketcher environment.

4. Choose the **Default Orientation** option from the **Saved view list** button of the **View** toolbar. The yellow colored arrow is displayed on the model, indicating the direction of extrusion.

5. In the dimension box present on the **Extrude** dashboard, enter a value of **8,** the model appears, as shown in Figure 4-58.

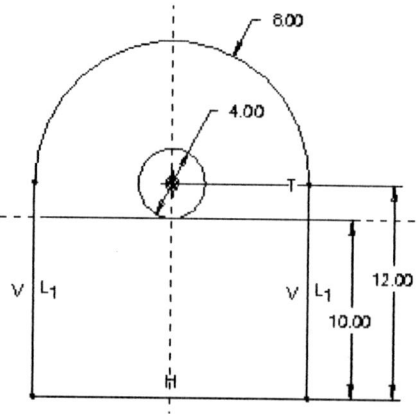

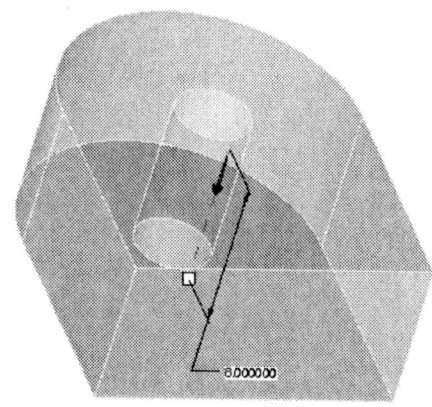

Figure 4-57 Sketch for the base feature with dimensions and constraints

Figure 4-58 Default orientation of the model and the arrow showing the direction of feature creation

Tip: *It is recommended to check the orientation of the base feature of a model when it is completed. To check whether the plane you specified for sketching was correct or not, choose the **Saved view list** button from the **View** toolbar. Choose the **FRONT** option from the drop-down list; the base feature will reorient in the graphics window such that you can view the front view of the base feature.*

Datums

6. Choose the **Build feature** button from the **Extrude** dashboard. The base feature is completed. You can use the middle mouse button to spin the model in order to view it from various directions.

Note

*When you choose the **Default Orientation** option from the **Saved view list** button, the orientation of the model is trimetric and not isometric. If you want the isometric view of the model to be displayed whenever you choose the **Default** option, then you have to use the **Environment** dialog box. The **Environment** dialog box is displayed when you choose **Tools** > **Environment** from the menu bar. From the dialog box, in the **Standard Orient** drop-down list, choose the **Isometric** option. Now, the default orientation will be set to isometric.*

Selecting the Sketching Plane for the Second Feature

The second feature is an extrude feature and will be created on the previous plane that was used to create the base feature.

1. Choose the **Extrude Tool** button from the **Base Features** toolbar. The **Extrude** dashboard is displayed below the graphics window.

2. Choose the **Placement** tab and from the slide-up panel choose the **Define** button. The **Sketch** dialog box is displayed.

3. Choose the **Use Previous** button from the **Sketch Plane** area in the **Sketch** dialog box. The yellow arrow is displayed in the graphics window, as shown in Figure 4-59.

When you choose the **Use Previous** button, the system selects the previous sketching plane that was used to create the base feature. This option is selected because the base feature and the second feature are on the same plane but have different depths of extrusion. If they had the same depth of extrusion, you could have drawn them on the same plane as a single feature. The **TOP** datum plane and its orientation is set automatically.

In Figure 4-59, the model is oriented in its default orientation. The model can be oriented in its default position by choosing the **Default** option from the **Saved view list** button.

4. Choose the **Sketch** button from the **Sketch** dialog box.

The system takes you to the sketcher environment.

Drawing the Sketch for the Second Feature

The second feature has a rectangular section that will be extruded to a depth of 14. To improve the clarity of the edges of the base feature, choose the **No Hidden** button from the **Model Display** toolbar before you start sketching the second feature.

1. Draw the sketch, as shown in Figure 4-60.

2. The sketch is automatically constrained and some weak dimensions are assigned to it. Add the required constraints and modify the weak dimensions, as shown in Figure 4-60.

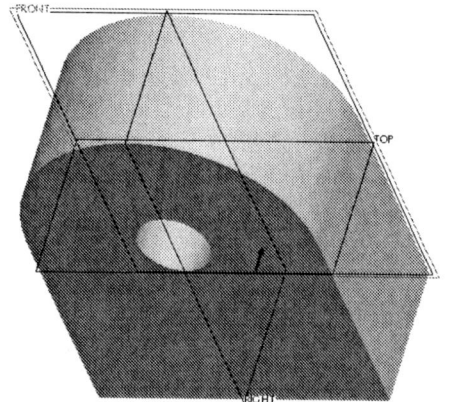

Figure 4-59 Arrow showing the direction of viewing the sketching plane

Figure 4-60 Sketch for the second feature

 **Tip**: *You can use the* **Create an entity from an edge** *button from the* **Sketcher Tools** *toolbar to use the bottom edge of the base feature. The edge of the base feature is required to complete the sketch for the second feature. Or else, draw an aligned line on the edge.*

3. Choose the **Continue with the current section** button to exit the sketcher environment. The **Extrude** dashboard is enabled below the graphics window. Choose the **Shading** button from the **Model Display** toolbar to view the shaded model.

4. Use the middle mouse button to orient the model, as shown in Figure 4-61. This orientation of the model gives you a better view of the sketch in the three-dimensional (3D) space. The yellow colored arrow is also displayed on the model, indicating the direction of extrusion.

5. Enter a value of **14** in the dimension box present on the **Extrude** dashboard and press ENTER. The second extruded feature is completed and its preview is displayed in the graphics window.

6. Choose the **Saved view list** button from the **View** toolbar. From the drop-down list choose the **Default Orientation** option.

7. Now, choose the **Build feature** button from the **Extrude** dashboard to confirm the feature creation. The model orients on the screen, as shown in Figure 4-62. You can also use the middle mouse button to spin the model.

Creating the Datum Plane for the Third Feature

A new datum plane is required to create the next feature. The datum plane will be created at an offset distance of 2 from the front face of the second feature. You need to turn on the display of datum planes if their display is off.

Datums

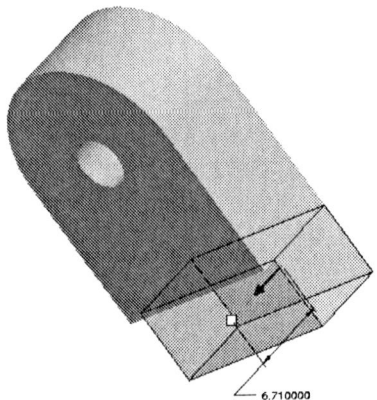

Figure 4-61 Arrow showing the direction of feature depth

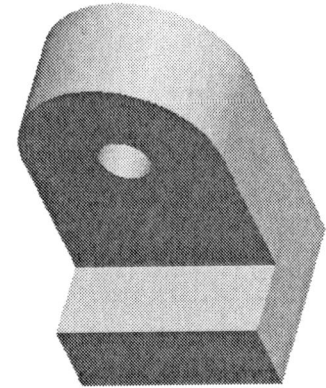

Figure 4-62 Second extruded feature with the base feature

1. Choose the **Datum Plane Tool** button from the **Datum** toolbar. The **DATUM PLANE** dialog box is displayed.

   Make sure that before opening the dialog box, there are no selections made. This can be confirmed from the **Status** bar that is below the graphics window. If there is a selection, you can remove it by left-clicking in the graphics window.

2. Using the left mouse button, select the front face of the second feature highlighted in Figure 4-63. The boundary of the selected front face is highlighted red in color.

   In the **DATUM PLANE** dialog box, under the **References** collector, the **Offset** constraint is displayed. This means that by the smart selection property of Pro/ENGINEER Wildfire 2.0, the **Offset** constraint is applied automatically to the datum plane you are going to create.

   Now, you need to specify the offset distance. If you enter a positive value, the datum plane will be created in the direction shown by the arrow and if you enter a negative value then the datum plane will be created in the direction opposite to that shown by the arrow.

3. In the **Translation** dimension box, under the **Offset** area of **DATUM PLANE** dialog box, enter a value of **-2** and press ENTER.

   The negative value is entered because the datum plane has to be created in the direction opposite to that shown by the yellow arrow. The datum plane named **DTM1** is created, as shown in Figure 4-64.

4. Choose the **OK** button.

Creating the Third Feature on DTM1

The datum plane **DTM1** is created and can be seen in the **Model Tree** as well as in the

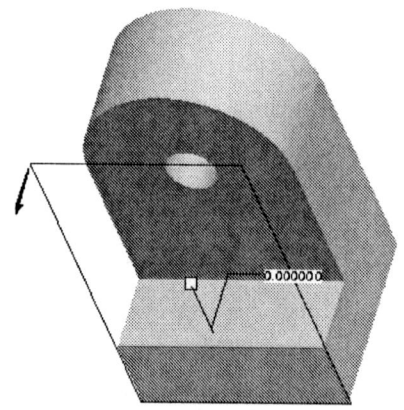
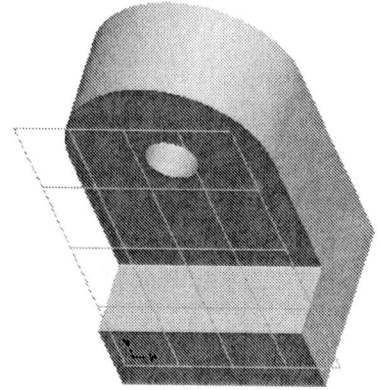

Figure 4-63 *The selected face and the arrow showing the positive direction of datum plane*

Figure 4-64 *The highlighted datum plane*

graphics window. The sketch of the next feature that will be extruded has to be created on the datum plane **DTM1**.

1. Choose the **Extrude Tool** button from the **Base Features** toolbar. The **Extrude** dashboard is displayed below the graphics window.

2. Choose the **Placement** tab and from the slide-up panel choose the **Define** button. The **Sketch** dialog box is displayed.

3. Select **DTM1** as the sketching plane for the third feature. A yellow arrow is displayed on the selected datum plane, as shown in Figure 4-65. This arrow shows the direction of viewing the sketching plane.

   The **TOP** datum plane and its orientation is selected by default.

4. Choose the **Sketch** button from the **Sketch** dialog box. The system takes you to the sketcher environment. Choose the **No Hidden** button from the **Model Display** toolbar.

5. Sketch the section for the third feature of the model and add constraints and dimensions to the sketch, as shown in Figure 4-66.

6. Choose the **Continue with current section** button to exit the sketcher environment.

   The **Extrude** dashboard is enabled below the graphics window. Turn the model display to **Shading**. Use the middle mouse button to orient the model, as shown in Figure 4-67. This orientation gives a better view of the model.

   Notice that the yellow arrow is pointing toward the direction of feature creation. But, you need to extrude the sketch in the opposite direction.

Datums

4-37

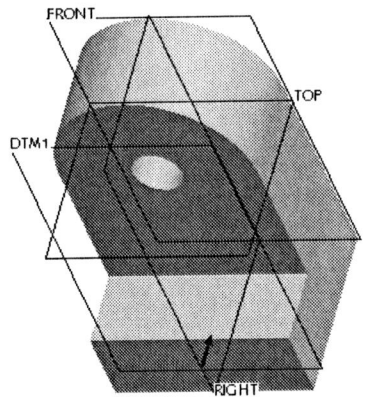

Figure 4-65 Arrow on **DTM1** showing the direction of viewing the sketching plane

Figure 4-66 Sketch with dimensions and constraints of the third feature

7. Choose the **Change depth direction of extrude to other side of sketch** button from the **Extrude** dashboard to change the direction of arrow.

8. Enter a value of **10** in the dimension box present on the **Extrude** dashboard.

   You can see the preview of the model after creating the third extruded feature.

9. Now, choose the **Build feature** button from the **Extrude** dashboard to confirm the feature creation.

   The default view that is displayed when you choose the **Default Orientation** option from the **Saved view list** button is shown in Figure 4-68.

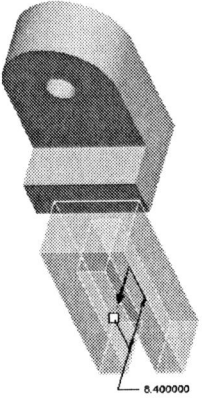
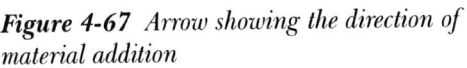

Figure 4-67 Arrow showing the direction of material addition

Figure 4-68 Model after creating the third

Selecting the Sketching Plane for the Cut Feature

A through cut will be created on the outer right face of the third feature. The circular section for the cut will be sketched and using the **Remove Material** button in the **Extrude** dashboard, the circular cut will be created. The sketching plane for the cut feature is shown in Figure 4-69.

Note

The circular cut feature can also be created using the Hole tool, which will be discussed in Chapter 5.

1. Choose the **Extrude Tool** button from the **Base Features** toolbar. The **Extrude** dashboard is displayed below the graphics window.

2. Choose the **Remove Material** button from the **Extrude** dashboard.

3. Choose the **Placement** tab and from the slide-up panel choose the **Define** button. The **Sketch** dialog box is displayed.

4. Select the face shown in Figure 4-69 for sketching.

5. Using the left mouse button select the **TOP** default datum plane and then select the **Top** option from the **Orientation** drop-down list.

6. Choose the **Sketch** button. The system takes you to the sketcher environment.

Sketching the Cut Feature

1. Turn the model display to **No Hidden**. Draw the sketch of the cut feature and add dimensions to it, as shown in Figure 4-70.

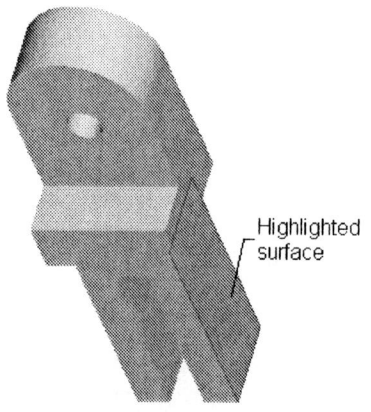

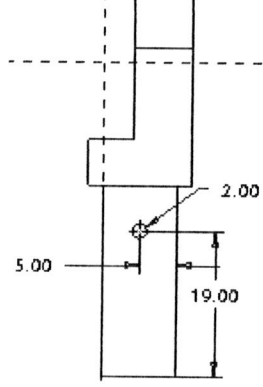

Figure 4-69 Sketching plane for hole feature

Figure 4-70 Sketch and dimensions for cut

2. Choose the **Continue with current section** button and turn the model display to **Shading**.

3. Choose the **Saved view list** button from the **View** toolbar and from the drop-down list choose the **Default Orientation** option. The model is displayed, as shown in Figure 4-71. Two yellow arrows also appear on the model. One arrow shows the direction of feature

Datums

creation and the other arrow shows the direction with respect to the sketch from where the material will be removed. The arrows should point, as shown in Figure 4-71.

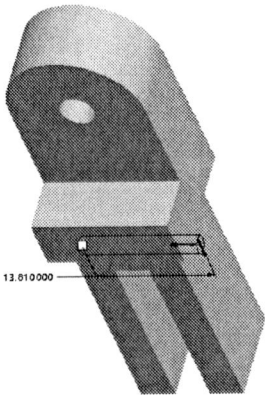

Figure 4-71 The two arrows on the cut feature

4. Choose the **Options** tab in the **Extrude** dashboard. The **Depth** slide-up panel is displayed.

5. From the **Side 1** drop-down list, choose the **Through All** option. The cut feature is completed and can now be previewed in the graphics window.

6. Choose the **Build feature** button from the **Extrude** dashboard. Turn the model display to shaded by choosing the **Shading** button from the **Model Display** toolbar. The trimetric view of the model is shown in Figure 4-72.

Saving the Model

1. Choose the **Save the active object** button from the **File** toolbar and save the model. The order of feature creation can be seen from the **Model Tree** shown in Figure 4-73.

Figure 4-72 Completed model for Tutorial 1

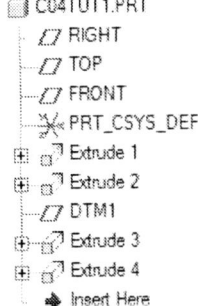

Figure 4-73 Model Tree for Tutorial 1

Tutorial 2

In this tutorial you will create the model shown in Figure 4-74. This figure also shows the front, top, and right-side views of the solid model. **(Expected time: 45 min)**

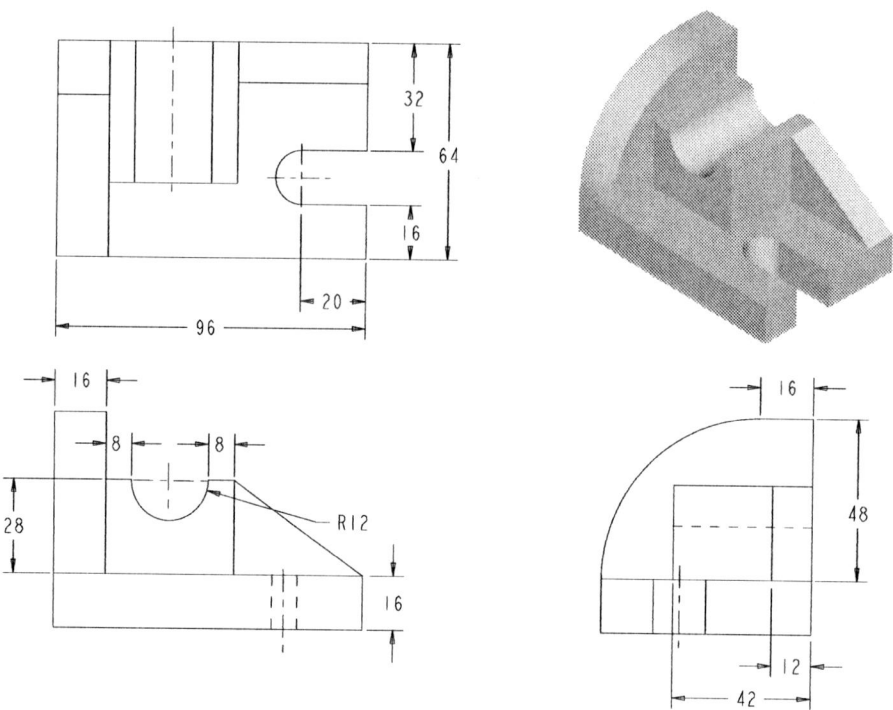

Figure 4-74 Top, front, right-side, and isometric views of the model

The following steps are required to complete this tutorial:

a. Examine the model and determine the number of features in it, refer to Figure 4-74.
b. Create the base feature, refer to Figures 4-75 and 4-76.
c. Create the second feature on the left face of the base feature, refer to Figures 4-77 through 4-81.
d. Create the third and fourth features on the same plane, but with different extrusion depths, refer to Figures 4-82 through 4-89.

After understanding the procedure for creating the model, you are now ready to create it. Set the working directory if required.

Starting a New Object File

1. Open a new part file and name it as **c04tut2**. The three default datum planes are displayed in the graphics window.

Datums

Selecting the Sketching Plane for the Base Feature

To create the sketch for the base feature, you need to first select the sketching plane for the base feature. In this model, you need to draw the base feature on the **TOP** datum plane. This is because the direction of extrusion of the base feature is perpendicular to the **TOP** datum plane.

1. Choose the **Extrude Tool** button from the **Base Features** toolbar. The **Extrude** dashboard is displayed below the graphics window.

2. Choose the **Placement** tab and from the slide-up panel choose the **Define** button. The **Sketch** dialog box is displayed.

3. Select the **TOP** datum plane as the sketching plane.

   A yellow arrow is displayed on the **TOP** datum plane, pointing in the direction of viewing the sketch. The **RIGHT** datum plane and its orientation is selected automatically.

4. Choose the **Sketch** button in the **Sketch** dialog box to enter the sketcher environment.

Specifying References

In the sketcher environment, the **References** dialog box is displayed at the top right corner of the screen. The status displayed in the **Reference status** area is **Fully Placed**. Exit the **References** dialog box by choosing the **Close** button from the dialog box.

Creating and Dimensioning the Sketch for the Base Feature

The sketch for the base feature consists of a rectangular shape with a slot, as shown in Figure 4-75. When this sketch is extruded, it will create the base feature with the slot, as shown in Figure 4-76.

1. Draw the sketch using various sketcher tools and add the required constraints and dimensions to it. Modify these dimensions, as shown in Figure 4-75.

2. Choose the **Continue with the current section** button to exit the sketcher environment.

3. Choose the **Default Orientation** option from the **Saved view list** button of the **View** toolbar.

   The default view of the model is displayed, but it does not fit on the screen. You may need to zoom for displaying the model properly in the graphics window. This gives you a better view of the sketch in the 3D space. A yellow arrow is also displayed on the model, indicating the direction of extrusion.

4. Enter a value of **16** in the dimension box present on the **Extrude** dashboard. The preview of the base feature is displayed in the graphics window.

5. Choose the **Build feature** button from the **Extrude** dashboard.

The base feature is completed and is shown in Figure 4-76. You can use the middle mouse button to spin the object to view it from various directions.

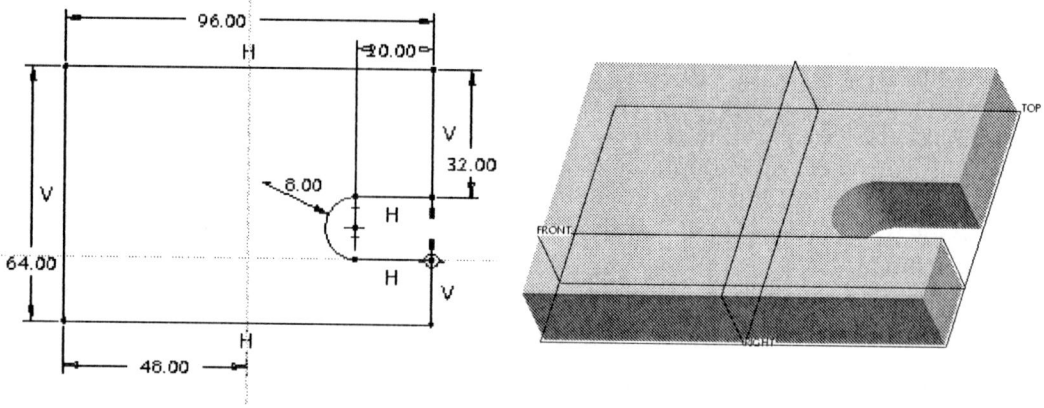

Figure 4-75 Sketch with dimensions and constraints for the base feature

Figure 4-76 Base feature of the model

Selecting the Sketching Plane for the Second Feature

The next feature is an extruded feature and will be created on the left face of the base feature. Therefore, you need to select the left face of the base feature as the sketching plane and then draw the sketch.

1. Choose the **Extrude Tool** button from the **Base Features** toolbar. The **Extrude** dashboard is displayed below the graphics window.

2. Choose the **Placement** tab and from the slide-up panel choose the **Define** button. The **Sketch** dialog box is displayed.

3. Using the middle mouse button, spin the model, as shown in Figure 4-77.

4. Now, select the left face of the base feature as the sketching plane.

    A yellow arrow that points in the direction of viewing the sketching plane appears on the left face of the base feature.

5. Choose the **Flip** button to flip the yellow arrow and to change the direction of viewing the sketching plane, as shown in Figure 4-77.

6. Select the top face of the base feature shown in Figure 4-78 and choose the **Top** option from the **Orientation** drop-down list.

    By selecting the top surface of the base feature, the model will be oriented in such a way that the highlighted planar surface will be at the top while sketching.

7. Choose the **Sketch** button from the **Sketch** dialog box to enter the sketcher environment.

Datums	4-43

Figure 4-77 Arrow showing the direction of viewing the model

Figure 4-78 Surface selected to be at the top

Creating and Dimensioning the Sketch for the Second Feature

The sketch for the second feature consists of two lines and an arc. The bottom edge of the sketch coincides with the top edge of the base feature. This sketch is extruded to a depth of 16. Before drawing the sketch, turn the model display to **No Hidden**.

1. Close the **References** dialog box that is displayed on the top right corner of the screen.

2. Draw the sketch using various sketcher tools, as shown in Figure 4-79. The sketch is dimensioned automatically and some weak dimensions are assigned to it.

3. Apply the constraints and modify the weak dimensions to the dimensions shown in Figure 4-79.

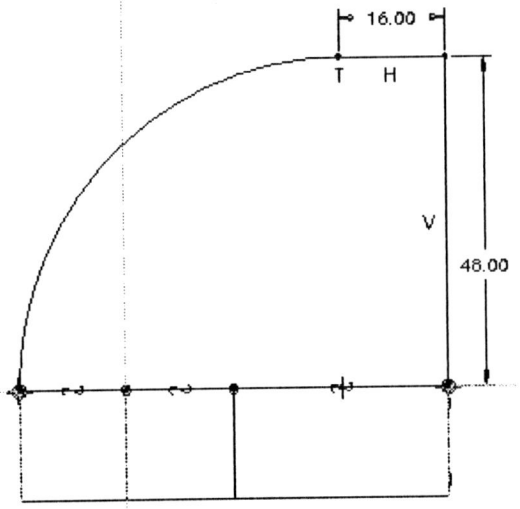

Figure 4-79 Sketch with dimensions and constraints for the second feature

4. After the sketch is complete, turn the model display to **Shading** and choose the **Continue with the current section** button.

   The **Extrude** dashboard is displayed below the graphics window. Use the middle mouse button to orient the model, as shown in Figure 4-80. A yellow arrow is also displayed on the model, indicating the direction of extrusion.

5. Type a value of **16** in the dimension box present in the **Extrude** dashboard and press ENTER. The second feature is completed and its preview is displayed on the screen.

6. Choose the **Build feature** button from the **Extrude** dashboard. The second feature is completed and is shown in Figure 4-81. You can use the middle mouse button to spin the model to view it from various directions.

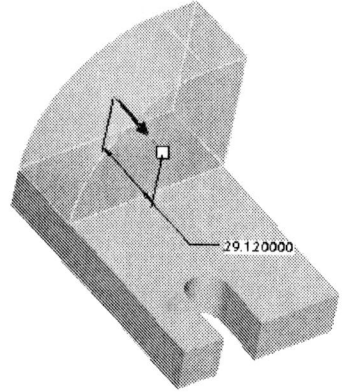

Figure 4-80 Arrow showing the direction of material addition

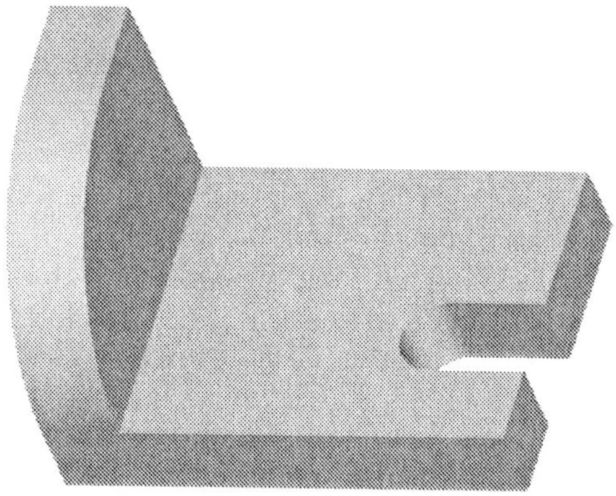

Figure 4-81 Model with the second extruded feature

Selecting the Sketching Plane for the Third Feature

The third feature is an extruded feature and will be created on the back planar surface of the base feature. Therefore, you need to define the back face of the base feature as the sketching plane and then draw the sketch.

1. Choose the **Extrude Tool** button from the **Base Features** toolbar. The **Extrude** dashboard is displayed below the graphics window.

2. Choose the **Placement** tab and from the slide-up panel choose the **Define** button. The **Sketch** dialog box is displayed.

3. Using the middle mouse button, spin the model and select the face of the base feature shown in Figure 4-82 as the sketching plane. A yellow arrow is displayed on the selected face.

4. Choose the **Flip** button to flip the yellow arrow, as shown in Figure 4-82. The arrow points in the direction of viewing the sketching plane.

5. Select the top face shown in Figure 4-83 and then select the **Top** option from the **Orientation** drop-down list. Now, the top face of the base feature is selected to be at the top while drawing the sketch.

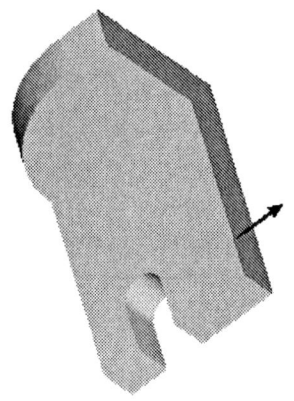

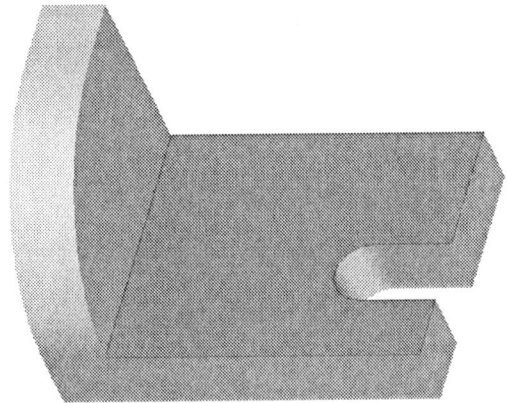

Figure 4-82 Arrow showing the direction of viewing the model

Figure 4-83 Face selected to be at the top

6. Choose the **Sketch** button to enter the sketcher environment.

Creating and Dimensioning the Sketch for the Third Feature

The sketch for the base feature consists of a rectangular section with a semicircular cut at the top. When this section is extruded, a feature with a semicircular slot will be created. Turn the display to **No Hidden**.

1. Draw the sketch using various sketcher tools, as shown in Figure 4-84.

2. The sketch is dimensioned automatically and some weak dimensions are assigned to it. Add the required constraints and modify the weak dimensions to the dimensions shown in Figure 4-84.

3. Choose the **Continue with the current section** button to exit the sketcher environment. The **Extrude** dashboard is enabled below the graphics window. Turn the display to **Shading**.

4. Use the middle mouse button to orient the model, as shown in Figure 4-85. This orientation of the model gives a better view of the sketch in 3D space. A yellow arrow is displayed on the model, indicating the direction of extrusion.

5. Type a value of **42** in the dimension box present in the **Extrude** dashboard and press ENTER. This feature is completed and its preview can be seen in the graphics window.

6. Choose the **Build feature** button from the **Extrude** dashboard to confirm the feature creation. The default view that is displayed when you choose the **Default Orientation** option from the **Saved view list** button is shown in Figure 4-86.

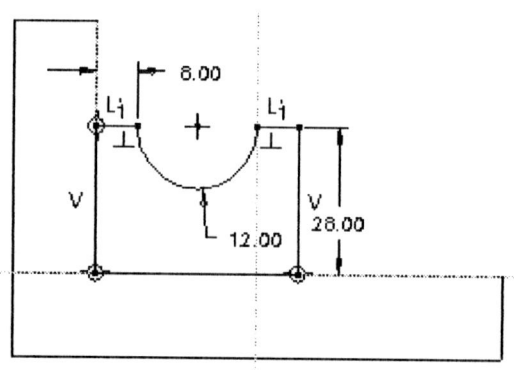

Figure 4-84 Sketch with dimensions and constraints

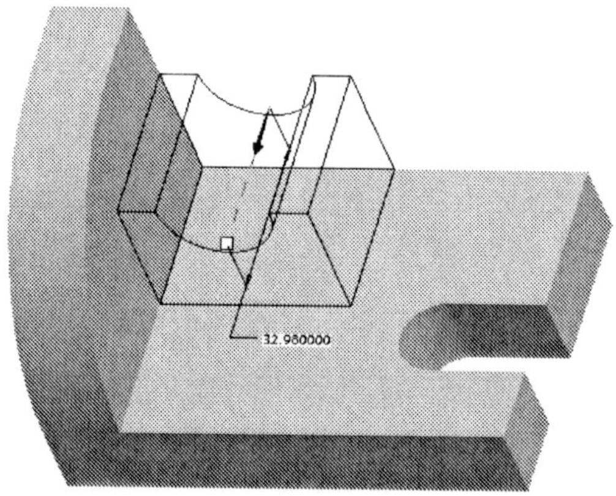

Figure 4-85 Arrow showing the direction of material addition

Selecting the Sketching Plane for the Last Feature

The last feature and the third feature are created on the same plane. But the depth of extrusion is different for both of them. This is the reason they are considered as separate features. Therefore, for sketching this feature, you can use the sketching plane that was used for creating the third feature.

1. Choose the **Extrude Tool** button and then invoke the **Sketch** dialog box.

2. Choose the **Use Previous** button from the **Sketch Plane** area in the **Sketch** dialog box. The yellow arrow is displayed in the graphics window, as shown in Figure 4-87.

   By specifying the **Use Previous** button, you are using the previous sketching plane that was defined for the previous feature. A yellow arrow is displayed on the face and points in the direction of viewing the sketch, as shown in Figure 4-87.

Figure 4-86 Model with the second extruded feature

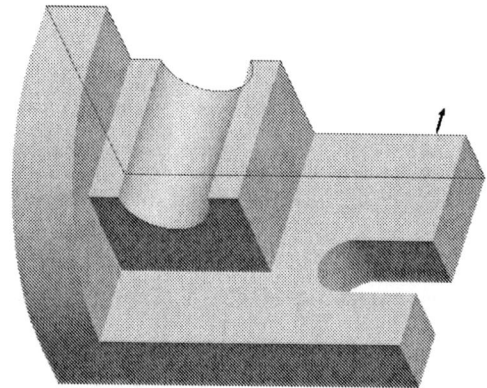

Figure 4-87 Arrow pointing in the direction of viewing the sketch plane

Datums

3. Choose the **Sketch** button to enter the sketcher environment.

Creating and Dimensioning the Sketch for the Last Feature

The sketch of this feature consists of three lines. The bottom edge of the sketch is aligned with the top edge of the base feature and the left edge is aligned with the right edge of the third feature. When this sketch is extruded, it creates a rib shape.

1. Draw the section sketch using various sketcher tools, as shown in Figure 4-88.

   The sketch is dimensioned automatically and some weak dimensions are assigned to it. You can add the required constraints to align the lines and points in the sketch with other features, as shown in Figure 4-88. In the figure, the constraint symbol displayed on the line indicates that the edges of the adjacent features are used to close the section.

2. After the sketch is complete, choose the **Continue with the current section** button. The **Extrude** dashboard is enabled. Using the middle mouse button orient the model, as shown in Figure 4-89.

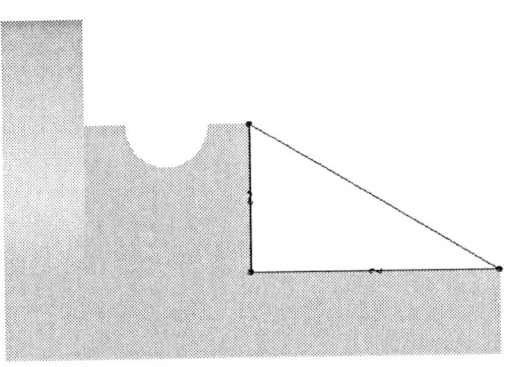
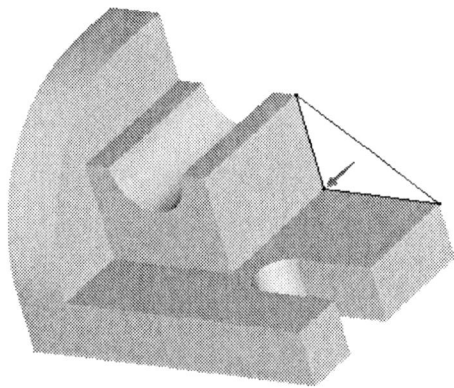

Figure 4-88 Sketch and constraints for the last feature

Figure 4-89 Arrow showing the direction of material addition

3. Enter a value of **12** in the dimension box present in the **Extrude** dashboard. The feature is completed and its preview can be seen in the graphics window.

4. Choose the **Build feature** button in the **Extrude** dashboard.

   The default view of the model, when you choose the **Default Orientation** option from the **Saved view list** button from the **View** toolbar, is shown in Figure 4-90.

Saving the Model

1. Choose the **Save the active object** button from the **File** toolbar and save the model. The order of feature creation can be seen from the **Model Tree** shown in Figure 4-91.

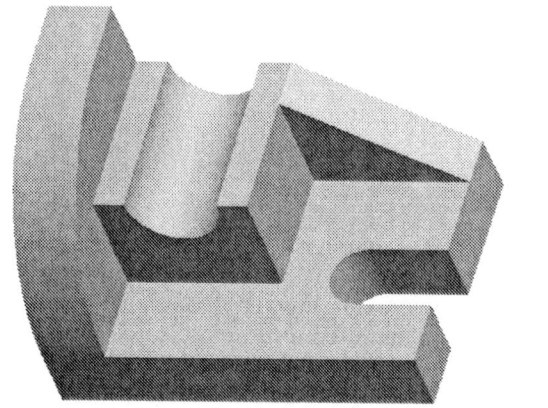

Figure 4-90 Completed model for Tutorial 2

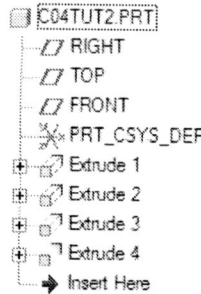

Figure 4-91 Model Tree for Tutorial 2

Tutorial 3

In this tutorial you will create the model shown in Figure 4-92. This figure also shows the front, top, and right-side views of the solid model.

(Expected time: 45 min)

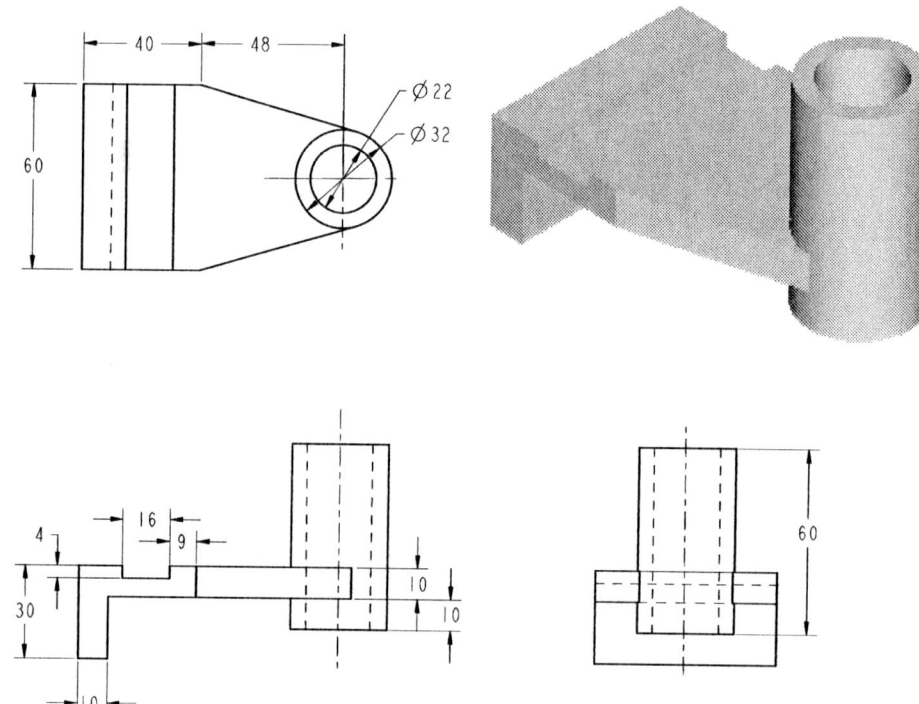

Figure 4-92 Top, front, right-side, and isometric views of the model

Datums

The following steps are required to complete this tutorial:

a. Examine the model and then determine the number of features in it, refer to Figure 4-92.
b. Create the base feature, refer to Figures 4-93 and 4-94.
c. Create the second feature, refer to Figures 4-95 through 4-98.
d. Create the hollow cylindrical feature on an offseted datum plane, refer to Figures 4-99 through 4-104.

Starting a New Object File

1. Set the working directory if required and open a new part file with the name **c04tut3**. The three default datum planes are displayed in the graphics window if the **Datum planes on/off** button is turned on.

Selecting the Sketching Plane for the Base Feature

To create the sketch for the base feature, you first need to select the sketching plane. In this model, you need to draw the base feature on the **FRONT** datum plane because the direction of extrusion is perpendicular to the **FRONT** datum plane.

1. Choose the **Extrude Tool** button from the **Base Features** toolbar.

2. Choose the **Placement** tab and from the slide-up panel choose the **Define** button. The **Sketch** dialog box is displayed.

3. Select the **FRONT** datum plane as the sketching plane.

   A yellow arrow is displayed on the **FRONT** datum plane and it points in the direction of viewing the sketch plane. The **RIGHT** datum plane and its orientation are selected by automatically.

4. Choose the **Sketch** button to enter the sketcher environment.

Specifying References

In the sketcher environment, the **References** dialog box is displayed. The status displayed in the **Reference status** area is **Fully Placed**. Close this dialog box.

Creating and Dimensioning the Sketch for the Base Feature

From the model, the section to be extruded for the base feature is evident. The section sketch is shown in Figure 4-93. When this sketch is extruded, it will create the base feature.

1. Draw the sketch using various sketcher tools, as shown in Figure 4-93.

2. The sketch is dimensioned automatically and some weak dimensions are assigned to it. Add the required constraints and modify the weak dimensions, as shown in Figure 4-93.

3. Choose the **Continue with the current section** button. The **Extrude** dashboard is displayed.

4. Choose the **Default Orientation** option from the **Saved view list** button of the **View** toolbar.

The model is reoriented and a yellow arrow is also displayed on it, indicating the direction of extrusion.

5. Type a value of **60** in the dimension box that is present on the **Extrude** dashboard and press ENTER.

6. Choose the **Build feature** button from the **Extrude** dashboard.

The base feature is completed, as shown in Figure 4-94. You can use the middle mouse button to spin the model to view it from various directions.

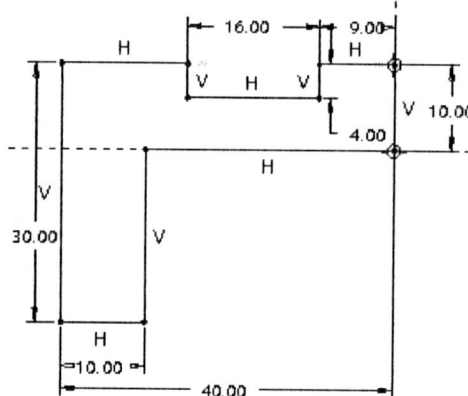

Figure 4-93 Sketch with dimensions and constraints for the base feature

Figure 4-94 Base feature of the model

Selecting the Sketching Plane for the Second Feature

The next feature is an extruded feature. The sketching plane for this feature is the top face of the base feature.

1. Choose the **Extrude Tool** button from the **Base Features** toolbar.

2. Choose the **Placement** tab and from the slide-up panel choose the **Define** button. The **Sketch** dialog box is displayed.

3. Select the top face of the base feature shown in Figure 4-95 as the sketching plane. A yellow arrow that points in the direction of viewing the sketch is displayed on the top face.

4. Choose the **Flip** button to flip the yellow arrow to point in the direction shown in Figure 4-95.

5. Select the **RIGHT** datum plane and then choose the **Right** option from the **Orientation** drop-down list.

6. Choose the **Sketch** button to enter the sketcher environment.

Creating and Dimensioning the Sketch for the Second Feature

The next feature to be created is an extrude feature. The section for the extrude feature is shown in Figure 4-96. Before drawing the sketch, turn the model display to **No Hidden**.

1. Close the **References** dialog box that is displayed on the top right corner of the screen.

2. Draw the sketch using various sketcher tools. In the sketch, draw a center line passing through the center of the arc, as shown in Figure 4-96. This center line helps in dimensioning the sketch. Add the required constraints and dimensions to the sketch, as shown in Figure 4-96. Before exiting the sketcher environment, turn the model display to **Shading**.

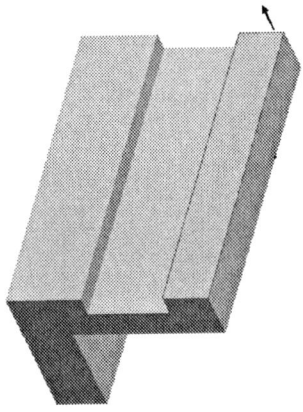 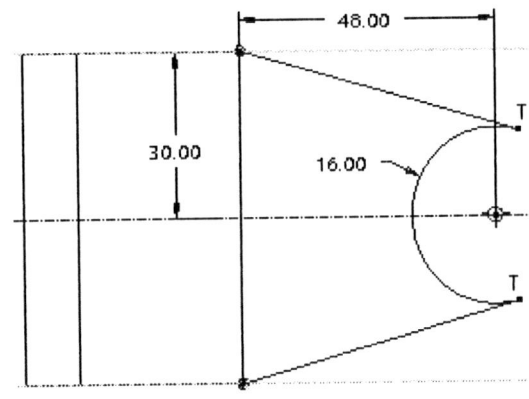

Figure 4-95 *Arrow pointing from the sketching plane in the direction of feature creation*

Figure 4-96 *Sketch with dimensions and constraints*

3. Choose the **Continue with the current section** button. The **Extrude** dashboard is enabled. Use the middle mouse button to orient the model, as shown in Figure 4-97.

4. Type **a value of 10** in the dimension box present on the **Extrude** dashboard and press ENTER.

5. Now, choose the **Build feature** button to confirm the feature creation. The default trimetric view of the extruded feature is shown in Figure 4-98.

Creating a Datum Plane for the Last Feature

To create the hollow cylindrical feature, you require a datum plane. This datum plane will be created at an offset distance of 10 from the bottom face of the second feature shown in Figure 4-99.

Note

The other method to create this feature is to select the top planar surface of the second feature as the sketching plane and extrude the sketch on both sides of the sketching plane. The depth of extrusion will be different at both the sides. If you use this method to create this cylindrical feature, you do not need to create a datum plane.

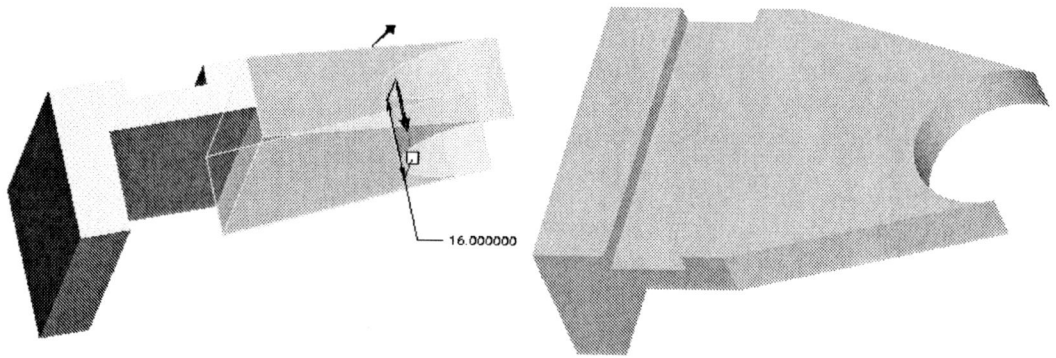

Figure 4-97 Arrow showing the direction of material addition

Figure 4-98 Model with the second extruded feature

1. Choose the **Datum Plane Tool** button from the **Datum** toolbar. The **DATUM PLANE** dialog box is displayed.

2. Spin the model using the middle mouse button and then using the left mouse button, select the bottom face of the second feature.

   As you select the face of the second feature, the **Offset** constraint is displayed in the **References** collector of the dialog box.

3. In the **Translation** dimension box, enter a value of **10**.

4. Choose the **OK** button from the **DATUM PLANE** dialog box.

   Datum plane **DTM1** is created, as shown in Figure 4-100 and will be selected as the sketching plane for creating the sketch.

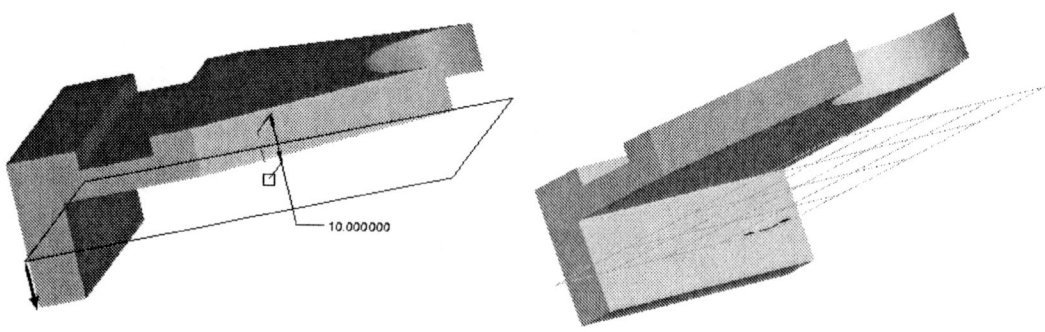

Figure 4-99 Creating the datum plane

Figure 4-100 Model after creating the datum plane

Datums

Selecting the Sketching Plane for the Last Feature

The plane **DTM1** will be selected as the sketching plane and the depth of extrusion will be defined from this plane.

1. Choose the **Extrude Tool** button from the **Base Features** toolbar.

2. Choose the **Placement** tab and from the slide-up panel choose the **Define** button. The **Sketch** dialog box is displayed.

3. Select **DTM1** as the sketching plane. The yellow arrow appears on the datum plane.

4. Choose the **Flip** button to reverse the direction of viewing the sketch. Now, the arrow points in the direction shown in Figure 4-101.

5. Select the **FRONT** datum plane and choose the **Bottom** option from the **Orientation** drop-down list.

6. Choose the **Sketch** button to enter the sketcher environment.

Creating and Dimensioning the Sketch for the Last Feature

The sketch for the hollow cylindrical feature will be drawn on the datum plane **DTM1**. The sketch for the hollow cylindrical feature consists of two concentric circles. Before drawing the sketch, turn the model display to **No Hidden**.

1. Close the **References** dialog box that is displayed on the top right corner of the screen.

2. Draw the sketch using various sketcher tools and add the required constraints and dimensions to it, as shown in Figure 4-102.

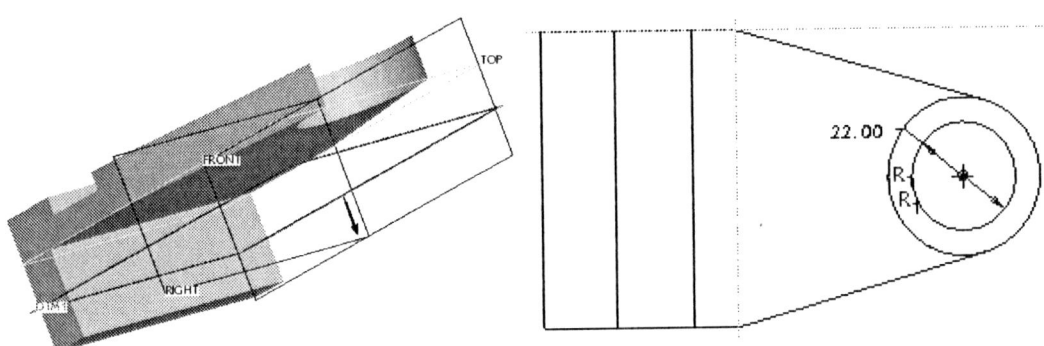

Figure 4-101 Direction of viewing the sketching plane

Figure 4-102 Sketch with dimensions and constraints

3. Choose the **Continue with the current section** button. The **Extrude** dashboard is displayed. Now, turn the model display to **Shading**.

Use the middle mouse button to orient the model, as shown in Figure 4-103. This view gives you a better view of the sketch in the 3D space.

4. Enter a value of **60** in the dimension box present on the **Extrude** dashboard.

5. Choose the **Build feature** button to confirm the feature creation. The default trimetric view of the complete model is shown in Figure 4-104.

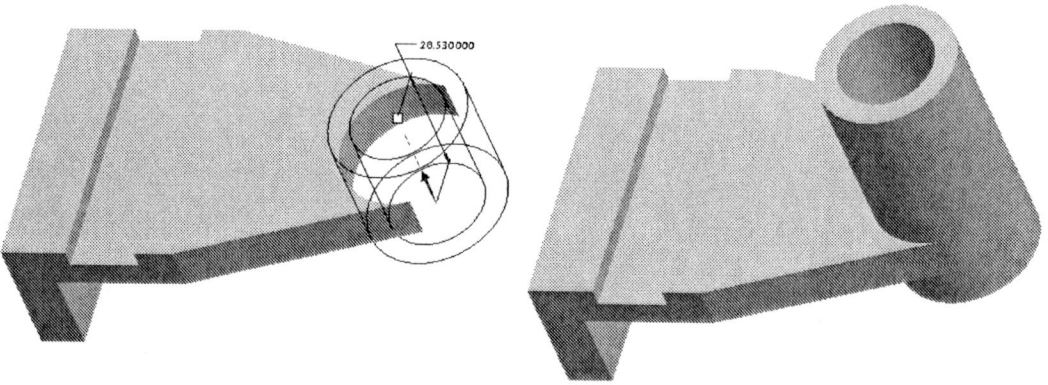

Figure 4-103 Preview of the feature

Figure 4-104 Final model of Tutorial 3

Saving the Model

1. Choose the **Save the active object** button from the **File** toolbar and save the model.

Self-Evaluation Test

Answer the following questions and then compare your answers with those given at the end of this chapter.

1. You can change the default names assigned to the datum planes. (T/F)

2. Datum points are also used to associate note in the drawings and attach datum targets. (T/F)

3. The constraint combination of creating an offset plane through cylinder is a valid combination. (T/F)

4. Generally all features in a model are created on a single sketching plane. (T/F)

5. A sketching plane can be selected on an existing face of a feature. (T/F)

6. While creating a datum feature, the _____ selected from the model help you to choose from the available constraints that you can apply to constrain the datum feature.

7. Datum axis is an _____ axis that is created in Pro/ENGINEER.

Datums

8. The trimetric view of the model is displayed when you choose the _____ option from the **Saved view list** button.

9. When you create a new object file, _____ default datum planes are displayed in the graphics window.

10. The default datum planes in the **Part** mode are named as _____, _____, and _____.

Review Questions

Answer the following questions:

1. Which one of the following is not a type of datums available in Pro/ENGINEER?

    (a) Axis (b) Plane
    (c) Circle (d) Curve

2. How many methods are available in Pro/ENGINEER to create a datum plane?

    (a) One (b) Two
    (c) Three (d) Four

3. Which one of the following dialog boxes is displayed when you choose the **Datum Point Tool** button from the **Datum** toolbar while extruding a section sketch?

    (a) **DATUM POINT** dialog box (b) **DATUM PLANE** dialog box
    (c) **DATUM AXIS** dialog box (d) None

4. Which one of the following combinations can be used to spin a model in the graphics window?

    (a) CTRL+ALT (b) middle mouse button
    (c) left mouse button (d) CTRL+right mouse button

5. Which one of the following menus in the menu bar is used to set the default orientation of the model to isometric?

    (a) **File** (b) **Insert**
    (c) **Tool** (d) None

6. Datum planes are considered as feature geometry and have mass and volume. (T/F)

7. To set the default orientation of the model to isometric, you have to use the **Environment** dialog box. (T/F)

8. Generally, the sketching plane for the base feature of any model is decided after viewing

the isometric view or the drawing views of the model. (T/F)

9. Datum planes are used as a reference for mirroring features, copying features, for creating a cross-section, as well as for orientation of references. (T/F)

10. Unlike the datum planes constraint options, all datum axes constraint options are stand-alone. (T/F)

Exercises

Exercise 1

Create the model shown in Figure 4-105. The dimensions, front view, and right-side view of the model are shown in Figure 4-106. **(Expected time: 45 min)**

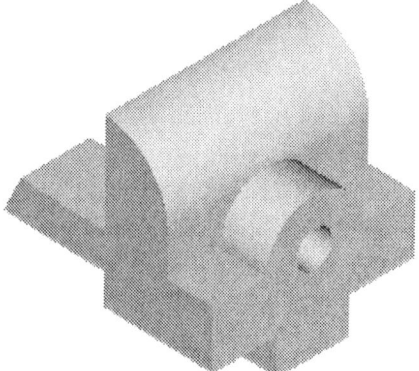

Figure 4-105 *Isometric view of the model*

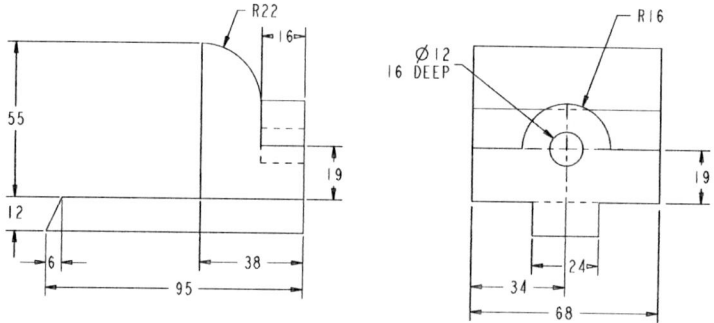

Figure 4-106 *Front and right-side views of the model*

Exercise 2

Create the model shown in Figure 4-107. The dimensions, and the front, top, right-side, and isometric views of the model are also shown in the figure. **(Expected time: 45 min)**

Datums 4-57

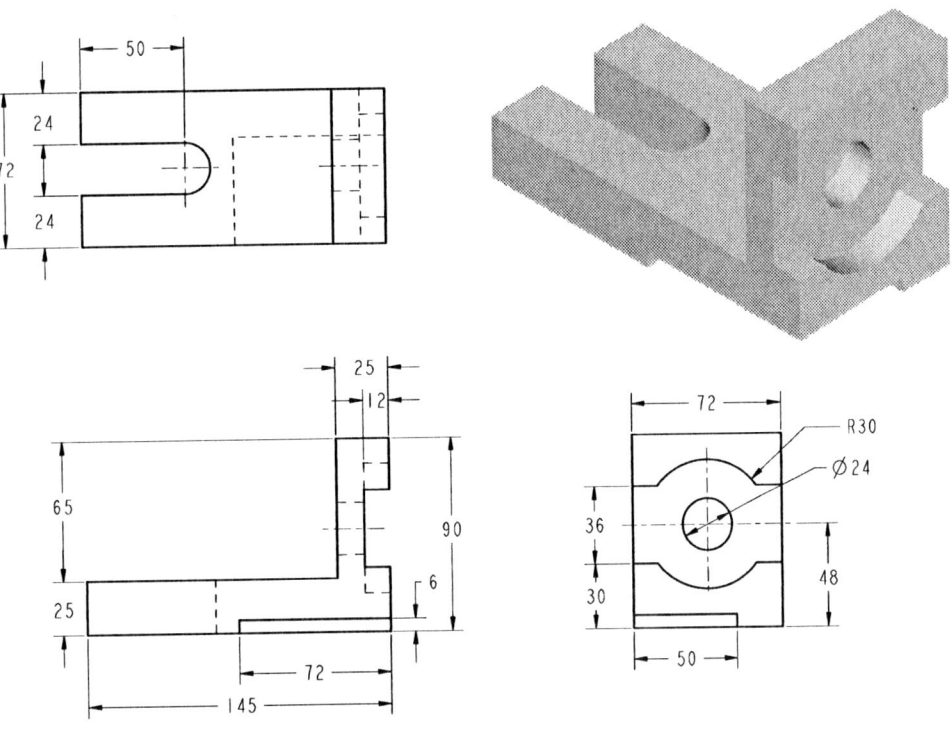

Figure 4-107 Top, front, right-side, and isometric views of the model

Exercise 3

Create the model shown in Figure 4-108. The top, front section, and two auxiliary views of the model are shown in Figure 4-109. **(Estimated time: 1 hr)**

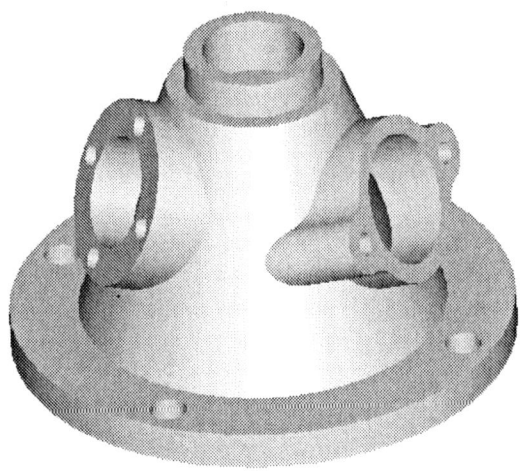

Figure 4-108 The 3D view of the model

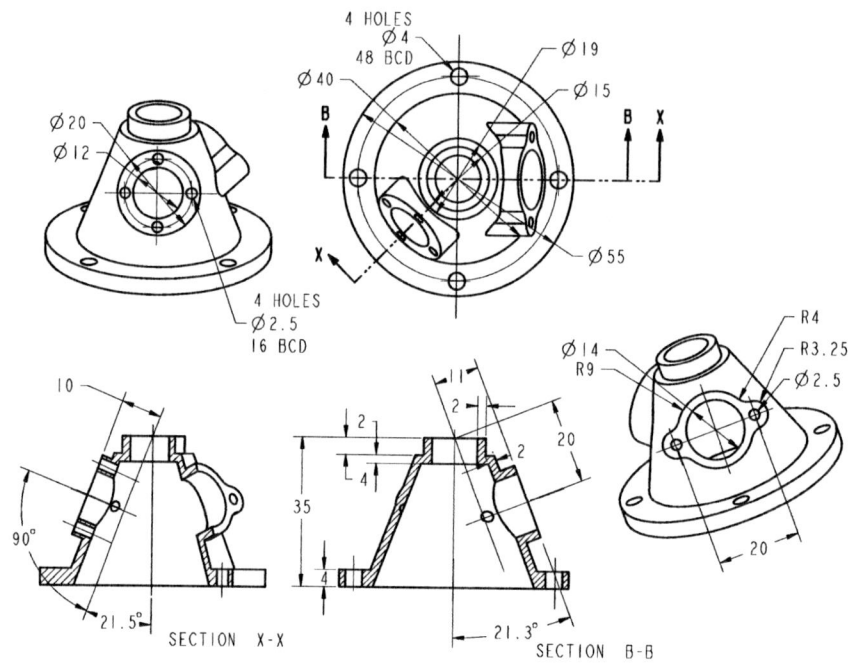

Figure 4-109 Top, front section, and two auxiliary views of the model

Hint to create the two side features in Exercise 3:

<u>Feature on the right</u>
1. Create an axis passing through the top face of the base feature and through the **RIGHT** datum plane.
2. Create a datum plane passing through the axis (created in the previous step) and at an angle of 21.5 degrees from the **RIGHT** datum plane.
3. Now, create an offset plane at a distance of 11 from the datum plane (created in the previous step). Select this datum plane as the sketching plane.

<u>Feature on the left</u>
1. Create a datum plane passing through the axis of revolution of the base feature and at an angle of 45 degrees from the **FRONT** datum plane.
2. Create a datum axis passing through the top face of the base feature and through the datum plane (created in the previous step).
3. Create a datum plane passing through datum axis (created in the previous step) and at an angle of 21.5 degrees from the datum plane (created in step1).
4. Create a datum plane offset to a distance of 10 from the datum plane (created in the previous step).

Exercise 4

Create the model shown in Figure 4-110. The dimensions, the left-side view, auxiliary view, front view, and isometric view of the model are also shown in the figure.

(Expected time: 45 min)

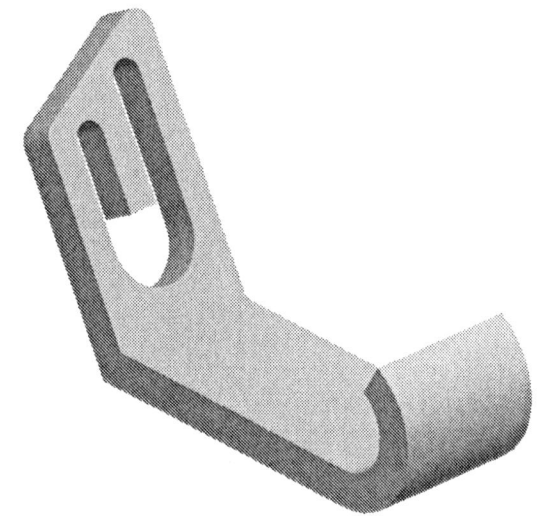

Figure 4-105 Isometric view of the model

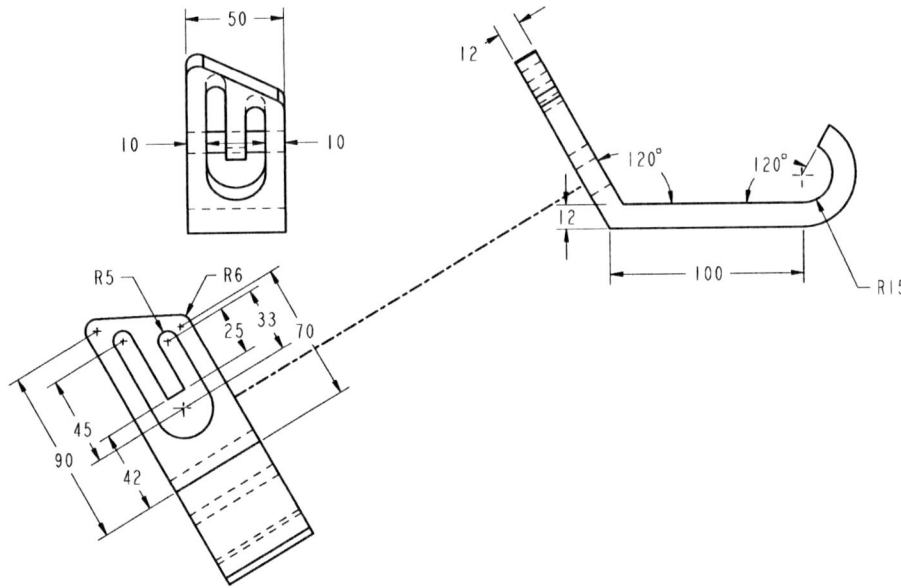

Figure 4-110 Left-side view, auxiliary view, front view, and isometric view of the model

Answers to the Self-Evaluation Test

1 - T, 2 - T, 3 - T, 4 - F, 5 - T, 6 - geometry, 7 - imaginary, 8 - trimetric, **Default Orientation**, 9 - three, 10 - **TOP, FRONT, RIGHT**.

Chapter 5

Options Aiding Construction of Parts-I

Learning Objectives

After completing this chapter you will be able to:
- *Create holes using the Hole dashboard.*
- *Create Round, Chamfer, and Rib.*
- *Edit features.*
- *Redefine, Reroute, and Reorder features.*
- *Suppress and delete features.*
- *Modify features.*

OPTIONS AIDING CONSTRUCTION OF PARTS

This chapter explains the feature creation tools provided in Pro/ENGINEER that help in creating a model and editing it. In this chapter you will also learn to create various types of holes that are required in most of the engineering designs. In previous chapters you learned to create holes using extrude cut, but in this chapter you will create holes using the **Hole** dashboard. Using the **Hole** dashboard, makes it easier to create and modify holes.

In this chapter you will also learn to create rounds, chamfers, and ribs. The options that are discussed in this chapter increase the efficiency of creating a design using Pro/ENGINEER.

CREATING HOLES

In engineering components, holes can be counterbore, countersink, tapered, or drilled. Pro/ENGINEER allows you to create all these types of holes. Pro/ENGINEER also provides industry standard holes that have standard dimensions. In Pro/ENGINEER Wildfire 2.0, holes are created using the **Hole** dashboard.

The Hole Dashboard

The **Hole** dashboard is displayed when you choose **Insert > Hole** from the menu bar or the **Hole Tool** button from the **Engineering Features** toolbar present in the **Right Toolchest**. You can create three types of holes using the **Hole** dashboard. The first type is straight hole, the second is sketched hole, and the third is standard hole.

Creating Straight Holes

Straight holes are the holes that have a circular cross-section with a constant diameter throughout the depth. They start at the placement plane and terminate at the user-defined depth or at the specified end surface. The **Hole** dashboard with **Create straight hole** button selected is shown in Figure 5-1. The options and tool buttons available in this dashboard are discussed next.

Figure 5-1 The **Hole** dashboard

Drop-down List

The drop-down list in the dashboard is used to specify the type of hole. The **Simple** option is selected by default. The other option available in this drop-down list is **Sketched**. The **Simple** option is used to draw straight holes. The **Sketched** option of creating holes is discussed later in the chapter.

Dimension Boxes

The first dimension box in the **Hole** dashboard is used to specify the diameter of the hole.

The second dimension box in the **Hole** dashboard is used to specify the depth of the

Options Aiding Construction of Parts-I

hole. The options to specify the depth of the hole are the same as those available in the **Extrude** dashboard.

Placement Tab

When you choose the **Placement** tab, the slide-up panel is displayed, as shown in Figure 5-2. This slide-up panel is displayed with the parameters that need to be specified to define a linear hole. The **Primary** collector in the slide-up panel is displayed in yellow color. This indicates that Pro/ENGINEER is prompting you to select the placement plane (primary reference). You can select the placement plane only when this collector is in yellow color. Once you have

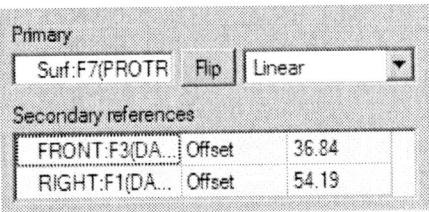

Figure 5-2 Placement tab slide-up panel

selected the primary reference, its name appears in the box. The primary reference can be a plane, an axis, or a point.

The **Flip** button is used to change the direction of hole creation with reference to the primary reference. When you choose this button, the direction of hole creation is reversed and you can view its preview in the graphics window. Remember that the hole will be created only if it is possible to create the hole in the specified direction.

The drop-down list to the right of the **Flip** button contains the options that determine the type of placement for the hole. You can use the options in this drop-down list after selecting the placement plane or primary reference. These options are discussed next.

> **Linear**. When you select this option, you need to specify the distances from two linear references. Generally, these linear references are the edges of the planar surface on the model, any two planar surfaces or axes, or a combination of any of these. Figure 5-3 shows a hole on the top face of the block. The procedure to create this hole using the **Hole** dashboard is as follows:

1. Invoke the **Hole** dashboard.
2. Choose the **Placement** tab to open the slide-up panel.
3. Select the top face of the model to place the hole. The preview of the hole is displayed.
4. Click in the **Secondary References** collector to make the secondary selections.
5. Select the left face or the top left edge of the model.
6. Use CTRL+left mouse button to select the front face or the top front edge of the model. Now, you have selected the two references that are needed to dimension the hole.
7. Specify the diameter of the hole in the dimension box present on the **Hole** dashboard.
8. From the depth flyout, select the **Drill to intersect with all surfaces** button.
9. Choose the **Build feature** button to exit the **Hole** dashboard.

Note
*When you have to select the secondary references, you need to click in the **Secondary References** collector and then select the references on the model.*

Note
The various dimensions of center of the hole from the selected references are displayed in the graphics window and also in the Secondary references collector. Click on the dimensions to modify them.

Radial. This option is used to create a hole that can be referenced to an axis. When you select this option, you need to select an axial reference and an angular reference to place the hole. The distance from the axis and the angle from a plane are entered in the boxes under the **Secondary References** collector. Figure 5-4 shows a radial hole on a curved surface. The following steps explain the procedure to create this hole.

1. Invoke the **Hole** dashboard.
2. Choose the **Placement** tab to open the slide-up panel.
3. Select the curved surface of the model to place the hole. The preview of the hole is displayed.
4. Click in the **Secondary References** collector to make the secondary selections.
5. Select the datum plane that is normal to curved surface to specify the angular value.
6. Use CTRL+left mouse button to select the top face or the top edge of the model. Now, you have selected the two references that were needed to dimension the hole.
7. Specify the diameter of the hole in the dimension box present on the **Hole** dashboard.
8. From the depth flyout, select the **Drill to intersect with all surfaces** button.
9. Choose the **Build feature** button to exit the **Hole** dashboard.

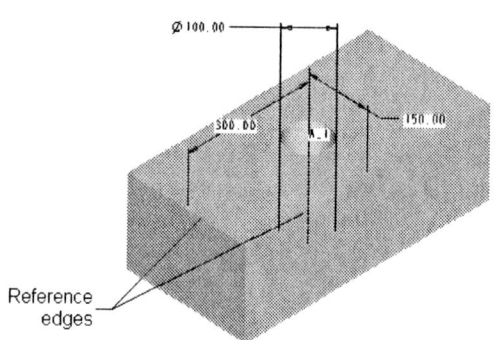

Figure 5-3 Linear dimensioning of hole   *Figure 5-4* Radial dimensioning of hole

 **Tip**: *Note that since you have selected the curved surface as the placement plane for the hole, therefore, you do not need to select the axis to specify the radial distance. However, if the placement plane is a plane surface, then you do need to select an axis to create a radial hole at some radius from the selected axis.*

Diameter. This option creates a diametrically placed hole. When you select this option, you need to select an axial reference and an angular reference to place the hole. Figure 5-5 shows a diameter hole on a plane. The following steps explain the procedure to create this hole.

Options Aiding Construction of Parts-I

1. Invoke the **Hole** dashboard.
2. Choose the **Placement** tab to open the slide-up panel.
3. Select the top face of the model to place the hole. The preview of the hole is displayed.
4. From the drop-down list, select **Diameter** option.
4. Click in the **Secondary References** collector to make the secondary selections.
5. Select the face shown in Figure 5-5 to specify the angular value.
6. Use CTRL+left mouse button to select the axis of the hole that is on the top face. Now, you have selected the two references that are needed to dimension the hole.
7. Specify the diameter of the hole and the angle in the **Secondary References** collector.
8. From the depth flyout, select the **Drill to intersect with all surfaces** button.
9. Choose the **Build feature** button to exit the **Hole** dashboard.

Coaxial. This option creates a hole coaxially. When you select this option, you need to select an axis. No dimensions are required to place a coaxial hole. Figure 5-6 shows a coaxial hole on a plane. The procedure to create this hole is explained next.

1. Invoke the **Hole** dashboard.
2. Choose the **Placement** tab to open the slide-up panel.
3. Select the top surface of the model to place the hole. The preview of the hole is displayed.
4. From the drop-down list, select the **Coaxial** option.
5. Click in the **Secondary References** collector to make the secondary selections.
6. Select the axis of revolution of the base feature.
7. Specify the diameter of the hole in the dimension box present on the **Hole** dashboard.
8. From the depth flyout, select the **Drill to intersect with all surfaces** button.
9. Choose the **Build feature** button to exit the **Hole** dashboard.

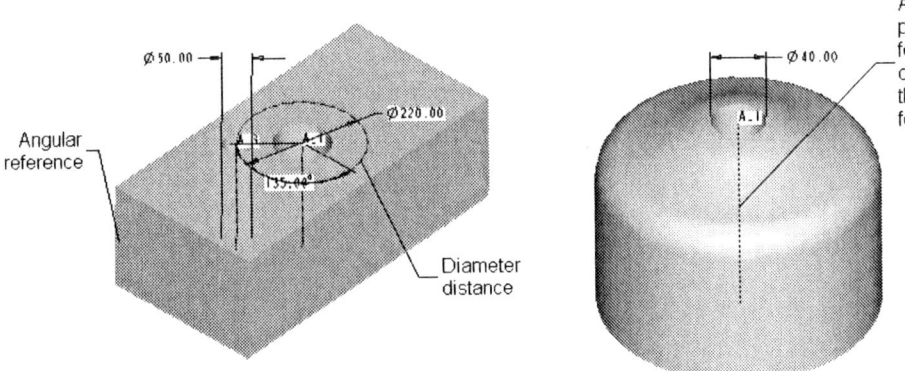

Figure 5-5 Diameter dimensioning of hole   *Figure 5-6 Coaxial hole*

Shape Tab

When you choose the **Shape** tab, the slide-up panel is displayed, as shown in Figure 5-7. This slide-up panel can be used to specify the depth and the diameter of the hole. The depth of the hole can be specified in two directions with respect to the placement plane. The options to specify the depth of the hole are similar to those that were available in the **Extrude** dashboard. In the slide-up panel, notice the sketch for the hole, which gives an

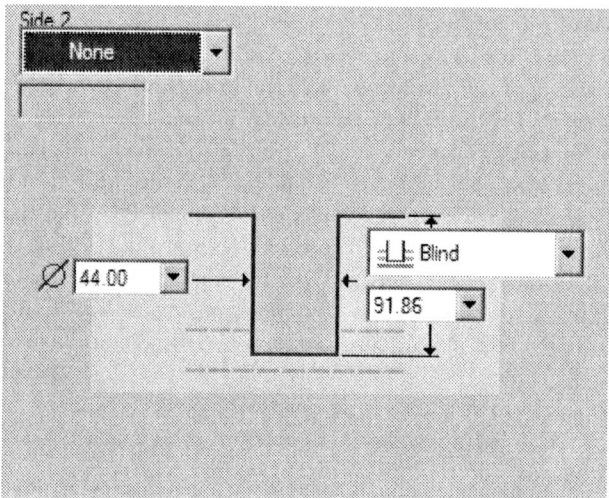

*Figure 5-7 The **Shape** slide-up panel*

idea of the shape of the hole and its dimensions. The sketch that is shown in the slide-up panel is showing the depth of the hole in the Side 1 direction.

The **Side 2** drop-down list is used to specify the depth of the hole in the direction opposite to Side 1. This means the hole can be created on both sides of the placement plane. By default, the **None** option is selected in this drop-down list.

Note
*The termination options in the **Shape** slide-up panel are similar to those discussed in the **Extrude** dashboard.*

When you specify the placement plane for a hole and then specify the two references in case of linear hole, you can notice the four drag handles on the hole. These four drag handles have four different functions to place the hole dynamically on the placement plane. Using these handles you can set the diameter of the hole, depth of the hole, and specify the two references to place the hole. Remember that these drag handles appear only when you have selected the primary reference or in other words the placement plane.

Tip: *The diameter of the hole and its depth in the Side 1 direction can be specified without using the slide-up panels. The options to specify diameter and depth in Side 1 direction are available on the Hole dashboard. Only when you need to specify the depth in the **Side 2** direction, you open the slide-up panel by choosing the **Shape** tab from the Hole dashboard.*

Creating Sketched Hole

The **Sketched** option allows you to sketch the cross-section for the hole that is revolved about a center axis. This option is used to draw custom shapes for the hole. When you choose the **Sketched** option from the first drop-down list in the **Hole** dashboard, the **Hole** dashboard appears, as shown in Figure 5-8. The tools and options in the dashboard are discussed next.

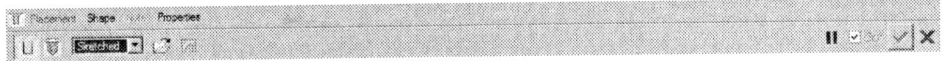

*Figure 5-8 The **Hole** dashboard*

Tip: *Remember that while placing any hole using the Hole dashboard, you need to define two things. First is the placement plane on which the hole feature will be created and the second is the dimensional references for all holes other than the Coaxial hole. For Coaxial hole, you need to select an axis.*

Activates Sketcher to create section Button

Choose the **Activates Sketcher to create section** button to open a new window with the sketcher environment. The cross-section for the hole is sketched using the normal sketcher tools available. While drawing the sketch, a center line must be drawn that acts as the axis of revolution for the section of hole. The sketched holes can be blind or through all, depending on the dimensions of its sketch.

When you complete the sketch for the hole and choose the **Continue with the current section** button to exit the sketcher environment, the **Hole** dashboard below the graphics window is enabled. Now, you need to specify the placement options to place the hole. The placement options are the same as discussed earlier.

Note
The placement options to place the hole can be specified after drawing the sketch for the hole or before drawing the sketch. The only difference is that when you specify the placement options after drawing the sketch, the preview of the hole is not available in the graphics window. This is because the primary reference is not selected yet. Therefore, it is recommended to specify the placement options before drawing the sketch for the hole.

Opens an existing sketched profile Button

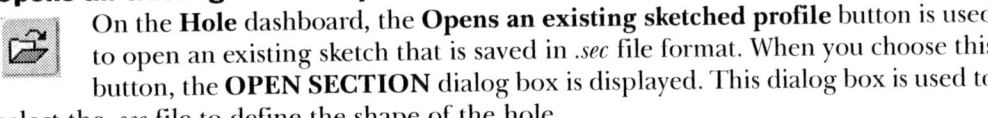
On the **Hole** dashboard, the **Opens an existing sketched profile** button is used to open an existing sketch that is saved in *.sec* file format. When you choose this button, the **OPEN SECTION** dialog box is displayed. This dialog box is used to select the *.sec* file to define the shape of the hole.

Placement Tab
The options in the slide-up panel that are displayed when you choose the **Placement** tab from the **Hole** dashboard are the same that were available in case of **Straight** hole. You can use the same types of placement options that were available when you created the **Straight** hole. These placement options were discussed earlier.

Shape Tab
The options in the slide-up panel that is displayed when you choose the **Shape** tab from the **Hole** dashboard are different than the options that were available in case of **Straight** holes. When you choose the **Shape** tab, the slide-up panel displays the preview of the sketch that you have drawn in the sketcher environment. You can pan and zoom the sketch in the slide-up panel.

Note
*If you choose the **Shape** tab before drawing the sketch for the hole, then the slide-up panel is blank and does not show a sketch.*

Standard Hole

The standard holes can be created by choosing the **Create standard hole** button from the **Hole** dashboard. The holes created using this button are based on industry standard fastener tables.

Note
Since the standard holes are based on industry standards and their dimensions are in millimeters (mm), therefore you need to select the template that measures the dimensions in mm. This can be done when you create a new file.

When you choose the **Create standard hole** button from the **Hole** dashboard, the dashboard appears, as shown in Figure 5-9. The tools and options available in the dashboard are discussed next.

*Figure 5-9 The **Hole** dashboard*

Drop-down Lists

The first drop-down list from the left in the dashboard is used to specify the type of thread on the hole. The **UNC** option is selected by default. The other options available in this drop-down list are **ISO** and **UNF**.

The second drop-down list in the **Hole** dashboard is used to specify the size of the screw that corresponds to the hole.

The third drop-down list in the **Hole** dashboard is used to specify the depth of the hole. The options in this drop-down list are similar to those available in the **Extrude** dashboard.

Adds tapping Button

This button is not enabled if the depth option is selected to specified depth value. As the name suggests, this button is used to tap a hole. Tapping is the process of cutting threads in the hole. In Pro/ENGINEER, these threads are cosmetic threads. This means that the threads created will not be visible in the hole. In the slide-up panel of the **Shape** tab, the preview of the cross-section of the tapped hole is displayed, as shown in Figure 5-10. In the preview of the section, the dimensions can be edited as required. The dotted lines represent the threading outside the hole. If you exit the **Taps the drilled hole** button, these dotted lines will not appear.

Adds countersink Button

As the name suggests, this button is used to create a countersink hole. A countersink hole has two diameters and the transition between the bigger diameter and the smaller diameter is in the form of a cone. In the slide-up panel of the **Shape** tab, the preview of the cross-section of the countersink hole is displayed, as

Options Aiding Construction of Parts-I

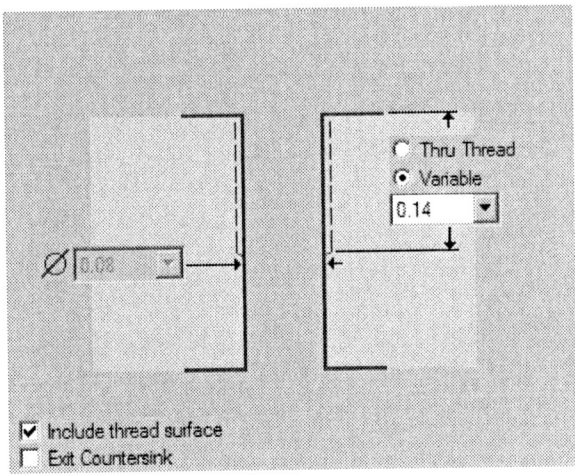

*Figure 5-10 The **Shape** slide-up panel*

shown in Figure 5-11. In the preview of the section, the dimensions can be edited as required.

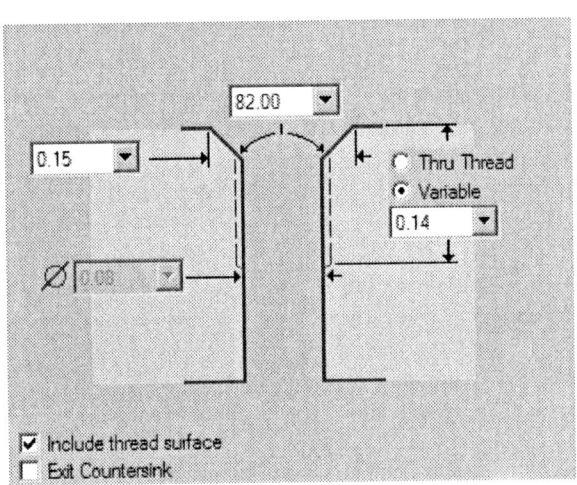

*Figure 5-11 The **Shape** slide-up panel*

Adds counterbore Button

As the name suggests, this button is used to create a counterbore hole. A counterbore hole is a stepped hole and has two diameters, a larger one and a smaller one. The larger diameter is called counter diameter and the smaller diameter is called drill diameter. In the slide-up panel of the **Shape** tab, the preview of the cross-section of the counterbore hole is displayed, as shown in Figure 5-12. In the preview of the section, the dimensions can be edited as required.

Placement Tab

The options in the slide-up panel that is displayed when you choose the **Placement**

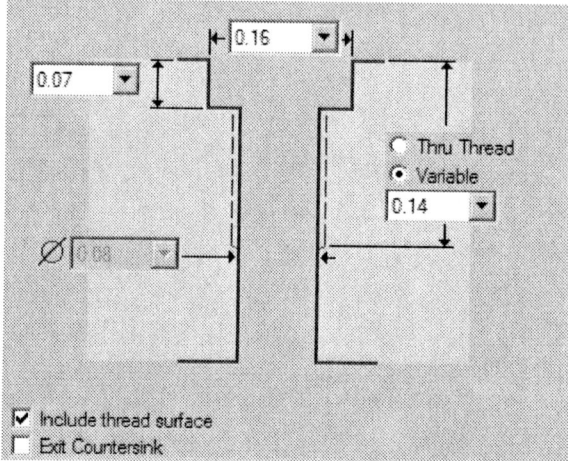

*Figure 5-12 The **Shape** slide-up panel*

option from the **Hole** dashboard are the same as those that were available in case of the **Straight** hole. You can use the same type of placement options that were available when you create the **Straight** hole.

Shape Tab
The options in the slide-up panel that is displayed when you choose the **Shape** tab from the **Hole** dashboard are different than those that were available in case of **Straight** holes. When you choose the **Shape** tab, the slide-up panel displays the preview of the hole. You can specify the dimensions for the hole in the respective dimension boxes.

Note Tab
This tab displays the specifications of the hole in the text format. This option is available only when you are creating a **Standard** hole.

Properties Option
This option displays the properties of the hole that you are creating.

Preview the Hole

The **Feature Preview** button is used to preview the hole created before confirming its creation. Changes and modifications in the hole parameters can be made once the hole is previewed.

Note

*The holes created using the **Hole** dashboard are parametric in nature and hence can be modified at anytime using the **Model Tree**. The method of modification using the **Model Tree** is discussed later in this chapter.*

Important Points to Remember While Creating Hole

The following points should be remembered while creating a hole:

Options Aiding Construction of Parts-I   5-11

1. While drawing the sketch of a hole, the sketch should have axis of revolution and at least one entity normal to it. Now, when you place this hole on the primary reference, then this normal entity is aligned with the plane (primary reference). If there are two entities normal to the axis, then the normal entity at the top of the sketch is aligned with the placement plane.

2. While creating a hole, the primary reference for placement of hole can be selected without choosing any option from the **Hole** dashboard. However, when you need to specify the secondary references, then you have to choose the **Placement** tab from the **Hole** dashboard. From the slide-up panel that is displayed, click in the **Secondary References** collector and when the **Primary** collector appears white in color, you can select the secondary references.

   Remember when the first reference is selected by using the left mouse button, then the second reference is selected by using CTRL+left mouse button.

3. It is recommended that if you are creating a **Standard** hole then the units of the model should be in mm. This can be achieved by using the appropriate template at the time of creating a new file or by setting the system of units.

4. An application of **Sketched** hole is a tapered hole. **Sketched** holes are always blind and have depth in one direction only.

5. You cannot select a convex or concave surface as the placement plane. To create hole on such a surface, you need to create a datum plane passing through the surface and then select this datum plane as the placement plane.

CREATING ROUNDS

The **Round Tool** creates a fillet or smooth rounded transition, between two adjacent surfaces, with either a circular or a conic profile. This feature creation tool can be invoked from the menu bar or from the **Engineering Features** toolbar in the **Right Toolchest**. The **Round** option can either add or remove material, depending on the edge references selected. You can select the edges or surfaces and create a round in a single set or define more than one set.

Figures 5-13 and 5-14 show some examples of rounds. From the figures, it is evident that the geometry of the rounds are tangent to the references selected. These references can be edges or surfaces. In Pro/ENGINEER, rounds are created using the **Round** dashboard.

Note
The round created after you select references on the model has circular cross-section and rolling shapes. In Pro/ENGINEER, there is more than one shape that a round can have.

Creating Basic Rounds

When you choose the **Round Tool** button from the **Engineering Features** toolbar or choose **Insert > Round** from the menu bar, the **Round** dashboard is displayed, as shown in Figure 5-15. You are prompted to select an edge or surface to create the round set.

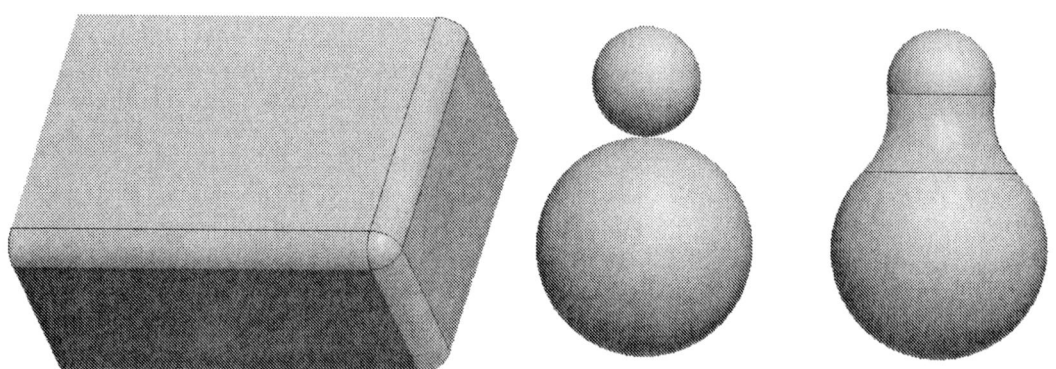

Figure 5-13 Round created on the edges   *Figure 5-14* Round created on surfaces

Figure 5-15 The **Round** dashboard

After you select an edge or surface, you are prompted to select an edge or a surface with the CTRL key to include in the same set. Use CTRL+left mouse button to select another edge or surface, if required. The preview of the round with a default radius value and drag handles appears on the model, as shown in Figure 5-16. Note that because the radius of all rounds in a set is equal, the drag handles are available on only one edge. You can drag the handles to dynamically specify the radius of the round or specify the radius in the **Round** dashboard. Once all parameters of the round are specified, choose the **Build feature** button to create it. The various options available in the **Round** dashboard are discussed next.

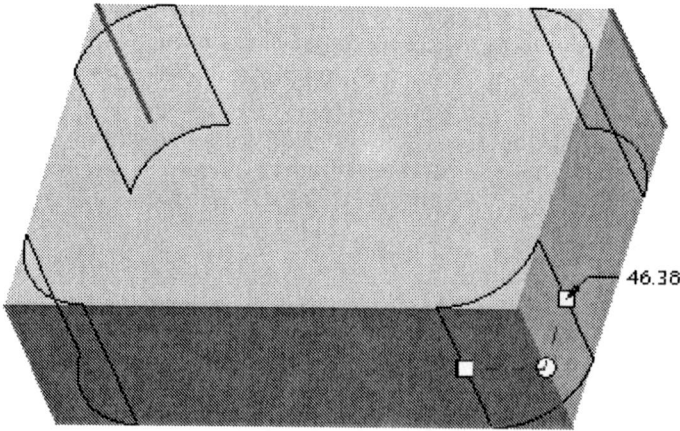

Figure 5-16 Preview of the round feature along with the radius of round and the drag handles

Options Aiding Construction of Parts-I

Tip: *You can create round on an edge by selecting the **Round Edges** option from the shortcut menu. Choose the **Geometry** option from the **Filter** drop-down list and then select an edge. Hold down the right mouse button and choose the **Round Edges** option from the shortcut menu. The **Round** dashboard is invoked and the preview of the round appears on the edge selected.*

Switch to set mode Button

When you invoke the **Round** dashboard, the **Switch to Set mode** button is chosen by default. This button is used to create rounds by defining sets. When this button is chosen, the dimension box to the right of this button is also displayed. The dimension box is used to specify the radius of round.

Tip: *The need for creating rounds by defining more than one set arises when you want to create rounds that have different radii. The rounds created by using this method will be a single feature and appear as a single feature in the **Model Tree**.*

Switch to transition mode button

The **Switch to transition mode** button is available only when you have defined at least one reference for the round. This button has primarily two applications. The first one is to define the shape of the round that is formed at a vertex and the other is to define a limit up to which the round will be created. Both these applications are discussed next.

If the round feature is being created on three intersecting edges, then this button is used to define the shape of the round that is created at the vertex. After choosing the **Round Tool** button press and hold the CTRL key down and select the three edges, as shown in Figure 5-17. Now, choose the **Switch to transition mode** button from the **Round** dashboard. The various transitions that are possible on the current selection are highlighted on the model, as shown in Figure 5-18. Select the transition that is displayed at the vertex. Now, from the drop-down list in the **Round** dashboard, select the desired option, which can be **Intersect**, **Corner Sphere** or **Patch**.

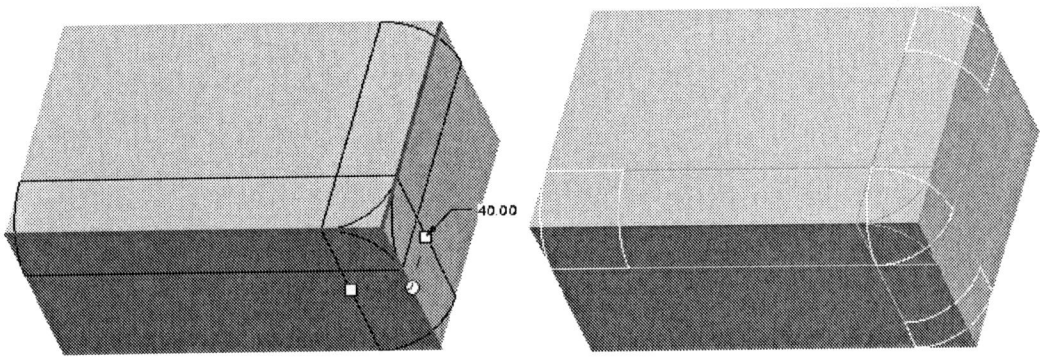

Figure 5-17 Edges selected for creating rounds   *Figure 5-18* Various possible transitions

The **Intersect** option merges the round in such a way that sharp edges are formed at the vertex, as shown in Figure 5-19.

The **Corner Sphere** option results in a smooth spherical merging of rounds at the vertex. If you increase the default radius value of the round feature by dragging the handle while creating the round using the **Corner sphere** option, handles are displayed on all selected edges. You can drag these handles to dynamically define the length of the fillet along the respective edges. The length values can also be defined in edit boxes that are displayed in the **Round** dashboard. Figure 5-20 shows the round created with the **Corner sphere** option.

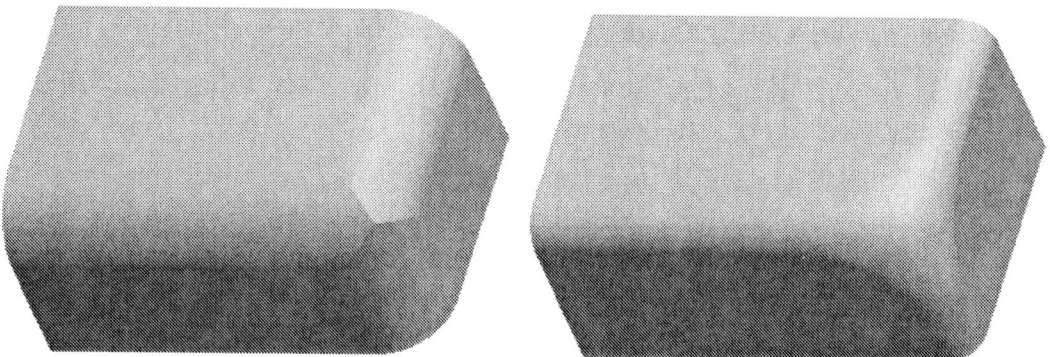

Figure 5-19 Round created using the **Intersect** option

Figure 5-20 Round created using the **Corner sphere** option

The **Switch to Transition mode** button also allows you to select a reference to stop the round on an edge. This means the round will be created on the selected edge up to the selected reference only. After selecting the edge for creating the round, choose the **Switch to transition mode** button to display the two transitions on either ends. Select any one of them to make it active. Now, select the **Stop at reference** option from the drop-down list available in the **Round** dashboard. Select a plane that will act as the stopping reference. Note that the edge on the side in which the transition is active will be retained and the round will be created on the other side. Figure 5-21 shows the example creating round up to a stopping reference. In this case, four edges have been filleted and the reference plane, which is highlighted by a mesh in the figure, is selected as the stopping reference.

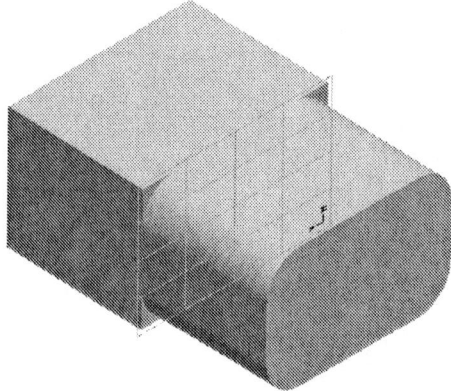

Figure 5-21 Transition rounds created on four edges and stopped by the highlighted reference plane

Options Aiding Construction of Parts-I

Note
At a time you can select only select one transition to make it active. The stopping references selected will be associated with the active transition only. Hence, if you want to define the stopping reference for more than one edge, then it has to be done individually for each active transition.

Sets Tab

When you choose the **Switch to set mode** button and then choose the **Sets** tab in the **Round** dashboard, the slide-up panel is displayed, as shown in Figure 5-23. All references and the geometry of the round are specified in this slide-up panel. The options in this slide-up panel are discussed next.

First Drop-down List

The first drop-down list that appears in the slide-up panel is used to specify the geometry of the round you need to create. This drop-down list contains the options, as shown in Figure 5-22. These options are discussed next:

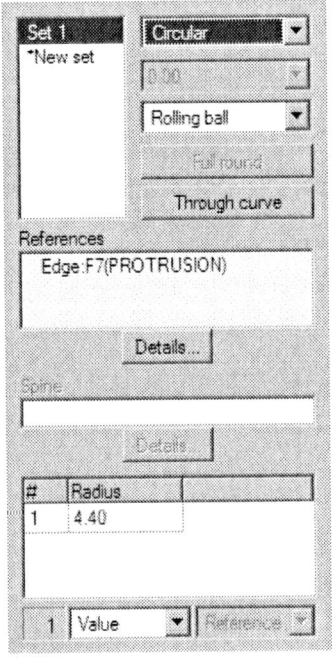

Figure 5-23 The Slide-up panel of the **Sets** tab

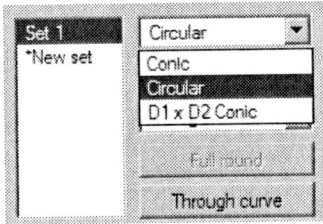

Figure 5-22 Drop-down list

Circular. When you choose this option from the drop-down list, the round that is created has a circular cross-section. This is the most widely used geometry type of rounds. The geometry is controlled by a radius value.

Conic. This type of round has a geometry of a conic. When you choose this option, the **Conic Parameter** drop-down list appears on the **Round** dashboard. The geometry of a conic round is controlled by two dimensions: the leg dimension and the conic parameter. The value of conic parameter can lie between 0.5 and 0.95.

In Figure 5-24, the round on the model that is on the left is created using a circular geometry and the round on the model that is on the right is created using a conic geometry.

D1xD2 Conic. The round created using this option has a geometry that can be controlled by three dimensions. Figure 5-25 shows the round that is created using the **D1xD2 Conic** option. This figure shows the two end legs dimensions and the middle dimension is the conic parameter. When you select this option to create a

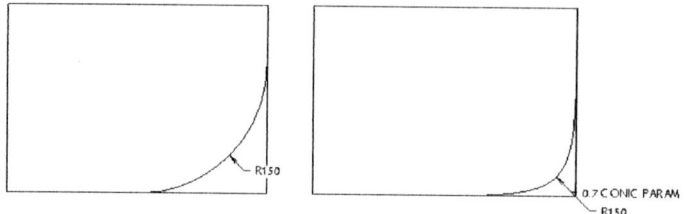

Figure 5-24 Preview of the round feature along with the radius of round

round, the three edit boxes are displayed on the **Round** dashboard. The **Reverse direction of conic legs** button is used to reverse the conic legs dimensions. Figure 5-26 shows the geometry of the conic round and the dimensions of the two legs after choosing this button.

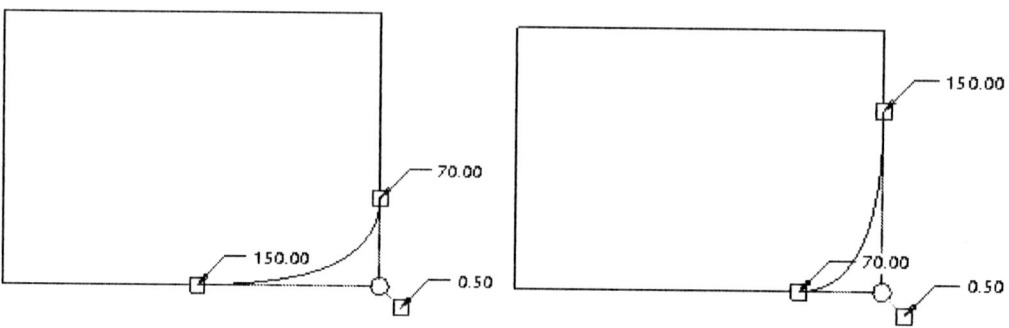

Figure 5-25 **D1xD2 Conic** *round*   *Figure 5-26* **D1xD2 Conic** *round after reversing*

Note
The rounds and chamfers are defined in sets when there is a variation in their dimensions. For example, if you want to create rounds on two sets of edges and the dimensions for both the rounds are different. In this case, you can define two sets of rounds; one set will define the rounds on one edge with some radius value and the second set will define the rounds on the other edge with a different radius value. The sets are also used in a similar way while creating chamfers.

Dimension Box
The edit box that is present below the first drop-down list is available only when you select the **Conic** or **D1xD2 Conic** options. This is because, this dimension box is used to enter the value for the sharpness of the conic round. The value lies between 0.5 and 0.95.

Second Drop-down List
The second drop-down list that is present in the slide-up panel is used to specify the method that Pro/ENGINEER will use to create the round. The options in this drop-down list are **Rolling Ball** and **Normal to spine**. These options are discussed next.

Rolling Ball. When you select this option to create a round, the round is created by rolling a spherical ball along the surfaces to which it will possibly stay tangent. This is the most common type of round that you will create.

Normal to spine. This option creates a round by sweeping an arc or a conic cross-section normal to the selected spine. Remember that in this case, the cross-section remains normal to the spine.

The area left to the drop-down lists in the slide-up panel lists the number of sets formed for creating rounds. This means you can define more than one set to create the rounds. Each set may have different values and references. To add a set, right-click in this area and choose the **Add** option from the shortcut menu or, you can just click on the **New set**. The set selected in this area, is the active set. It is recommended to fully define one set of round and then define the other set.

Full Round Button

The **Full Round** button creates a complete round between two selected edges or two planar or non-planar surfaces. This button is available only after selecting two references and not available when you choose the **Conic** or **D1xD2 Conic** option or the **Normal to spine** option. The following steps explain the procedure to create a full round by selecting two edges.

1. Invoke the **Round** dashboard.
2. Select the first edge and then use CTRL+left mouse button to select the second edge. Note that the edges selected should have a surface between them that can be converted into a round.
3. Make sure the **Circular** and **Rolling ball** options are selected from the first and third drop-down lists respectively.
4. Choose the **Full Round** button from the slide-up panel that is displayed when you choose the **Sets** tab from the **Round** dashboard. The round is created.

The following steps explain the procedure to create a full round by selecting the two surfaces.

1. Invoke the **Round** dashboard.
2. Select the first surface and then use CTRL+left mouse button to select the second surface. Note that the edges selected should have a surface between them that can be converted into a round, for example, the top and bottom faces of a rectangular block.
3. If the **Full Round** button is not chosen automatically then select it from the slide-up panel that is displayed when you choose the **Sets** tab from the **Round** dashboard. You are prompted to select a surface to replace with a full round.
4. Notice that in the slide-up panel, the **Driving surface** display box turns yellow in color. This indicates that now you can select the driving surface (surface to replace with a full round).

Figure 5-27 shows the round created using the **Full Round** option between the edge on the upper and the lower faces of the base of the model.

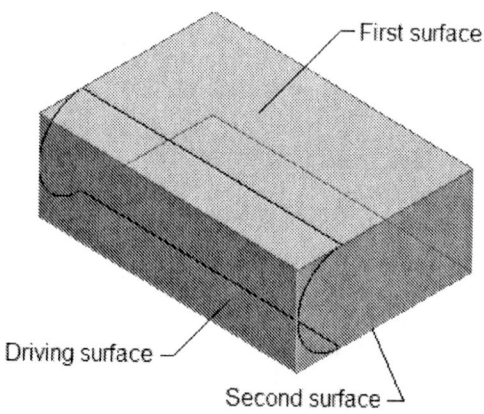

*Figure 5-27 Surfaces selected for **Full round** option*

Through Curve Button
The **Through Curve** button creates a round whose radius is specified by an existing curve.

References Collector
The **References** collector displays the references selected on the model. These references can be deleted by selecting them in the collector and then holding down the right mouse button to display a shortcut menu. Choose the **Remove** or **Remove All** option from this shortcut menu.

Spine Collector
The **Spine** collector displays the spine selected on the model. This collector is active only when you select the **Normal to spine** option from the second drop-down list. The **Spine** collector changes to **Driving surface** collector when you select two surfaces and then choose the **Full Round** button.

The area under the **Spine** collector is used to define the dimension values of the rounds.

Transitions Tab
When you choose the **Transition** tab, the slide-up panel is displayed, as shown in Figure 5-28. This slide-up panel displays the transitions if they are present in a round. For example, in Figure 5-21, there are four transitions (each at the four edges).

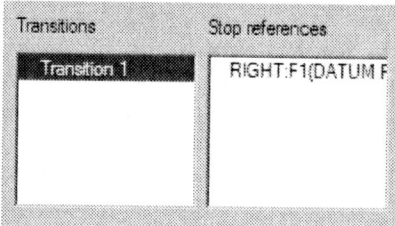

*Figure 5-28 The **Transitions** slide-up panel*

Pieces Tab

When you choose the **Pieces** tab, the slide-up panel is displayed. It provides the possible locations of the round in the form of pieces. For example, consider a case where you select two faces to fillet. One of the faces has a cut feature such that it is having two disjoint surfaces, as shown in Figure 5-29. When you move the cursor over any piece under the **Pieces** area, that piece is highlighted in cyan color on the model. This tab is useful in resolving ambiguity which arises if there exists multiple placement location for rounds with the current selection of references. This happens if two surfaces are intersecting each other at more than one places. From the pieces listed in the **Pieces** area, you can select the pieces to exclude by selecting the **Excluded** option from the drop-down list that is displayed when you click on the **Included** field, as shown in Figure 5-30.

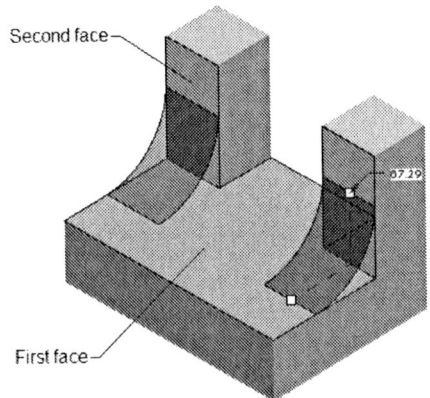

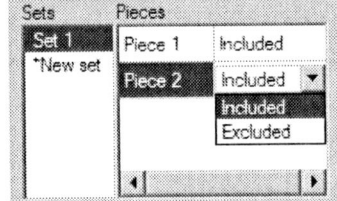

Figure 5-29 Faces selected for creating round   *Figure 5-30 The Pieces slide-up panel*

Options Tab

When you choose the **Options** tab, the slide-up panel is displayed that has two radio buttons. These radio buttons allow you to make the round as a surface or as a solid.

Properties Tab

When you choose this option, the slide-up panel is displayed. This slide-up panel displays the feature identity. Click on the **i** button to read the information about the feature. Click on the sash on the right of the **Feature info** window to close it.

Creating a Variable Radius Round

A variable radius round is the one in which the radius of the round varies through the length of the edge. For example, refer to Figure 5-31, which shows a constant radius round and Figure 5-32, which shows a variable radius round. To create a variable radius round, select the edge. When the preview of the round appears on the edge, hold down the right mouse button any where in the drawing window to display a shortcut menu. Choose the **Make variable** option from this shortcut menu. Now, in the slide-up panel that is displayed when you choose the **Sets** tab, the two default values of the radius are available, which can be modified as per requirement. You can drag the handles displayed at both ends of the selected edge to dynamically modify the radius of round.

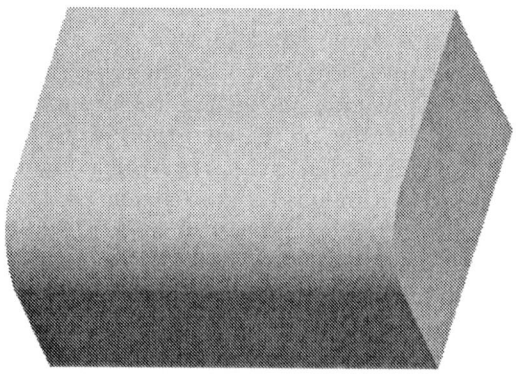
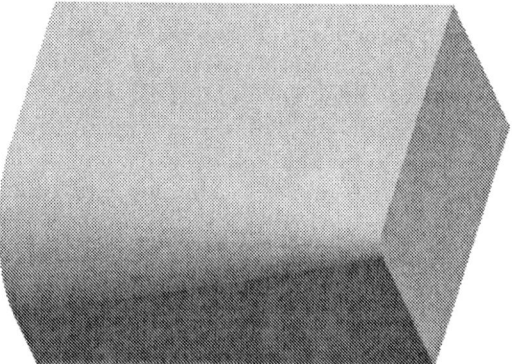

Figure 5-31 Constant radius round   *Figure 5-32* Variable radius round

Note
You can also create a variable radius round by selecting two surfaces. In this case, the edge that is common to the two surfaces is converted into a round.

You are also allowed to add additional points between the two end points and define variable radius in these additional points. To define additional points, choose the **Sets** tab to invoke the slide-up panel. Bring the cursor to the area in the slide-up panel where the radius value and the location of the points are displayed and right-click to invoke the shortcut menu. Choose the **Add radius** option; one additional point appears in the same area of the slide-up panel and also on the selected edge. This point by default is placed on the selected edge by a ratio value with respect to end point. You can select the **Reference** option from the drop-down list present under this area and select a location on the edge to place the point. The location of the point can also be modified by dragging the circular handle displayed on the edge.

Note
Rounds can be created on one edge, chain of edges, between two surfaces, an edge and surface, and between two edges.

Points to Remember While Creating Rounds
The following points should be remembered while creating rounds:

1. To avoid conflict due to parent-child relationship, other features should not be referenced to the edges created by rounds or to the round surface.
2. Preferably, rounds should be added end. This means, the rounds should be the last features in any solid model.
3. The regeneration time of the model increases if the number of round features in it are increased.
4. Use the shortcut menus extensively while creating rounds to reduce the modeling time.

Tip: *Rounds and chamfers are used in components to reduce the stress concentration at the sharp corners and edges. Hence, they reduce the chances of failure of a component under a specified loading condition.*

Options Aiding Construction of Parts-I

CREATING CHAMFERS

The **Chamfer tool** is used to bevel the selected edges and corners as per some specified parameters. Pro/ENGINEER creates two type of chamfers. The first is the **Corner** chamfer and the second is the **Edge** chamfer. Figure 5-33 shows the two types of chamfers. To create chamfer on a corner, choose **Insert > Chamfer > Corner** from the menu bar. To create a chamfer on an edge, choose **Insert > Chamfer > Edge** from the menu bar. You can also choose the **Chamfer Tool** button from the **Engineering Features** toolbar to create the edge chamfers.

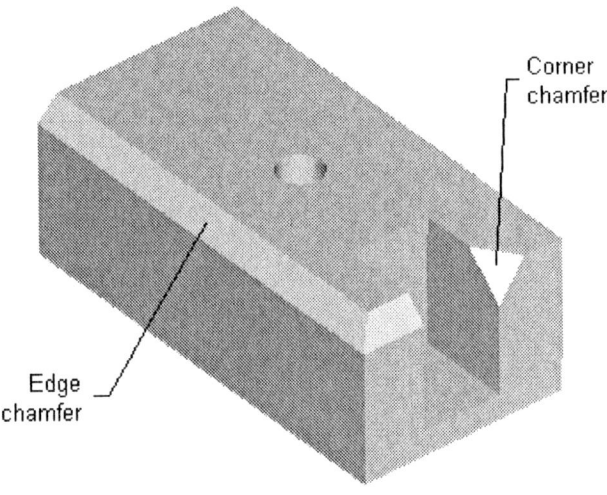

Figure 5-33 Different types of chamfers

Corner Chamfer

A **Corner** chamfer creates a beveled surface at the intersection of three edges. When you choose the **Corner Chamfer** option from menu bar, the **SELECT** dialog box and the **CHAMFER (CORNER)** dialog box is displayed (Figure 5-34) and you are prompted to select a corner to chamfer. Select one of the edges that form the corner, the **PICK/ENTER** menu is displayed, as shown in Figure 5-35. Note that while selecting the edge, click close to the corner that is to be chamfered. The options in the **PICK/ENTER** menu are used to specify the chamfer distance and are discussed next.

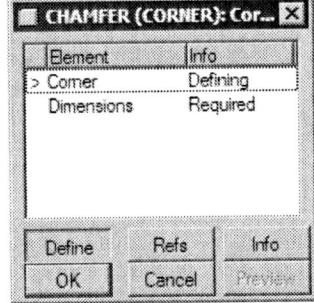

Figure 5-34 The **CHAMFER (CORNER)** dialog box

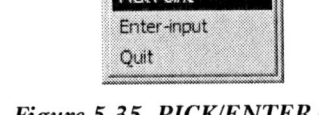

Figure 5-35 PICK/ENTER menu

Pick Point

The **Pick Point** option is used to select a point on the highlighted edge. The selected point denotes the chamfer distance on that edge from the corner. Once the point is selected, the next edge gets highlighted in blue and you are prompted to select a point on it. This step is repeated for the third edge also. Once points have been selected on all three edges, the **PICK/ENTER** menu disappears, indicating that the required information to create a corner chamfer is completed. Choose **Preview** button from the **CHAMFER (CORNER)** dialog box to preview the chamfer and then choose the **OK** button to create it.

Enter-input

When you use the **Enter-input** option, the **Message Input Window** is displayed and you are prompted to enter the length dimension along the highlighted edge. After doing so, the next edge will be highlighted in green. Enter the value in the Message input window. After you set the parameters for the second edge, the third edge is highlighted. Similarly, specify the value for the third edge. The **PICK/ENTER** menu disappears, choose **Preview** from the **CHAMFER (CORNER)** dialog box to preview the chamfer and then choose the **OK** button to create it.

Note

*When you are creating the corner chamfer, notice that only the **Edge** filter is available in the **Filter** drop-down list in the **Status** bar. As mentioned earlier, the filters available with this version of Pro/ENGINEER narrow the entities available for selection. Thus, you can easily select the reference entities on the model.*

Edge Chamfer

An **Edge** chamfer creates a beveled surface along the selected edge. When you choose **Insert > Chamfer > Edge Chamfer** from the menu bar or choose the **Chamfer Tool** button from the **Engineering Features** toolbar, the **Chamfer** dashboard is displayed, as shown in Figure 5-36 and you are prompted to select any number of edges to create chamfer. The options in this menu are discussed next.

*Figure 5-36 The **Chamfer** dashboard*

Switch to set mode Button

When you invoke the **Chamfer** dashboard, the **Switch to set mode** button is chosen by default. This button is used to create chamfers by defining either one set or more than one sets. When this button is chosen, the drop-down list to the right of this button in the **Chamfer** dashboard is also displayed. This drop-down list is used to specify the type of chamfer. The options in the drop-down list are discussed next.

D x D

The **D x D** option creates a chamfer such that the chamfer is created at an equal distance from the selected edge on both the faces connected by the edge. In Figure 5-37, the dimension value 10 is the value of D, where the distance D is measured from the corner.

D1 x D2

The **D1 x D2** option creates a chamfer at two user-defined distances from the selected edge. In Figure 5-38, the dimension value 10 is the value of D1 and the dimension value 16 is the value of D2. The values for D1 and D2 can be entered from the **Chamfer** dashboard. You can choose the **Interchange distance dimensions of the chamfer** button available on the right of the D2 edit box to interchange the two dimensions.

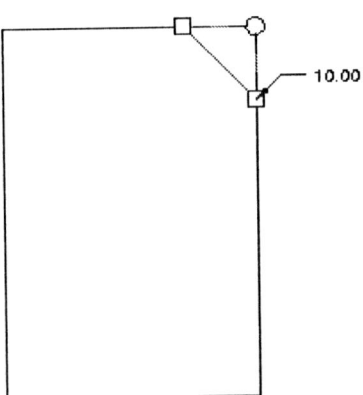

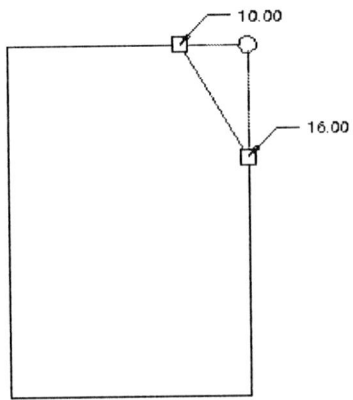

Figure 5-37 Creating chamfers using the *D x D* option

Figure 5-38 Creating chamfers using the *D1 x D2* option

Angle x D

The **Angle x D** option creates a chamfer at a user-defined distance from the selected edge and at a user-defined angle measured from a specified surface. In Figure 5-39, the angle is 60 and the value of D is 15.

45 x D

The **45 x D** option creates a chamfer at the intersection of two perpendicular surfaces. The chamfer created is at an angle of 45-degree from both the surfaces and at a distance D from the edge along each surface. Once the chamfer is created, the distance D can be modified. Figure 5-40 shows this type of chamfer where the value of D is 25.

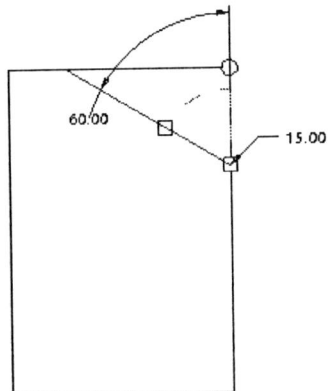

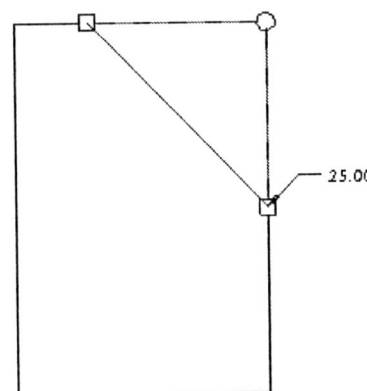

Figure 5-39 *Angle x D* option

Figure 5-40 *45 x D* option

Switch to transition mode Button

The use of **Switch to transition mode** button in the Chamfer dashboard is similar to the one available in the **Round** dashboard. This button will be available only when you have defined at least one reference for the chamfer. Using this button, you can define the shape of the chamfer that is formed at a vertex and also to define a limit up to which the chamfer will be created. Both these applications are discussed next.

Note

*After entering into the transition mode, you cannot change the value of **D**. Hence, it is recommended to set the value of **D** prior to specifying the transition. In case, it is required to change the value of **D** after setting the transition then choose the **Switch to set mode** button to return to the set mode and change the value of **D**.*

If the chamfer feature is being created on three intersecting edges, then this button is used to define the shape of the chamfer that is created at the vertex. After choosing the **Chamfer Tool** button, select the three edges. Now, choose the **Switch to transition mode** button from the **Chamfer** dashboard; the various transitions that are possible on the current selection are highlighted on the model, as shown in Figure 5-41. Select the transition that is displayed at the vertex. Now, from the drop-down list in the **Chamfer** dashboard, select the desired option, which can be **Default (Intersect)**, **Patch**, or **Corner Plane**.

The **Default (Intersect)** option merges the chamfer in such a way that sharp edges are formed at the vertex, as shown in Figure 5-42.

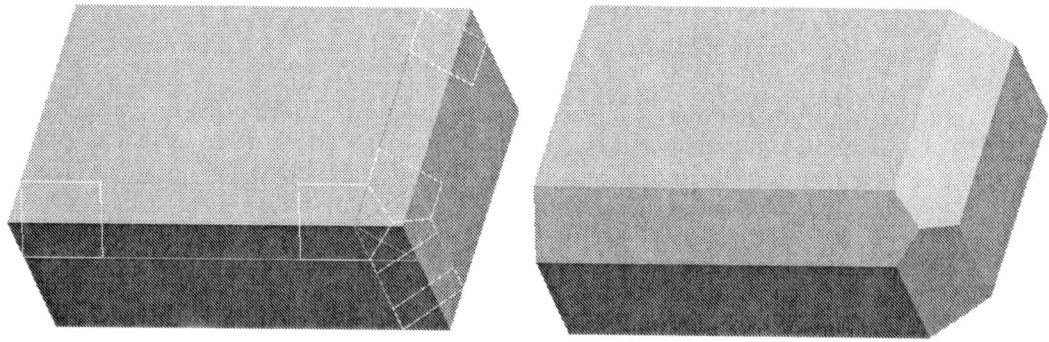

Figure 5-41 Various possible transitions   *Figure 5-42* Chamfer created using the **Default (Intersect)** option

The **Patch** option results in a smooth transition of chamfer by creating a fillet at the vertex with respect to a selected surface. After selecting this option from the drop-down list, the **Optional surface** collector is displayed to its right. Click once in this collector to bring it into the selection mode. Now, select one of the surfaces that form the selected edges. The resulting chamfer created by selecting the top surface of the model as the **Optional surface** is shown in Figure 5-43. Note that the fillet radius is displayed on the selected surface and can be dynamically modified by dragging the handle displayed on the preview or by entering a value in the **Radius** edit box displayed in the **Chamfer** dashboard.

The **Corner Plane** option can be used only when three chamfers are converging at a vertex. By using this option a plane is created at the point of merger, as shown in Figure 5-44. This plane is in the shape of a triangle and its size depends on the specified value of **D**.

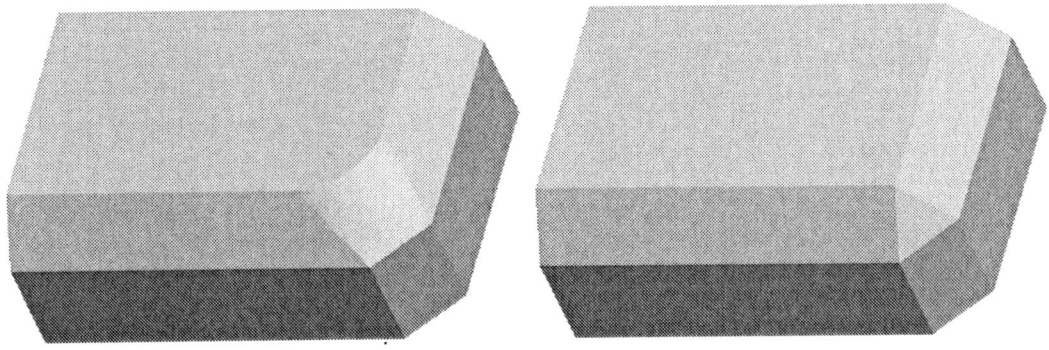

Figure 5-43 Chamfer created using the **Patch** *option*

Figure 5-44 Chamfer created using the **Corner Plane** *option*

The **Switch to transition mode** button also allows you to select a reference to stop the chamfer on an edge. As discussed earlier, when you choose this button, the transitions that are possible on the current selection are highlighted on the model. The highlighted piece that you select is not chamfered up to the selected reference. The reference to stop the transition can be selected after you select the **Stop at Reference** option from the drop-down list present on the **Chamfer** dashboard. This means the chamfer will be created on the selected edge starting from the opposite end and up to the selected reference only, as shown in Figure 5-45. The reference plane is highlighted by a mesh in the figure.

Sets Tab

When you choose the **Sets** tab in the **Chamfer** dashboard, the slide-up panel is displayed, as shown in Figure 5-46. All references and dimensions of the chamfer can be specified in this slide-up panel. The options in this slide-up panel are discussed next:

Set 1 Display Area

This display area displays the number of sets created. To add a set, right-click to invoke a shortcut menu and choose the **Add** option from it or just click on **New set**. It is recommended that you define an additional set only when you have defined the first set. You can switch among the sets by selecting them from this area. The set that is highlighted in this area can be previewed in the graphics window.

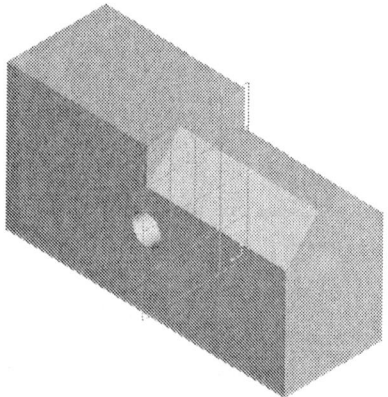

Figure 5-45 Transition chamfer

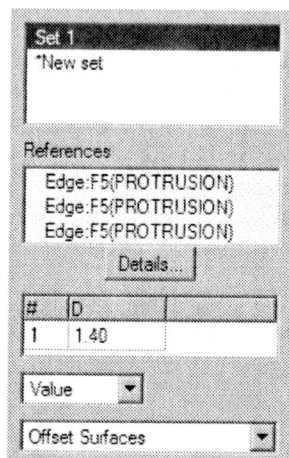

Figure 5-46 The Slide-up panel of the Sets tab

References collector

The **References** collector displays the edges selected to chamfer. An edge can be removed from the selection set by, right-clicking to invoke a shortcut menu and choosing the **Remove** option from it.

The area below the **References** collector is used to specify the dimension values. The dimension values can also be specified on the dashboard.

Transitions Tab

When you choose the **Transitions** tab, the slide-up panel is displayed, as shown in Figure 5-47. This slide-up panel displays the transitions, if they are present in a chamfer. For example, in Figure 5-45, there are two transitions (each at the two edges).

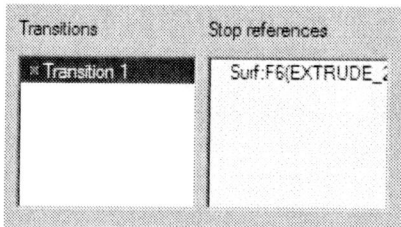

Figure 5-47 The Transitions slide-up panel

Pieces Tab

The use of **Pieces** tab is the same as discussed in the **Round** dashboard. You can use it to exclude selected pieces in case there exists some ambiguity while creating chamfers.

Options Tab
When you choose the **Options** tab, the slide-up panel is displayed, which has two radio buttons. These radio buttons allow you to make the chamfer as a surface or as a solid.

Properties Tab
When you choose this option, the slide-up panel is displayed, which shows the feature identity.

UNDERSTANDING RIBS
Ribs are defined as thin wall-like structures used to bind the joints together so that they do not fail under an increased load. In Pro/ENGINEER, the section for the rib is sketched as an open section and can be extruded equally in both directions of the sketch plane or on either side. The procedure of creating a rib is similar to that of creating a extrusion.

In Pro/ENGINEER, you can create two types of ribs: rotational ribs and straight ribs. Rotational ribs are constructed on cylindrical parts and straight ribs are created on planar faces. Figure 5-48 shows a rotational rib and Figure 5-49 shows a straight rib. The options available for the creation of these ribs are the same. The creation of these ribs depends on the geometry of the feature on which the rib is created.

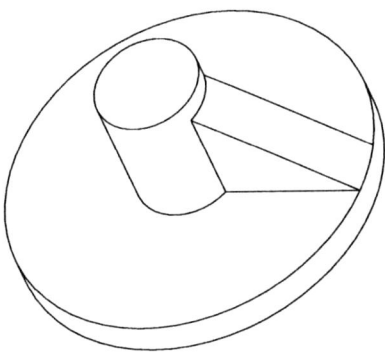
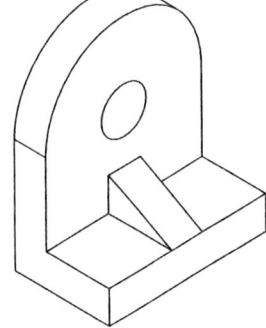

Figure 5-48 Rotational rib feature   *Figure 5-49 Straight rib feature*

Creating Ribs

Ribs are created using the **Rib** dashboard shown in Figure 5-44. This dashboard is displayed when you choose **Insert > Rib** from the menu bar or choose the **Rib Tool** button from the **Engineering Features** toolbar.

*Figure 5-50 The **Rib** dashboard*

To create a rib, you first need to select a sketching plane and then draw the geometry for the rib feature in the sketcher environment. The sketch for the rib feature should be an open entity.

The options and the tools available in the **Rib** dashboard are discussed next.

References Tab

When you choose the **References** tab, the slide-up panel is displayed, as shown in Figure 5-51. In the slide-up panel, the **Sketch** collector shows **Select 1 item**. This indicates that the sketch for the rib feature is not drawn. To draw the sketch, choose the Define button. The sketch dialog box is displayed using which you can define the sketch plane and the references.

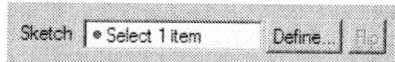

*Figure 5-51 The **References** slide-up panel*

 The **Flip** button is used to flip the side where the material should be added. This button will be available only when you draw the sketch and exit the sketcher environment.

Change the thickness option between both sides, side 1, and side 2 Button

 This button is available on the **Rib** dashboard only when the sketch of the rib is drawn and you exit the sketcher environment. You have an option to extrude the sketch on either sides of the sketching plane or symmetrically to both sides of the sketching plane.

 Note
Because the sketch of the rib can be extruded on both sides of the sketching plane, therefore, while creating a rib, you must always select an appropriate sketching plane such that it lies at the center of rib feature

The procedure to create a Rib feature using the **Rib** dashboard is discussed next.

1. Invoke the **Rib** dashboard by choosing the **Rib Tool** button form the Engineering Features toolbar or choose **Insert > Rib** form the menu bar.
2. From the **Rib** dashboard, choose the **References** tab to display the slide-up panel and choose the **Define** button from it.
3. Select the sketch plane and specify the reference in the **Sketch** dialog box. Now, create an open sketch that defines the rib profile. Make sure that the endpoints of the sketch are aligned with the edges of the model.
4. When you exit the sketcher environment, a yellow arrow is displayed in the graphics window. This arrow shows the direction of material addition and should point toward the geometry of the model and not away form it. The direction of the arrow can be reversed by using the **Flip** button available in the slide-up panel that is displayed on selecting the **References** tab.
5. If the arrow direction is correct, then you can preview the rib feature on the graphics window. The rib thickness can be dynamically modified by dragging the handles displayed on the rib or by entering a value in the edit box available in the **Rib** dashboard.

Options Aiding Construction of Parts-I

6. Now, use the **Change the thickness option between both sides, side 1, and side 2** button to change the direction of extrusion. You can create the rib feature either symmetrically about the sketching plane or on either side of it. Every successive click on this button will change the position of the rib feature with respect to the sketching plane.
7. After specifying all parameters for the rib feature, choose the **Build feature** button.

EDITING FEATURES OF A MODEL

Editing is one of the most important aspects of the product design. Most of the designs require editing either during their creation or after they are created. As mentioned earlier, Pro/ENGINEER is a parametric and feature-based solid modeling software. Hence, the features constituting a model can be individually edited. For example, Figure 5-52 shows a cylindrical part that has six counterbore holes created at some bolt circle diameter (BCD).

Now, in case you have to edit the features such that the six holes are to be converted into eight holes and the counterbore holes are to be converted into countersink holes, as shown in Figure 5-53, you just need to use two operations. The first operation will convert six holes into eight holes by using the **Edit** option in the shortcut menu that is displayed when you select the hole from the **Model Tree** and hold down the right mouse button. The second operation will open the **Hole** dashboard and convert the counterbore holes into countersink holes.

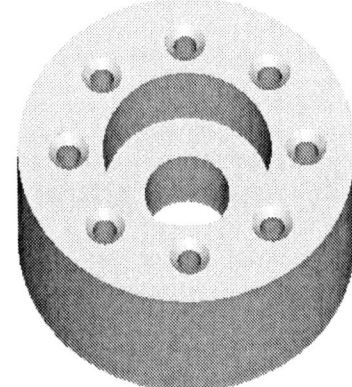

Figure 5-52 Six counterbore holes   *Figure 5-53* Eight countersink holes

Similarly, you can also edit the datums and the features referenced to these datums. Because there exists a parent-child relationship between the two, therefore, the child feature is also modified when the parent feature is modified. For example, if you have created a feature using a datum plane that is at some offset distance, the feature will be automatically repositioned when the offset value of the datum plane is changed. The following methods explain how to edit the features in Pro/ENGINEER.

Editing Definition or Redefining Features

Editing the definition of features allows you to make changes in the parameters that were used to create it. You can also modify the sketches of the sketched features by editing their definition. To invoke the **Edit Definition** option, there are three methods available in Pro/ENGINEER Wildfire 2.0. These methods are listed next.

1. Select the feature that has to be redefined from the graphics window; the selected feature is highlighted in red. Now, hold down the right mouse button. A shortcut menu is displayed; choose the **Edit Definition** option from this menu.

   Sometimes, it becomes difficult to select the desired feature from the model to redefine it. Therefore, you can use the filters available in the **Filter** drop-down list on the **Status** bar.

   Tip: *Suppose you have created a cube and the display is selected to be* **Shaded**. *Now, you need to select the top face of the cube. By default, in the* **Filter** *drop-down list the* **Smart** *option is selected. Select the cube on the top face; the whole cube turns red in color. Again click on the top face; it gets selected and is highlighted in red.*

2. You can also use the **Model Tree** to redefine a feature. Select the feature that has to be redefined from the **Model Tree**; the selected feature is highlighted in red in the graphics window. Now, right-click on the feature listed in the **Model Tree**. A shortcut menu is displayed; choose **Edit Definition** from this menu.

3. Select the feature to redefine from the graphics window or from the **Model Tree** and then choose **Edit > Definition** from the menu bar. Remember that the **Definition** option in the **Edit** menu is available only when a feature has been selected.

Reordering Features

Reordering the features is defined as the process of changing the sequence of feature creation in a model. Sometimes after creating a model, it may be required to change the order in which the features of the model were created. A feature can be placed before or after another feature. For this purpose, the **Model Tree** or the **Menu Manager** is used.

If you use the **Model Tree**, you just have to drag and drop the feature you want to reorder below or above the destination feature.

To invoke the **Menu Manager**, choose **Edit > Feature Operations** from the menu bar, as shown in Figure 5-54. It was used extensively in the prior releases of Pro/ENGINEER Wildfire but now, is only used mainly to perform operations on features.

After choosing the **Reorder** option from the **FEAT** menu of **Menu Manager**, the **SELECT FEAT** submenu is displayed, as shown in Figure 5-55, and you are prompted to select the features that you want to reorder. Select the feature to be reordered from the **Model Tree** and then choose the **Done** option from the **SELECT FEAT** submenu or use the middle mouse button to confirm the selection. The **Message** area displays the possible positions where the selected feature can be inserted.

Now, select **BEFORE** or **AFTER** from the **REORDER** submenu, as shown in Figure 5-56. Depending upon this selection the feature to be reordered will be placed before or after another selected feature. Select another feature and choose the **Done** option from the **Menu Manager** to complete the reordering process.

Options Aiding Construction of Parts-I

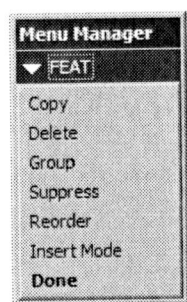

Figure 5-54 *Menu Manager*

Figure 5-55 *SELECT FEAT submenu*

Figure 5-56 *REORDER submenu*

The example that is discussed here will explain why and when the need for reordering of features in a model arise. Consider the model shown in Figure 5-57. It consists of a rectangular pattern of three columns and two rows. Now, a shell feature is created on this model, which removes the top and the front face, as shown in Figure 5-58.

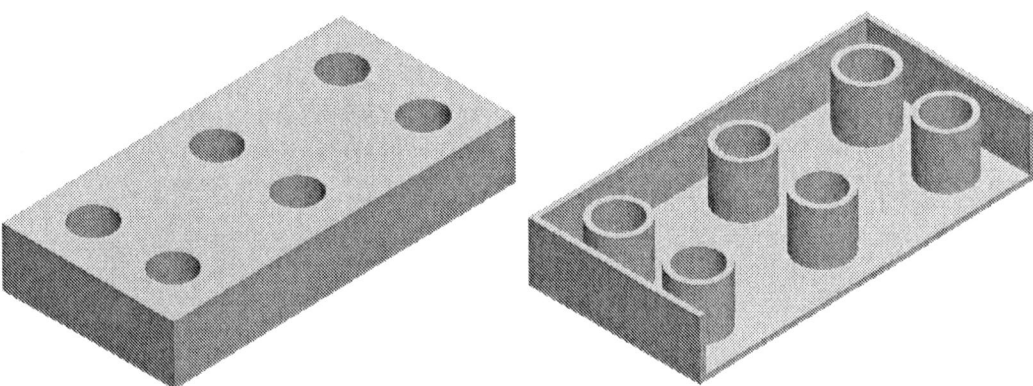

Figure 5-57 *Model with pattern*

Figure 5-58 *Model after creating the shell feature*

In the Figure 5-58, it is evident that the material equal to the wall thickness of the shell feature is added around the hole features. If this is not desired, the model has to be modified and here the need for reordering the feature arises. You will reorder the features such that the shell feature is inserted before the hole feature and its pattern. When you reorder the features, all features will be automatically adjusted in the new order. The model after reordering the hole feature is shown in Figure 5-59.

Rerouting Features

Rerouting of features is required when there is a need of changing its sketch plane. While doing so you can also redefine the references that were selected while creating the sketch. For rerouting a feature, select it from the **Model Tree** or from the graphics area. Now choose **Edit > References** from the menu bar. Note that the **References** option is only available after a feature has been selected.

Figure 5-59 Model after reordering the features

After choosing the **References** option, the system prompts whether to roll back the model. If you select **Yes** then all features created after the selected feature will be suppressed and they will not be visible on the model. The suppressed features will be unsuppressed after the rerouting is completed. If **No** is selected, all features of the model will remain active while the selected feature is being rerouted. The desired selection can be made by choosing the appropriate button on the right of the Message area.

Now, the **Menu Manager** is displayed and also a surface is highlighted in the graphics window. You are prompted to select an alternate surface. The highlighted surface can be the sketch plane or a reference that was selected while creating the feature. You can select another plane or surface to replace the highlighted surface. Once a selection is made another surface gets highlighted for replacement. This process will continue till all references are replaced. Finally, the Menu Manger will be automatically closed and the model will be regenerated as per the new references.

Note

While rerouting a feature, if you do not want to replace a highlighted surface then select the same surface again as the alternate surface.

Suppressing Features

Once the feature is suppressed, it will neither be displayed in the graphics window nor in the drawing views. Note that the feature is not deleted, only its visibility is turned off. You can anytime resume the feature by unsuppressing it using the **Model Tree** or using **Edit > Resume** from the menu bar. As soon as you unsuppress the feature, it will be displayed in the graphics window as well as in the drawing views. When a model has many features, then suppressing some features decreases its regeneration time. To suppress a feature, right-click on it in the **Model Tree** or select it in the graphics window and then press and hold down the right mouse button to display the shortcut menu. Choose **Suppress** from the shortcut menu to suppress the selected feature. You can also suppress a feature using the **Menu Manager**. To unsuppress the feature, select it in the **Model Tree** and invoke the shortcut menu, select the **Resume** option from it.

Tip: *By default, the suppressed features are not displayed in the **Model Tree**. To display the suppressed features, choose **Settings** > **Tree Filters** from the **Model Tree**. The **Model Tree Items** dialog box is displayed. Select the **Suppressed Objects** check box and choose the **OK** button. The suppressed features will now be displayed in the **Model Tree** with a small black box displayed on its name.*

Note
If the feature that is suppressed has some child features, they will also be suppressed. When you select such a feature, the system prompts you to confirm the suppression of the highlighted features in the graphics window.

Deleting a feature

The feature that is not required can be deleted from the model. Right-click on the feature in the **Model Tree** or in the graphics window to invoke the shortcut menu. From this menu, choose the **Delete** option. Once the feature to be deleted is selected, it gets highlighted along with its child features. The system confirms before deleting the selected feature. If you confirm the deletion, the feature will be deleted along with its child features.

Modifying a feature

Once a feature is created, you can still modify the feature by modifying its dimensions. This editing operation reflects the parametric nature of Pro/ENGINEER. Right-click on the feature in the **Model Tree** or in the graphics window to invoke the shortcut menu. From this menu, choose the **Edit** option. Now, the dimensions of the feature are displayed on the graphics window. These dimensions are the same that were defined while sketching or creating the feature.

In some cases, for example, rounds, holes, and so on, the dimensions of these features were defined as parameters. Hence, these dimensions are also displayed when you select them to modify. To modify the dimensions, double-click on the dimension in the graphics window. The dimension is converted into an edit box. Type the required value for the selected dimension and press ENTER. Once you modify a dimension, you also need to regenerate the feature. Choose the **Regenerates Model** button from the **Top Toolchest**.

If you want to modify a pattern, it is recommended that you select the instance other than the original one. This is because if you select the original feature, the incremental dimensions are not displayed and so you cannot modify the incremental values of the pattern. Patterns are discussed in Chapter 6.

Tip: *You can double-click on a feature in the graphics window to display its dimensions. These dimensions can be modified by double-clicking on them and then entering a new value in the displayed edit box.*

Dynamically Modifying the Sketch of a Feature

If you want to modify the sketch of a feature without entering the sketcher environment, then you need to modify it dynamically. To dynamically modify a sketch, follow these steps:

1. Select the feature preferably by using the **Features** filter from the **Filter** drop-down list.

2. When the feature turns red in color, hold down the right mouse button to invoke the shortcut menu.

3. From the shortcut menu, choose the **Edit** option. The dimensions of the sketch appear on the feature.

4. Select the sketch and hold down the right mouse button to invoke the shortcut menu. Choose the **Edit** option from this shortcut menu.

5. Now, bring the cursor close to the sketch. You will notice that the cursor changes to a pointer. The pointer symbol represents that you can select an entity from the sketch and drag it to modify the shape of the sketch.

6. Modify the shape of the sketch by dragging its entities.

7. Choose the **Regenerates Model** button from the **Top Toolchest** to apply the modifications.

TUTORIALS

Tutorial 1

Create the model shown in Figure 5-60. The dimensions, and the front, top, and left-side views of the model are also shown in the figure. **(Expected time: 45 min)**

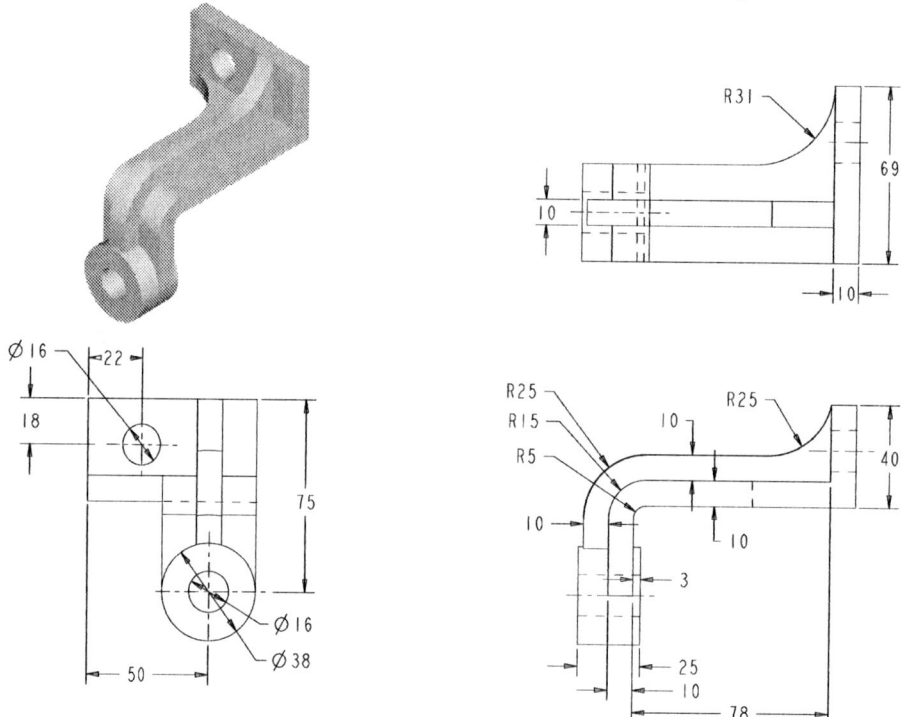

Figure 5-60 *Isometric shaded, left-side, front, and top views of the model*

Options Aiding Construction of Parts-I

The following steps are required to complete this model:

a. Examine the model and determine the number of features in it, refer to Figure 5-60.
b. Create the base feature on the **FRONT** datum plane, refer to Figures 5-61 and 5-62.
c. Create the second extrude feature on the **TOP** datum plane, refer to Figures 5-63 and 5-64.
d. Create the third extrude feature on the front planar surface of the second feature, refer to Figures 5-65 and 5-66.
e. Create the non symmetrically extruded cylindrical feature on the front face of the third feature, refer to Figures 5-67 and 5-68.
f. Created hole feature that is coaxial to the cylindrical feature, refer to Figure 7-70.
g. Create the round features, refer to Figures 5-71 and 5-72.
h. Create the last feature, which is the rib, refer to Figure 5-73 and 5-74.

After understanding the procedure for creating the model, you are now ready to create it. When Pro/ENGINEER session is started, the first task is to set the working directory. Because this is the first tutorial of this chapter, you need to create a folder named *c05*, if it does not exist. Choose the **New Directory** button in the **Select Working Directory** dialog box and create a directory named *c05* at *C:\ProE-WF-2.0*.

Starting a New Object File

1. Open a new part file and name it *c05tut1*. The three default datum planes are displayed in the graphics window. The **Model Tree** is also displayed on the graphics window. Close the **Model Tree** by clicking on the sash.

Creating the Base Feature

To create the sketch for the base feature, you first need to select the sketching plane for it. In this model, you need to draw the base feature on the **FRONT** datum plane because the direction of extrusion of this feature is normal to this plane.

1. Choose the **Extrude Tool** button from the **Base Features** toolbar; the **Extrude** dashboard is displayed.

2. Choose the **Placement** tab to display the slide-up panel and choose the **Define** button; the **Sketch** dialog box is displayed.

3. Select the **FRONT** datum plane as the sketching plane. The **RIGHT** datum plane and its orientation are selected by default.

4. Select the **TOP** datum plane and then choose the **Top** option from the **Orientation** drop-down list.

5. Choose the **Sketch** button from the **Sketch** dialog box to enter the sketcher environment.

6. Once you enter the sketcher environment, create the sketch of the base feature and apply constraints and dimensions, as shown in Figure 5-61. Note that in the sketch, the bottom line of the rectangular section coincides with the **TOP** datum plane.

As evident from the sketch of the base feature shown in Figure 5-61, the **RIGHT** datum plane is located at a dimension of 50 from the left edge because later in the tutorial, the rib feature will be created on this plane. The distance of 50 can be calculated from Figure 5-60. As mentioned earlier, the sketch for a rib is by default extruded to both sides of the sketching plane. Therefore, when you create the sketch for the rib feature on the **RIGHT** datum plane, and it will be extruded on both the sides.

7. After the sketch is completed, choose the **Continue with the current section** button to exit the sketcher environment.

   The **Extrude** dashboard is enabled and appears below the graphics window. All model attributes that are selected by default are accepted to create the model.

8. Enter **10** as the depth in the dimension box on the **Extrude** dashboard.

9. Choose the **Default Orientation** option from the **Saved view list** drop-down list and then choose the **Build feature** button from the **Extrude** dashboard to create the base feature.

Creating the Second Feature

The second feature is also an extruded feature and will be created on the **TOP** datum plane. Therefore, you need to define the **TOP** datum plane as the sketching plane.

1. Choose the **Extrude Tool** button from the **Base Features** toolbar; the **Extrude** dashboard is displayed.

2. Choose the **Placement** tab and from the slide-up panel, choose the **Define** button; the **Sketch** dialog box is displayed.

3. Select the **TOP** datum plane as the sketching plane.

4. Select the front face of the base feature shown in Figure 5-62 and then select the **Bottom** option from the **Orientation** drop-down list.

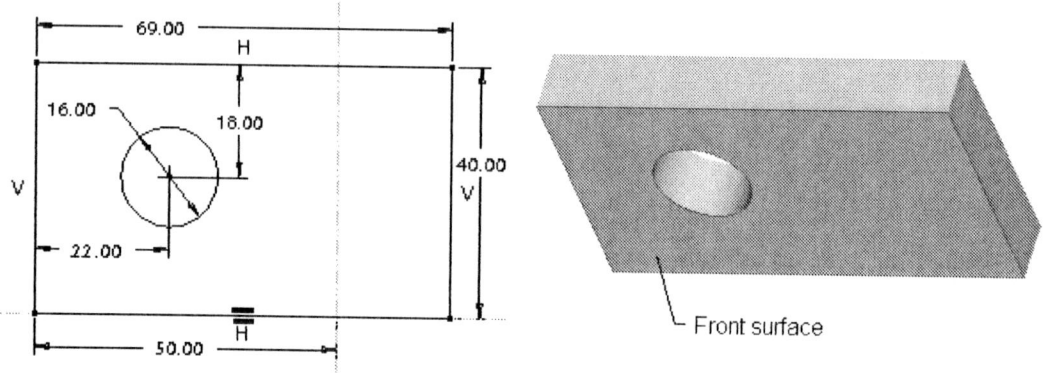

Figure 5-61 Sketch with dimensions and constraints for the base feature

Figure 5-62 Front face of the base feature selected to be at the bottom while sketching

Options Aiding Construction of Parts-I 5-37

5. Choose the **Sketch** button from the **Sketch** dialog box to enter the sketcher environment. After entering the sketcher environment, turn the model display to **No Hidden**.

Note
*Before you start sketching a section for any feature, it is recommended that you turn the model display to **No Hidden** and turn off the datum plane display using the **Datum planes on/off** button. This is done to improve the clarity of the graphics window while sketching.*

6. Create the sketch for the second feature and apply the constraints and dimensions, as shown in Figure 5-63.

   You can use the **Create an entity from an edge** button from the **Sketcher Tools** toolbar to use the edge of the base feature. The edge of the base feature is required to close the section for the second feature. However if you do not use the edge, you will have to draw a line to close the sketch. You will then have to align it with the bottom edge and add the constraint and dimensions, as shown in Figure 5-63.

Note
If you do not close the section loop by drawing a line or using the edge of the base feature, a yellow arrow is displayed when you exit the sketcher environment. Using this arrow you can specify the direction in which the material will be added.

7. After completing the sketch, turn the model display to **Shading** and choose the **Continue with the current section** button. The **Extrude** dashboard is enabled below the graphics window.

8. Enter a value of **10** in the dimension box present on the **Extrude** dashboard and press ENTER.

9. Choose the **Build feature** button from the **Extrude** dashboard. The second feature is completed and the shaded default trimetric view is shown in Figure 5-64.

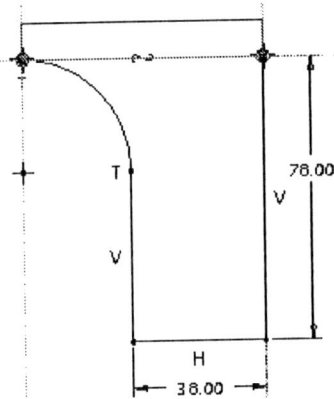

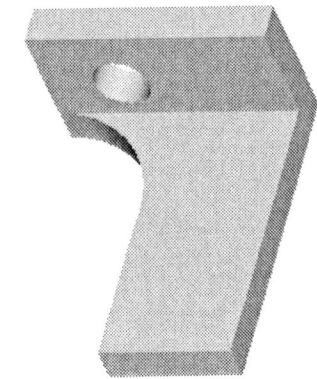

Figure 5-63 Sketch with dimensions and constraints for the second feature

Figure 5-64 The default trimetric view of the completed second feature and the base feature

Creating the Third Feature

The sketch of the third feature will be drawn on the front planar surface of the second feature and will be extruded to the given depth.

1. Choose the **Extrude Tool** button from the **Base Features** toolbar.

2. Choose the **Placement** tab and from the slide-up panel, choose the **Define** button; the **Sketch** dialog box is displayed.

3. Select the face of the second feature shown in Figure 5-65 as the sketching plane.

4. Select the top face of the second feature as the reference and then select the **Top** option from the **Orientation** drop-down list only if they are not selected by default.

5. Choose the **Sketch** button from the **Sketch** dialog box to enter the sketcher environment.

6. Once you enter the sketcher environment, turn the model display to **No Hidden**. Create the sketch for the third feature and apply constraints and dimensions, as shown in Figure 5-66.

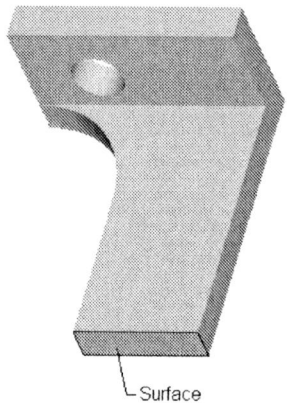
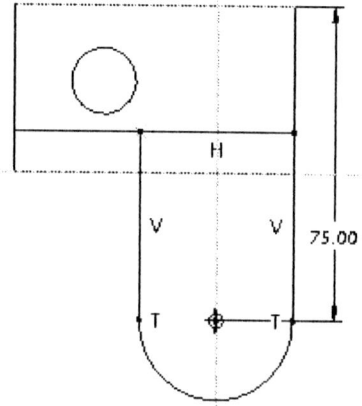

Figure 5-65 Planar surface selected as the sketching plane for the third feature

Figure 5-66 Sketch for the third feature with dimensions and constraints

Note
While drawing the sketch for any feature, it is recommended that you first apply the required constraints and then dimension the sketch.

7. Choose the **Continue with the current section** button. The **Extrude** dashboard is enabled below the graphics window.

8. Enter a value of **10** in the dimension box present on the **Extrude** dashboard.

9. Choose the **Build feature** button from the **Extrude** dashboard. Turn the model display to **Shading**. The third feature is completed. You can use the middle mouse button to spin the model to view it from various directions.

Options Aiding Construction of Parts-I

Creating the Fourth Feature

The fourth feature of the model is an extruded feature and the sketch of this feature is drawn on the front planar surface of the third feature.

1. Choose the **Extrude Tool** button from the **Base Features** toolbar.

2. Choose the **Placement** tab and from the slide-up panel, choose the **Define** button; the **Sketch** dialog box is displayed.

3. Select the front face shown in Figure 5-67 as the sketch plane. The extrusion of the cylindrical feature will be on both sides of the front face.

   The attributes of the feature, like extruding to both sides, and the depth of extrusion will be specified after the sketch of the feature is drawn.

4. Select the top face of the second feature as reference and then select the **Top** option from the **Orientation** drop-down list, if they are not selected by default.

5. Choose the **Sketch** button from the **Sketch** dialog box to enter the sketcher environment.

6. Once you enter the sketcher environment, turn the model display to **No Hidden**.

7. Draw the circular section for the fourth feature, as shown in Figure 5-68. Choose the **Create concentric circle** button from the **Sketcher Tools** toolbar to draw the sketch for this feature. Select the arc using the left mouse button. As you move the mouse, the yellow rubber-band circle changes its size. Move the cursor close to the arc, the cursor snaps to the arc. Use the left mouse button and select a point on the arc. You will notice that the equal radius constraint is applied to the sketch. Now, exit this tool.

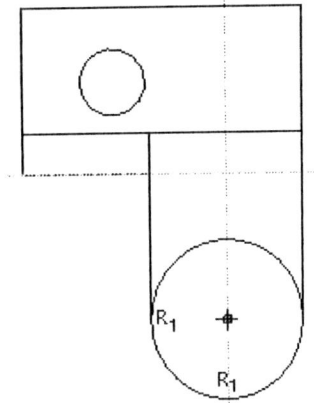

Figure 5-67 Front face   *Figure 5-68* Sketch for the fourth feature

8. After the sketch is completed, choose the **Continue with the current section** button. The **Extrude** dashboard is enabled below the graphics window.

9. Choose the **Options** option from the **Extrude** dashboard. From the slide-up panel, choose the **Blind** option from the **Side 2** drop-down list.

10. In the dimension box that is on the right of the **Side 1** drop-down list, enter a value of **12**.

11. In the dimension box that is on the right of the **Side 2** drop-down list, enter a value of **13**.

12. Turn the model display to **Shading** and choose the **Build feature** button from the **Extrude** dashboard to create the forth feature.

Creating the Hole Feature

The hole feature will be created using the **Hole** dashboard. The hole will be placed coaxial to the fourth cylindrical feature.

Note
*Because the hole will be created coaxially, you will be required to select the axis of the circular feature. You may need to turn on the axis display if it is turned off. Choose the **Datum axes on/off** button from the **Datum Display** toolbar to turn on the datum axis display.*

1. Choose the **Hole Tool** button from the **Engineering Features** toolbar. The **Hole** dashboard is displayed, in which, the **Create straight hole** button is selected by default.

2. Choose the **Placement** tab from the **Hole** dashboard; the slide-up panel is displayed.

3. Select the front face of the cylindrical feature, as shown in Figure 5-69, to place the hole.

   As you select the front face of the cylindrical feature, the preview of the hole is displayed in the graphics window. Now, you need to specify the reference for the hole placement.

4. From the drop-down list present in the slide-up panel, select the **Coaxial** option. Now, you have specified the type of hole you need to create.

5. Click in the **Secondary References** collector so that the **Primary** collector turns white in color.

6. Select the axis of the cylindrical feature from the graphics window.

7. In the diameter dimension box on the **Hole** dashboard, type **16** and press ENTER.

8. From the depth flyout in the **Hole** dashboard, select the **Drill to intersect with all surfaces** button.

9. Choose the **Build feature** button from the **Hole** dashboard. The hole is created and the trimetric view of the shaded model with the hole is shown in Figure 5-70.

Creating the Two Round Features

Now, the two round features will be created. Both the rounds have different radii and will be created by defining two sets.

Options Aiding Construction of Parts-I

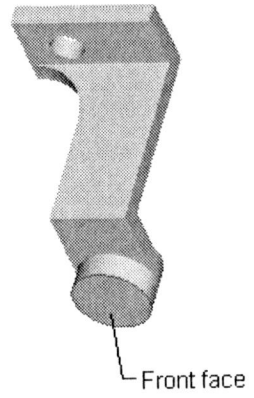
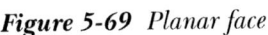

Figure 5-69 Planar face

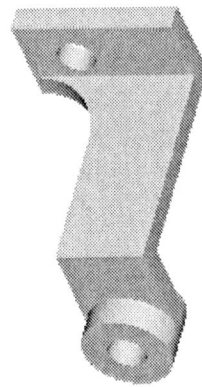

Figure 5-70 The default trimetric view

1. Choose the **Round Tool** button from the **Engineering Features** toolbar; the **Round** dashboard is displayed.

2. Choose the **Sets** tab to display the slide-up panel. Let this slide-up panel remain open so that you can view the selections you make on the model.

3. Spin the model and select one of the faces shown in Figure 5-71. Press the CTRL key and then select the face.

   The preview of the round is created on the edge that connects the two selected faces. A default value of the radius is also displayed.

4. Double-click on the default radius value that is displayed on the preview of the round. In the edit box that appears, type a value of **5** and press ENTER. The first round is created.

5. After spinning the model, select one of the faces shown in Figure 5-72 and then use

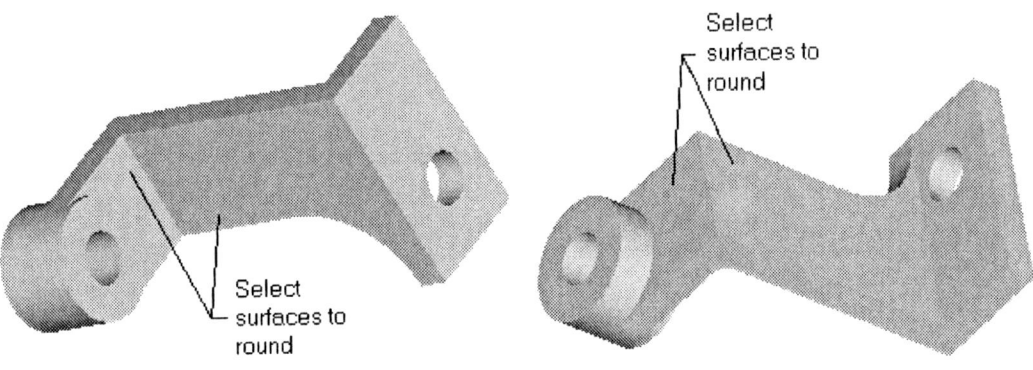

Figure 5-71 Faces for creating round of radius 5

Figure 5-72 Faces for creating round of radius 15

CTRL+left mouse button to select the second face. You will notice that in the slide-up panel, **Set 1, Set 2** appears. This indicates that second set is defined.

The preview of the round is created on the edge that connects the two selected faces. A default value of the radius is also displayed.

6. Double-click on the default radius value that is displayed on the preview of the second round. In the edit box that appears, type a value of **15** and press ENTER.

7. Choose the **Build feature** button from the **Round** dashboard.

The second round is created. In the **Model Tree** the two rounds created appear as one feature.

Creating the Rib Feature

Some advance planning that was done while drawing the sketch for the base feature will help you to draw the rib feature now. The location for the **RIGHT** datum plane was calculated while sketching the base feature. The section for the rib feature will be drawn on the **RIGHT** datum plane.

1. Choose the **Rib Tool** button from the **Engineering Features** toolbar; the **Rib** dashboard is displayed.

2. A rib feature is always sketched from the side view. Now, turn on the display of datum planes from the **Datum Display** toolbar.

3. Choose the **References** tab and from the slide-up panel, choose the **Define** button; the **Sketch** dialog box is displayed.

4. Select the **RIGHT** datum plane as the sketching plane from the graphics window.

5. Select the **TOP** datum plane as reference and then select the **Top** option from the **Orientation** drop-down list, only if they are not selected by default.

6. Choose the **Sketch** button to enter the sketcher environment. When you enter the sketcher environment, the **References** dialog box is displayed and the status displayed is **Fully Placed**. Close the **References** dialog box.

You will need to turn the model display to **No Hidden** before drawing the sketch for the rib feature.

7. Draw the open sketch for the rib feature and apply the required constraints and dimensions, as shown in Figure 5-73.

8. After the sketch is completed, choose the **Continue with the current section** button.

A yellow arrow is displayed on the sketch pointing in the direction of material addition.

Options Aiding Construction of Parts-I   5-43

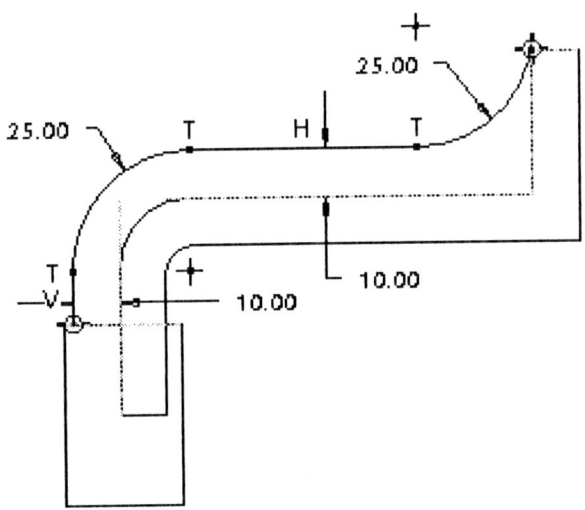

Figure 5-73 Open sketch of the rib feature with dimensions and constraints

Because the section for the rib feature is open, Pro/ENGINEER allows you to specify the direction where the material should be added.

9. Enter **10** in the dimension box present on the **Rib** dashboard.

10. Choose the **Build feature** button from the **Rib** dashboard to exit this feature creation tool.

All features in the model have been created and the model is now complete. The trimetric shaded view of the completed model is shown in Figure 5-74.

Saving the Model

1. Choose the **Save the active object** button from the **File** toolbar and save the model. The order of feature creation can be seen from the **Model Tree** shown in Figure 5-75.

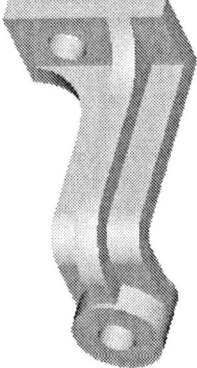

Figure 5-74 The default trimetric view of the model

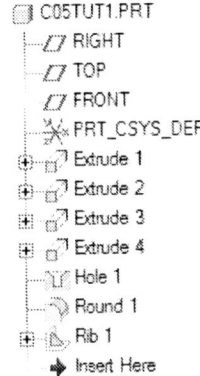

Figure 5-75 Model Tree for Tutorial 1

Tutorial 2

In this tutorial you will create the model shown in Figure 5-76. The dimensions of the model are shown in Figure 5-77.

(Expected time: 30 min)

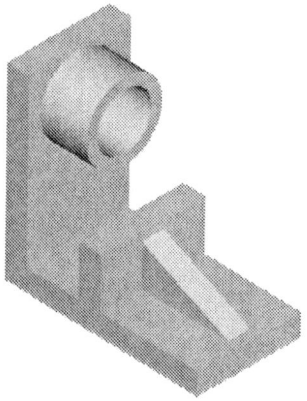

Figure 5-76 Isometric view of the solid model

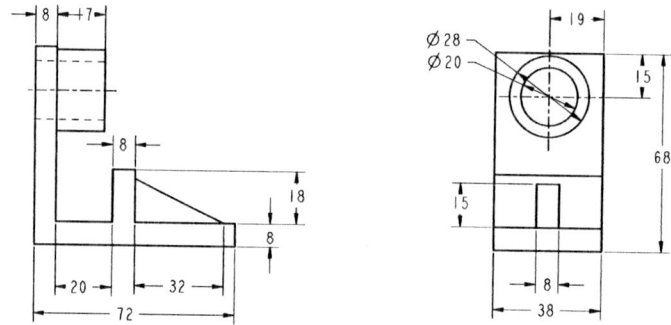

Figure 5-77 Front view and the right-side view of the model for Tutorial 2

The following steps are required to complete this model:

a. Examine the model and determine the number of features in it, refer to Figure 5-77.
b. Create the base feature, refer to Figures 5-78 and 5-79.
c. Create the cylindrical feature, refer to Figures 5-80 and 81.
d. Create the hole feature coaxially on the cylindrical feature, refer to Figure 5-83.
e. Create the rib feature, refer to Figures 5-84 and 5-85.

After understanding the procedure for creating the model, you are now ready to draw it. The working directory has already been selected in Tutorial 1 and therefore you do not need to select it again. However, if there is a need of changing the working directory, choose **File > Set Working Directory** and then select *c05* in the **Select Working Directory** dialog box.

Starting a New Object File

1. Open a new part file and name it as *c05tut2*. The three default datum planes and the **Model Tree** appears in the graphics window. Exit the **Model Tree**. The datum planes will not appear if they were previously turned off.

Creating the Base Feature

1. Choose the **Extrude Tool** button from the **Base Features** toolbar.

2. Choose the **Placement** tab and from the slide-up panel choose the **Define** button. The **Sketch** dialog box is displayed.

3. Select the **FRONT** datum plane for drawing the sketch of the base feature.

4. Select the **TOP** datum plane from the graphics window and then select the **Top** option from the **Orientation** drop-down list.

5. Choose the **Sketch** button from the **Sketch** dialog box to enter the sketcher environment.

6. Once you enter the sketcher environment, create the sketch of the base feature and apply constraints and dimensions, as shown in Figure 5-78.

7. After the sketch is complete, choose the **Continue with the current section** button and exit the sketcher environment.

   The **Extrude** dashboard is enabled and appears below the graphics window.

8. Enter a depth of **38** in the dimension box that is present on the **Extrude** dashboard.

   The default trimetric view of the base feature is shown in Figure 5-79.

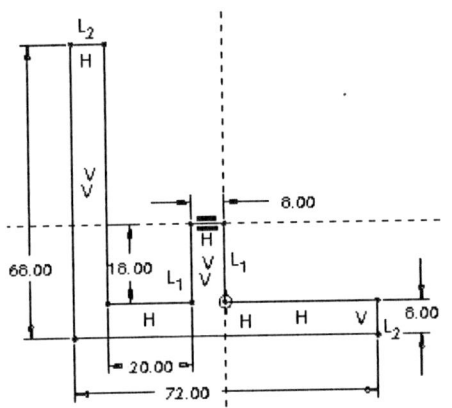
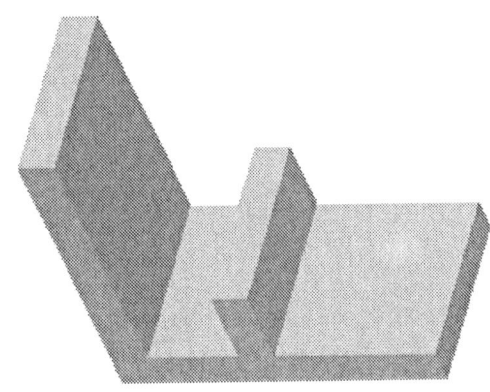

Figure 5-78 Sketch with dimensions and constraints for the base feature

Figure 5-79 The default trimetric view of the base feature

9. Choose the **Build feature** button from the **Extrude** dashboard to exit the feature creation tool.

Note

*In this tutorial the base feature can also be created by extruding it on both sides of the sketching plane, that is the **FRONT** datum plane. This would reduce the model creation by one feature. This is because, as evident from the model, later in the tutorial you will have to create a datum plane to create the rib feature. However, to make you familiar with creating datum planes on-the-fly, the base feature is extruded on one side of the sketching plane.*

Creating the Second Feature

The second feature is a cylindrical feature that is sketched on the planar surface of the base feature that is highlighted in Figure 5-80.

1. Choose the **Extrude Tool** button from the **Base Features** toolbar.

2. Choose the **Placement** tab and from the slide-up panel choose the **Define** button. The **Sketch** dialog box is displayed.

3. Select the face of the base feature shown in Figure 5-80 as the sketching plane.

4. Select the **TOP** datum plane from the graphics window and then select the **Top** option from the **Orientation** drop-down list.

5. Choose the **Sketch** button from the **Sketch** dialog box to enter the sketcher environment. After entering the sketcher environment, turn the model display to **No Hidden**.

6. Draw the sketch for the second feature. Apply and modify the dimensions, as shown in Figure 5-81.

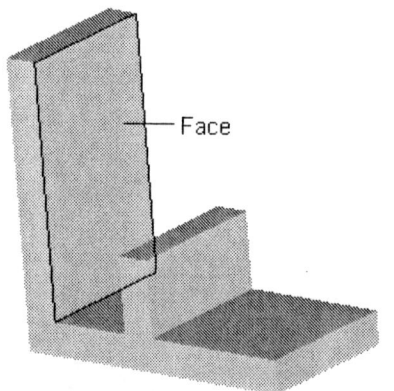
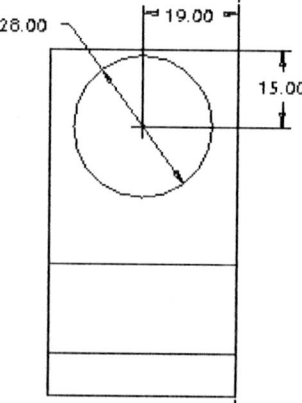

Figure 5-80 The face of the base feature selected as the sketching plane

Figure 5-81 Sketch and dimensions for the second feature

7. Turn the model display to **Shading**. Exit the sketcher environment by choosing the **Continue with the current section** button. The **Extrude** dashboard is enabled and appears below the graphics window.

8. Enter a value of **17** in the dimension box present on the **Extrude** dashboard.

Options Aiding Construction of Parts-I

9. Choose the **Build feature** button from the **Extrude** dashboard to exit the feature creation tool. The trimetric view of the model with the second feature is shown in Figure 5-82.

Creating the Hole Feature

The hole feature will be created using the **Hole** dashboard. The coaxial hole will be created on the cylindrical feature. The axis of the cylindrical feature will be used as axial reference to create the coaxial hole.

1. Choose the **Insert > Hole Tool** from the menu bar. The **Hole** dashboard is displayed. The **Create straight hole** button in the **Hole** dashboard is selected by default.

2. Choose the **Placement** tab from the **Hole** dashboard. The slide-up panel is displayed.

3. Select the front face of the cylindrical feature to place the hole.

   As you select the front face of the cylindrical feature, the preview of the hole is displayed in the graphics window. Now, you need to specify the reference for the hole placement.

4. From the drop-down list present in the slide-up panel, select the **Coaxial** option. You have specified the type of hole you need to create.

5. Click in the **Secondary References** collector; the **Primary** collector turns white in color.

6. Select the axis of the cylindrical feature from the graphics window.

7. In the diameter dimension box on the **Hole** dashboard, enter **20** and press ENTER.

8. From the depth flyout on the **Hole** dashboard, select the **Drill to intersect with all surfaces** button.

9. Choose the **Build feature** button from the **Hole** dashboard. The hole is created and the trimetric view of the shaded model with the hole is shown in Figure 5-83.

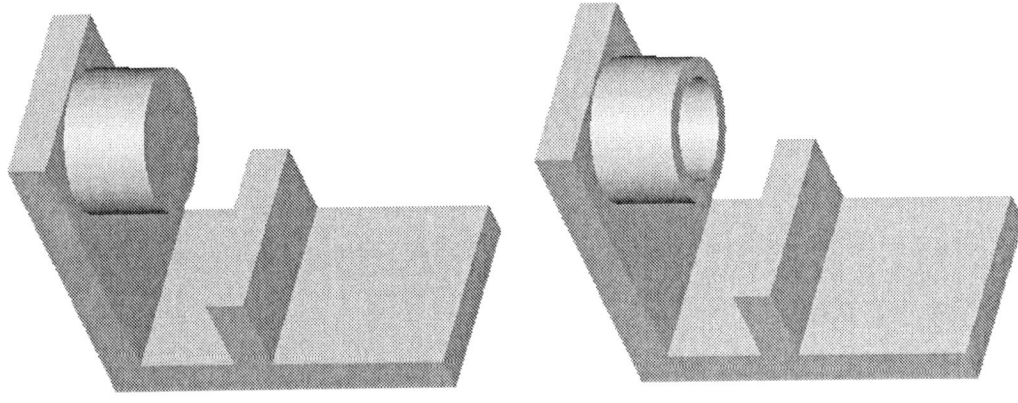

Figure 5-82 Model with the second feature   *Figure 5-83* Model with the hole feature

Creating the Rib Feature

To create the rib feature, a datum plane is created on-the-fly. As mentioned earlier, rib features are always drawn from the side view.

1. Choose the **Rib Tool** button from the **Engineering Features** toolbar.

2. Choose the **References** tab and from the slide-up panel choose the **Define** button. The **Sketch** dialog box is displayed.

3. Choose the **Datum Plane Tool** button from the **Datum** toolbar. The **DATUM PLANE** dialog box is displayed. You may need to move the **Sketch** dialog box to bring the **Datum Plane Tool** button into view.

4. Select the axis of the hole, press the CTRL key and select the **FRONT** datum plane. You can view your selections in the **References** collector.

5. Click the **FRONT** reference in the **References** collector. A drop-down list appears to the right of the reference. Select the **Parallel** option from this drop-down list.

6. Choose **OK** from the **DATUM PLANE** dialog box to exit it.

A datum plane that passes through the selected axis and is parallel to the **FRONT** datum plane will be created. This datum plane is selected automatically as the sketching plane. Now, you need to select the reference plane.

7. Select the **TOP** datum plane and then select the **Top** option from the **Orientation** drop-down list.

8. Choose the **Sketch** button; the system takes you to the sketcher environment.

9. Close the reference dialog box. Draw the open sketch of the rib feature and apply the dimensions, as shown in Figure 5-84. Exit the sketcher environment by choosing the **Continue with the current section** button. The **Rib** dashboard is enabled and appears below the graphics window.

10. Choose the **Flip** button on the **References** slide-up panel to flip the arrow. The preview of the rib is displayed on the model.

11. Enter a value of **8** in the edit box present on the **Rib** dashboard. This value is the thickness of the rib. Choose the **Build feature** button.

The trimetric view of the complete model with the rib feature is shown in Figure 5-85.

Saving the Model

1. Choose the **Save the active object** button from the **File** toolbar to save the model and then close the active window.

Options Aiding Construction of Parts-I 5-49

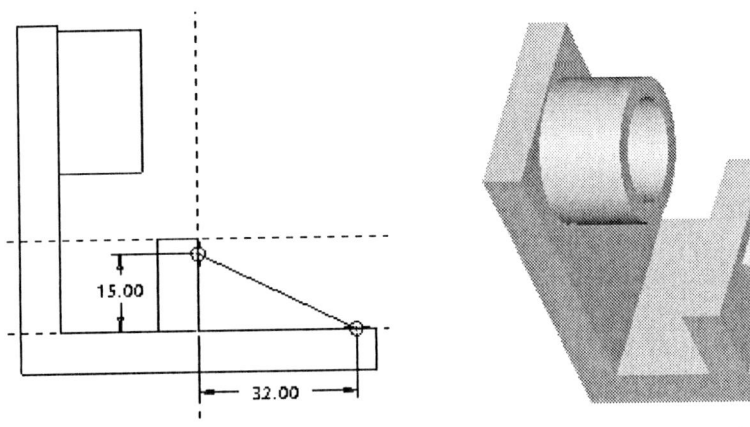

Figure 5-84 Sketch for the rib feature with the model display set to **No Hidden**

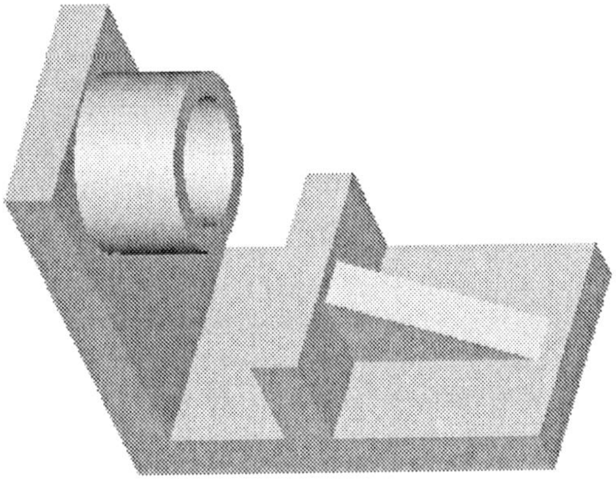

Figure 5-85 The default trimetric view of the final model

Tutorial 3

In this tutorial you will create the model shown in Figure 5-86. The solid model, dimensions, and the front and the right-side views are also shown in this figure. **(Expected time: 45 min)**

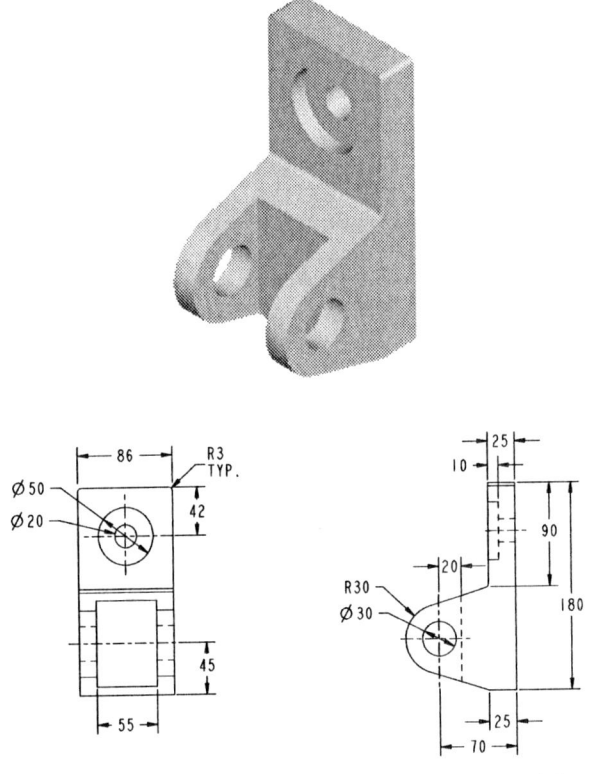

Figure 5-86 The isometric, front, and right-side views of the model

The following steps are required to complete this model:

a. Examine the model and determine the number of features in it, refer to Figure 5-86.
b. Create the base feature, refer to Figures 5-87 and 5-88.
c. Create the cut feature, refer to Figures 5-89 and 5-90.
d. Create a counterbore hole, refer to Figures 5-91 through 5-93.
e. Create the round, refer to Figures 5-94 and 5-95.

Starting a New Object File

1. Open a new part file and name it as **c05tut3**. The three default datum planes and the **Model Tree** appear in the graphics window if they were not turned off previously.

Creating the Base Feature

1. Choose the **Extrude Tool** button from the **Base Features** toolbar.

2. Choose the **Placement** tab and from the slide-up panel choose the **Define** button. The **Sketch** dialog box is displayed.

3. Select the **RIGHT** datum plane as the sketching plane.

4. Select the **TOP** datum plane from the graphics window and then select the **Top** option from the **Orientation** drop-down list.

5. Choose the **Sketch** button from the **Extrude** dashboard to enter the sketcher environment.

6. Once you enter the sketcher environment, create the sketch of the base feature and apply the dimensions and constraints, as shown in Figure 5-87.

7. Exit the sketcher environment by choosing the **Continue with the current section** button. The **Extrude** dashboard is enabled and appears below the graphics window.

8. Enter a value of **86** in the dimension box present on the **Extrude** dashboard.

9. Choose the **Build feature** button to exit the feature creation tool. The base feature is completed. The default trimetric view of the base feature is shown in Figure 5-88.

Creating the Second Feature

The second feature is an extruded cut feature. This cut feature is created on a datum plane that passes through the center of the base feature.

1. Choose the **Extrude Tool** button from the **Base Features** toolbar.

2. Choose the **Remove material** button.

3. From the depth flyout, choose the **Extrude on both sides of sketch plane by half the specified depth value in each direction** button.

Options Aiding Construction of Parts-I

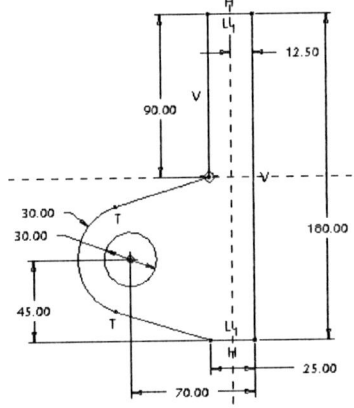 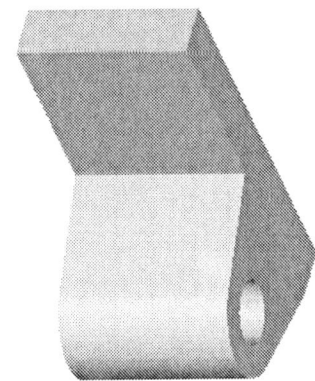

Figure 5-87 Sketch with dimensions and constraints for the base feature

Figure 5-88 The default trimetric view of the base feature

4. Choose the **Placement** tab and from the slide-up panel choose the **Define** button. The **Sketch** dialog box is displayed.

5. Choose the **Datum Plane Tool** button from the **Datum** toolbar. The **DATUM PLANE** dialog box is displayed.

6. Select the **RIGHT** datum plane. In the **Translation** edit box that appears in the dialog box, enter a value of **43**. This value is half of the width of the base feature.

7. Choose **OK** to exit the **DATUM PLANE** dialog box.

8. Select the **FRONT** datum plane and then select the **Left** option from the **Orientation** drop-down list.

9. Choose the **Sketch** button. The system takes you to the sketcher environment.

10. Draw the sketch for the cut feature and apply the dimensions and constraints, as shown in Figure 5-83.

    In Figure 5-83, some dimensions appear light in color. These dimensions are weak dimensions and it is not important to convert them into strong dimensions. These dimensions are not important for the creation of this feature. However, the geometry for the cut should be similar to that shown in Figure 5-89.

11. After the sketch is completed, exit the sketcher environment. The **Extrude** dashboard is enabled and appears below the graphics window.

12. Enter a depth of **55** in the dimension box present on the **Extrude** dashboard. The system will accept this depth symmetrical to the sketching plane.

13. Choose the **Build feature** button to exit the feature creation tool. The cut feature is

completed now and the default trimetric view of the cut feature along with the base feature is shown in Figure 5-90.

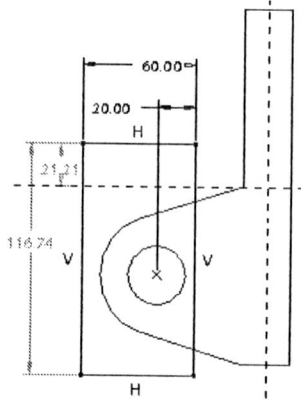

Figure 5-89 *Sketch with dimensions and constraints for the cut feature*

Figure 5-90 *The default trimetric view of the model*

Creating the Hole Feature

The hole is created using the **Hole** dashboard.

1. Choose the **Hole Tool** button from the **Engineering Features** toolbar. The **Hole** dashboard is displayed.

2. From the drop-down list present on the **Hole** dashboard, select the **Sketched** option.

3. Choose the **Activates Sketcher to create section** button to enter the sketcher environment.

4. In the sketcher environment, draw the sketch of the counterbore hole, as shown in Figure 5-91. After drawing the sketch, exit the sketcher environment.

5. Select the front face of the base feature to place the hole. The preview of the hole appears on the selected face.

6. Choose the **Placement** tab from the **Hole** dashboard. The slide-up panel is displayed.

7. Click in the **Secondary References** collector so that the **Primary** collector turns white in color.

8. Select the two references, as shown in Figure 5-92 for dimensioning. Note that the second reference should be selected using CTRL+left mouse button.

9. Modify the dimensions of the references by entering their values in the **Placement** slide-up panel. The center of hole is at the distance of **42** from the top edge and **43** from the right edge. After viewing the preview of the hole you may have to specify negative value of any one of the distances.

Options Aiding Construction of Parts-I 5-53

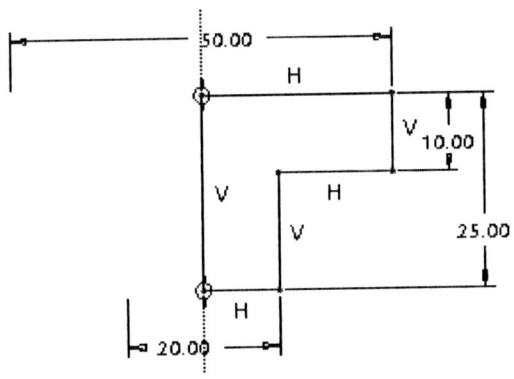

Figure 5-91 Cross-section of the hole with dimensions and constraints

Figure 5-92 Edges to be selected for placing hole

10. Choose the **Build feature** button to exit this feature creation tool. The hole will be created, as shown in Figure 5-93.

Note
When you select a feature, entity, or item from the graphics window and right-click to invoke the shortcut menu, the right mouse button should be held down until the shortcut menu is displayed. Once the shortcut menu is displayed, the right mouse button can be released.

Creating the Round Feature

The round is created using the **Simple** option.

1. Choose the **Round Tool** button from the **Engineering Features** toolbar. The **Round** dashboard is displayed.

2. Using the CTRL key, select the edges that are shown in Figure 5-94. The selected edges turns red in color.

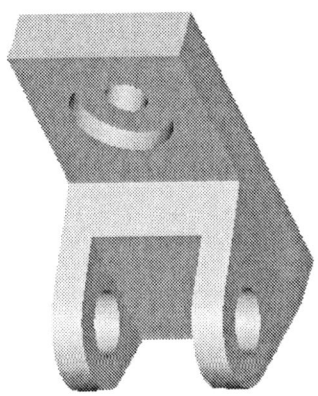
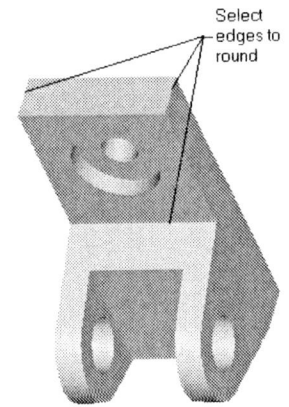

Figure 5-93 The default trimetric view

Figure 5-94 Edges to be selected to round

3. Enter **3** in the dimension box that is present on the **Round** dashboard or double-click on the default value displayed on the preview of the round and enter the radius value in the displayed edit box. The round is created and its preview appears on the model.

4. Choose the **Build feature** button to exit the round feature creation tool. The trimetric view of the final model with round feature created is shown in Figure 5-95.

Saving the Model

1. Choose the **Save the active object** button from the **File** toolbar and save the model.

   The order of feature creation can be seen from the **Model Tree** shown in Figure 5-96.

Figure 5-95 Final model for Tutorial 3

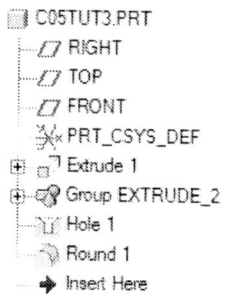
Figure 5-96 Model Tree for Tutorial 3

Tutorial 4

Create the model shown in Figure 5-91. The dimensions, the top view, the front section view, and the right section view of the model are shown in Figure 5-92.   **(Expected time: 1 hr)**

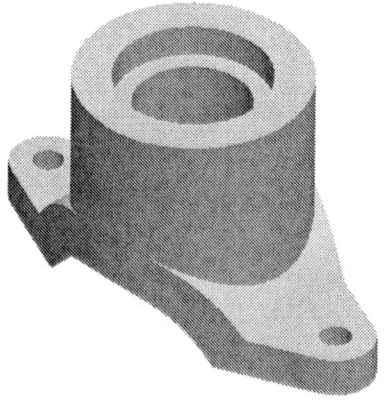
Figure 5-97 Isometric view of the model

Options Aiding Construction of Parts-I   5-55

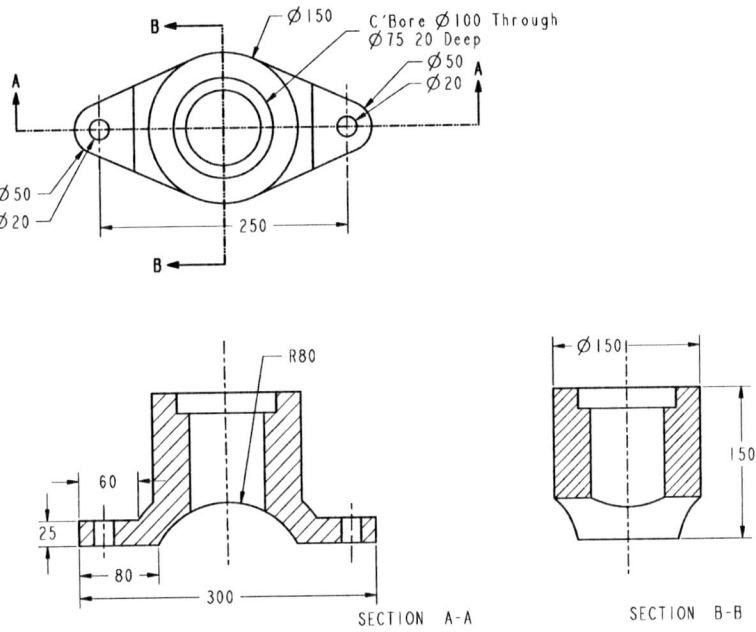

Figure 5-98 Top view, front section view, and right section view of the model

The following steps are required to complete this model:

a. Examine the model and determine the number of features in it, refer to Figure 5-97.
b. Create the base feature, refer to Figures 5-99 and 5-100.
c. Create the cut feature, refer to Figures 5-101 through 5-103.
d. Create the cylindrical feature on an offseted datum plane, refer to Figure 5-104 through 5-106.
e. Create a counterbore hole coaxially on the cylindrical feature, refer to Figures 5-107 and 5-108.
f. Create straight holes on the top face of base feature, refer to Figures 5-109 and 5-110.

Starting a New Object File

1. Open a new part file and name it as *c05tut4*. The three default datum planes are displayed in the graphics window if the **Datum planes on/off** button is turned on. If the **Model Tree** is open then close it by selecting the sash on its right edge.

Selecting the Sketching Plane for the Base Feature

In this model, you need to draw the base feature on the **FRONT** datum plane because the direction of extrusion of the base feature is perpendicular to the **FRONT** datum plane.

1. Choose the **Extrude Tool** button from the **Base Features** toolbar.

2. Choose the **Placement** tab and from the slide-up panel choose the **Define** button. The **Sketch** dialog box is displayed.

3. Select the **FRONT** datum plane as the sketching plane. As you select the sketching plane, the **RIGHT** datum plane and its orientation **Right** are set automatically.

4. Choose the **Sketch** button to enter the sketcher environment.

Specifying References

In the sketcher environment, the **References** dialog box is displayed. The status displayed in the **Reference status** area is **Fully Placed**. Close this dialog box.

Creating and Dimensioning the Sketch for the Base Feature

From the model, the section to be extruded for the base feature is not evident. Therefore, you need to visualize the sketch for the base feature. The sketch is shown in Figure 5-93. When this sketch is extruded, it will create the base feature.

1. Draw the section sketch using various sketcher tools, as shown in Figure 5-99.

2. The sketch is dimensioned automatically and some weak dimensions are assigned to it. Add the required constraints and modify the weak dimensions, as shown in Figure 5-99.

3. Choose the **Continue with the current section** button. The **Extrude** dashboard is displayed.

4. Choose the **Default Orientation** option from the **Saved view list** button of the **View** toolbar.

    The default view is displayed which gives you a better view of the sketch in the 3D space. A yellow arrow is also displayed on the model, indicating the direction of extrusion.

5. Enter a value of **150** in the dimension box that is present on the **Extrude** dashboard.

6. Choose the **Build feature** button from the **Extrude** dashboard.

    The base feature is completed, as shown in Figure 5-100. You can use the middle mouse button to spin the model to view it from various directions.

Creating the Second Feature

The next feature is an extruded cut feature. The sketching plane for this feature is the bottom face of the base feature. To get the required shape that is at the base of the model, you need to cut the base feature in such a way such that you get the required shape. This sketch and its dimensions can be referred to from the top view of the model shown in Figure 5-101. Before drawing the sketch, change the model display to **No Hidden**.

1. Choose the **Extrude Tool** button from the **Base Features** toolbar.

Options Aiding Construction of Parts-I 5-57

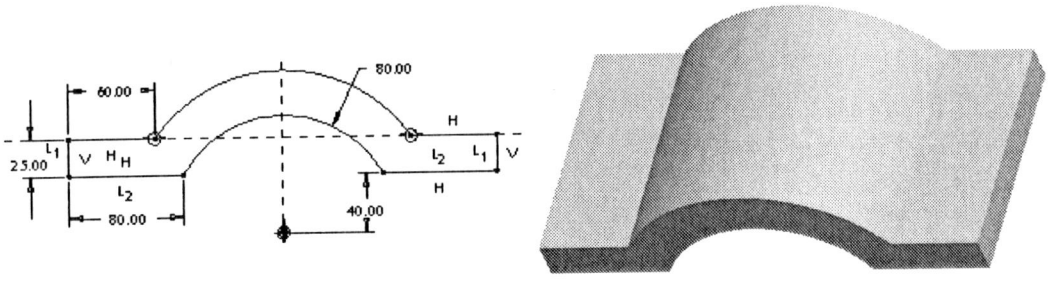

Figure 5-99 Sketch with dimensions and constraints

Figure 5-100 Default view of the base feature

2. Choose the **Remove Material** button from the Extrude dashboard.

3. Choose the **Placement** tab and from the slide-up panel choose the **Define** button. The **Sketch** dialog box is displayed.

4. Select the bottom face of the base feature as the sketching plane.

5. Select the **RIGHT** datum plane and from the **Orientation** drop-down list, choose the **Right** option.

6. Choose the **Sketch** button to enter the sketcher environment.

7. Draw the sketch of the cut feature, and apply constraints and dimensions, as shown in Figure 4-101.

8. Choose the **Continue with the current section** button to exit the sketcher environment. The **Extrude** dashboard is displayed below the graphics window.

9. Choose the **Shading** button from the **Model Display** toolbar.

10. Choose the **Default Orientation** option from the **Saved view list** drop-down list. The model orients in its default orientation.

11. Choose the **Change material direction of extrude to other side of sketch** button from the **Extrude** dashboard. The yellow arrow points in the direction, as shown in Figure 5-102.

12. Choose the **Options** tab in the **Extrude** dashboard to display the slide-up panel. From the **Side 1** drop-down list, choose the **Through All** option. The model can be previewed in the graphics window.

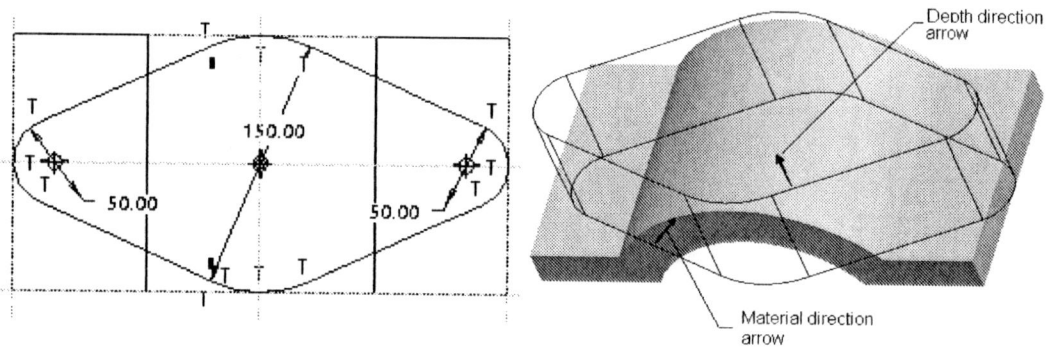

Figure 5-101 Sketch with dimensions and constraints for the cut feature

Figure 5-102 Two arrows and the preview of the cut feature

13. Choose the **Build feature** button from the **Extrude** dashboard to complete the feature creation. The model after creating the cut feature is shown in Figure 5-103.

Creating the Sketching Plane for the Third Feature

You need to create a datum plane to create the third feature. The datum plane will be at a distance of 150 from the bottom face of the model.

1. Choose the **Datum Plane Tool** button from the **Datum** toolbar. The **DATUM PLANE** dialog box is displayed.

2. Spin the model using the middle mouse button and then using the left mouse button, select the bottom face of the model.

   When you select the bottom face of the model, the **Offset** constraint is displayed in the **References** collector of the dialog box. Note that the yellow arrow is pointing in the opposite direction. Therefore, you need to enter a negative value of the offset distance.

3. In the **Translation** dimension box, type a value of **-150** and press ENTER.

4. Choose the **OK** button from the **DATUM PLANE** dialog box.

   Datum plane **DTM1** is created, as shown in Figure 5-104 and will be selected as the sketching plane for creating the sketch.

Creating the Third Feature

The plane **DTM1** will be selected as the sketching plane and the depth of extrusion will be given from this plane. The third feature is cylindrical in shape and its outer edge is tangent to the edge of the base feature.

1. Choose the **Extrude Tool** button from the **Base Features** toolbar.

Options Aiding Construction of Parts-I

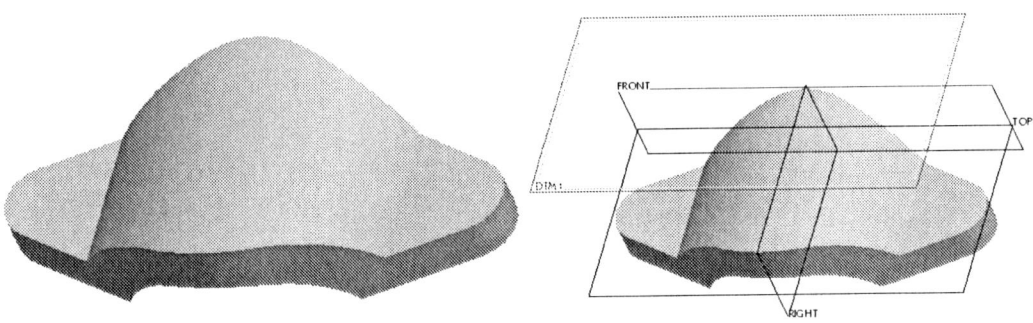

Figure 5-103 Model after creating the cut feature   *Figure 5-104* DTM1 created at an offset distance

2. Choose the **Placement** tab and from the slide-up panel choose the **Define** button. The **Sketch** dialog box is displayed.

3. Select **DTM1** as the sketching plane. The yellow arrow appears on the datum plane.

4. Select the **FRONT** datum plane and then from the **Orientation** drop-down list, choose the **Top** option.

5. Choose the **Sketch** button to enter the sketcher environment.

6. Draw a concentric circle for the cylindrical feature, as shown in Figure 5-105.

7. Choose the **Continue with the current section** button to exit the sketcher environment.

8. Choose the **Default Orientation** option from the **Saved view list** drop-down list. The model orients in its default orientation.

9. Choose the **Options** tab in the **Extrude** dashboard to display the slide-up panel. From the **Side 1** drop-down list, choose the **To Next** option. The model can be previewed in the graphics window.

10. Choose the **Build feature** button from the **Extrude** dashboard to complete the feature creation. The model after creating the cut feature is shown in Figure 5-106.

Creating the Counterbore Hole

The fourth feature is a counterbore hole that can be easily created using the **Hole** feature creation tool.

1. Choose the **Hole Tool** button from the **Engineering Features** toolbar. The **Hole** dashboard is displayed.

2. From the drop-down list present on the **Hole** dashboard, select the **Sketched** option.

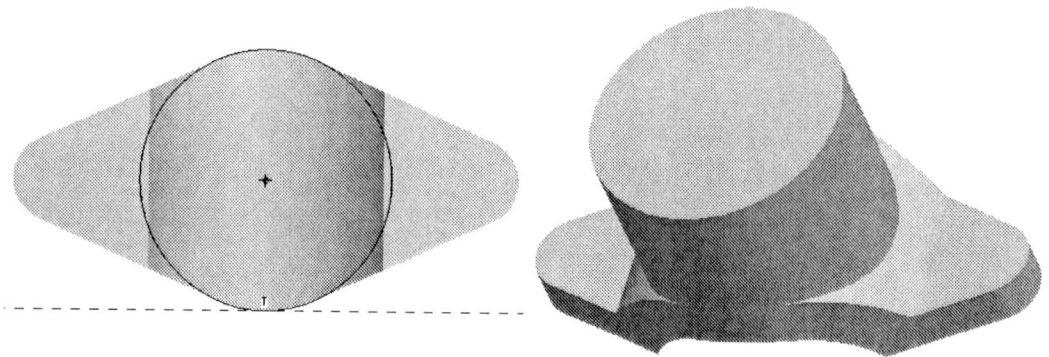

Figure 5-105 Sketch for the cylindrical feature

Figure 5-106 Model after creating the cylindrical feature

3. Choose the **Activates Sketcher to create section** button to enter the sketcher environment.

4. In the sketcher environment, draw the sketch of the counterbore hole, as shown in Figure 5-107. After drawing the sketch, exit the sketcher environment.

5. Select the top face of the cylindrical feature to place the hole. The preview of the hole appears on the selected face.

6. Choose the **Placement** tab from the Hole dashboard. The slide-up panel is displayed.

7. Select the **Coaxial** option from the drop-down list in the slide-up panel.

8. Click in the **Secondary References** collector so that the **Primary** collector turns white in color.

9. Select the axis of the cylindrical feature to which the hole will be coaxial. The preview of the hole can be viewed on the model.

10. Choose the **Build feature** button to exit the **Hole** dashboard. The counterbore hole is shown in Figure 5-108.

Creating the Datum Axis for the Hole features

The last feature is a pair of two holes that are on the top face of the base feature. The two holes will be placed coaxially with the two axes that you need to create.

1. Choose the **Datum Axis Tool** button from the **Datum** toolbar. The **DATUM AXIS** dialog box is displayed.

2. Select the curved surface shown in Figure 5-109 to create the datum axis.

Options Aiding Construction of Parts-I 5-61

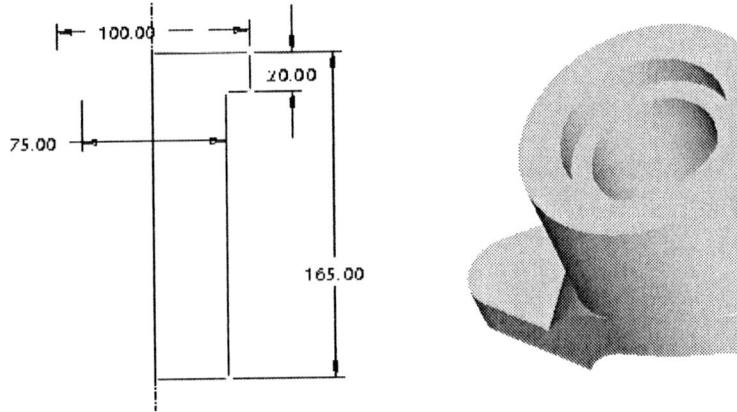

Figure 5-107 Sketch with dimensions

Figure 5-108 Model with counterbore hole

3. Similarly, create a datum axis on the left of the model. Figure 5-110 shows the model after creating the two datum axes.

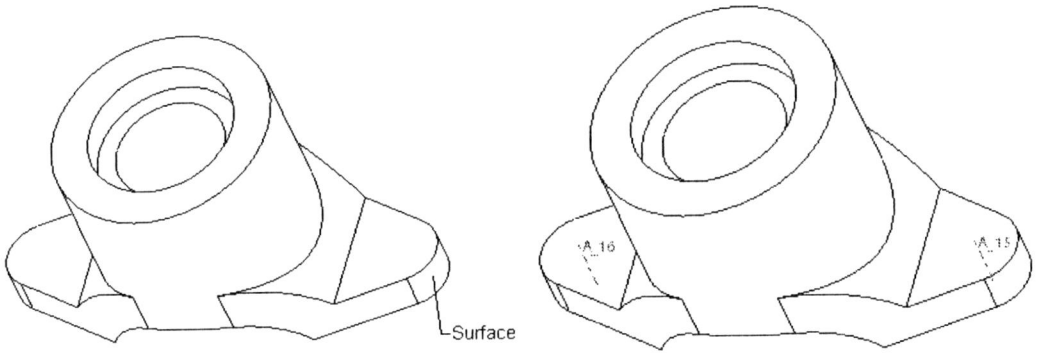

Figure 5-109 Surface to select for creating the datum axis

Figure 5-110 The two datum axis

Creating Holes

To create the two coaxial holes, you need to invoke the **Hole** dashboard twice. This is because only one hole can be created at a time by using the **Hole** dashboard.

1. Choose the **Hole Tool** button from the **Engineering Features** toolbar. The **Hole** dashboard is displayed.

2. Select the top face of the base feature. The preview of the straight hole appears on the selected face.

3. Choose the **Placement** tab from the **Hole** dashboard. The slide-up panel is displayed.

4. Select the **Coaxial** option from the drop-down list.

5. Click in the **Secondary References** collector so that the **Primary** collector turns white in color.

6. Select the datum axis to which the hole will be coaxial.

7. Enter the diameter of hole as **20** in the dimension box present on the **Hole** dashboard.

8. Choose the **Build feature** button to exit the **Hole** dashboard.

9. Again, invoke the **Hole** dashboard and similar to the first hole create the second hole.

The two holes on the base feature are shown in Figure 5-111.

Saving the Model

1. Choose the **Save the active object** button from the **File** toolbar and save the model. The order of feature creation can be seen from the **Model Tree** shown in Figure 5-112.

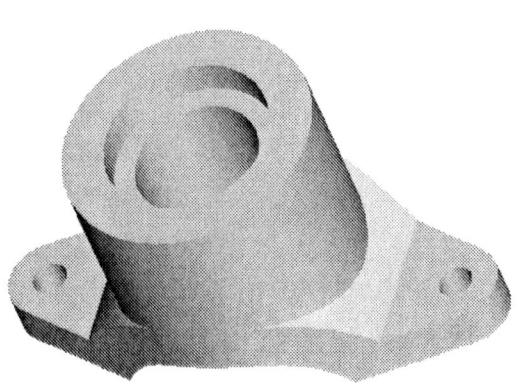

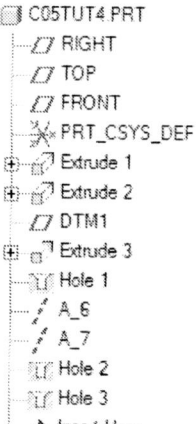

Figure 5-111 Model after creating the two holes

Figure 5-112 Model Tree *of Tutorial 4*

Self-Evaluation Test

Answer the following questions and then compare your answers with those given at the end of this chapter.

1. A hole created using the **Hole** dashboard is parametric in nature. (T/F)

2. A hole cannot be created on both sides of the sketching plane or the placement plane. (T/F)

3. In Pro/ENGINEER, holes can also be sketched. (T/F)

4. The **Full Round** button in the **Sets** slide-up panel allows you to enter the radius of round. (T/F)

Options Aiding Construction of Parts-I

5. The **Model Tree** is used extensively in Pro/ENGINEER for editing a model. (T/F)

6. The rib feature is always created from the _____ view.

7. A _____ hole is a stepped hole and has two diameters, a larger one and a smaller one.

8. Straight holes are the holes that have circular cross-section having _____ diameter throughout the depth.

9. _____ are defined as thin wall-like structures used to bind the joints together so that they do not fail under an increased load.

10. By default, the suppressed features are not displayed in the _____.

Review Questions

Answer the following questions:

1. If the feature that is suppressed has some child features then they will also be suppressed. (T/F)

2. While in the sketcher mode, the constraints to a sketch should be applied before the dimensions. (T/F)

3. The chamfers created in Pro/ENGINEER are parametric in nature. (T/F)

4. The **Model Tree** can be used to redefine a feature. (T/F)

5. When you redefine a rib feature, the _____ dashboard is displayed.

6. The sketch of a rib feature can be extruded to _____ side(s) of the sketching plane.

   (a) one   (b) both
   (c) either   (d) All of the above

7. _____ is a stepped hole with two diameters.

   (a) Counterbore   (b) Countersink
   (c) Straight   (d) None

8. Which of the following option of the **Menu Manager** is used to change the sequence in which the features are created?

   (a) **Feature**   (b) **Modify**
   (c) **Regenerate**   (d) **Reorder**

9. At the intersection of how many edges does a **Corner** chamfer create a beveled surface?

(a) One (b) Two
(c) Three (d) None

10. Rounds and chamfers are used in engineering components to reduce the _____ on the corners.

(a) Stress concentration (b) Tension
(c) Conduction (d) None

Exercises

Exercise 1

Create the model shown in Figure 5-113. The dimensions and front and top views of the model are shown in Figure 5-114.   **(Expected time: 45 min)**

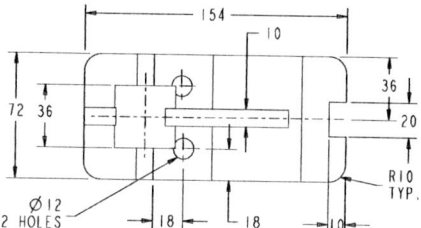

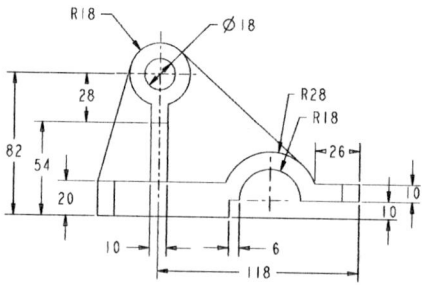

Figure 5-113 Isometric view of the model

Figure 5-114 Top and front views of the model; hidden lines are suppressed for clarity

Exercise 2

Create the model shown in Figure 5-115. The dimensions and front and right-side views of the model are shown in Figure 5-116. **(Expected time: 30 min)**

Figure 5-115 Isometric view of the model for Exercise 2

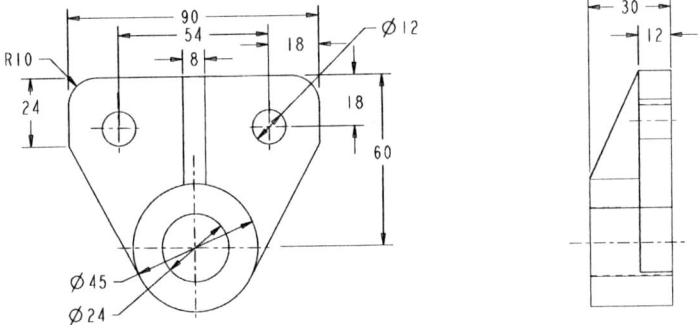

Figure 5-116 Front view and the right-side view of the model

Exercise 3

Create the model shown in Figure 5-117. The dimensions and front and right-side views of the model are shown in Figure 5-118. **(Expected time: 30 min)**

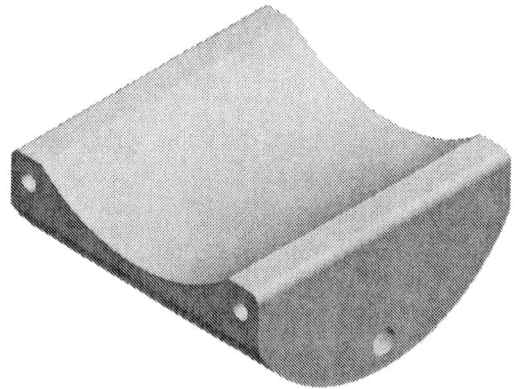

Figure 5-117 Isometric view of the model for Exercise 3

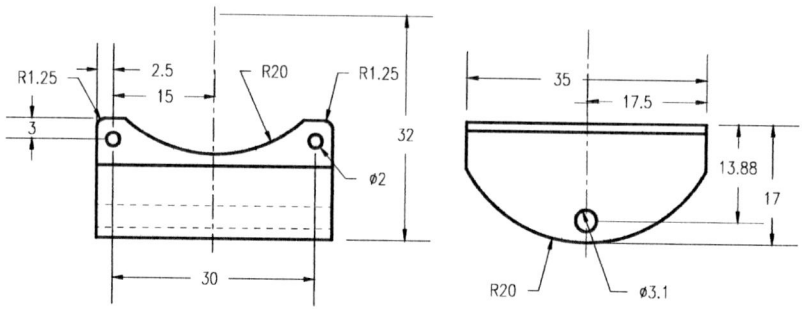

Figure 5-118 Front view and the right-side view of the model

Answers to the Self-Evaluation Test

1 - T, **2** - F, **3** - T, **4** - F, **5** - T, **6** - side, **7** - counterbore, **8** - constant, **9** - Ribs, **10** - **Model Tree**.

Chapter 6

Options Aiding Construction of Parts-II

Learning Objectives

After completing this chapter you will be able to:
- *Create Dimension pattern.*
- *Create Direction pattern.*
- *Create Axis pattern.*
- *Create Fill pattern.*
- *Create Reference pattern.*
- *Create Table-driven pattern.*
- *Control the size of the pattern instances using constrains in sketcher environment.*
- *Use the Copy option.*
- *Use the Move option.*
- *Use the Mirror option.*
- *Use the Mirror Tool button.*
- *Create section of a model.*

In this chapter you will learn about various methods of duplicating the existing features. In Pro/ENGINEER Wildfire 2.0, you can duplicate a feature by using the following methods:

- **Pattern**
- **Copy**
- **Mirror**

You will also learn to create sections in solid models. A solid model is sectioned for viewing the profile of its cross section and also for creating section drawing views.

CREATING FEATURE PATTERNS

Patterns are one or two dimensional incremental array of features created from a single feature called the parent feature or the leader. When a pattern is created, the leader also becomes a part of the pattern. When you pattern a feature, you need to specify the total number of features to be created, including the one that is being patterned and the increment in the dimensions if required.

Uses of Patterns

Patterns are very helpful in solid modeling as they speed up the model creation process. The uses of patterns in solid modeling are discussed next.

1. Patterns create multiple copies of a feature, and hence save time that would otherwise be spent in creating the features individually.

2. All instances in a pattern, including the parent feature, act as a single feature. Therefore, they can be easily suppressed or mirrored.

3. All instances in a pattern are related parametrically. Hence, you can modify the number of instances in a pattern, the spacing between the instances, and other pattern related parameters.

4. If the dimensions of the parent feature are modified, then the dimensions of the child features are also modified.

Creating Patterns

In Pro/ENGINEER, patterns are created by choosing the **Pattern Tool** button from the **Edit Features** toolbar present in the **Right Toolchest**. This button is object-action tool. This means that this button is enabled in the **Edit Features** toolbar only when you have selected the feature to be patterned.

To invoke the pattern tool from the menu bar, choose **Edit > Pattern**. The **Pattern** dashboard is displayed, as shown in Figure 6-1. The options and the tools available on the dashboard depend on the type of pattern you are creating. There are six types of patterns that can be created in Pro/ENGINEER, they are, **Dimension, Direction, Axis, Fill, Table**, and **Reference** patterns. The options and tools available in the Pattern dashboard are discussed next.

Options Aiding Construction of Parts-II

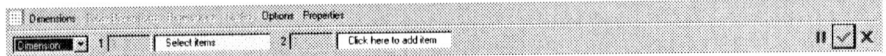

Figure 6-1 The **Pattern** *dashboard*

Drop-down list

In the drop-down list present on the **Pattern** dashboard, the **Dimension** option is selected by default. The other options in this drop-down list are **Direction**, **Axis**, **Fill**, **Table**, and **Reference**. When you select the options from this drop-down list, the options and tools available in the **Pattern** dashboard change accordingly. The types of patterns that can be created in Pro/ENGINEER are discussed next:

Dimension Patterns

In dimension patterns, the existing dimensions of the parent feature are used to create a pattern. This pattern can be created in either one direction or in two directions. When you select the option to create the pattern in the second direction, all instances that were created in the first direction can also be created in the second direction. You need to specify the increment value for the instances, which can be either positive or negative. Once you have specified the increment value in a direction, the system creates the specified number of instances (including the parent feature) in that direction. Figure 6-2 shows a hole that is patterned and the patterned hole is shown in Figure 6-3.

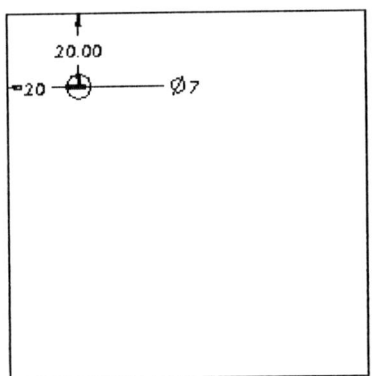

Figure 6-2 Part with a hole

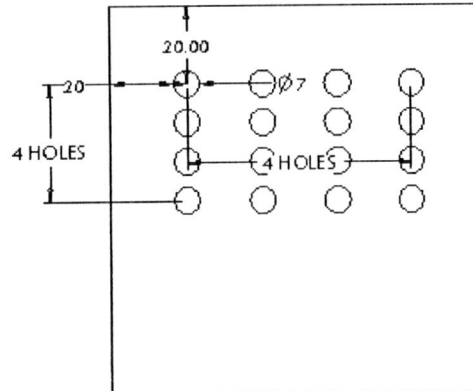

Figure 6-3 Hole patterned in two directions

To create a dimension pattern, use the following steps:

1. Select the feature to be patterned and then choose the **Pattern Tool** button.
2. Click the dimension in the first direction. The edit box is displayed; enter the increment in the dimension where you need to place the second instance of the pattern.
3. In the **1** edit box present on the dashboard, enter the number of instances. The instances in the first direction are specified and now you need to specify the number of instances and the increment in dimension in the second direction. Note that the dimension in the second direction should be selected only if you need to create instances in the second direction.

4. Click on **Click here to add item** in the **2** collector. It displays **Select items**. Now, click the dimension in the second direction; the edit box appears. Enter the dimension increment in the edit box.
5. In the **2** edit box, enter the number of instances needed in the second direction. You can view the black dots where the instances of the pattern will be placed.
6. Choose the **Build feature** button to exit the tool and to view the pattern.

Direction Patterns

The **Direction** option is used to create pattern in the specified direction. The direction is specified by selecting an edge, a plane, or a linear curve. This type of pattern is different from the **Dimension** pattern in the way that this type of pattern does not use the dimensions of the feature to be patterned. When you select this option from the drop-down list, you will be prompted to select the reference for the first direction. An arbitrary dimension that is not the dimension of the parent feature is displayed in the specified direction. After specifying the first direction you can specify the second direction in which you need to place the instances.

Note that the direction of pattern can be reversed by selecting the **Flip** button present on the **Pattern** dashboard. Figure 6-4 shows the linear curve that is selected to create the directional pattern. The direction is indicated by the arrow. The preview of the pattern instances are shown by black dots. The pattern is created, as shown in Figure 6-5.

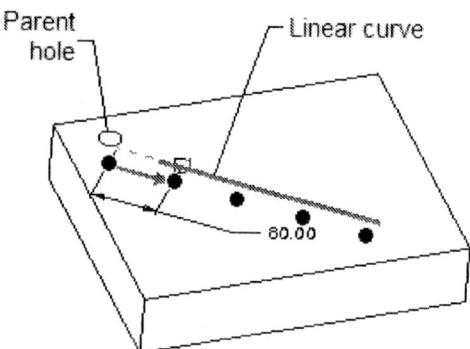
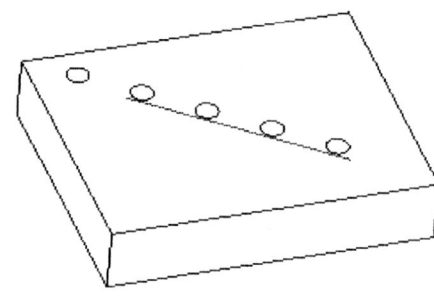

Figure 6-4 Linear curve selected to create a direction pattern

Figure 6-5 Direction pattern

Note
*The reference that you select for specifying the direction of a **Direction** pattern should belong to a feature created before the feature to be patterned, otherwise, the desired reference will not be highlighted for selection.*

Axis Patterns

The **Axis** option is used to create rotational patterns. When you choose this option the **Pattern** dashboard is displayed, as shown in Figure 6-6. To create this type of pattern an axis is required. This axis can be a datum axis or the axis of a feature, for example, an

Options Aiding Construction of Parts-II

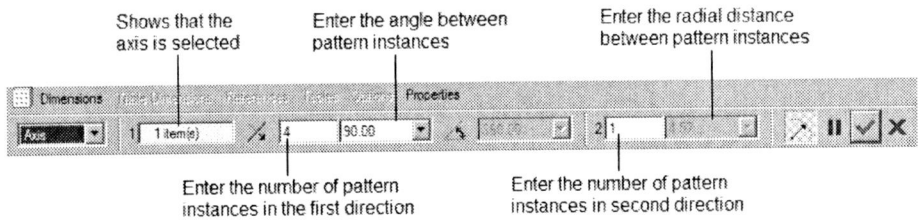

*Figure 6-6 The **Pattern** dashboard with the **Axis** option selected*

axis formed by a revolve feature, like a hole. The parameters that you need to specify in order to create an axis pattern are shown in Figure 6-7.

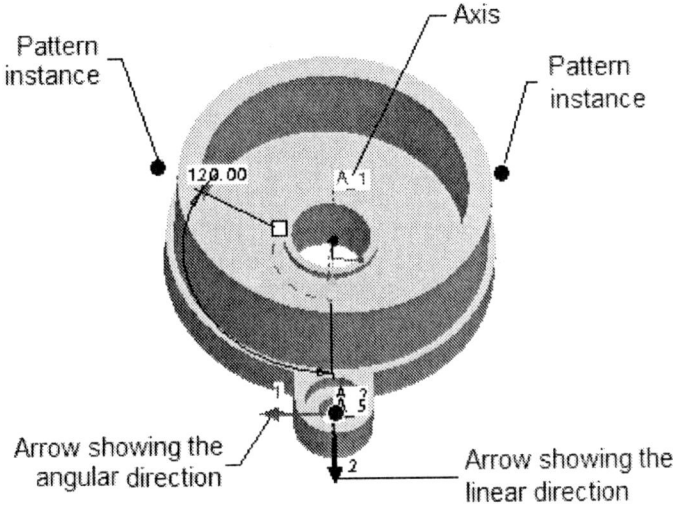

Figure 6-7 Parameters for the axis pattern

Note
Remember that for creating the axis pattern of a feature, the reference axis must be created before the feature that is to be patterned.

To create an axis pattern, use the following steps:

1. From the drop-down list present on the **Pattern** dashboard, select the **Axis** option. You are prompted to select an axis.
2. Turn on the display of axis and select the axis.
3. Enter the required angle between each instance and the then number of pattern instances in the edit box.

The instances in the first direction are specified and now you can specify the number of instances and the increment in dimension in the second direction. Note that the number of instances and dimension in the second direction should be specified only if you need to create instances in the second direction. If you only need to create the rotational pattern, as shown in Figure 6-8 then you just need to create instances in the first direction. The instances in the second direction along with the instances in the first direction is shown in Figure 6-9.

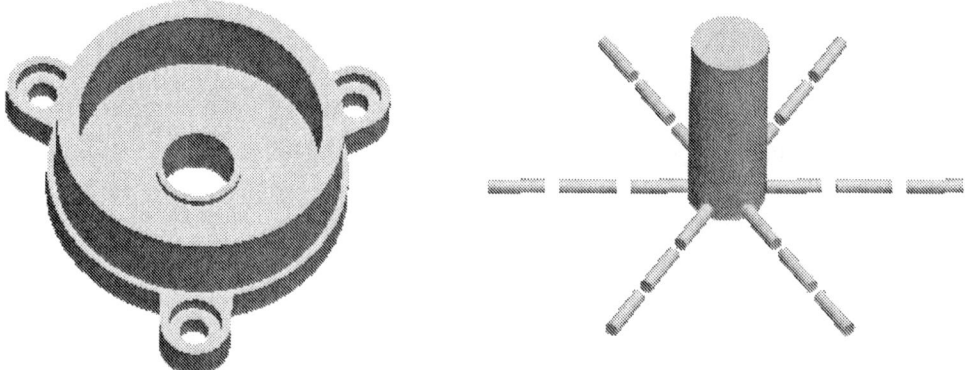

Figure 6-8 Rotational pattern created using the axis

Figure 6-9 Pattern created in both circular and linear directions

4. Enter the number of pattern instances in the **2** edit box for the second direction and then enter the radial distance between pattern instances in the edit box. The pattern is created and now you can view the black dots where the instances will be placed.
5. Choose the **Build feature** button to exit the tool and view the pattern.

Fill Patterns

Fill patterns are used to fill the sketched area by a selected feature. This is the easiest and fastest method to create pattern in the sketched area. When you select the **Fill** option from the drop-down list, the **Pattern** dashboard appears, as shown in Figure 6-10.

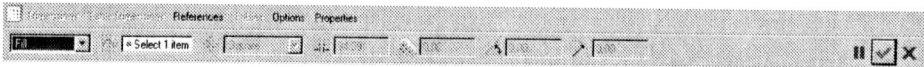

Figure 6-10 The **Pattern** dashboard

The options and tools available in this dashboard are discussed next.

References tab. When you choose this tab the slide-up panel is displayed in which the **Define** button is available. Choose the **Define** button to invoke the **Sketch** dialog box. This dialog box is used to sketch the area that will be filled by the pattern instances. Select the sketching plane and its orientation and then enter the sketcher environment to draw the sketch.

Drop-down list. This drop-down list is available only when the area to be filled is defined. The options in this drop-down list are used to specify the shape of the fill pattern. The options are **Square, Diamond, Triangle, Circle, Curve**, and **Spiral**. Generally, it is advised that the sketch should be similar to the shape of the fill pattern you need. Figures 6-11 through 6-16 show the patterns of various shapes.

Options Aiding Construction of Parts-II 6-7

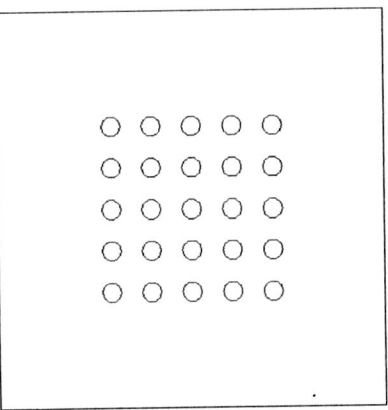

Figure 6-11 *Fill pattern in **Square** shape on the square sketched area*

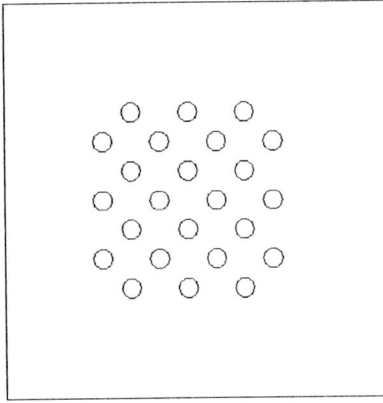

Figure 6-12 *Fill pattern in **Diamond** shape on the square sketched area*

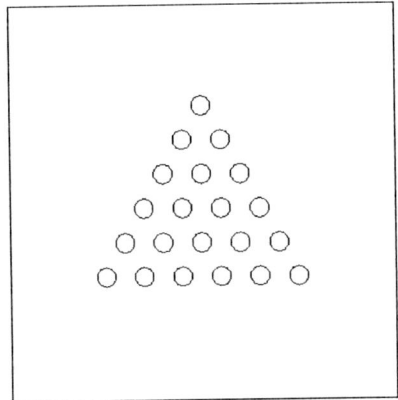

Figure 6-13 *Fill pattern in **Triangle** shape on the sketched triangle*

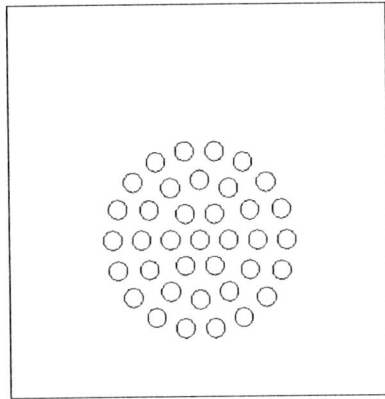

Figure 6-14 *Fill pattern in **Circle** shape on the sketched circle*

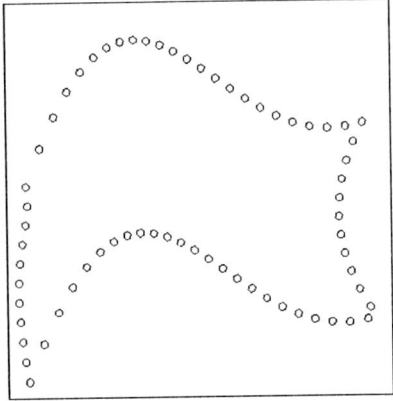

Figure 6-15 *Fill pattern in **Curve** shape on the sketched curve*

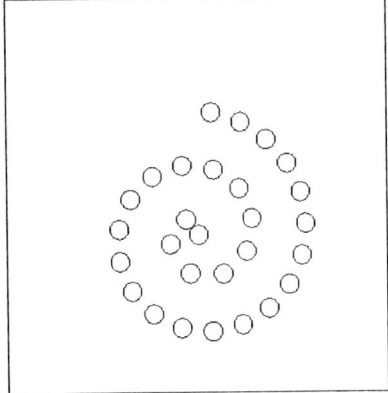

Figure 6-16 *Fill pattern in **Spiral** shape on the sketched circle*

Dimension boxes. The dimension boxes available on the dashboard are used to specify the different parameters of the fill pattern. You can specify the gap between the two members of the pattern, the distance of all members (instances) from the sketched boundary, rotation angle of the pattern, and radial distance between instances in case of circular and spiral shapes.

Tip: *When you create a Fill pattern, you can select the member of the pattern that you want to remove from the pattern. This can be done before exiting the Pattern dashboard. To remove the member of a pattern, select it in the preview of the Fill pattern. The selected member turns white, suggesting that it is removed. To resume the removed member of the pattern, select it once again.*

Reference Patterns

In the reference patterns, an existing pattern is referenced to create a new pattern. In this type of pattern, the parent feature of the new pattern should be referenced to the parent feature of the existing pattern.

Figure 6-17 shows a rib feature that is referenced to the parent hole feature. In this figure, the parent rib feature is created on a plane that was created on-the-fly while creating the rib feature. This plane passes through the axis of the parent hole feature. Hence, a relationship is built between the parent rib feature and the parent hole feature. Note that a group is formed when the rib feature is created using a datum on-the-fly. For creating the pattern of the rib feature, you have to select the whole group from the **Model Tree**. By selecting only the rib feature, you cannot create its reference pattern.

To create the reference pattern of the rib feature, select its group from the **Model Tree** and choose the **Reference Tool** button from the **Edit Features** toolbar to invoke the **Pattern** dashboard. The Reference option is selected by default in the Pattern dashboard. Click on the **Build feature** button, the pattern of the rib feature is created without specifying the increment in dimensions. The Figure 6-18 shows the rib feature patterned using the reference of the hole pattern.

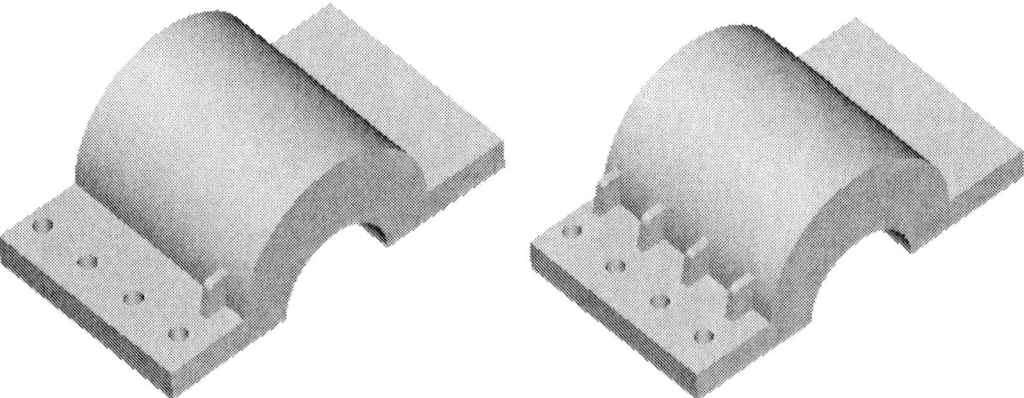

Figure 6-17 Rib feature referenced to the hole feature   *Figure 6-18* Rib feature is reference patterned

Options Aiding Construction of Parts-II 6-9

Note
*If you create the datum plane passing through the axis of the hole using the **Datum Plane Tool** button and then create the rib feature on it, the rib feature will not be patterned with reference to the holes. This is because the rib does not have any direct relation with the hole. The rib has relation with the datum plane and the datum plane has a relation with the hole. However, since the datum plane has direct relation with the hole, you can pattern the datum planes with reference to the holes using the **Reference** option.*

Tip: *If you do not want to create datum on-the-fly and still want to create a reference pattern, as shown in Figure 6-18 then you need to create a datum plane and a rib feature on this plane. Group the datum plane and the rib feature. After grouping the two, select the group from the **Model Tree** and choose the **Pattern Tool** button. Now the reference pattern can be created .*

Table Patterns

Table-driven patterns are created by defining a table. In this table you need to specify the dimensions of the instances from the edge or faces from where the leader of the pattern is referenced (leader of a pattern is the feature that is selected to create a pattern). When you select the **Table** option from the drop-down list, the dashboard appears, as shown in Figure 6-19.

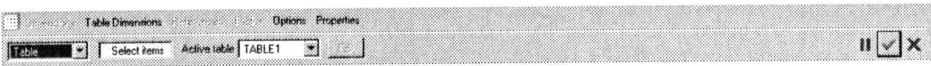

Figure 6-19 The **Pattern** *dashboard*

In Figure 6-19, the **Active Table** collector shows the table that is active. You can create more than one table to create a pattern. You can choose the **Edit** button to display the **Pro/TABLE** shown in Figure 6-20. This button will be available only after you have selected at least one dimension.

To create the pattern shown in Figure 6-21, follow the steps given next:

1. Create a solid protrusion of dimension 120x100x15.
2. Create a hole of diameter **5** on the top face of the base feature at a distance of 20 from the left edge and 20 from the bottom edge.
3. Select the hole feature and choose the **Pattern Tool** button from the **Edit Features** toolbar. The **Pattern** dashboard is displayed and the dimensions of the hole feature appear on the hole.
4. Select the **Table** option from the drop-down list present on the dashboard.
5. Select the dimension **20** that is from the left edge and then use CTRL+left mouse button to select the dimension **20** that is from the bottom edge.
6. Choose the **Edit** button to display the **Pro/TABLE**.
7. In column **C1** enter 1 under **idx** (index number). The number 1 signifies the first instance. To enter the value 1, click under **idx**. The cell is highlighted, enter 1.
8. Toward the right of **1**, in column **C2** enter the distance along the first dimension and in column **C3** enter the distance along the second dimension.

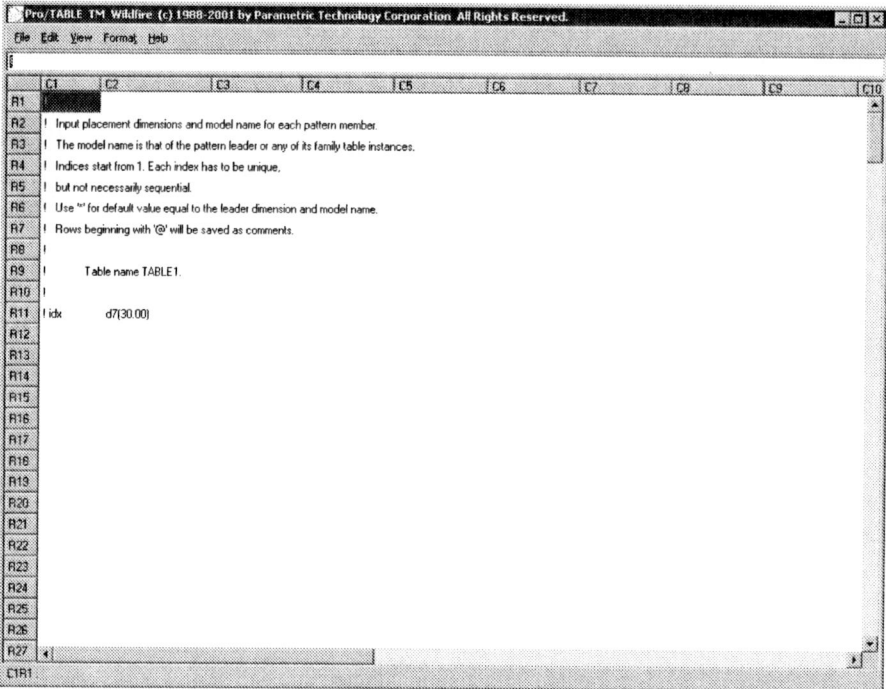

Figure 6-20 Pro/TABLE

9. Similarly, under the rows that are below **idx**, enter the dimensions of other instances of the holes, as shown in Figure 6-22. Remember that distance of each hole is measured from where the leader is dimensioned.

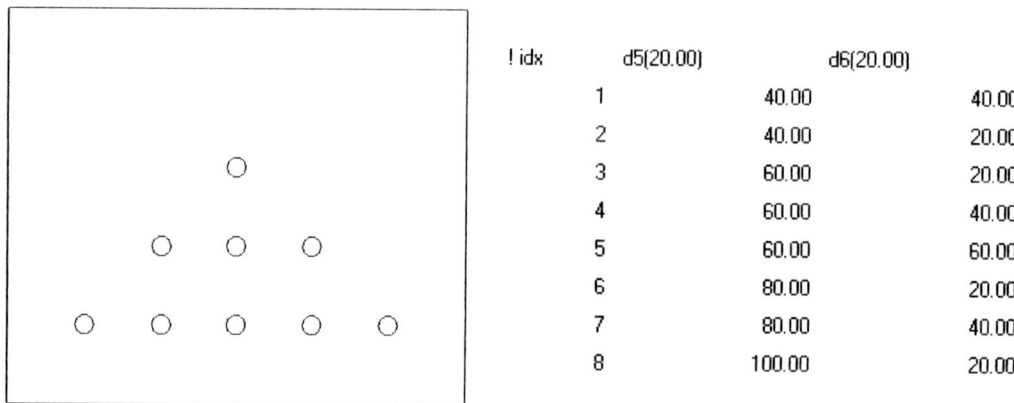

Figure 6-21 Table-driven pattern *Figure 6-22 Coordinate values of instances*

10. Choose **File > Save** and then choose the **Exit** option to exit the **Pro/TABLE**.
11. Choose the **Build feature** button to exit the feature creation tool.

Generally, table-driven patterns are used when the incremental distances between the instances of the pattern are nonuniform. This pattern is also useful when the coordinate locations of the instances are known.

 **Tip**: *Pattern tables can be created from scratch by picking the dimensions and filling in all values. In most cases, it is easier and less time-consuming to create a dimensional pattern that is similar to what you need and then convert it to a table.*

Options Tab

When you choose the **Options** tab, the slide-up panel is displayed, as shown in Figure 6-23. The options in this slide-up panel are discussed next:

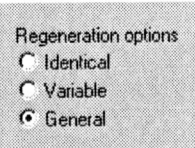

Figure 6-23 **Options** *slide-up panel*

Identical Pattern

The **Identical** option is used to create an identical pattern. You need to select at least one incremental dimension to pattern the feature. Depending on the incremental dimension selected, the resulting pattern will be linear or rotational. A linear pattern is created when the driving dimension is linear and a rotational pattern is created when the driving dimension is angular. You can enter a positive or a negative value as the increment in a pattern dimension. All instances of a pattern that are created using this option are identical in size and geometry. This is the reason the patterns created using this option are known as the **Identical** patterns. Figure 6-24 shows a hole feature on the base feature and Figure 6-25 shows holes patterned linearly.

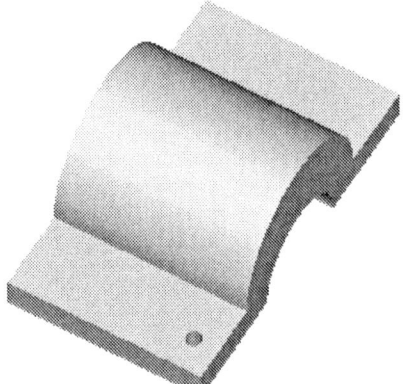 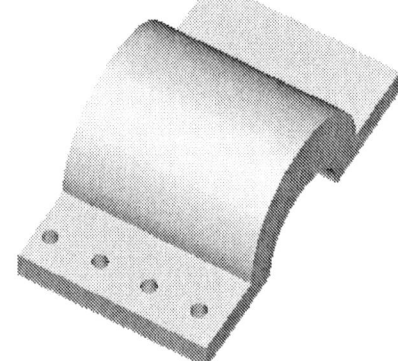

Figure 6-24 Hole on the base feature   *Figure 6-25* Hole patterned on the base feature

Similarly, Figure 6-26 shows a hole feature on the base feature and Figure 6-27 shows the rotational pattern of the hole feature.

Figure 6-26 Hole on the base feature   *Figure 6-27* Rotational pattern of hole feature

As evident from Figures 6-25 and 6-27, all instances in the identical patterns are placed on the same placement surface and no feature intersects the edges of the placement surface, any other instance, or any other feature other than the placement surface. Note that it is not possible to pattern the hole feature shown in Figure 6-25 on the right flap using the identical patterns. However, you can use the **General** option to pattern the hole on the right flap of the model shown in Figure 6-25.

Variable Pattern

The **Variable** type of pattern is used when the instances vary in size. In this type of pattern the instances can be placed on different surfaces and can also intersect with the edges of the placement surface. The feature shown in Figure 6-28 is patterned using the **Variable** option. The patterned feature is shown in Figure 6-29. In Figure 6-29, the length and the diameter of the rod varies in all instances.

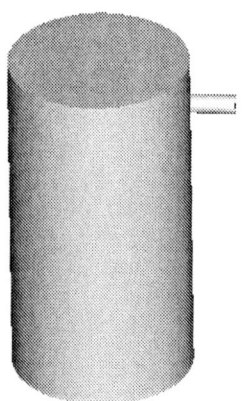 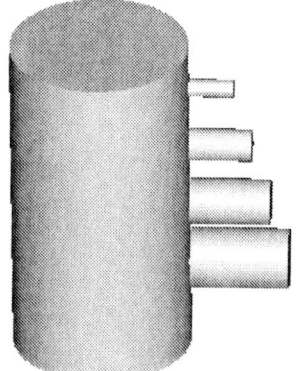

Figure 6-28 A rod on the base feature   *Figure 6-29* Varying pattern of the rod

General Pattern

The most complex patterns can be created using the **General** pattern. This option is used to create patterns in which the instances touch each other and intersect with other instances or the edges of the surface. This option of creating patterns is also used when instances intersect with the base feature and the intersection is not visible. The hole shown in Figure 6-30 is patterned using the **General** option. The pattern is shown in Figure 6-31.

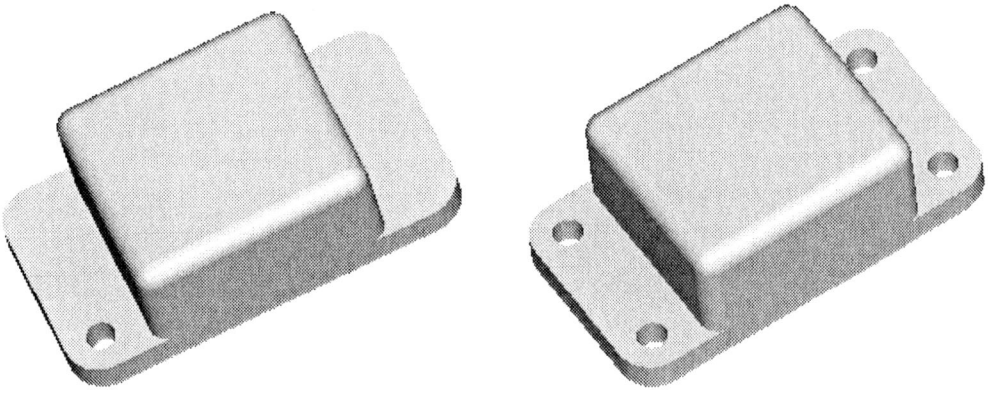

Figure 6-30 Hole on the base feature   *Figure 6-31 General pattern*

Note
*If the feature that you need to pattern has formed a group and is listed in the **Model Tree** as a group feature, then you need to select the group from the **Model Tree** to pattern the desired feature.*

*A feature forms a group when it has used other features to form itself. For example, if you create datum on-the-fly and use it as a sketch plane then after the feature is created, it is listed in the **Model Tree** as a group feature. This group comprises of the feature, the datum plane, and the sketch of the feature.*

Deleting a Pattern
You can delete a pattern by selecting it from the **Model Tree**. Select the pattern feature from the **Model Tree** and right-click to invoke the shortcut menu. Choose the **Delete Pattern** option from the shortcut menu. However, note that the parent feature (leader) is not deleted when you delete the pattern by using the **Delete Pattern** option even if it was selected along with other instances for deletion.

If you want to delete the pattern along with the parent feature, choose the **Delete** option from the shortcut menu. The system highlights the pattern in the graphics window and confirms the deletion of the pattern from the user.

COPYING FEATURES
The **Copy** option is used for copying and mirroring the selected features. This option reduces the time required in the model creation and also helps in maintaining its design intent. The **Copy** option is available in the **FEAT** menu in the **Menu Manager**. To invoke the **Menu Manager**, choose **Edit > Feature Operations** from the menu bar.

When you choose the **Copy** option from the **FEAT** menu, the **COPY FEATURE** submenu appears with various options, as shown in Figure 6-32. The options of this submenu are explained next.

New Refs

The **New Refs** option is used to copy a feature by varying the dimension values and by selecting new references. The references can be edges, axes, or placement planes. Select the **Done** option from the **Menu Manager** after other desired options have been selected in it. You are prompted to select the feature to be copied. After doing so, select the **Done** option; the **GP VAR DIMS** menu is displayed, as shown in Figure 6-33. Using this menu, you can select the dimensions that are to be varied while copying a feature with new references. After making the selection, select the **Done** option to enter the modified dimensions values. Now, the **WHICH REF** menu is displayed, as shown in Figure 6-34.

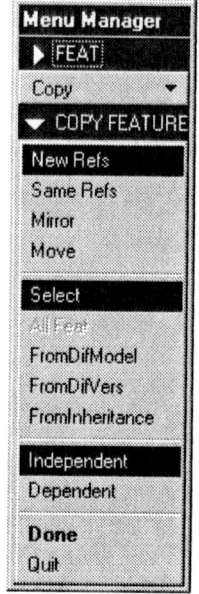

Figure 6-32 COPY FEATURE submenu

You can copy a feature using the **New Refs** option by two methods. You can keep the same dimensional or placement reference for the copied feature as that of the original feature by using the **Same** option in the **WHICH REF** menu. This means that you can use the same edge or surface as the references for the copied part. The other possibility is that you can use new references for the copied feature. This can be achieved by using the **Alternate** option in the **WHICH REF** menu. This provides you with a greater flexibility to copy the features.

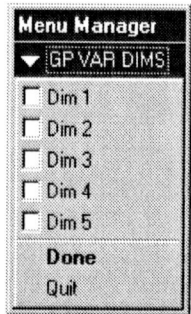

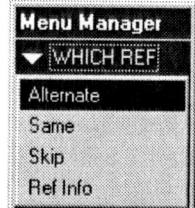

Figure 6-33 GP VAR DIMS menu   *Figure 6-34 WHICH REF menu*

Same Refs

When you use the **Same Refs** option, you have the flexibility to vary the dimensions of the copied features, but the dimensional and placement references of the copied feature are the same as for the source feature. You only need to select the dimensions that will be varied while copying the feature.

Mirror

The **Mirror** option is used to copy a feature by mirroring it about a specified datum plane or a face. When you invoke this option, you are prompted to select the features to be mirrored. Once you select the features to be mirrored, you are prompted to select a plane or create a datum about which the features will be mirrored. As soon as you select a datum plane or a planar surface, the selected feature will be mirrored. This option not only reduces the model completion time, but it also helps in maintaining the symmetry in the features of a model. Figure 6-35 shows a rib feature that is to be mirrored and the datum plane about which the rib feature will be mirrored. Figure 6-36 shows the model after mirroring the rib feature and turning off the visibility of the datum plane.

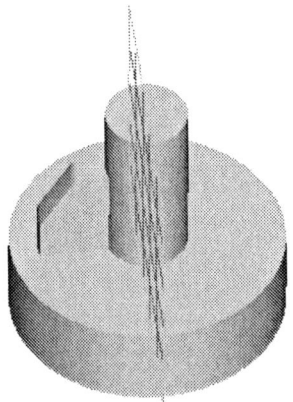 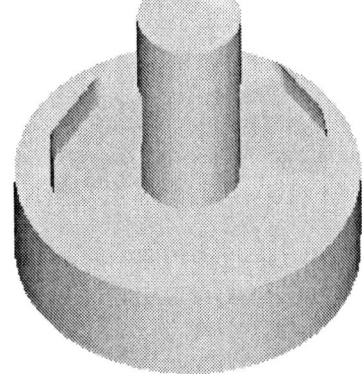

Figure 6-35 Rib feature and the mirror plane   *Figure 6-36* Mirrored rib feature

Tip: *To mirror a feature at an angle of 90 degrees to the parent feature, create a datum plane at an angle of 45 degrees to the parent feature and then use this datum plane to mirror the feature.*

Move

The **Move** option is used to copy features by translating or rotating them. When you invoke this option, you will be prompted to select the features to be translated. To invoke the **Move** option, choose **Move > Select > Independent > Done** from the **COPY FEATURE** menu. The **SELECT FEAT** menu is displayed and you are prompted to select the features to be translated. After selecting the features and choosing the **Done** option from the **SELECT FEAT** menu, you are prompted to define the movements by combination of translation and rotation. The **MOVE FEATURE** submenu is displayed, as shown in Figure 6-37. The selected feature can be translated or rotated using the options in this submenu.

Figure 6-37 MOVE FEATURE submenu

Translating the Features

You can select a feature from the graphics window or from the **Model Tree** to copy it while translating. You need to specify a perpendicular plane to define the direction in which the feature will be copied and the offset distance. You can also select multiple features to copy.

Rotating the features

You can select a feature from the graphics window or from the **Model Tree** and then rotate it about an axis, edge, datum curve, or coordinate system. Using this option you can select multiple features to copy.

Select

The **Select** option provides the flexibility of choosing the features to be copied. When you use this option, you need to select the features either from the graphics window or from the **Model Tree**. You can select any number of features to copy.

All Feat

The **All Feat** option is available only when you copy a feature using the **Mirror** or the **Move** option. When you use this option, all features created are copied. Remember that you need to specify a coordinate system while using this option.

FromDifModel Option

The **FromDifModel** (From different model) option is used to copy a feature from a different model. This option is available only when you are using the **New Refs** option of the **COPY FEATURE** submenu. This is due to change in the references required to copy from one model to another. Therefore, all references will be new.

FromDifVers Option

You can copy features from a different version of the current model using the **FromDifVers** (From different versions) option. This option is available only when you are using the **New Refs** and the **Same Refs** options of the **COPY FEATURE** submenu.

Independent Option

The **Independent** option specifies that the dimensions of the copied features are independent of the dimensions of the parent feature. The features that you copy from a different model or different versions are independent by default. The dimensions of such copied features have no relation with those of the original feature. This is the reason the **Dependent** option is not available while using the **FromDifModel** or the **FromDifVers** options.

Dependent Option

The **Dependent** option specifies that the dimensions of the copied features are dependent on the dimensions of the parent feature. Therefore, if you make any modification in the section of the parent feature, the changes are automatically reflected in the copied feature.

MIRRORING A GEOMETRY

Choose the **Mirror Tool** button from the **Edit Features** toolbar. This button is only available after a feature has been selected. After selecting this button, the **Mirror** dashboard is displayed and you are prompted to select a datum plane about which the model will be mirrored. You can select any datum plane or a planar surface as the mirroring plane. Now, choose the **Build Feature** button to create the mirrored feature.

Options Aiding Construction of Parts-II 6-17

This option is slightly different from the **Copy > Mirror** option. When you use the **Copy > Mirror** option, you can select features to be copied. But, by using the **Mirror Tool** button, the whole model can be mirrored about a specified datum plane. The model can be selected by selecting its name that is available at the top of the **Model Tree**. All features in the mirrored model are related to the parent model. Any modification in the parent model are reflected in the mirrored model. Figure 6-38 shows a model and a datum plane for mirroring the model and Figure 6-39 shows the resulting mirrored model.

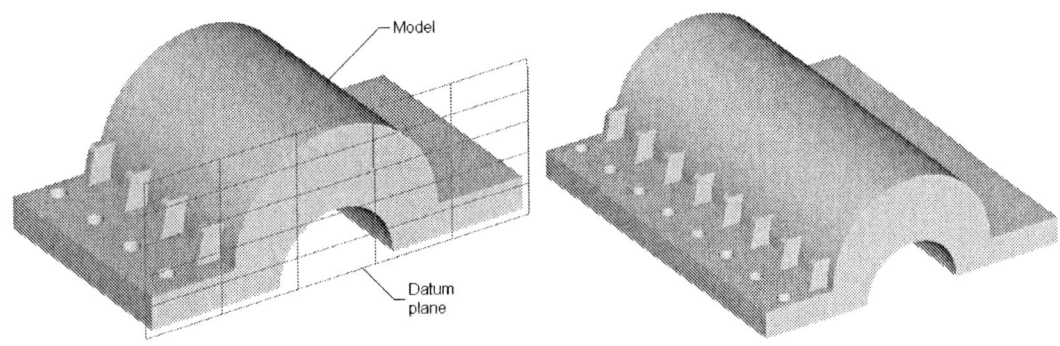

Figure 6-38 Model and a plane   *Figure 6-39* Resulting mirrored model

CREATING A SECTION OF THE SOLID MODEL

The complex geometry of a model sometimes needs sectioning so that you can see the inner sections of various features that are not visible from outside. These sections are also used for showing sections in the drawing views.

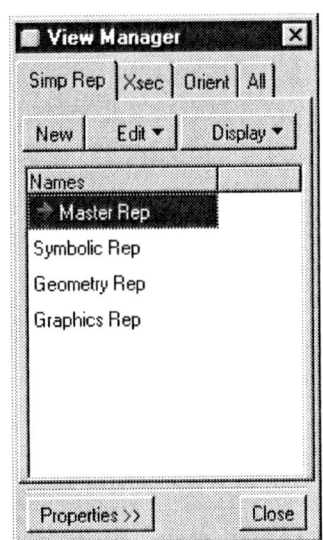

Choose **View > View Manager** from the menu bar. The **View Manager** dialog box is displayed, as shown in Figure 6-40.

Choose the **New** button from the dialog box. The **Rep0001** is displayed in the **Names** display box. Press ENTER. This the default name given to the section that is going to be created. If required, you can enter another name here. The dialog box disappears and then appears again. Select **Rep0001** and then right-click. From the shortcut menu choose the **Redefine** option. The **EDIT METHOD** menu is displayed, as shown in Figure 6-41.

In this menu the methods of representing a model are listed. Here, you will learn the **Work Region** method of representing the model.

Figure 6-40 View Manager dialog box

Figure 6-41 EDIT METHOD menu

Work Region Method

This method of representing a model is very similar to removing material (creating cut) from a model. The following steps explain the procedure to create the section of the model shown in Figure 6-42.

Figure 6-42 Solid model

1. Choose the **Work Region** option from the **EDIT METHOD** menu. The **SOLID OPTS** submenu is displayed, as shown in Figure 6-43.
2. Choose **Extrude > Solid > Done** from the **SOLID OPTS** submenu. The **Extrude** dashboard is displayed.
3. Choose the **Placement** tab and from the slide-up panel choose the **Define** button. The **Sketch** dialog box is displayed.
4. Select the top face of the plate as the sketch plane and select the **RIGHT** datum plane to be at the right while drawing the sketch.
5. When you enter the sketcher environment, draw the section lines, as shown in Figure 6-44. This section is an open loop. The dimensions are not of importance to create this section.
6. Exit the sketcher environment and select the **Extrude to intersect with all surface** button from the depth flyout. The yellow arrow is pointing in the reverse direction.
7. Choose the **Change material direction of extrude to other side of sketch** button to reverse the direction of arrow.
8. Choose the **Build feature** button from the **Extrude** dashboard. The sectioned model is shown in Figure 6-45.

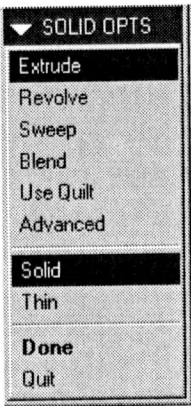

Figure 6-43 SOLID OPTS submenu

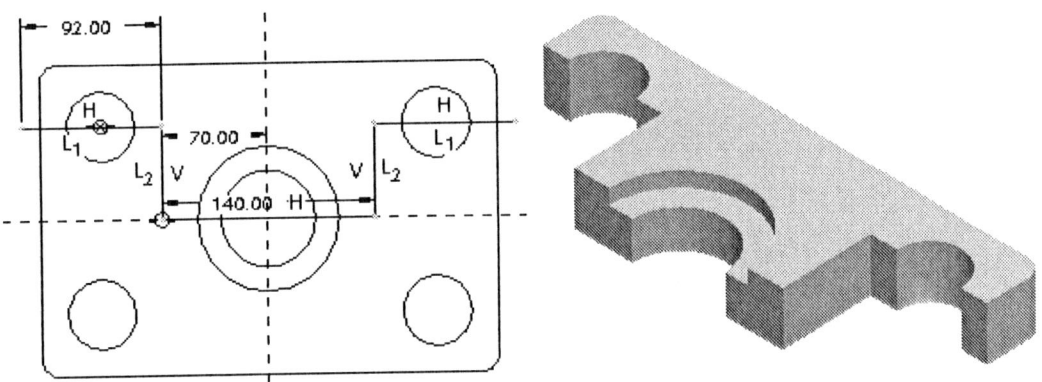

Figure 6-44 Sketch of the section lines   *Figure 6-45 Sectioned model*

To resume the view of the complete model, invoke the **View Manager**. Right-click on **Master Rep** in the Names display box to invoke the shortcut menu and select the **Set Active** option from it. Close the View Manager once the complete view of the model is resumed.

Note
*The created section is saved with the name that is entered in the **Names** display box while creating it. You can create multiple sections and save them with different names. To view a particular section, right-click on its name in the **Names** display box and select the **Set Active** option from the shortcut menu.*

TUTORIALS

Tutorial 1

In this tutorial you will create the model shown in Figure 6-46. This figure also shows the top, front, and right-side views of the model.   **(Expected time: 30 min)**

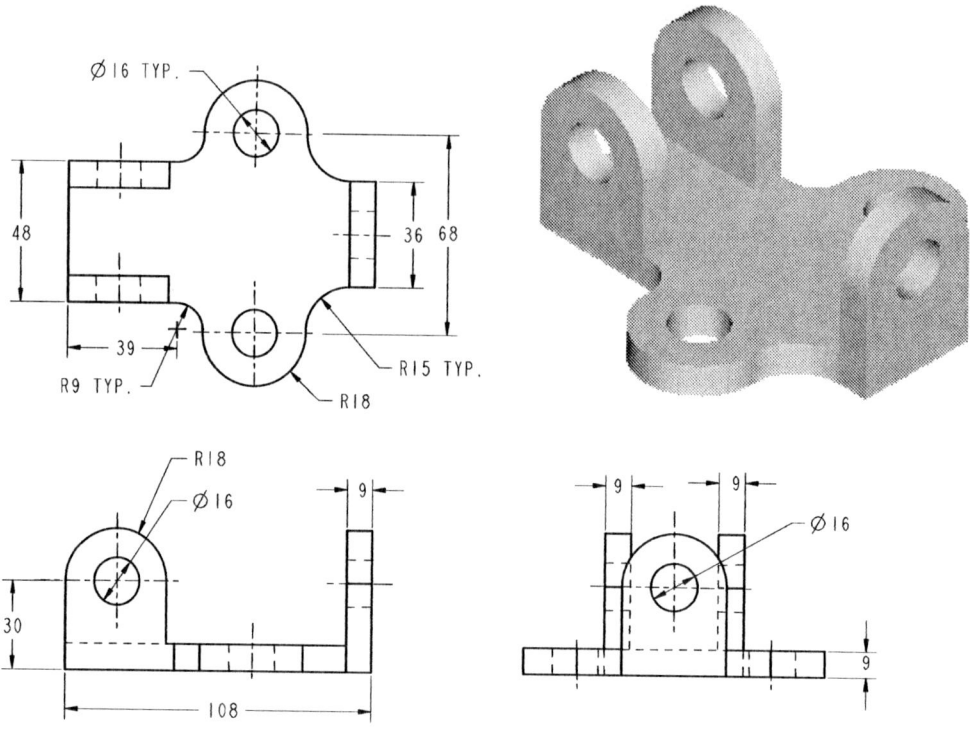

Figure 6-46 Top, front, isometric, and right-side views of the model

The following steps are required to complete this tutorial:

a. Examine the model and determine the number of features in it. The model is composed of four features, refer to Figure 6-46.
b. Create the base feature on the TOP datum plane, refer to Figures 6-47 and 6-48.
c. Create the second feature on the right face of the base feature, refer to Figures 6-49 through 6-51.
d. Create the third feature by copying the second feature, refer to Figures 6-52 and 6-53.
e. Create the forth feature by mirroring the third feature, refer to Figures 6-54.

After understanding the procedure for creating the model, you are now ready to create it. When Pro/ENGINEER session is started, the first task is to set the working directory. Since this is the first tutorial of this chapter, you need to create a folder named *c06*, if it does not exist and set it as **Working Directory**.

Options Aiding Construction of Parts-II
6-21

Starting a New Object File

1. Open a new part file and name it *c06tut1*. The three default datum planes are displayed in the graphics window. The **Model Tree** is also displayed in the graphics window. Close the **Model Tree** by clicking on the sash present on the right edge of the **Model Tree**.

Creating the Base Feature

To create the sketch for the base feature, you need to select the **TOP** datum plane as the sketching plane.

1. Choose the **Extrude Tool** button from the **Base Features** toolbar. The **Extrude** dashboard is displayed below the graphics window.

2. Choose the **Placement** tab and from the slide-up panel choose the **Define** button. The **Sketch** dialog box is displayed.

3. Select the **TOP** datum plane as the sketching plane. As you select the sketching plane, the **RIGHT** datum plane and its orientation **Right** are set automatically.

4. Choose the **Sketch** button to enter the sketcher environment.

5. Once you enter the sketcher environment, create the sketch of the base feature and apply constraints and dimensions, as shown in Figure 6-47.

   Note that in the sketch, the bottom half of the sketch is mirrored to create the top half of the sketch. This is evident from the constraints of symmetry applied to the sketch in Figure 6-47. Also, as evident from the sketch of the base feature shown in Figure 6-47, the **RIGHT** datum plane is located at a dimension of 18 from the left edge because later in the tutorial, this datum plane will be used as a reference to copy a feature.

6. After the sketch is complete, choose the **Continue with the current section** button to exit the sketcher environment.

   The **Extrude** dashboard is enabled and is displayed below the graphics window.

7. Enter **9** as the value of the depth in the dimension box present on the **Extrude** dashboard.

8. Choose the **Build feature** button from the **Extrude** dashboard.

   The base feature is completed and now the second feature will be created. The default trimetric view of the base feature is shown in Figure 6-48.

   Note
   *The two holes are integrated in the base feature. These holes are sketched while drawing the sketch for the base feature. Hence, the base feature is created as one single feature that includes the two holes. The other method is to create the two holes separately on the base feature using the **Hole** dashboard. In the method to create the features separately, the total features created will be three.*

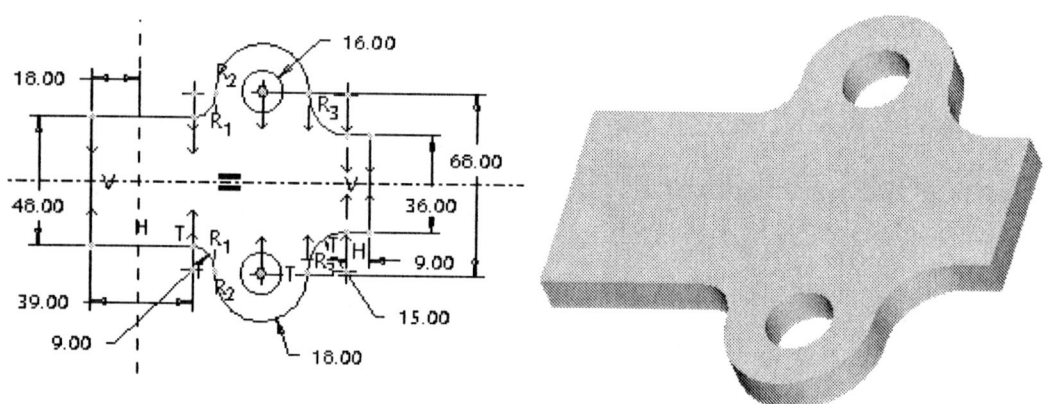

Figure 6-47 Sketch of the base feature

Figure 6-48 The default trimetric view of the base feature

Creating the Second Feature

The second feature is also an extruded feature and is created on the right face of the base feature. Therefore, you need to select the right face as the sketching plane. Note that the sketch of this feature has to be created keeping in mind some important steps. The feature may not get copied later on if you do not use the following steps in creating the sketch of this feature.

1. Choose the **Extrude Tool** button from the **Base Features** toolbar. The **Extrude** dashboard is displayed below the graphics window.

2. Choose the **Placement** tab and from the slide-up panel choose the **Define** button. The **Sketch** dialog box is displayed.

3. Select the right face of the base feature as the sketching plane.

4. Choose the **Flip** button to reverse the direction of the yellow arrow.

5. Select the **TOP** datum plane and then select the **Top** option from the **Orientation** drop-down list.

6. Choose the **Sketch** button to enter the sketcher environment.

   After entering the sketcher environment, turn the model display to **No Hidden**.

7. Choose the **Create 2 point lines** button from the **Sketcher Tools** toolbar and draw the right vertical line starting from the point shown in Figure 6-49. Notice that the endpoint of the right vertical line is not aligned with the edge on the top face of the base feature.

8. Next, draw the horizontal line in continuation with the first line, as shown in Figure 6-49. Notice that since the endpoint of the right vertical line is not aligned with the edge on the top face of the base feature, the horizontal line is also not aligned with that edge.

Options Aiding Construction of Parts-II 6-23

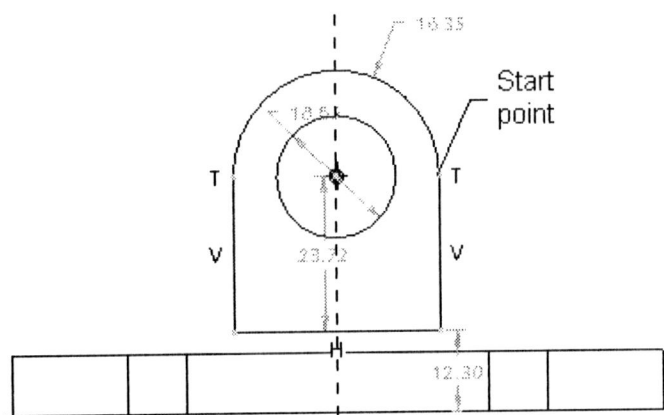

Figure 6-49 Sketch of the second feature with weak dimensions

9. Next, draw the left vertical line in continuation with the horizontal line and then complete the sketch by drawing the arc and the circle, see Figure 6-49.

10. Choose the **Impose sketcher constraints on the section** button to display the **Constraints** dialog box. Choose the **Create same points, points on entity or collinear constraint** button and select the bottom horizontal line of the sketch. Now, select the edge on the top face of the base feature as the second entity to apply the constraint. Notice that the sketch will extend such that the horizontal line is now aligned with the top face.

11. If the center points of the arc and the circle are not aligned with the **FRONT** datum plane, align them.

12. Add the dimensions and then modify them, as shown in Figure 6-50. After completing the sketch, turn the model display to **Shading** and choose the **Continue with the current section** button. The **Extrude** dashboard is enabled.

13. Enter **9** as the value of the depth in the dimension box present on the **Extrude** dashboard.

14. Choose **Build feature** from the **Extrude** dashboard. The default shaded trimetric view of the model after creating the second feature is shown in Figure 6-51.

Creating the Third Feature

You can create the third feature using two methods. The first method is to draw the sketch of the third feature by defining a sketching plane and then extruding it to the given depth. The second method is to copy the second feature by defining new references. This is because the third feature is similar in geometry and dimensions to the second feature of the model. In this tutorial you will use the second method to create the third feature.

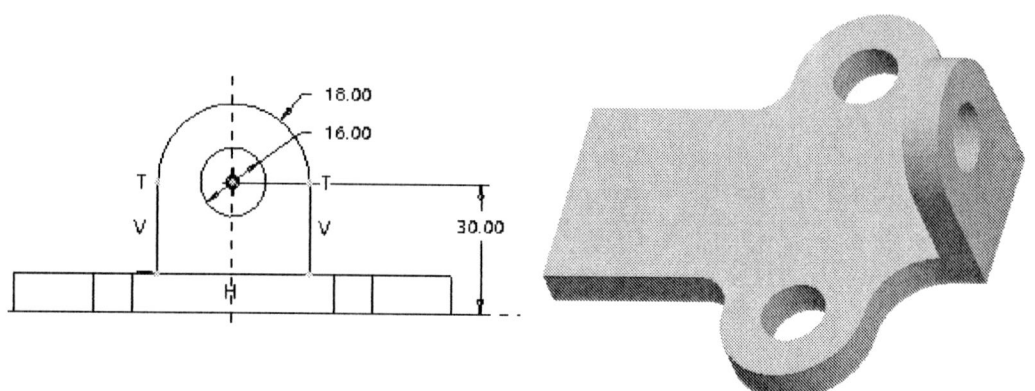

Figure 6-50 Sketch of the second feature   *Figure 6-51* Model after creating the second feature

1. Invoke the **Menu Manager** by choosing **Edit > Feature Operations** from the menu bar.

2. Choose **FEAT > Copy** from the **Menu Manager**. The **COPY FEATURE** submenu is displayed.

3. Choose **New Refs > Select > Dependent** from the **COPY FEATURE** submenu and choose **Done**. The **SELECT FEAT** submenu is displayed and you are prompted to select the feature to be copied.

4. Select the second feature from the graphics window. The selected feature is highlighted with red boundary.

5. Choose **Done** from the **SELECT FEAT** submenu. The **GP VAR DIMS** menu is displayed. Since you do not need to vary any dimension of the source feature, you can proceed further without selecting any dimension.

6. Choose **Done** from the **GP VAR DIMS** menu. The **WHICH REF** menu is displayed and you are prompted to select a sketching plane reference corresponding to the highlighted surface.

7. Select the face shown in Figure 6-52 from the graphics window as the surface on which the copied feature will be placed. The **TOP** datum plane is highlighted and you are prompted to select a horizontal sketcher reference corresponding to the highlighted surface.

8. Choose the **Same** option from the **WHICH REF** menu. The **FRONT** datum plane is highlighted and you are prompted to select section dimensioning reference corresponding to the highlighted surface.

9. Select the **RIGHT** datum plane. The edge of the top face of the base feature is highlighted and you are prompted to select section dimensioning reference corresponding to the highlighted surface.

Options Aiding Construction of Parts-II   6-25

10. Choose the **Same** option from **WHICH REF** menu. A green arrow is displayed and you are prompted to select the upward direction of the horizontal plane for protrusion.

11. Choose **Okay** from the **DIRECTION** submenu and then choose **Done** from the **GRP PLACE** menu. The feature is copied at the required location on the model, as shown in Figure 6-53.

12. Choose **Done** from the **Menu Manager** to complete feature creation.

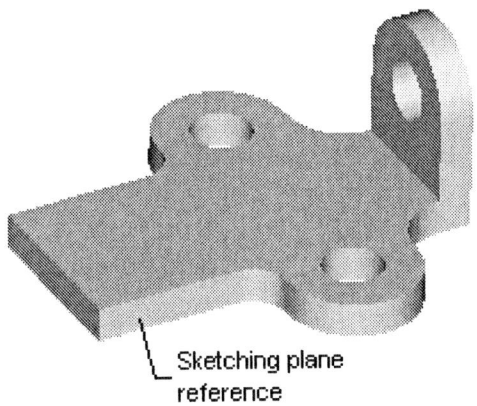

Figure 6-52 Sketching plane reference for third feature

Figure 6-53 The model after copying the feature with new references

Creating the Fourth Feature

The fourth feature can be created by sketching and extruding it to the given depth. You can also create this feature by placing a mirrored copy of the third feature at the required location. In this tutorial you will use the second method because it consumes less time.

1. Choose **FEAT > Copy** from the **Menu Manager**. The **COPY FEATURE** submenu is displayed.

2. Choose **Mirror > Select > Dependent > Done** from the **COPY FEATURE** submenu. The **SELECT FEAT** submenu is displayed and you are prompted to select the features to be mirrored.

3. Select the third feature from the graphics window and choose **Done** from the **SELECT FEAT** submenu. You are prompted to select a plane or create a datum to mirror about.

4. Select the **FRONT** datum plane as the mirror plane. The third feature is mirrored about the **FRONT** datum plane. The trimetric view of the final model is shown in Figure 6-54.

5. Choose **Done** from the **Menu Manager** to complete feature creation.

6. Choose the **Save the active object** button from the **File** toolbar and save the model. The order of feature creation can be seen from the **Model Tree** shown in Figure 6-55. Note that the feature id numbers in your model may be different from the ones shown in this figure.

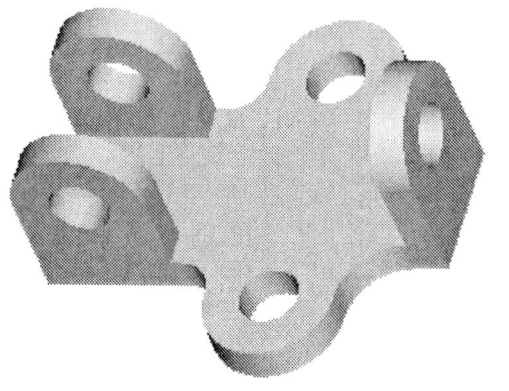

Figure 6-54 Default trimetric view of the model

Figure 6-55 Model Tree for Tutorial 1

Tutorial 2

In this tutorial you will create the model shown in Figure 6-56. The dimensions of the model are given in the top view and the front view of the model shown in Figure 6-57.

(Expected time: 45 min)

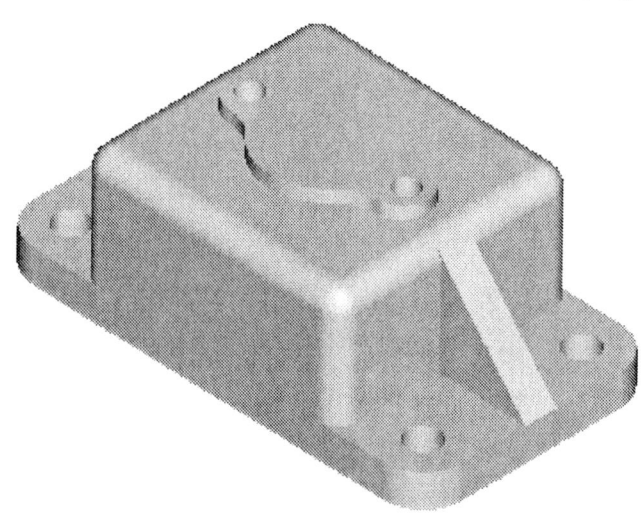

Figure 6-56 Solid model for Tutorial 2

The following steps are required to complete this tutorial:

a. Examine the model and determine the number of features in it. The model is composed of nine features, refer to Figure 6-56.
b. Create the base feature on the TOP datum plane, refer to Figures 6-58 and 6-59.
c. Create the round features, refer to Figures 6-60 and 6-61.
d. Create the hole feature and then pattern it, refer to Figures 6-62 and 6-63.

Options Aiding Construction of Parts-II

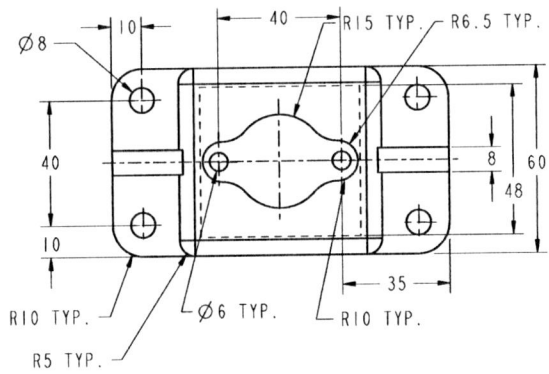

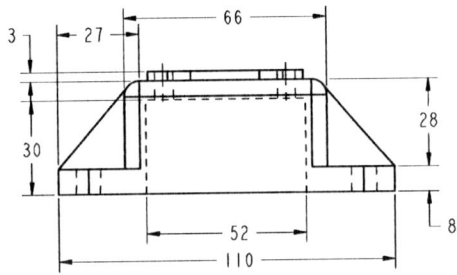

Figure 6-57 Top view and front view of the model

e. Create the rib feature on the front datum plane, refer to Figures 6-64 and 6-65.
f. Mirror the rib feature, refer to Figure 6-66.
g. Create the extrude feature on the top planar face of the base feature, refer to Figures 6-67 and 6-68.
h. Create the cut feature on the bottom planar face of the base feature, refer to Figures 6-69 and 6-70.
i. Create the hole feature on the top face of the model and then copy it, refer to Figures 6-71 and 6-72.

After understanding the procedure for creating the model, you are now ready to draw it. If required set the **Working Directory** to c06 folder.

Starting a New Object File

1. Open a new part file and name it *c06tut2*. The three default datum planes are displayed in the graphics window.

Creating the Base Feature

To create the sketch for the base feature, you first need to select the sketching plane for the base feature. In this model, you need to draw the base feature on the **FRONT** datum plane because the direction of extrusion of this feature is perpendicular to it. The base feature will be created symmetric to the **FRONT** datum plane.

1. Choose the **Extrude Tool** button from the **Base Features** toolbar. The **Extrude** dashboard is displayed below the graphics window.

2. Choose the **Placement** tab and from the slide-up panel choose the **Define** button. The **Sketch** dialog box is displayed.

3. Select the **FRONT** datum plane as the sketch plane.

4. Select the **TOP** datum plane and then select the **Top** option from the **Orientation** drop-down list.

5. Choose the **Sketch** button to enter the sketcher environment.

6. Once you enter the sketcher environment, create the sketch of the base feature and apply constraints and dimensions, as shown in Figure 6-58.

   In the sketch, you should note that since the base feature is symmetrical, therefore, a vertical center line is drawn and then the right half of the sketch is mirrored to create the left half. This is evident from the constraints of symmetry applied to the sketch in Figure 6-58. These constraints appear as arrow symbols in the sketch.

   As evident from the sketch of the base feature shown in Figure 6-58, the **TOP** datum plane is aligned with the bottom line segment.

7. After the sketch is complete, choose the **Continue with the current section** button to exit the sketcher environment.

   The **Extrude** dashboard is enabled and appears below the graphics window. Now, you need to extrude the sketch symmetrically to both sides of the sketching plane in order to create the base feature symmetrical to the **FRONT** datum plane. This is because, later in the tutorial, you will need to select the default datum planes as mirror planes for mirroring features. Hence, you do not need to create datum planes.

8. Choose the **Extrude on both sides of sketch plane by half the specified depth value in each** button from the depth flyout present on the **Extrude** dashboard.

9. Enter a depth of **60** in the dimension box present on the **Extrude** dashboard.

10. Choose the **Build feature** button from the **Extrude** dashboard to exit the feature creation tool.

    The base feature is completed and now the second feature will be created. The default trimetric view of the base feature is shown in Figure 6-59.

Creating the Second Feature

The second feature is a round feature of radius 5.

Options Aiding Construction of Parts-II

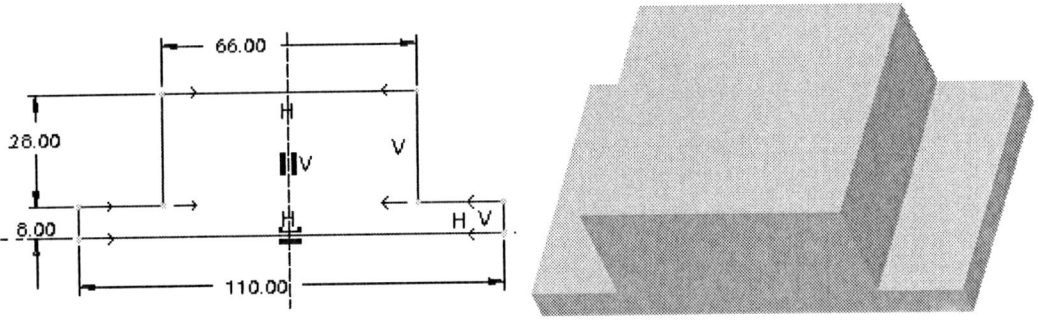

Figure 6-58 Sketch with dimensions and constraints for the base feature

Figure 6-59 Default trimetric view of the base feature

1. Choose the **Round Tool** button from the **Engineering Features** toolbar. The **Round** dashboard is displayed.

2. Choose the **Sets** tab to display the slide-up panel. Let this slide-up panel remain open so that you can view the selections you make on the model.

3. Select one of the edges shown in Figure 6-60. To select the second edge and the subsequent edges you need to press the CTRL key and then select the edge.

   The preview of the round is created on the edges that you have selected. A default value of the radius is also displayed.

4. Double-click on the default radius value that is displayed on the preview of the round. In the edit box that appears, enter a value of **5**. The first set of rounds is created.

   Now, you need to create the second set of rounds.

5. After spinning the model, select the four vertical edges of the bottom portion of the base feature to round. Remember to use the CTRL+left mouse button to select the second and subsequent edges.

   You will notice that in the slide-up panel, **Set 1**, **Set 2** appears. This indicates that second set is defined. The preview of the round is created on the edges. A default value of the radius is also displayed.

6. Double-click on the default radius value that is displayed on the preview of the round. In the edit box that appears, enter a value of **10**.

7. Choose the **Build feature** button from the **Round** dashboard to create the round feature.

The round feature is completed. The default trimetric view of the round feature is shown in Figure 6-61. In the **Model Tree** the two rounds created appear as one feature.

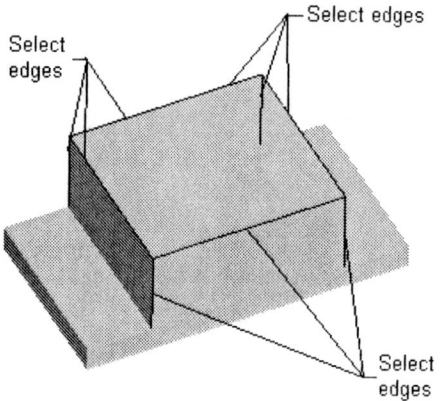

Figure 6-60 *Edges to be selected to round*

Figure 6-61 *The default trimetric view of the base feature with two sets of round*

Creating the Third Feature

The third feature is a through hole and will be created using the **Hole** dashboard.

1. Choose the **Hole Tool** button from the **Engineering Features** toolbar. The **Hole** dashboard is displayed. The **Create straight hole** button is selected by default.

2. Create the hole, as shown in Figure 6-62 by specifying the placement parameters. The placement parameters can be referred from Figure 6-57.

Creating a Pattern of the Hole Feature

As evident from Figure 6-56, you need to create four instances of the hole. The first instance is created by using the **Hole** dashboard and you can use the **Pattern Tool** button to create the remaining three instances. You will create a rectangular pattern of the hole feature. You can also create all holes using the **Hole** dashboard and specify the placement parameters for each of them. But to save time, it is recommended that you create a pattern of the hole.

1. Select the hole feature and then choose the **Pattern Tool** button from the **Edit Features** toolbar.

    The **Pattern** dashboard is displayed. You are prompted to select the dimensions to vary in the first direction.

2. Make sure that the **General** option is selected in the **Options** slide-up panel.

    You cannot use the **Identical** option to create the rectangular pattern of the hole feature. This is because when you use the **Identical** option to pattern, the pattern cannot intersect the base feature on which the hole is created. If the top portion of the base feature is created as a separate feature, then the hole can be patterned using the **Identical** option.

3. Select the dimension value **10** from the graphics window. Since the two dimensions that are displayed on the graphics window are both 10, therefore, select the dimension 10 that is along the shorter side of the base feature. After you select the dimension in the first direction, the edit box is displayed.

4. Enter a value of **40** in the edit box.

5. Hold down the right mouse button to display a shortcut menu. Select the **Direction 2 Dimensions** option from the shortcut menu.

6. Select the dimension value **10** that is along the longer side of the base feature. After you select the dimension in the second direction, the edit box is displayed.

   Note that in the edit boxes present on the dashboard, the number of instances, **2**, is specified by default.

7. Enter a value of **90** in the edit box.

8. Choose the **Build feature** button from the **Pattern** dashboard.

   The rectangular pattern of holes is displayed, as shown in Figure 6-63. You can use the middle mouse button to spin the model and to display the model, as shown in Figure 6-63.

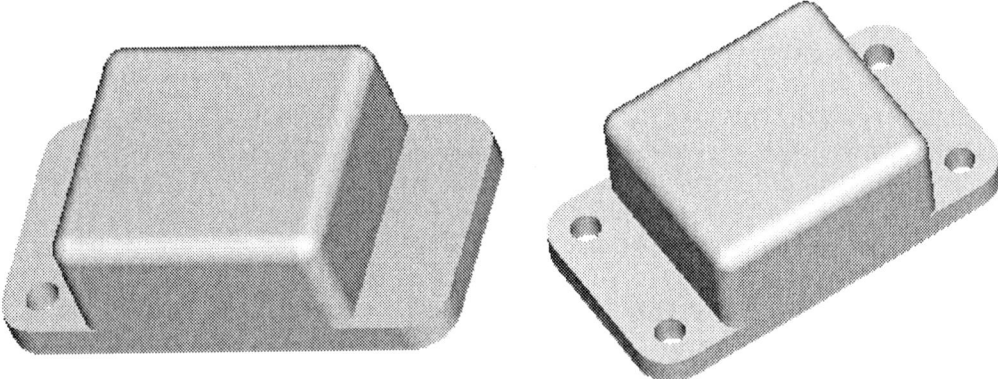

Figure 6-62 The hole feature on the base feature

Figure 6-63 Model after creating the hole pattern

Note
If you have multiple holes to be created on a model, it is recommended that you create their pattern, if possible. This is because when you assemble bolts in these holes in the Assembly mode, it becomes very easy to assemble them using the reference pattern.

Creating the Rib Feature

The sketch of the rib feature will be created on the **FRONT** datum plane and the required thickness will be applied to it. This is the fourth feature of the model.

1. Choose the **Rib Tool** from the **Engineering Features** toolbar. The **Rib** dashboard is displayed.

2. Choose the **References** tab and from the slide-up panel choose the **Define** button. The **Sketch** dialog box is displayed.

3. Select the **FRONT** datum plane as the sketching plane.

4. Select the **TOP** datum plane from the graphics window and then select the **Top** option from the **Orientation** drop-down list.

5. Choose the **Sketch** button to proceed to the sketcher environment.

6. Draw the sketch of the rib feature, as shown in Figure 6-64.

Note

As evident from Figure 6-64, the top end of the inclined line in the sketch is aligned with the curve and the tangent constraint is applied to the line and the curve. Similarly, the bottom end of the inclined line is also aligned with the two edges. This is the reason there are no dimensions in the sketch and the sketch for the rib feature is fully constrained.

7. Exit the sketcher environment.

8. From the **References** slide-up panel, choose the **Flip** button.

9. Specify the rib thickness as 8 in the dimension box present on the **Rib** dashboard. Choose the **Build feature** button to create the rib feature.

The rib feature is shown in Figure 6-65. You can use middle mouse button to spin the model in the orientation, as shown in Figure 6-65.

Creating a Copy of the Rib Feature

A copy of the rib feature will be created, as shown in Figure 6-66. This copy feature is the fifth feature of the model. The other method to create the same is to create the sketch of the rib feature on a sketching plane. But this method will consume a lot of time. Therefore, the rib feature will be copied about the **RIGHT** datum plane.

1. Choose **Edit > Feature Operations** from the menu bar. The **Menu Manager** is displayed.

2. Choose **FEAT > Copy** from the **Menu Manager**. The **COPY FEATURE** submenu is displayed.

Options Aiding Construction of Parts-II

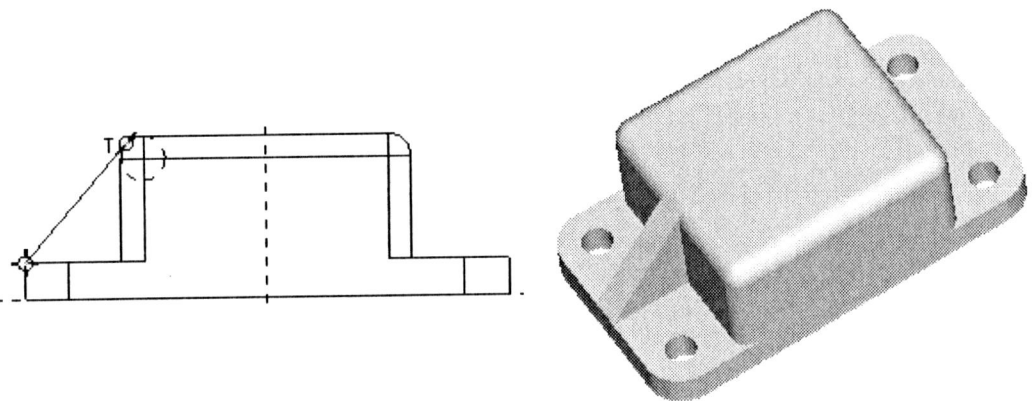

Figure 6-64 Fully constrained sketch for the rib   *Figure 6-65* Model after creating the rib feature

3. Choose **Mirror** > **Select** > **Independent** > **Done** from the **COPY FEATURE** submenu. The **SELECT FEAT** submenu is displayed and you are prompted to select the features to be mirrored.

4. Select the rib feature from the graphics window and choose **Done** from the **SELECT FEAT** submenu. You are prompted to select a plane or create a datum plane to mirror about.

5. Select the **RIGHT** datum plane as the mirror plane. The selected feature is mirrored about the **RIGHT** datum plane, as shown in Figure 6-66.

Figure 6-66 Model after mirroring the rib

6. Choose **Done** from the **FEAT** menu to exit the **Menu Manager**.

Creating the Protrusion Feature

The sixth feature is an extruded feature that will be created on the top face of the base feature.

1. Choose the **Extrude Tool** button from the **Base Features** toolbar.

2. Choose the **Placement** tab and from the slide-up panel choose the **Define** button. The **Sketch** dialog box is displayed.

3. Select the top face of the base feature as the sketching plane.

4. Select the **RIGHT** datum plane and then select the **Right** option from the **Orientation** drop-down list.

5. Choose the **Sketch** button to enter the sketcher environment.

6. After you enter the sketcher environment, draw the sketch of the extruded feature, as shown in Figure 6-67.

   Here in the sketch, you should note that the tangent and equal radii constraints are applied. Also the center of the top arc and the bottom arc coincides with the intersection of the two datum planes.

7. After the sketch is complete, choose the **Continue with the current section** button to exit the sketcher environment.

   The **Extrude** dashboard is enabled and appears below the graphics window.

8. Enter a depth of **3** in the dimension box present on the **Extrude** dashboard. Choose the **Build feature** button from the **Extrude** dashboard. The extruded feature is completed and the default trimetric view is shown in Figure 6-68.

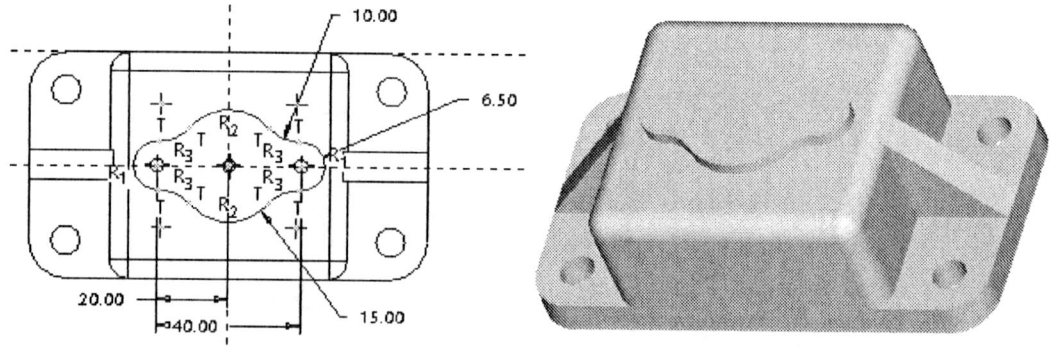

Figure 6-67 Sketch with dimensions and constraints for the extruded feature

Figure 6-68 The default trimetric view after creating the extruded feature

Creating the Cut Feature

You need to create an extruded cut on the bottom planar surface of the base feature and this is the seventh feature of the model.

1. Choose the **Extrude Tool** button form the **Base Features** toolbar.

2. Select the **Remove Material** button from the **Extrude** dashboard.

3. Choose the **Placement** tab and from the slide-up panel choose the **Define** button. The **Sketch** dialog box is displayed.

4. Select the bottom face of the base feature as the sketching plane.

5. Select the **RIGHT** datum plane and then select the **Right** option from the **Orientation** drop-down list.

6. Choose the **Sketch** button to enter the sketcher environment.

7. After you enter the sketcher environment, draw the sketch and dimension it, as shown in Figure 6-69.

8. After completing the sketch, choose the **Continue with the current section** button. The **Extrude** dashboard is enabled and appears below the graphics window.

9. Enter a value of **30** in the dimension box present on the **Extrude** dashboard.

10. Choose the **Build feature** button. You can spin the model using the middle mouse button. The cut feature is shown in Figure 6-70.

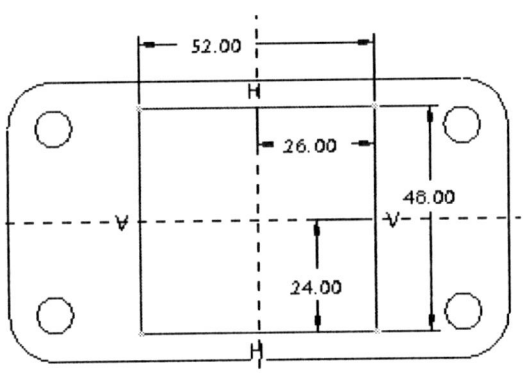

Figure 6-69 Sketch for the cut feature with dimensions

Figure 6-70 Model after creating the cut feature

Creating the Hole

Next, a through hole will be created on the top extruded feature using the **Hole** dashboard.

1. Invoke the **Hole** dashboard, the **Create straight hole** button is selected by default in it. Specify the placement parameters of the hole as given in Figure 6-56 and create the hole. The model after creating the hole is shown in Figure 6-71.

Creating a Copy of the Hole

You need to create a mirror copy of the hole, as shown in Figure 6-72 and this will be the ninth feature of the model. The hole feature will be mirrored about the **RIGHT** datum plane.

1. Invoke the **Menu Manager** using the **Edit** menu. Choose **FEAT > Copy** from the **Menu Manager**. The **COPY FEATURE** submenu is displayed.

2. Choose **Mirror > Select > Independent > Done** from the **COPY FEATURE** submenu. The **SELECT FEAT** submenu is displayed and you are prompted to select the features to be mirrored.

3. Select the previous hole feature from the graphics window and choose **Done** from the **SELECT FEAT** submenu. You are prompted to select a plane or create a datum plane to mirror about.

4. Select the **RIGHT** datum plane as the mirror plane and choose **Done**. The hole feature is mirrored about the **RIGHT** datum plane, as shown in Figure 6-72.

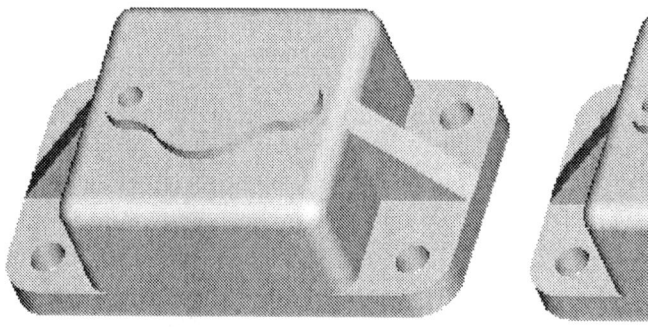

Figure 6-71 Hole feature on the extruded feature   *Figure 6-72* The copied hole feature

Saving the Model

You have to save the model because you may need it later.

1. Choose the **Save the active object** button from the **File** toolbar and save the model.

   The order of feature creation can be seen from the **Model Tree** shown in Figure 6-73.

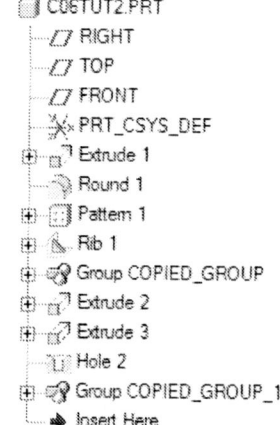

Figure 6-73 Model Tree for Tutorial 2

Tutorial 3

In this tutorial you will create the model shown in Figure 6-74. This figure also shows the top view, front view, and isometric view of the model. **(Expected time: 30 min)**

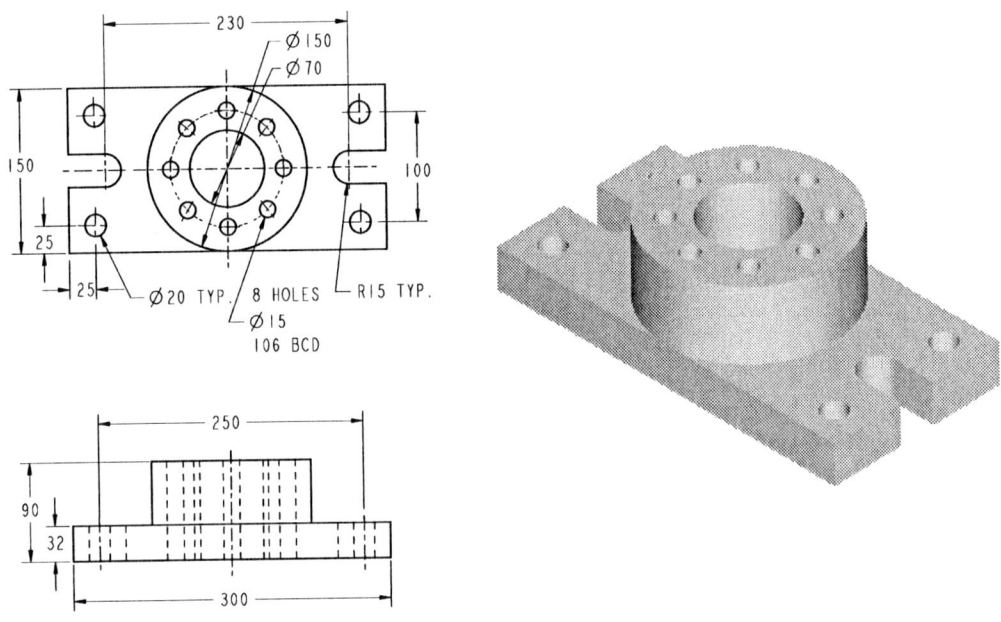

Figure 6-74 Top, front, and isometric views of the model

The following steps are required to complete this tutorial:

a. Examine the model and determine the number of features in it. The model is composed of five features, refer to Figure 6-74.
b. Create the base feature on the **TOP** datum plane, refer to Figures 6-75 and 6-76.
c. Create the cylindrical extrude feature, refer to Figures 6-77 and 6-78.
d. Create the hole feature coaxially with the cylindrical feature, refer to Figure 6-79.
e. Create the hole feature on the top planar surface of the base feature, and then pattern it refer to Figures 6-80 and 6-81.
f. Create the hole feature on the top planar surface of the cylindrical feature, refer to Figure 6-82.
g. Create a rotational pattern of this hole, refer to Figure 6-83.

After understanding the procedure for creating the model, you are now ready to draw it. If required set the **Working Directory** to c06 folder.

Starting a New Object File

1. Open a new part file and name it *c06tut3*. The three default datum planes are displayed in the graphics window.

Creating the Base Feature

To create the sketch for the base feature, you first need to select the sketching plane for the base feature. In this model, you need to draw the base feature on the **TOP** datum plane. This is because the direction of extrusion of this feature is perpendicular to the **TOP** datum plane.

1. Choose the **Extrude Tool** button from the **Base Features** toolbar.

2. Choose the **Placement** tab and from the slide-up panel choose the **Define** button. The **Sketch** dialog box is displayed.

3. Select the **TOP** datum plane as the sketching plane.

4. Select the **RIGHT** datum plane and then select the **Right** option from the **Orientation** drop-down list, if it is not selected by default.

5. Choose the **Sketch** button to enter the sketcher environment.

6. Once you enter the sketcher environment, create the sketch of the base feature and apply constraints and dimensions, as shown in Figure 6-75.

7. After the sketch is complete, choose the **Continue with the current section** button to exit the sketcher environment.

   The **Extrude** dashboard is enabled and appears below the graphics window.

8. Enter a depth of **32** in the dimension box that is present on the **Extrude** dashboard and choose **Build feature** from the dashboard.

   The base feature is completed and now the second feature will be created. The default trimetric view of the base feature is shown in Figure 6-76.

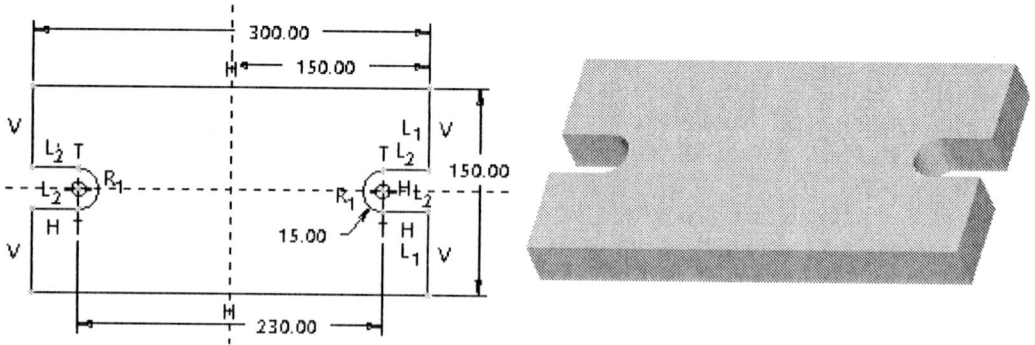

Figure 6-75 Sketch with dimensions and constraints for the base feature

Figure 6-76 Default trimetric view of the base feature

Options Aiding Construction of Parts-II

Creating the Second Feature

The second feature is also an extruded feature and will be created on the top face of the base feature. Therefore, you need to define the top face as the sketching plane for the second feature.

1. Choose the **Extrude Tool** button from the **Base Features** toolbar.

2. Choose the **Placement** tab and from the slide-up panel choose the **Define** button. The **Sketch** dialog box is displayed.

3. Select the top face of the base feature as the sketching plane.

4. Select the **RIGHT** datum plane and then select the **Right** option from the **Orientation** drop-down list, if it is not selected by default.

5. Choose the **Sketch** button to enter the sketcher environment.

6. Create the sketch for the second feature and dimension it, as shown in Figure 6-77.

7. After creating the sketch, choose the **Continue with the current section** button. The **Extrude** dashboard is enabled.

8. Enter a value of **58** in the dimension box present on the **Extrude** dashboard and choose the **Build feature** button.

The second feature is completed and the default trimetric view is shown in Figure 6-78.

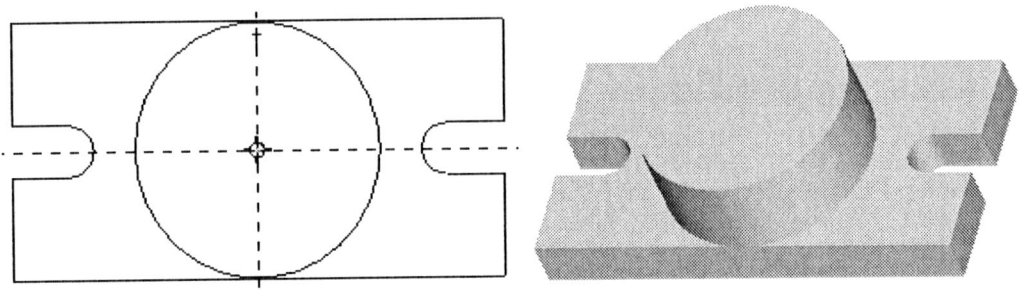

Figure 6-77 Sketch of the cylindrical feature with diameter of the cylinder

Figure 6-78 Default trimetric view of the cylindrical feature

Creating the Third Feature

The third feature is a through hole that is coaxial to the cylindrical feature. The hole feature will be created using the **Hole** dashboard.

1. Choose the **Hole Tool** button from the **Engineering Features** toolbar. The **Hole** dashboard is displayed. The **Create straight hole** button in the **Hole** dashboard is chosen by default.

2. Create a hole of diameter **70**, as shown in Figure 6-79 by specifying the placement parameters.

Creating the Fourth Feature

The fourth feature is a through hole and will be created on the top planar surface of the base feature. The hole feature will be created using the **Hole** dashboard.

1. Choose the **Hole Tool** button from the **Engineering Features** toolbar. The **Hole** dashboard is displayed. The **Create straight hole** button in the **Hole** dashboard is selected by default.

2. Create the hole, as shown in Figure 6-80 by specifying the placement parameters.

Figure 6-79 Coaxial hole on the cylindrical feature

Figure 6-80 Hole on the base feature

Patterning the Hole Feature

Next, you need to create a rectangular pattern of the hole feature that is created on the base feature. You can also create the individual holes, but this will consume a lot of time and the number of features in the **Model Tree** will increase. Hence, it is recommended that you create a rectangular pattern of the hole feature.

1. Select the hole feature and then choose the **Pattern Tool** button from the **Edit Features** toolbar.

   The **Pattern** dashboard is displayed. You are prompted to select the dimensions to vary in the first direction.

2. Select the **Identical** option from the **Options** slide-up panel.

   Here, you can use the **Identical** option because the feature on which the pattern is created is not intersecting the pattern.

Options Aiding Construction of Parts-II 6-41

3. Select the dimension value **25** from the graphics window. Since the two dimensions that are displayed on the graphics window are both 25, hence select the dimension 25 that is along the shorter side of the base feature. After you select the dimension in the first direction, the edit box is displayed.

4. Enter a value of **100** in the edit box.

5. Hold down the right mouse button to display the shortcut menu. Choose the **Direction 2 Dimensions** option from the shortcut menu.

6. Select the dimension value **25** that is along the longer side of the base feature. After you select the dimension in the second direction, the edit box is displayed.

7. Enter a value of **250** in the edit box. Note that in the edit boxes present on the dashboard, the number of instances, **2**, is specified by default.

8. Choose the **Build feature** button from the **Pattern** dashboard. The rectangular pattern of holes will be displayed, as shown in Figure 6-81.

Creating a Hole on the Cylindrical Feature

The hole on the cylindrical feature will created diametrically using the **Hole** dashboard.

1. Choose the **Hole Tool** button from the **Engineering Features** toolbar. The **Hole** dashboard is displayed. The **Create straight hole** button in the **Hole** dashboard is chosen by default.

2. Choose the **Placement** tab from the **Hole** dashboard to display the slide-up panel.

3. Select the top face of the cylindrical feature as the placement plane.

4. From the drop-down list in the slide-up panel, select the **Diameter** option.

5. Click in the **Secondary References** collector and now select the axis of the cylindrical feature.

6. Enter a value of **106** in the second dimension box present on the right of the **Diameter** option in the **Secondary References** display box.

7. Use CTRL+left mouse button and select the **FRONT** datum plane from the graphics window. Enter a value of **90** in the dimension box present in the **Secondary References** collector.

8. Enter a value of **20** as the diameter of the hole in the diameter dimension box present on the **Hole** dashboard.

9. Select the **Drill to intersect with all surfaces** option from the depth flyout available in the **Hole** dashboard.

10. Choose the **Build feature** button from the **Hole** dashboard. The hole is created, as shown in Figure 6-82.

Figure 6-81 Rectangular pattern of the hole feature

Figure 6-82 Diametrical hole on the cylindrical feature

Creating the Rotational Pattern of the Hole Feature

Creating the remaining holes individually on the cylindrical feature will consume a lot of time. Therefore, you need to create a rotational pattern of the hole feature.

1. Select the hole feature and then choose the **Pattern Tool** button from the **Edit Features** toolbar.
   The **Pattern** dashboard is displayed. You are prompted to select the dimensions to vary in the first direction.

2. Select the **Identical** option from the **Options** slide-up panel.

3. The dimensions of the hole feature are displayed in the graphics window. Select the angular dimension **90**. The edit box is displayed.

4. Enter a value of **45** in the edit box and press ENTER. Now you need to specify the number of instances of the hole feature in the pattern.

5. Enter **8** in the **1** edit box and press ENTER. Choose the **Build feature** button. The rotational pattern is created, as shown in Figure 6-83.

Saving the Model

You have to save the model because you may need it later.

1. Choose the **Save the active object** button from the **File** toolbar and save the model.

   The order of feature creation can be seen from the **Model Tree** shown in Figure 6-84.

Options Aiding Construction of Parts-II 6-43

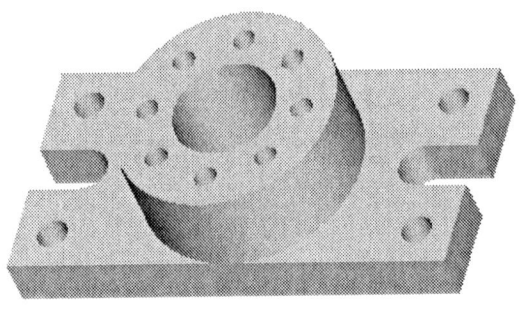

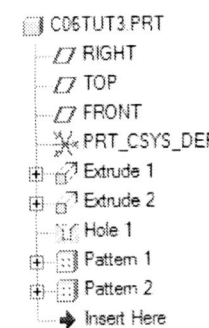

Figure 6-83 The complete model

Figure 6-84 Model Tree for Tutorial 3

Tutorial 4

In this tutorial you will create the model of the cylinder head shown in Figure 6-85. Figure 6-86 shows the top view and the front section view of the model. **(Expected time: 45 min)**

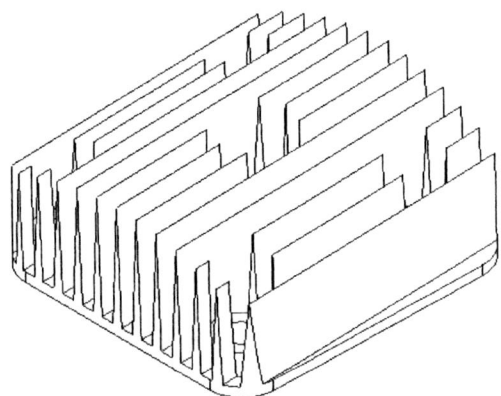

Figure 6-85 Isometric view of the model

The following steps are required to complete this tutorial:

a. Examine the model and determine the number of features in it. The model is composed of twelve features, refer to Figure 6-85.
b. Create the base feature on the **TOP** datum plane, refer to Figures 6-87 and 6-88.
c. Create the round features on the vertical edges of the base feature, refer to Figure 6-89.
d. Create the cylindrical feature on the bottom face of the base feature, refer to Figure 6-90.
e. Create the revolve cut feature on a plane passing through the center of the cylindrical feature, refer to Figures 6-91 and 6-92.
f. Create the fifth feature for the fin that later will be patterned, refer to Figures 6-93 through 6-95.

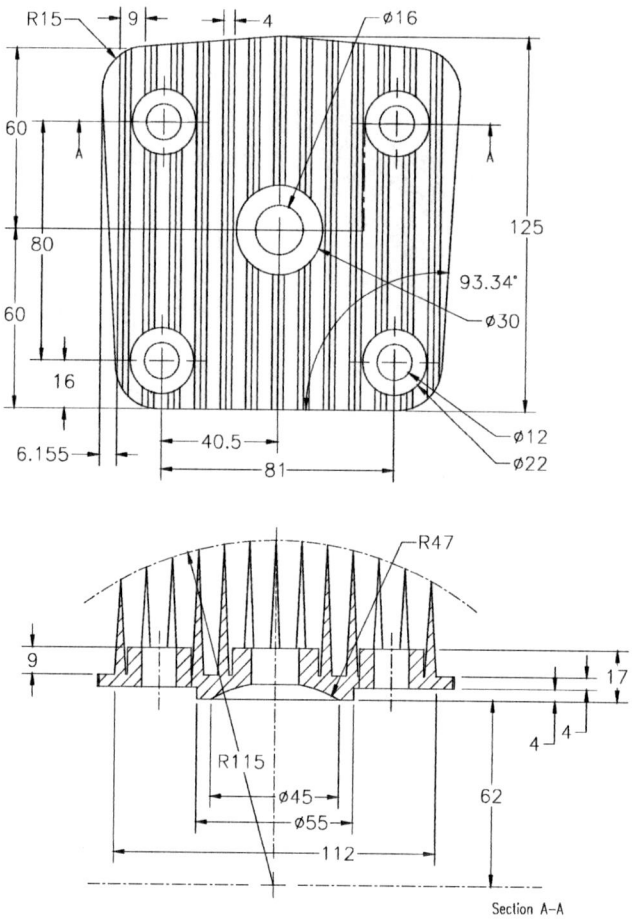

Figure 6-86 Top view and the front section view of the model

g. Create the cut feature that will remove the projections of the fins that are projecting out of the base feature, refer to Figure 6-96.
h. Create a circular cut feature on the top face of the base feature, refer to Figures 6-67 and 6-98.
i. Create a cylindrical feature on the top face of the base feature, refer to Figure 6-99.
j. Create the ninth feature is as a cut feature, refer to Figures 6-100 and 6-101. This feature is reference patterned to create its other instances, refer to Figure 6-102.
k. Create the tenth feature as a cylindrical feature on the top face of the base feature, refer to Figure 6-103.
l. Create the referenced pattern of the tenth feature, refer to Figure 6-104.
m. The eleventh feature is a coaxial hole. After creating the feature, create a reference pattern of this hole, refer to Figure 6-105.
n. Create the last feature as the coaxial hole that will be created on the eighth feature, refer to Figure 6-106.

Options Aiding Construction of Parts-II

After understanding the procedure for creating the model, you are now ready to draw it. If required set the **Working Directory** to c06 folder.

Starting a New Object File

1. Open a new part file and name it *c06tut4*. The three default datum planes are displayed in the graphics window.

Creating the Base Feature

In this model, you need to draw the sketch of the base feature on the **TOP** datum plane. The sketch is a polygon. The right half of the polygon will be drawn first and then it will be mirrored about the centerline to create the complete polygon.

1. Choose the **Extrude Tool** button from the **Base Features** toolbar.

2. Choose the **Placement** tab and from the slide-up panel choose the **Define** button. The **Sketch** dialog box is displayed.

3. Select the **TOP** datum plane as the sketching plane.

4. Select the **RIGHT** datum plane and then select the **Right** option from the **Orientation** drop-down list, if it is not selected automatically.

5. Choose the **Sketch** button to enter the sketcher environment.

6. Once you enter the sketcher environment, create the sketch of the base feature and apply constraints and dimensions, as shown in Figure 6-87.

7. After the sketch is complete, choose the **Continue with the current section** button to exit the sketcher environment.

   The **Extrude** dashboard is enabled and appears below the graphics window.

8. Enter a depth of **4** in the dimension box that is present on the **Extrude** dashboard and choose **Build feature** from the dashboard.

   The base feature is completed and now the second feature will be created. The default trimetric view of the base feature is shown in Figure 6-88.

Creating the Second Feature

The second feature is rounds of radius 15.

1. Choose the **Round Tool** button from the **Engineering Features** toolbar. The **Round** dashboard is displayed.

2. Choose the **Sets** tab to display the slide-up panel. Let this slide-up panel remain open so that you can view the selections you make on the model.

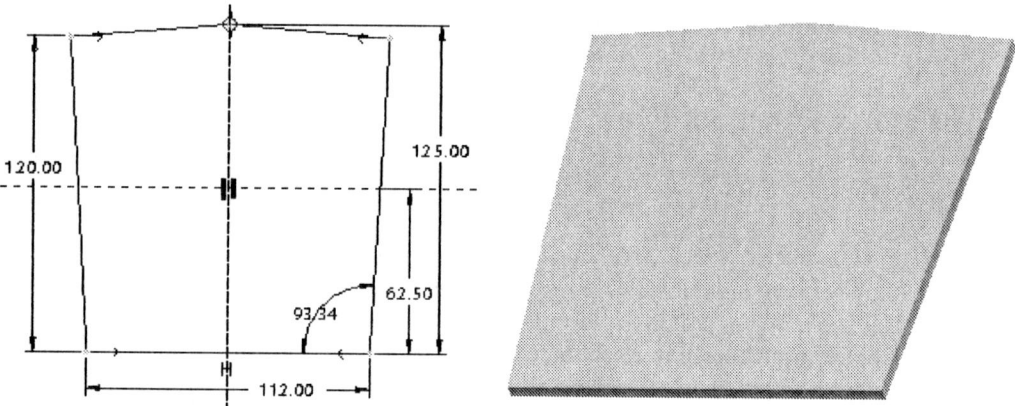

Figure 6-87 Sketch of the base feature with dimensions and constraints

Figure 6-88 Default trimetric view of the base feature

3. Select one of the four vertical edges on the corners. To select the second edge and the subsequent edges you need to press the CTRL key and then select the edge. Refer to Figure 6-86 for the selection of edges.

   The preview of the round is created on the edges that you have selected. A default value of the radius is also displayed.

4. Double-click on the default radius value that is displayed on the preview of the round. In the edit box that appears, enter a value of **15**. You can also enter the radius value in the dimension box available in the **Round** dashboard.

5. Choose the **Build feature** button to create the rounds.

   The default trimetric view of the model after creating the rounds is shown in Figure 6-89.

Creating the Third Feature

The third feature is a cylindrical feature and will be created on the bottom face of the base feature.

1. Invoke the **Extrude** dashboard and using the **Sketch** dialog box select the bottom face of the base feature as the sketch plane.

2. Enter the sketcher environment and draw the sketch of the cylindrical feature. Refer to Figure 6-86 for drawing the sketch.

3. Exit the sketcher environment and enter **4** as the depth of extrusion.

4. Exit the **Extrude** dashboard.

   The cylindrical feature is completed and is shown in Figure 6-90.

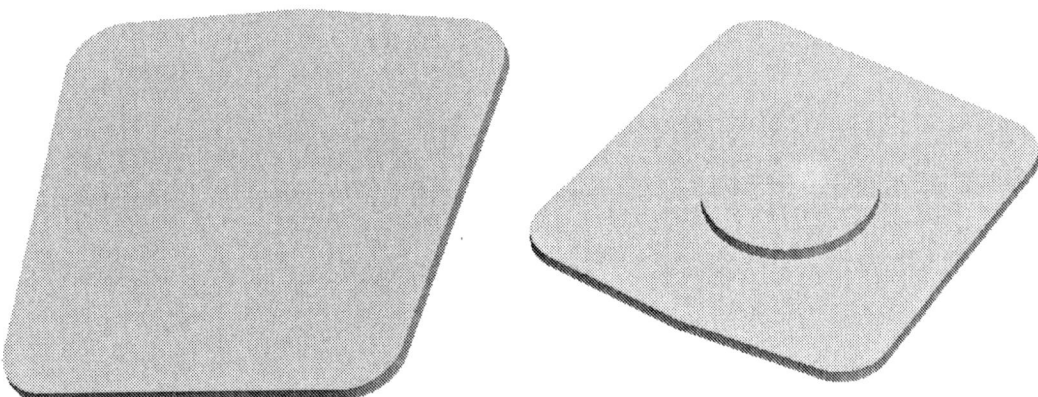

Figure 6-89 Model with the rounds

Figure 6-90 Model after creating the cylindrical feature at the bottom face

Creating the Fourth Feature

The fourth feature is a revolve cut that will be created on the datum plane passing through the center of the cylindrical feature.

1. Invoke the **Revolve** dashboard and select the **Remove Material** button.

2. Choose the **Placement** tab and from the slide-up panel choose the **Define** button. The **Sketch** dialog box is displayed. Select the **FRONT** datum plane as the sketch plane.

3. Select the **RIGHT** datum plane and then select the **Right** option from the **Orientation** drop-down list.

4. Once you enter the sketcher environment, create the sketch of the revolve cut feature and apply constraints and dimensions, as shown in Figure 6-91.

5. After the sketch is complete, exit the sketcher environment.

   The **Revolve** dashboard is enabled and appears below the graphics window. The value for angle of revolution, **360**, is selected by default.

6. Choose **Build feature** from the dashboard.

   The revolve cut feature is created and is shown in Figure 6-92.

Creating the Fifth Feature

The fifth feature is the fin and its pattern. The sketch of the fin will be created on the **FRONT** datum plane. The pattern of the fin feature will be created by controlling the dimensions of the sketch and constraining it.

This type of pattern can only be created if dimensions of the feature that is to be patterned is proper. In the sketch of the feature, a construction entity must exist. This

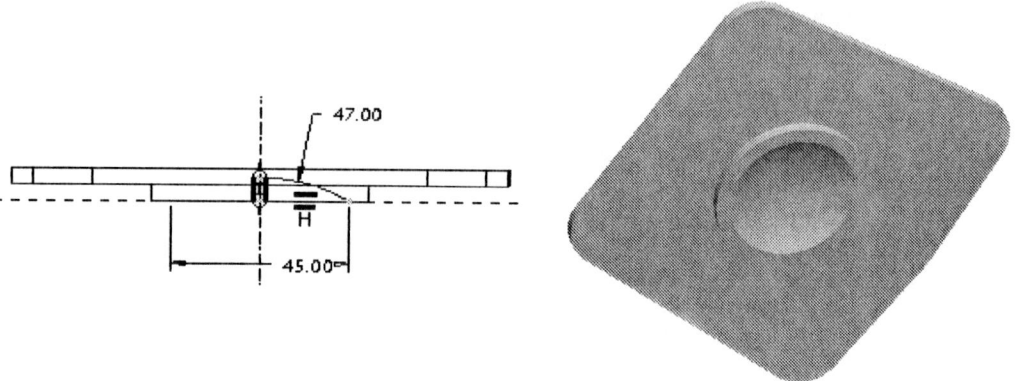

Figure 6-91 Sketch of the revolve feature with dimensions and constraints

Figure 6-92 Revolve cut feature

construction entity can be a circle, an arc, or a line. The construction entity is used as the driving entity for the pattern that will be created later. The pattern instances are created in such a way that they follow the geometry of the construction entity.

1. Invoke the **Extrude** dashboard and using the **Sketch** dialog box select the **FRONT** datum plane as the sketch plane.

2. Select the **RIGHT** datum plane and then select the **Right** option from the **Orientation** drop-down list, if they are not selected by default.

3. Choose the **Sketch** button to enter the sketcher environment.

4. Once you enter the sketcher environment, draw the arc shown in Figure 6-93 by using the **Create an arc by picking its center and endpoints** button.

 Note
In the sketch of the fin feature, you need to draw the sketch such that later you can control the size of the instances of the pattern. You will use the dimension 6.155 to create the pattern. Also, the vertical height of the fin feature is not specified because in that case the fin feature will not slide along the construction arc and will be locked with the top construction arc.

5. Select the arc and then hold down the right mouse button to invoke the shortcut menu. From the shortcut menu, choose the **Construction** option. The arc is converted to a construction arc. Create the sketch of the fin feature and apply constraints and dimensions, as shown in Figure 6-93.

6. After the sketch is complete, exit the sketcher environment.

7. Select the **Extrude on both sides of sketch plane by half the specified depth value in each** button from the depth flyout that is present on the **Extrude** dashboard.

Options Aiding Construction of Parts-II 6-49

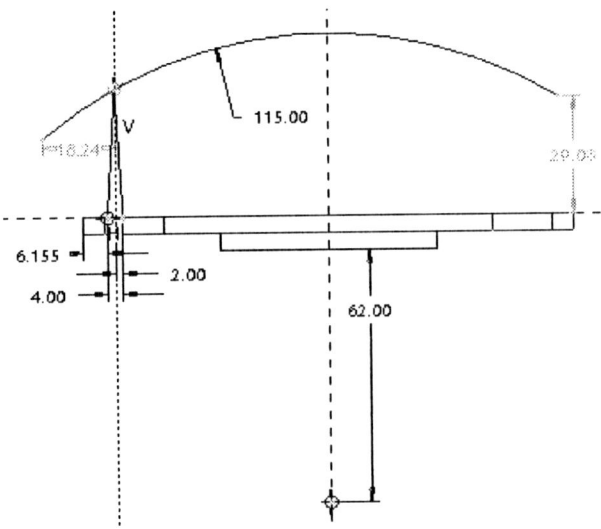

Figure 6-93 Sketch for the fin feature with two weak dimensions

8. Enter a value of **135** in the dimension box present on the dashboard.

9. Choose **Build feature** from the dashboard.

   The default trimetric view of the fin feature is shown in Figure 6-94. Now, you need to pattern the fin feature to create the remaining 12 fins.

10. Select the fin and invoke the **Pattern** dashboard. The dimensions of the fin feature are displayed in the graphics window.

11. Select the value **6.155**; the edit box appears. Enter a value of **9** in this edit box.

12. Enter a value of **13** in the dimension box present on the dashboard. These are the number of instances of the fin feature.

13. Choose the **Build feature** button to exit the **Pattern** dashboard. The model after patterning the fin feature is shown in Figure 6-95.

Creating the Sixth Feature

The sixth feature is an extrude cut that will be created on the bottom face of the base feature. This cut will remove the fins that are projecting out of the base feature.

1. Invoke the **Extrude** dashboard and select the **Remove Material** button.

2. Select the bottom face of the base feature as the sketch plane.

3. Select the **RIGHT** datum plane and then select the **Right** option from the **Orientation** drop-down list if they are not selected by default.

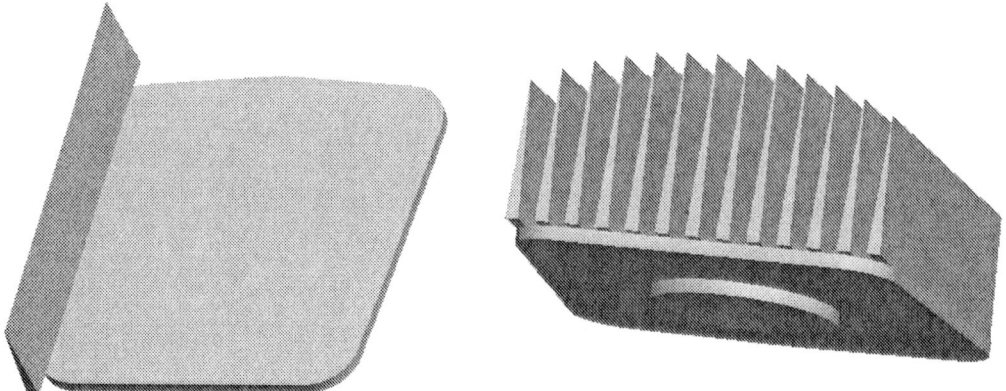

Figure 6-94 Fin created on the base feature   *Figure 6-95* Pattern of the fin feature

4. Once you enter the sketcher environment, use the edge of the base feature to create the sketch that will be later extruded to create the cut.

5. After the sketch is complete, exit the sketcher environment.

   The **Extrude** dashboard is enabled and appears below the graphics window.

6. Select the **Extrude to intersect with all surface** button from the depth flyout.

7. Choose the **Flip** button to reverse the direction of yellow arrow. This yellow arrow points in the direction from where the material will be removed.

8. Choose the **Build feature** button from the dashboard. The default trimetric view of the model after the cut feature is created is shown in Figure 6-96.

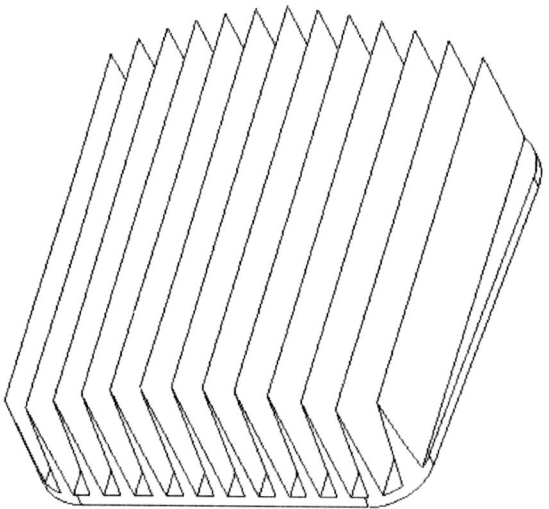

Figure 6-96 Model after trimming the fins by creating the cut

Options Aiding Construction of Parts-II   6-51

Creating the Seventh Feature

The seventh feature is a is an extrude cut that will be created on the top face of the base feature.

1. Invoke the **Extrude** dashboard and select the **Remove Material** button.

2. Select the top face of the base feature as the sketch plane.

3. Select the **RIGHT** datum plane and then select the **Right** option from the **Orientation** drop-down list, if they are not selected by default.

4. Once you enter the sketcher environment, draw the sketch of the circular cut feature, as shown in Figure 6-97.

5. After the sketch is complete, exit the sketcher environment.
   The **Extrude** dashboard is enabled and appears below the graphics window.

6. Select the **Extrude to intersect with all surface** button from the depth flyout.

7. If required, change the depth direction of extrude such that the fins of the model are cut.

8. Choose the **Build feature** button from the dashboard. The top view of the model after creating the circular cut is shown in Figure 6-98.

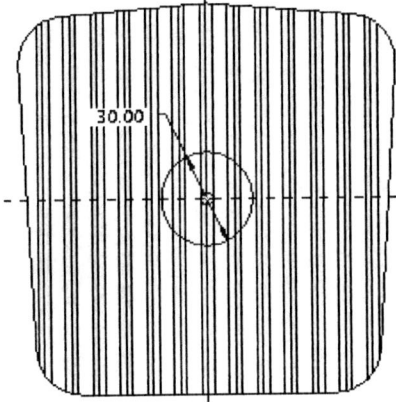

Figure 6-97 Sketch of the cut feature

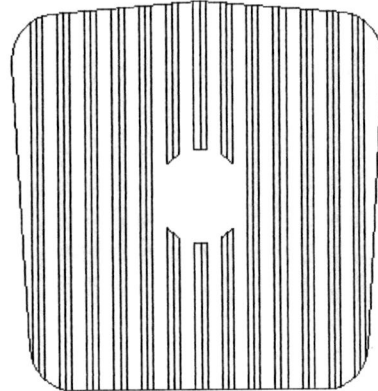

Figure 6-98 Model after creating cut

Creating the Eighth Feature

The eighth feature is a protrusion feature that will be created on the top face of the base feature.

1. Invoke the **Extrude** dashboard and select the top face of the base feature as the sketch plane.

2. Select the **RIGHT** datum plane and then select the **Right** option from the **Orientation** drop-down list, if they are not selected by default.

3. Once you enter the sketcher environment, use the edge of the cut feature to create the sketch of the protrusion feature.

4. After the sketch is complete, exit the sketcher environment.

5. Enter a value of **9** in the dimension box present on the **Extrude** dashboard.

6. Choose the **Build feature** button from the dashboard. The model after creating the cylindrical feature is shown in Figure 6-99.

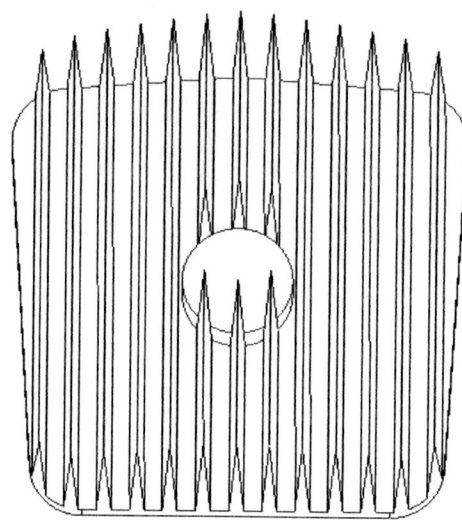

Figure 6-99 Model after creating the cylindrical feature

Creating the Ninth Feature

The ninth feature is a cut feature and it will be patterned to create the other instances. The cut feature is created on the top face of the base feature.

1. Invoke the **Extrude** dashboard and select the **Remove Material** button.

2. Select the top face of the base feature as the sketch plane.

3. Select the **RIGHT** datum plane and then select the **Right** option from the **Orientation** drop-down list.

4. Once you enter the sketcher environment, draw the sketch of the circular cut feature, as shown in Figure 6-100.

5. After the sketch is complete, exit the sketcher environment. The **Extrude** dashboard is enabled and appears below the graphics window.

Options Aiding Construction of Parts-II   6-53

6. Select the **Extrude to intersect with all surface** button from the depth flyout.

7. Change the depth direction of extrude such that the fins of the model are cut.

8. Choose the **Build feature** button from the dashboard. The model after creating the circular cut is shown in Figure 6-101.

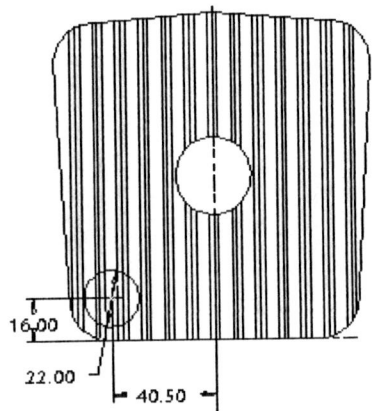

Figure 6-100 Sketch of the cut feature   *Figure 6-101* Model after creating cut

Now, the cut feature will be patterned to create its other instances.

9. Select the cut feature from the **Model Tree** and then choose the **Pattern Tool** button from the **Edit Features** toolbar. Select the **Dimension** option if it is not selected by default in the drop-down list.

   The **Pattern** dashboard is displayed. You are prompted to select the dimensions to vary in the first direction.

10. Select the **General** option from the **Options** slide-up panel if it is not selected.

11. Select the dimension value **40.5** from the graphics window. After you select the dimension in the first direction, the edit box is displayed.

12. Enter a value of **-81** in the edit box.

13. Hold down the right mouse button to display a shortcut menu. Choose the **Direction 2 Dimensions** option from the shortcut menu.

14. Select the dimension value **16**. After you select the dimension in the second direction, the edit box is displayed.

15. Enter a value of **80** in the edit box. Note that in the edit boxes present on the dashboard, the number of instances, **2**, is specified by default.

16. Choose the **Build feature** button from the **Pattern** dashboard. The rectangular pattern of cut features is displayed, as shown in Figure 6-102.

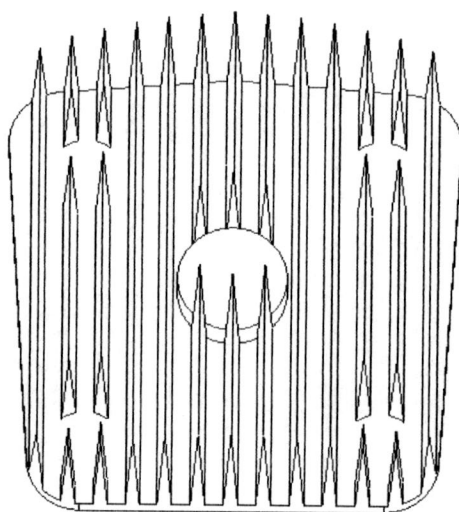

Figure 6-102 *Model after patterning the cut feature*

Creating the Tenth Feature

The tenth feature is a extrude feature and it will be patterned to create the other instances. The protrusion feature is created on the top face of the base feature.

1. Invoke the **Extrude** dashboard.

2. Select the top face of the base feature as the sketch plane.

3. Select the **RIGHT** datum plane and then select the **Right** option from the **Orientation** drop-down list, if they are not selected by default.

4. Once you enter the sketcher environment, draw the sketch of the protrusion feature using the edges of the cut feature.

5. After the sketch is complete, exit the sketcher environment.

6. Enter a value of **9** in the dimension box present on the dashboard.

7. Choose the **Build feature** button from the dashboard. The model after creating the protrusion feature is shown in Figure 6-103.

    Now, a reference pattern of the protrusion feature will be created in order to create its other instances.

8. Select the extrude feature from the **Model Tree** and then choose the **Pattern Tool** button from the **Edit Features** toolbar.

Options Aiding Construction of Parts-II

The **Reference** option is selected by default in the drop-down list present on the **Pattern** dashboard.

9. Choose **Build feature** from the **Pattern** dashboard. The model after creating the pattern is shown in Figure 6-104.

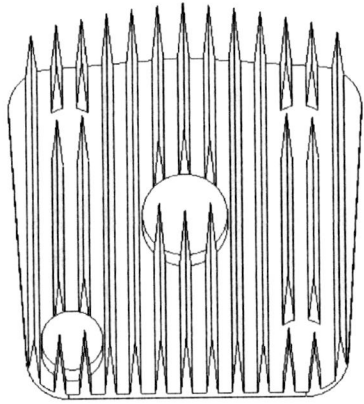

Figure 6-103 Model after creating the cylindrical feature

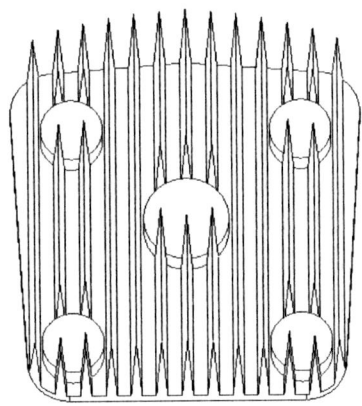

Figure 6-104 Model after creating the reference pattern of the cylindrical feature

Creating the Eleventh Feature

The eleventh feature is a hole that is created on the top face of the tenth feature and later it is reference patterned to create its other instances.

1. Create a hole of diameter **12** on the top face of the protrusion feature created on the lower left corner.

    Note
   Make sure that the primary reference while creating the hole is the top face of the protrusion feature and the secondary reference is the axis. If you take the axis as the primary reference, the pattern of holes may not be what is desired.

2. Pattern the feature using the **Reference** option. The model after creating the pattern of the hole feature is shown in Figure 6-105.

Creating the Last Feature

The last feature is a hole of diameter **16** that is created on the top face of the eighth feature. This hole is coaxial with the cylindrical feature on which it is being created. The model after creating the hole feature is shown in Figure 6-106.

Saving the Model

You have to save the model because you may need it later.

1. Choose the **Save the active object** button from the **File** toolbar and save the model. The order of feature creation can be seen from the **Model Tree** shown in Figure 6-107.

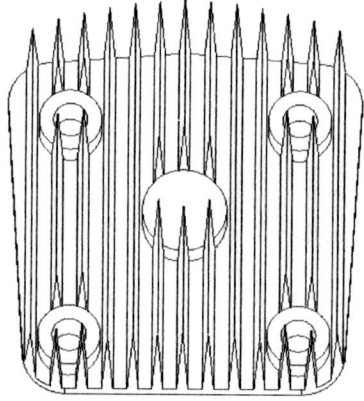

Figure 6-105 Model after creating the reference pattern of the hole feature

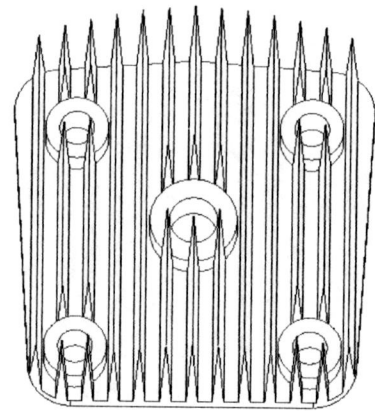

Figure 6-106 Model after creating the hole on the cylindrical feature

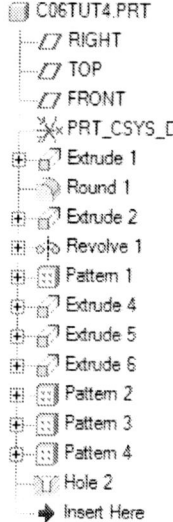

Figure 6-107 Model Tree of Tutorial 4

Creating a section view of the model

1. Choose **View > View Manager** from the menu bar. The **View Manager** dialog box is displayed.

2. Choose the **New** button from the dialog box. **Rep0001** is displayed in the **Names** display box.

3. Press ENTER. The dialog box disappears and then appears again.

4. Select **Rep0001** and then right-click. From the shortcut menu choose the **Redefine** option. The **EDIT METHOD** menu is displayed.

5. Choose the **Work Region** option from the **EDIT METHOD** menu. The **SOLID OPTS** submenu is displayed.

Options Aiding Construction of Parts-II   6-57

6. Choose **Extrude > Solid > Done** from the **SOLID OPTS** submenu. The **Extrude** dashboard is displayed and from it invoke the **Sketch** dialog box.

7. Select the bottom most planar face of the model as the sketch plane and select the **RIGHT** datum plane to be at the right while drawing the sketch.

8. Choose the **Flip** button to reverse the direction of yellow arrow. You need to draw the sketch from the side of the sketch plane where the fins are facing. This is because when you draw the sketch, you have to make sure that the section lines are not cutting the fins.

9. When you enter the sketcher environment, draw the section lines, as shown in Figure 6-108. This section is an open loop. The dimensions are not of importance to create this section.

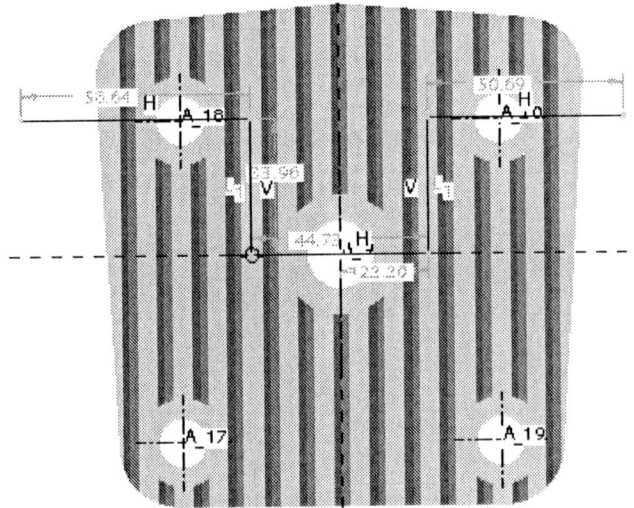

Figure 6-108 Sketch of the section lines with weak dimensions

10. Exit the sketcher environment and select the **Extrude to intersect with all surface** button from the depth flyout. The yellow arrow is pointing in the reverse direction.

11. Choose the **Change material direction of extrude to other side of sketch** button to reverse the direction of arrow.

12. Choose the **Change depth direction of extrude to other side of sketch** button to reverse the direction extrude.

13. Choose the **Build feature** button from the **Extrude** dashboard. The sectioned model is shown in Figure 6-109.

14. Choose the **Done/Return** option from the **EDIT METHOD** menu and close the **View Manager**.

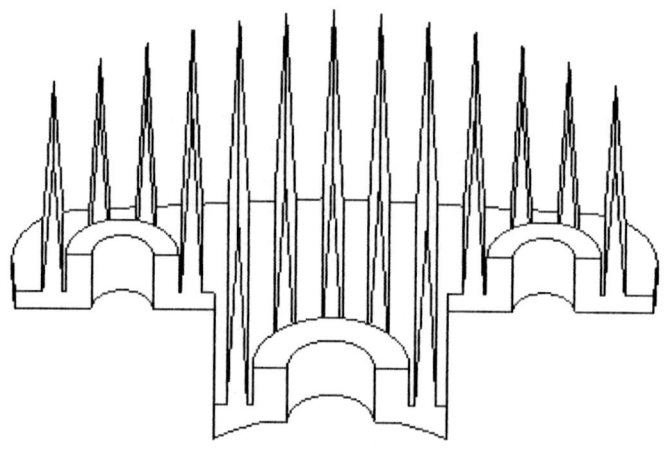

Figure 6-109 Sectioned model

Self-Evaluation Test

Answer the following questions and then compare your answers with the answers given at the end of this chapter.

1. When a pattern is created, the leader or the parent feature also becomes a part of the pattern. (T/F)

2. Once a pattern is created, all instances in the pattern including the parent feature act as a single feature. (T/F)

3. In Reference pattern, an existing pattern is referenced to create a new pattern. (T/F)

4. Using the **Pattern** dashboard, you can create only linear patterns. (T/F)

5. If you select a pattern feature from the **Model Tree** and right-click to display the shortcut menu and then choose the **Delete** option, the whole pattern is deleted including the leader. (T/F)

6. The _____ option is used to delete a pattern leaving the leader feature.

7. The _____ option is used to mirror an entire geometry about a plane.

8. The _____ option from the **Pattern** dashboard is used to create a pattern in which the dimensions of the instances can be varied.

9. The first feature in a pattern is called _____.

Options Aiding Construction of Parts-II

10. The _____ menu is used to select the dimensions that you want to vary from the leader while copying.

Review Questions

Answer the following questions:

1. Which of the following options from the **Pattern** dashboard are used to create patterns in which the instances touch each other, and intersect with other instances or the edges of the surface?

    (a) **Identical** (b) **Variable**
    (c) **General**   (d) None

2. Which of the following datums is required to create a rotational pattern?

    (a) Graph (b) Curve
    (c) Axis   (d) Point

3. Which of the following options from the **Pattern** dashboard cannot be used to create a pattern that intersects an edge of a feature on which the pattern has to be created?

    (a) **Identical** (b) **Variable**
    (c) **General**   (d) None

4. Which of the following options from the **Pattern** dashboard is used to create a pattern that has all instances of different size?

    (a) **Identical** (b) **Variable**
    (c) **General**   (d) None

5. Which of the following options in the **COPY FEATURE** submenu creates a parent-child relationship between the copied feature and the source feature?

    (a) **FromDifModel** (b) **FromDifVers**
    (c) **Independent**   (d) **Dependent**

6. You can mirror features using the datum planes or planar surfaces. (T/F)

7. To create a rotational pattern, you should specify an angular increment. (T/F)

8. The option **New Refs** is in the **COPY FEATURE** submenu. (T/F)

9. When you copy a feature using the **New Refs** option, then it is not related with the original feature. (T/F)

10. The features copied from different models or from different versions are always independent. (T/F)

Exercises

Exercise 1

Create the model shown in Figure 6-110. The top view, front view, right-side view, detailed, and sectioned views of the model are shown in Figure 6-111. **(Expected time: 45 min)**

Figure 6-110 Solid model for Exercise 1

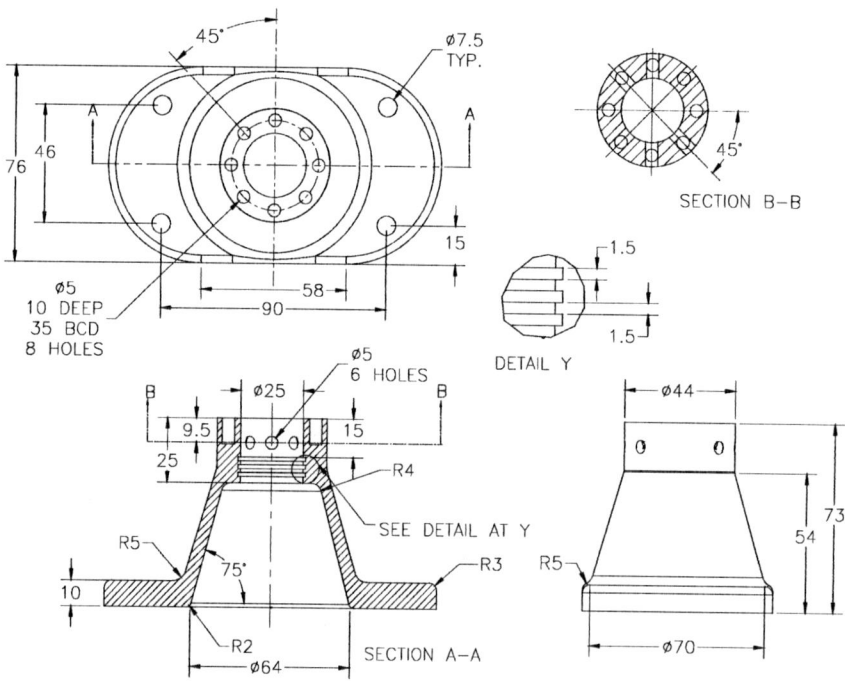

Figure 6-111 Top, front, right-side, detailed, and sectioned views of the model

Options Aiding Construction of Parts-II

Exercise 2

Create the model shown in Figure 6-112. The figure also shows the top view, front view, and the right-side view of the model. **(Expected time: 30 min)**

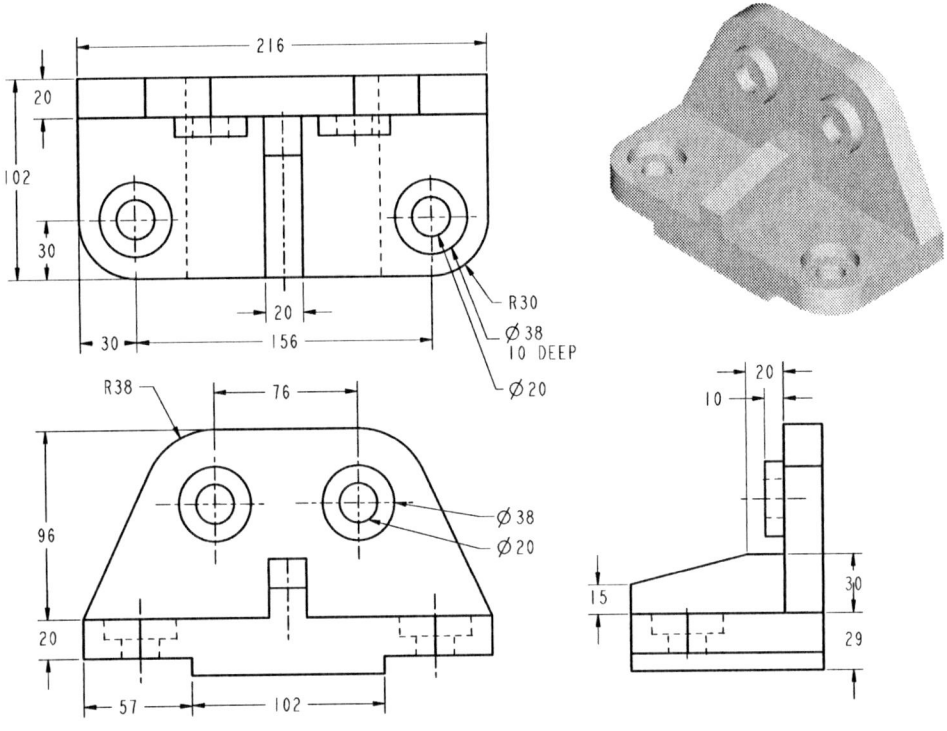

Figure 6-112 Top, front, right-side, and isometric views of the

Exercise 3

Create the model shown in Figure 6-113. The top view, front view, and the isometric view is shown in the figure. **(Expected time: 30 min)**

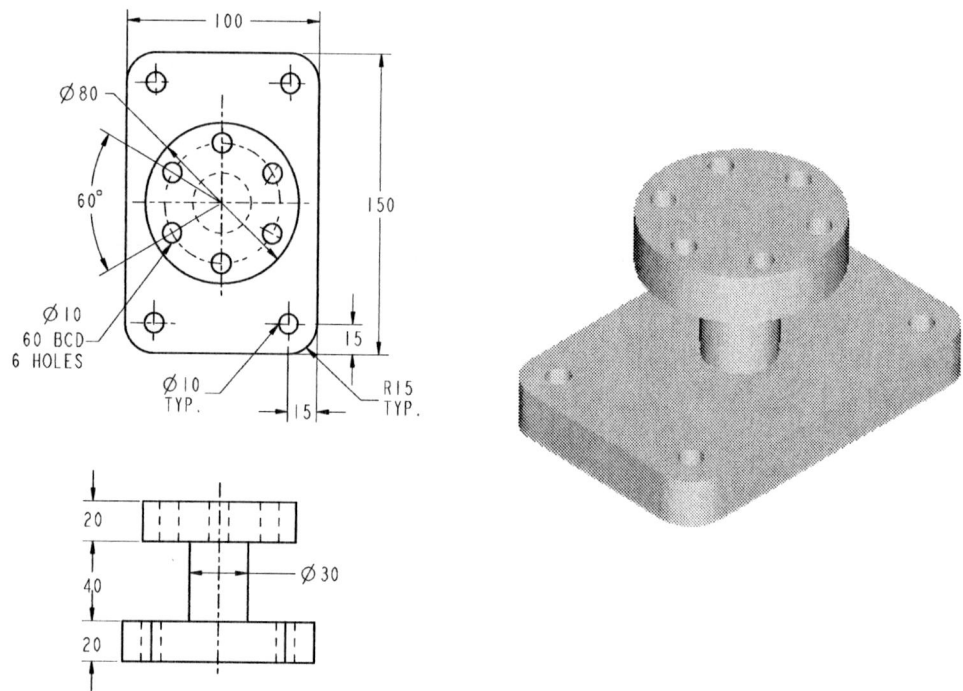

Figure 6-113 Top view, front view, and the isometric view of the model

Exercise 4

Create the model shown in Figure 6-114. The front and sectioned right-side views are shown in the Figure 6-115.   **(Expected time: 1 hr)**

Note

*To create the three protrusion features on the outer cylindrical surface, you will have to create a datum on-the-fly using the **Datum Plane Tool** button to orient the sketch view. Also, when you enter the sketcher environment, you need to select the central axis of the model for defining the references using the **References** dialog box. As you select the central axis, the message in the **Reference status** area of the **References** dialog box will be **Fully Placed**. This suggests that you do not require any datum plane to define the references. Use this central axis to dimension the sketch. If you do not select the central axis to define the references using the **References** dialog box, the resulting feature may not get patterned.*

You will learn about the aligned section views and total aligned section views in Chapter 10.

Options Aiding Construction of Parts-II

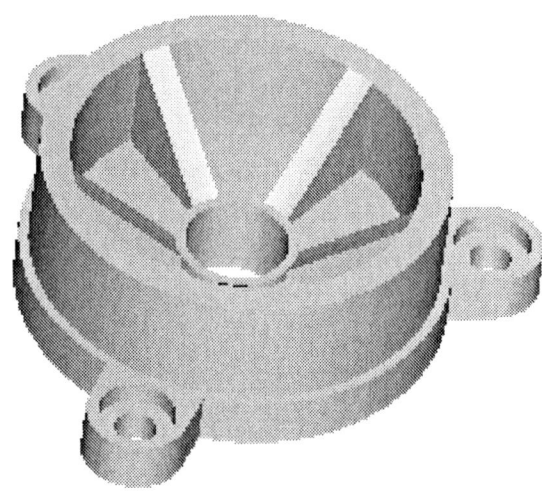

Figure 6-114 Solid model for Exercise 4

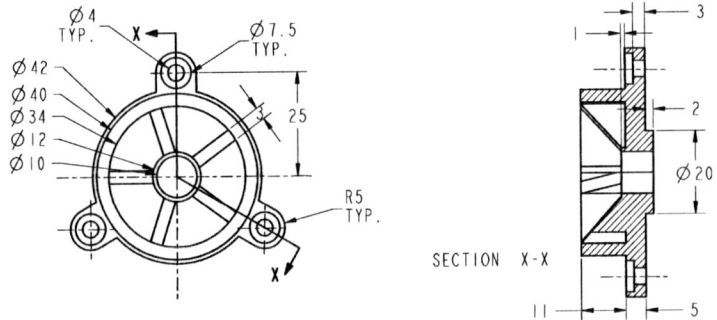

Figure 6-115 Top view and the total aligned sectioned view of the model

Answers to the Self-Evaluation Test
1 - T, 2 - T, 3 - T, 4 - F, 5 - T, 6 - **Delete Pattern**, 7- **Mirror Tool**, 8 - **Variable**, 9 - leader, 10 - **GP VAR DIMS**.

Chapter 7

Advanced Modeling Tools-I

Learning Objectives

After completing this chapter you will be able to:
- *Create sweep features*
- *Create features using sweep cut.*
- *Create parallel, rotational, and general blend.*
- *Use blend vertex in blend features.*
- *Create shell features.*
- *Create datum curves.*
- *Create draft features.*

OTHER PROTRUSION OPTIONS

The **Extrude Tool** and **Revolve Tool** buttons available in the **Base Features** toolbar were discussed in Chapter 3, and the **Sweep** and **Blend** options are discussed in this chapter. As mentioned earlier, Protrusion and Cut are the two basic options available in Pro/ENGINEER that are used to create a feature.

Note

All options that are available for creating a cut are similar to those that are available for creating a protrusion. Remember, that a cut is performed on an existing feature and therefore, the Cut option is available only when at least a base feature exists in the graphics window.

In this chapter you will also learn about the tools of solid modeling that make the creation of a complex model easy. In the next section you will learn about sweep features.

SWEEP FEATURES

The **Sweep** option extrudes a section along a defined trajectory. The order of operation is to first create a trajectory and then a section. Trajectory is a path along which a section is swept. The trajectory for a sweep feature can be either sketched or selected. The **Sweep** option of protrusion is similar to the **Extrude** option. The only difference being that in the case of the **Extrude** option, the feature is extruded in a direction normal to the sketching plane, but in the case of the **Sweep** option, the section is swept along the sketched or selected trajectory. The trajectory can be open or closed. Normal sketching tools are used for sketching the trajectory. The cross section of the swept feature remains constant throughout the sweep.

Note

Some important points to remember while drawing a trajectory and a section for a sweep feature are discussed later in this chapter.

Creating Swept Protrusions

The **Sweep** option can be used for adding material as well as for removing material, that is, for protrusion as well as for cut features. You can choose the **Sweep** option from the menu bar. In the **Insert > Sweep > Protrusion** option that is discussed here, material defined by the section is added in the specified path. The **SWEEP TRAJ** (Sweep Trajectory) menu appears, as shown in Figure 7-1. The options in this menu are discussed next.

Figure 7-1 SWEEP TRAJ menu

Sketch Traj Option

The **Sketch Traj** (Sketch Trajectory) option is used when you want to sketch the trajectory for the sweep feature. This is the most commonly used option for defining the trajectory. As mentioned earlier, the trajectory can be open or closed. There are some limitations for using closed or open trajectory with closed or open section. These limitations are discussed in the next section. When you choose the **Sketch Traj** option, you are prompted to select a sketching plane. The sketching plane you select will be parallel to the screen when you sketch the

Advanced Modeling Tools-I

trajectory. Figure 7-2 shows how the section is swept along the sketched trajectory, and Figure 7-3 shows the shaded image of the swept feature.

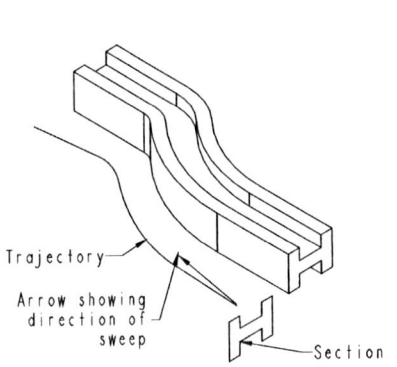

Figure 7-2 Sweep along the sketched trajectory

Figure 7-3 Shaded image

The following points specify the combinations of trajectories and sections that are possible/not possible to create.

1. Open section and open trajectory are not possible.

2. Closed section and open trajectory are possible.

3. If the sketched trajectory is a closed loop then after you exit the sketcher environment, the **ATTRIBUTES** menu is displayed, as shown in Figure 7-4. There are two options that are available: **Add Inn Fcs** (Add inner faces) and **No Inn Fcs** (No inner faces).

Figure 7-4 ATTRIBUTES menu

For **Add Inn Fcs**, only open sections are possible, as shown in Figure 7-5. The shaded image of the corresponding swept feature is shown in Figure 7-6. These two figures explain the **Add Inn Fcs** option.

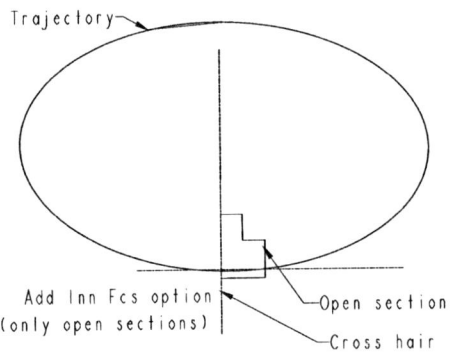

*Figure 7-5 **Add Inn Fcs** option*

Figure 7-6 Shaded image

For **No Inn Fcs**, only closed sections are possible, as shown in Figure 7-7. The shaded image of the corresponding swept feature is shown in Figure 7-8.

Figure 7-7 No Inn Fcs option

Figure 7-8 Sweep feature

Select Traj Option

The **Select Traj** (Select Trajectory) option allows you to select a trajectory in the graphics window. The trajectory to be selected can be an existing edge or a datum curve. Creation of datum curves will be discussed later in the chapter. When you choose this option from the **SWEEP TRAJ** menu, the **CHAIN** menu is displayed. The **CHAIN** menu is also discussed later in this chapter.

Figures 7-9 and 7-10 show two examples of selecting the edges of the base feature and then using these as a trajectory to sweep. The corresponding swept features are also shown.

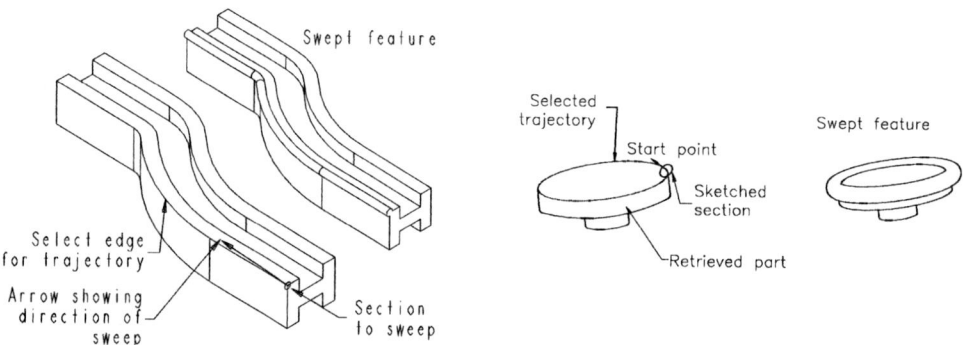

Figure 7-9 Sweep along the selected trajectory

Figure 7-10 Sweep along the selected trajectory

Sketching a Trajectory Aligned to an Existing Geometry

When one end of the sketched trajectory is aligned to the adjacent geometry of the existing feature, Pro/ENGINEER provides two options. The first option is to merge the ends of the sweep feature with the adjacent geometry and the second option is to leave the ends of the sweep feature free. These options are available in the **ATTRIBUTES** menu that is displayed after you complete the sketch of the trajectory and choose the **Continue with the current section** button. The **ATTRIBUTES** menu is shown in Figure 7-11.

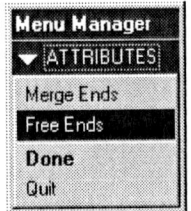

Figure 7-11 ATTRIBUTES menu

The **ATTRIBUTES** menu is displayed only when the trajectory is aligned to an edge or surface of the feature that already exists in the graphics window. This means that the **ATTRIBUTES** menu does not appear if the sweep feature you are drawing is the base feature of a model or, in other words, if there is no adjacent geometry to which the trajectory can be merged. The options available in the **ATTRIBUTES** menu are discussed next.

Merge Ends
The **Merge Ends** option merges the end of a sweep feature to the surface to which the end of the trajectory is aligned. For this option the trajectory should be aligned to the adjacent geometry.

Free Ends
The **Free Ends** option leaves the sweep feature partially attached to the adjacent feature even if the end of the trajectory is aligned to the adjacent geometry.

Figure 7-12 show the **Merge Ends** and **Free Ends** options. In the figure shown below, the trajectory is aligned with the adjacent geometry in both the cases. Figure 7-13 shows the corresponding shaded image.

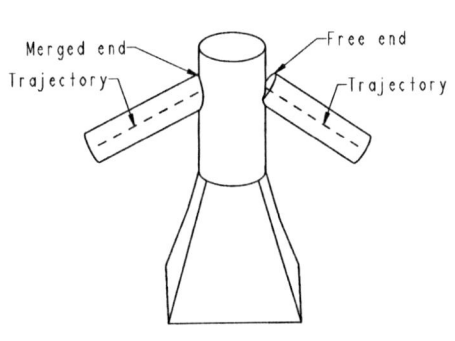

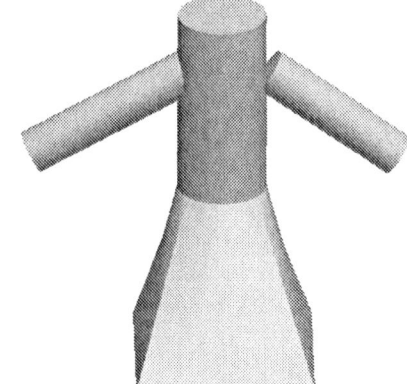

Figure 7-12 Merge Ends and Free Ends   *Figure 7-13 Shaded image*

Creating Sweep Feature by Selecting a Trajectory
When you choose the **Select Traj** option from the **SWEEP TRAJ** menu, the **CHAIN** menu is displayed, as shown in Figure 7-14. You can use the **CHAIN** menu only if you have created a feature that will be used to select the trajectory. The options in this menu are used to select a trajectory. These options are discussed next.

Tip: *The following points should be remembered while creating a sweep feature:*

1. *Similar to other sketched features the trajectory of the sweep feature is also sketched after selecting a sketching plane.*

2. *The section for the sweep trajectory is sketched using the normal sketcher tools when the sketch trajectory option is selected.*

3. *At bends in a trajectory, the radius of the bend should be proportionate to the cross section to be swept to avoid overlapping. If the section size is large and the radius of the curve or bend is small, overlapping takes place and the sweep feature will not be created. Therefore, make sure that the ratio of the size of the section to the size of the trajectory is appropriate.*

One By One

The **One By One** option of the **CHAIN** menu is selected by default. Using this option you can select an edge or curve individually, one by one. You have to hold the **CTRL** key down while selecting the edges. When you select an edge, it is highlighted in red. Before selecting the edge, make sure that the **Select** option in the **CHAIN** menu is highlighted. The edge once selected and confirmed by choosing **OK** button from the **Select** dialog box or by using the middle mouse button can also be unselected by selecting the **Unselect** option from the **CHAIN** menu.

Tangnt Chain

Using this option, you can select an edge or edges tangent to the selected edge. When you select an edge, all edges tangent to the selected edge are highlighted. If the selected edge is not tangent to any other edge then the function of this option is the same as that of the **One By One** option. The difference being that you can select only one edge in the case of the **Tangnt Chain** option.

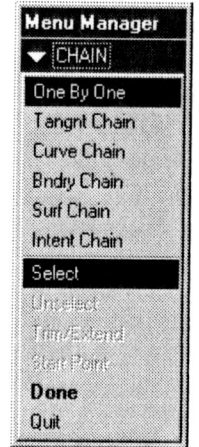

Figure 7-14 CHAIN menu

Curve Chain

You can select a chain of curves by using the **Curve Chain** option.

Bndry Chain

The **Bndry Chain** (Boundary Chain) option is used only for surface features. You can define a chain by selecting a quilt and using its one-sided edges. If the quilt has more than one loop, select a specific loop to define the chain.

When you select the edge of the quilt, it is highlighted in blue and the **CHOOSE** menu is displayed with the options, as shown in Figure 7-15.

Figure 7-15 CHOOSE menu

Advanced Modeling Tools-I

Surf Chain

Using the **Surf Chain** option, you can define a chain by selecting a surface and using its edges. If the surface has more than one loop, then you are prompted to specify a loop to define the chain. When a surface is selected, the **CHAIN OPT** menu is displayed, as shown in Figure 7-16. Choose either **Select All** or **From-To** from the **CHAIN OPT** menu.

Figure 7-16 CHAIN OPT menu

Intent Chain

The **Intent Chain** option is used to select multiple edges. When a section is extruded, the edges formed by the extrusion consists of intent chains. The intent chains can either be the edges of the section or the edges of the extruded surface. The edges selected should form a closed loop.

Creating Thin Sweep Protrusion

The **Sweep > Thin Protrusion** option creates a thin sweep feature with a specified thickness. This option is similar to the **Extrude > Thin** option that was discussed in Chapter 3. In case of thin features, a certain thickness has to be specified. The thickness is specified on one side of the section or symmetrically to both the sides of the section. The resultant sweep is similar to the solid sweep created with a section comprising of two closed loops at some offset distance. Figure 7-17 shows the sections that can be used to create the model shown in Figure 7-18.

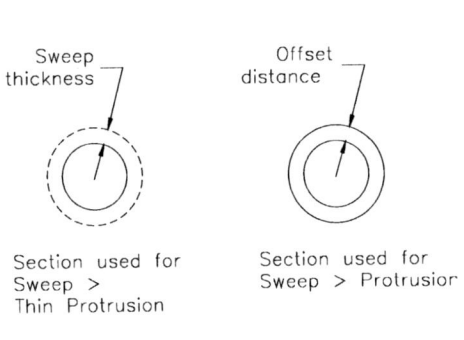

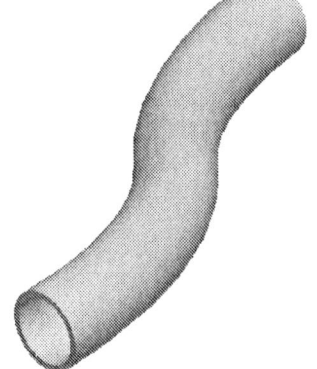

Figure 7-17 Two possible options to create the same model

Figure 7-18 Same model created using different sections

Creating a Sweep Cut

To create a **Sweep Cut** feature, the procedure to be followed is the same as that in **Sweep Protrusion**. The only difference is that in case of cut features, the material is removed from an existing feature. The **Cut** option can be invoked by choosing **Insert > Sweep > Cut** from the menu bar. Cut can be a solid swept cut or thin swept cut. Figure 7-19 shows trajectories for the **Sweep Cut** feature. Figure 7-20 shows the shaded model of an open trajectory sweep cut and a closed trajectory sweep cut.

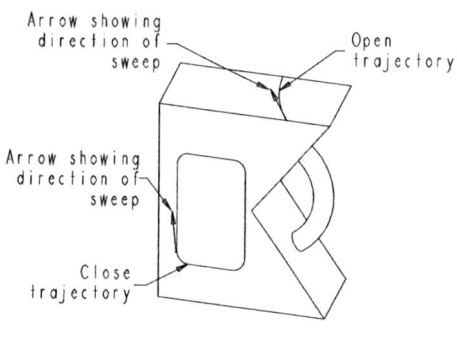

Figure 7-19 Trajectories for **Sweep Cut**

Figure 7-20 Shaded model with sweep cuts

BLEND FEATURES

Blend features are composed of two or more sections that are joined through transitional faces at their edges so as to form a continuous feature. The number of entities in each section that creates the blend feature should be the same. For example, you cannot blend a circle with a rectangle. This is because a rectangle is composed of four entities and a circle of one. It can be achieved only if the circle is divided into four entities.

In Pro/ENGINEER, the **Blend** feature is of two types; **Protrusion** and **Cut**. The **Blend** option is used where the feature to be created has varying cross sections. To invoke this option, choose **Insert > Blend > Protrusion** or **Insert > Blend > Cut** from the menu bar.

When you choose the **Blend > Protrusion** option, the **BLEND OPTS** submenu is displayed, as shown in Figure 7-21. The options in this menu are discussed next.

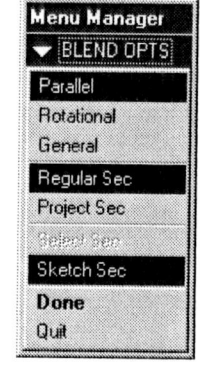

Figure 7-21 **BLEND OPTS** submenu

Parallel Option

Parallel blends have sections that are drawn parallel to each other with a specified distance between them.

After choosing **Parallel > Regular Sec > Sketch Sec > Done** from the **BLEND OPTS** submenu, the **ATTRIBUTES** menu is displayed, as shown in Figure 7-22. The options in this menu are discussed next.

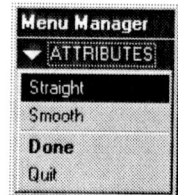

Figure 7-22 **ATTRIBUTES** menu

Straight Option

The **Straight** option is used to connect the vertices of all sections in a blend feature with straight lines.

Smooth Option

The **Smooth** option is used to connect the vertices of all sections in a blend feature with curves.

Figure 7-23 shows three sections that are used to create the blend feature. Figures 7-24 and 7-25 show the parallel blend features with straight edges and smooth edges respectively, created using the sections shown in Figure 7-23.

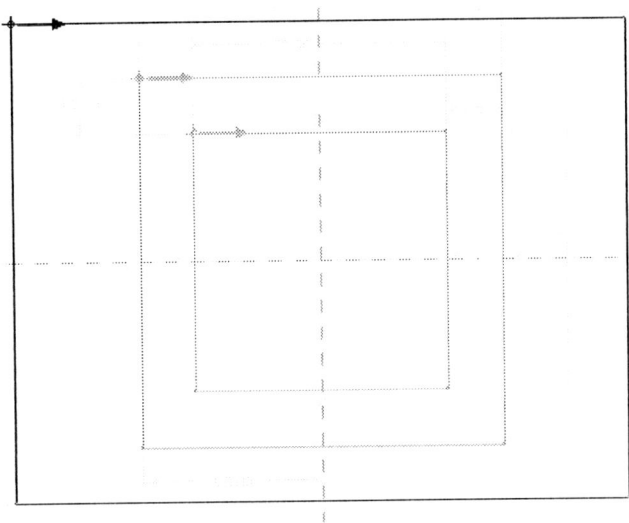

Figure 7-23 Parallel sections

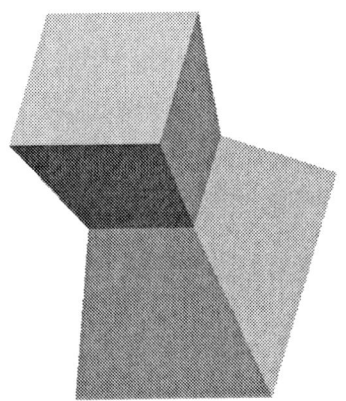   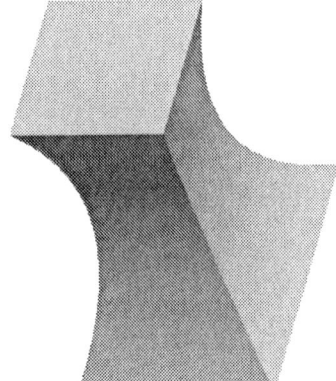

Figure 7-24 Parallel blend with straight edges   *Figure 7-25 Parallel blend with smooth edges*

 Note
While drawing a section, the start point of all sections should be in the same direction in order to avoid twisted blend features.

Tip: *The following points should be remembered while creating a **Parallel** blend feature:*

1. *After completing first section, choose **Sketch** > **Feature Tools** > **Toggle Section** to proceed for drawing the second section. You can also hold down the right mouse button and choose the **Toggle Section** option from the shortcut menu that appears. If you choose the **Continue with the current section** button before drawing the second section, the system prompts you to use the **Toggle Section** option to continue with the second section.*

2. *Active section appears in cyan color and the other sections appear in gray.*

3. *All sections in a blend feature must have the same number of entities. However, you can blend a point with any section irrespective of the number of entities.*

4. *System prompts for depth between subsequent sections after completion of all sections in the blend.*

5. *By default, the start point of any entity that is drawn to define a section is considered as the **Start Point** of the section. To change the **Start Point** of a section, select the point to be defined as the **Start Point** and hold down the right mouse button to display the shortcut menu. Choose the **Start Point** option from this shortcut menu to change the start point.*

6. *After defining the sections in a blend feature and before choosing the **Continue with the current section** button, the sections can be modified by using the **Toggle Section** option.*

Rotational Blend

The rotational blends have sections that are rotated about the Y-axis up to a maximum of 120-degree and the distance between two sections is measured from the coordinate system. Between each section, an angle called the **rotational blend angle** has to be defined. In this type of blend, each section has its own user-defined coordinate system. If the rotational blend angle entered between the two sections is equal to 0-degree, then the **Rotational** blend option functions the same way as the **Parallel** blend option.

Note that all nonparallel blends can be open or close. Therefore, after choosing **Rotational > Regular Sec > Sketch Sec > Done** from the **BLEND OPTS** submenu, the **ATTRIBUTES** menu is displayed, as shown in Figure 7-26. The options in this menu are discussed next.

Open option

The **Open** option is used when the blend feature to be created has to be kept open.

Closed option

The **Closed** option is used to create a closed blend feature. In this type of blend feature,

Advanced Modeling Tools-I 7-11

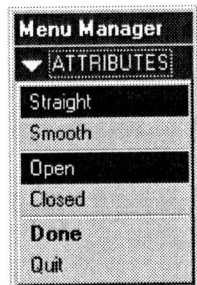

Figure 7-26 ATTRIBUTES menu

 **Tip**: *It is recommended that the closed blend features should have at least three sections.*

Pro/ENGINEER closes the feature by automatically blending the last section with the first section. Figure 7-27 shows the sections used to create a rotational smooth blend feature. The three default datum planes can also be seen. From Figure 7-27, it is evident that two sections are used to create the blend feature and that these sections are at an angle of 45-degree. It is also evident from the figure that the second section is dimensioned from the coordinate system that was defined in the first section. Figure 7-28 shows the shaded model of the same blend feature.

The following steps explain, in brief, the procedure to create the blend feature shown in Figure 7-28.

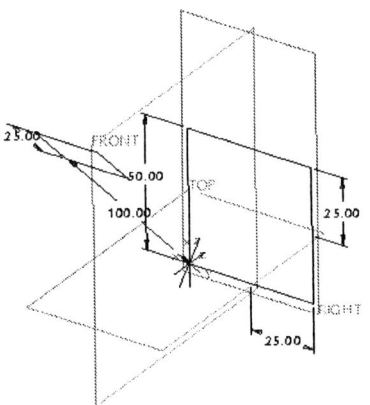

Figure 7-27 The two sections with dimensions and the default datum planes used to create the blend feature shown in the adjacent figure

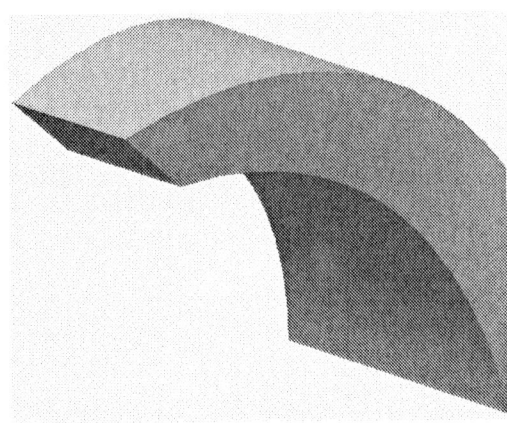

Figure 7-28 Shaded model of rotational open blend feature

1. Invoke the **Blend** option and from the **BLEND OPTS** menu choose **Rotational > Regular Sec > Sketch Sec > Done**. The **ATTRIBUTES** menu is displayed.

2. Choose **Smooth > Open > Done** from the **ATTRIBUTES** menu. You are prompted to select a plane to sketch.
3. Select the **RIGHT** datum plane as the sketch plane and orient the **FRONT** datum plane to be on top.
4. After selecting the sketch plane and its orientation, draw the first section that is a square of side 50. The start point on this square should be at the lower right corner.
5. Place the coordinate system at the lower left corner of the square.
6. Choose the **Continue with the current section** button to proceed for drawing the next section. Accept the default Y-axis rotation angle of 45-degrees for section 2.
7. Draw the second section that is a square of size 25.
8. Place the coordinate system for this section at a distance of 100 toward the left of the lower left corner of the square. Remember that the start point of this section should also lie at the lower right corner of the rectangle.
9. Exit the sketcher environment. The **Message Input Window** appears and you are prompted that do you want to continue to draw the next section.
10. Choose the **No** button. Now, you can see the preview of the feature that you have created and then exit the feature creation tool.

General Option

Using the **General** option, sections are translated and rotated about the x, y, and z axes. The sections are aligned using the user-defined coordinate system. The coordinate system has to be manually placed in every section sketch that constitutes the blend feature.

USING BLEND VERTEX

As mentioned earlier, each section of the blend feature must have an equal number of entities. However, you can use the **Blend Vertex** option if the number of entities in all sections are not equql. For example, to create a blending between a rectangle and a triangle, add blend vertex on a point other than the start point of the triangle. The two vertices of the triangle will be blended with the two vertices of the rectangle and the blend vertex in the triangle will be blended with the remaining two vertices in the rectangle, as shown in Figure 7-29.

To add a blend vertex in a sketch, select the point where you want to place the blend vertex. The selected point is highlighted in red. Choose **Sketch > Feature Tools > Blend Vertex**. The blend vertex is placed at the selected point.

Note
The **Blend Vertex** can be used only in either the first or last section of a blend feature.

SHELL OPTION

The **Shell** option scoops out the material from the model and at the same time removes the selected faces, leaving behind a thin model with some specified wall thickness. The **Shell** option can be invoked by choosing the **Shell Tool** from the **Engineering Features** toolbar. You can also choose this option from the menu bar by choosing **Insert > Shell**. The **Shell** dashboard is displayed, as shown in Figure 7-30.

Advanced Modeling Tools-I

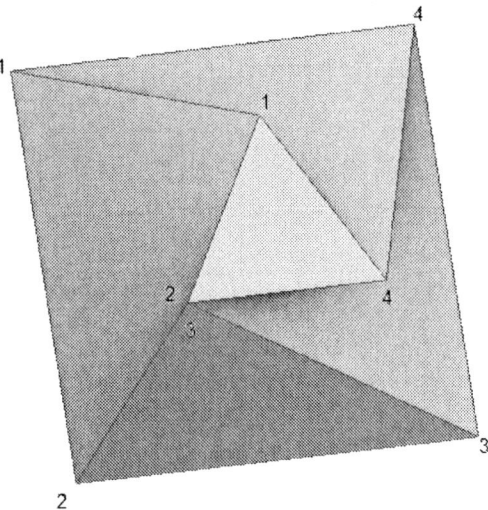

Figure 7-29 Blending a square with a triangle using the blend vertex

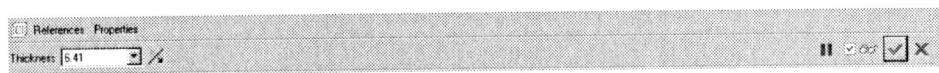

Figure 7-30 The **Shell** dashboard

Note
*The **Shell** option is used on existing models and hence this option is available only when a model exists in the graphics window.*

Using the **Shell** dashboard you can create two types of shell.

1. Constant thickness shell
2. Variable thickness shell

Creating Constant Thickness Shell

The constant thickness shell is a shell that has a uniform thickness on all four faces of the model. The following steps explain the procedure to create a constant thickness shell.

1. Invoke the **Shell** dashboard.
2. Select the top face of the model to remove, as shown in Figure 7-31. The selected face will be removed from the model, leaving the specified thickness from the boundary of the selected face.
3. Enter the thickness value of the shell in the dimension box present on the dashboard.
4. Choose the **Feature Preview** button from the dashboard to preview the shell feature.
5. Choose the **Build feature** button from the **Shell** dashboard to exit it.

The shell is created on the selected face, as shown in Figure 7-32. The thickness of the shell is uniform on all faces of the model.

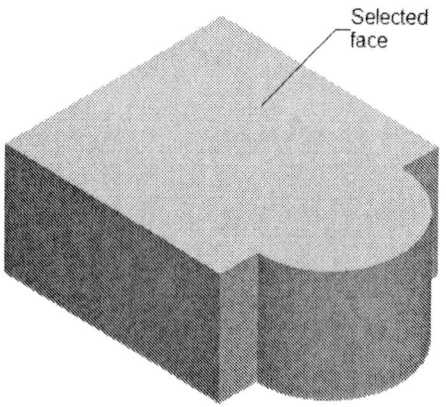

Figure 7-31 Highlighted surface to shell   *Figure 7-32* Shell created on the selected surface

Note
*When you invoke the **Shell** dashboard, the **Surface** filter is selected by default.*

Creating Variable Thickness Shell

The variable thickness shell is a shell that has different thickness values assigned to adjacent faces. The following steps explain the procedure to create a variable thickness shell.

1. Invoke the **Shell** dashboard.
2. Select the top face of the model to remove, as shown in Figure 7-33.
3. Choose the **References** tab to invoke the slide-up panel. In the slide-up panel, there are two collectors. The **Removed Surfaces** collector shows the surface id of the face that you have selected to remove. The **Non Default Thickness** collector shows the surfaces that you will select for creating the variable thickness shell.
4. Click in the **Non Default Thickness** collector and now select the adjacent faces that are shown in Figure 7-33. To select the second and the successive faces you need to use CTRL+left mouse button.
   Now you need to specify the different thickness values with reference to the selected face.
5. Enter the thickness values in the dimension boxes that are present on the right side of the surfaces in the **Non Default Thickness** collector.

The shell is created on the selected faces and with thickness values assigned to them, see Figure 7-34.

Note
The thickness value entered can be positive or negative. If the value entered is positive, the material is removed, leaving the shell thickness inside the boundary of the selected face. But when the value entered is negative, the shell thickness is added outside the boundary of the selected face.

Figure 7-35 shows the model whose top surface is selected to be removed in order to create a shell. Figure 7-36 shows the model after shelling. Notice that the shell thickness is left on the selected face and the remaining material is removed.

Advanced Modeling Tools-I

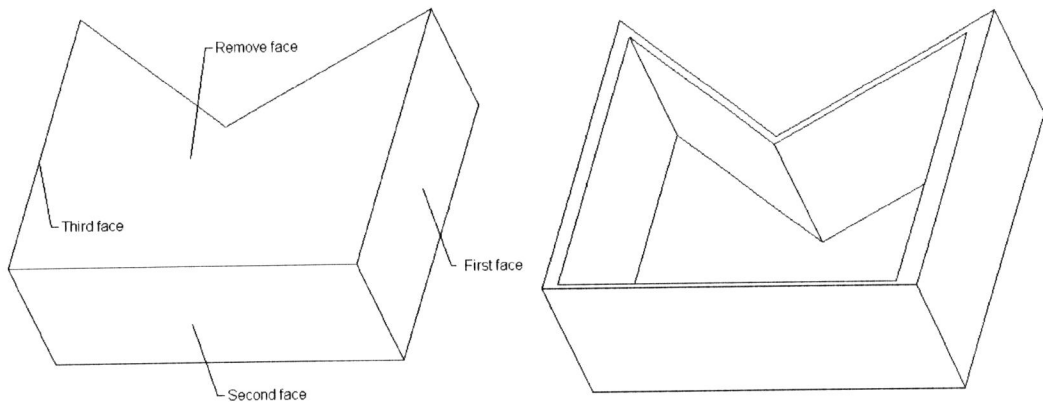

Figure 7-33 Faces of the model to select for variable thickness and the top face to remove

Figure 7-34 Variable thickness shell

Figure 7-35 Solid model

Figure 7-36 Model after shelling the top face

DATUM CURVES

Datum curves are useful in creation of advanced solid and surface features such as the sweep trajectories to create a sweep feature. A datum curve is considered as a feature and is displayed in the **Model Tree**. There are various tools and options to create a datum curve. You can use the **Insert a datum curve** button and **Sketched Datum Curve Tool** button from the **Datum** toolbar to create a datum curve. You can also choose **Intersect**, **Project**, and **Wrap** options in the **Edit** menu to create a datum curve.

Creating Datum Curve Using the Insert a datum curve Button

 The **Insert a datum curve** button is used to create datum curves. When you choose this button from the **Right Toolchest**, the **CRV OPTIONS** menu is displayed, as shown in Figure 7-37. The options in this menu are discussed next.

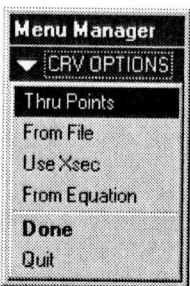

Figure 7-37 CRV OPTIONS menu

Thru Points

The **Thru Points** option creates a datum curve by selecting the existing datum points or vertices. The resulting datum curve may be a spline curve or can have a user-defined radii.

When you choose **Done** from this menu, the **CONNECT TYPE** menu is displayed, as shown in Figure 7-38. The options in this menu are discussed next.

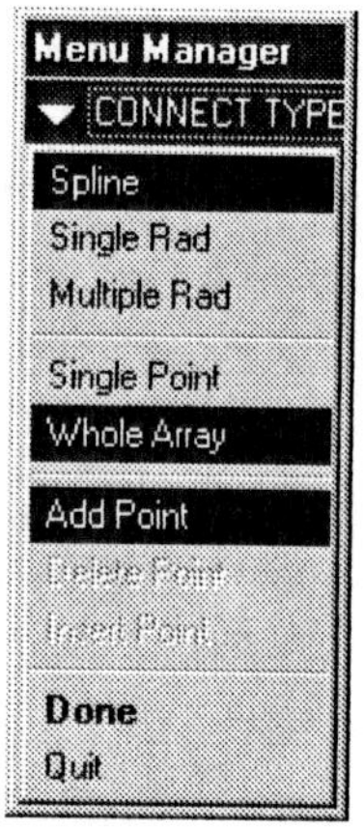

Spline

The **Spline** option creates a datum curve in the form of a spline and passing through the selected datum points or vertices.

Single Rad

The **Single Rad** option creates a datum curve that has a constant radius at the bends. This option is called single radius because the system prompts you to specify a radius value at the bend. The bend is the location on the datum curve that lies between two datum points or vertices.

Figure 7-38 CONNECT TYPE menu

Multiple Rad

As the name suggests, the **Multiple Rad** option allows you to specify radius at every bend that occurs on a datum curve. Using this option you can either specify the same radius that was specified at the previous bend or a new radius value.

Single Point

The **Single Point** option is used to select the datum points individually. This option is provided because if the datum points are created such that they are a single feature, then all of them may be selected.

Whole Array

The **Whole Array** option is used to select all datum points that act as a single feature.

From File

The **From File** option is used to import a datum curve from IGES, VDA, *.ibl file formats. You can save a model that is in any of the above-mentioned formats and export the geometry in the form of a curve. When you open the exported file to create a datum curve, the geometry of the model is converted to a datum curve.

Use Xsec

The **Use Xsec** option is used to create a datum curve that has the geometry of an existing cross section. The creation of sections is discussed in Chapter 6.

From Equation

The **From Equation** option is used to create datum curves by defining equations using the coordinate systems. When you choose this option, you are prompted to select a coordinate system. Select the default coordinate system from the graphics window. The **SET CSYS TYP** menu is displayed, as shown in Figure 7-39.

Menu Manager
SET CSYS TYP
Cartesian
Cylindrical
Spherical

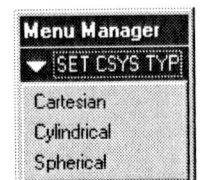

Figure 7-39 SET CSYS TYP *menu*

This menu provides the types of coordinate system that you can choose from. Choose the coordinate system depending on the equation you want to use to create a datum curve. If you choose the **Cartesian** option from the menu, the **rel.ptd - Notepad** window is displayed, as shown in Figure 7-40.

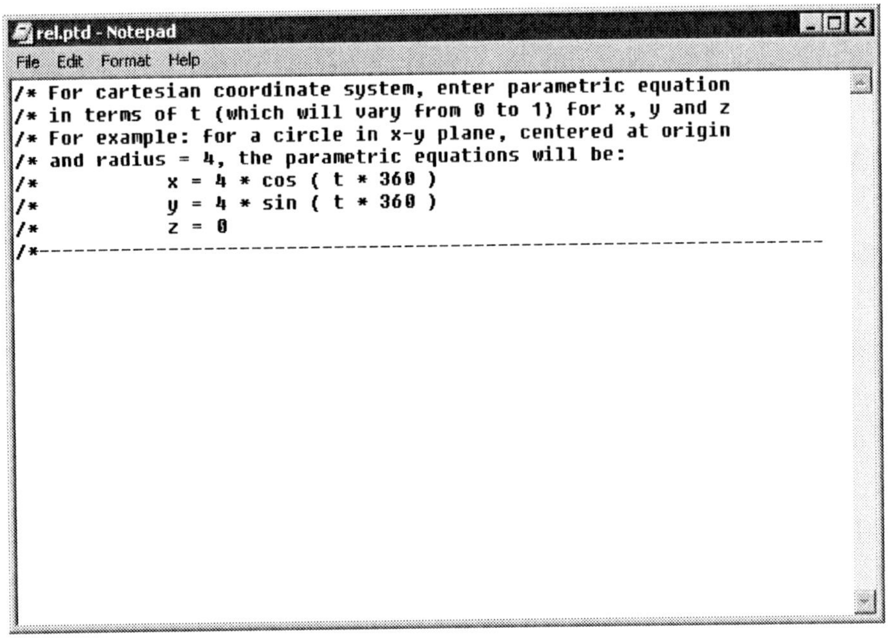

rel.ptd - Notepad
File Edit Format Help
/* For cartesian coordinate system, enter parametric equation
/* in terms of t (which will vary from 0 to 1) for x, y and z
/* For example: for a circle in x-y plane, centered at origin
/* and radius = 4, the parametric equations will be:
/* x = 4 * cos (t * 360)
/* y = 4 * sin (t * 360)
/* z = 0
/*---

Figure 7-40 *Notepad window*

Using this notepad you can define the equations. The notepad shows the instructions that should be followed while writing an equation in the notepad. These instructions vary and depend on the type of coordinate system you have selected. After you have entered the equations in the notepad, save the file and then exit the notepad.

Creating a Datum Curve in Spiral Shape

The following steps explain the procedure to create a spiral-shaped datum curve.

1. Choose the **Cylindrical** option from the **SET CSYS TYP** menu. The notepad is displayed. In the notepad the parametric equations to create a circle are given as follows:

   r = 4
   theta = t * 360
   z = 0

   In the above equations, the variable t varies from 0 to 1, r is the radius of the circle, and z is the third equation that is set equal to 0. Now, to understand the given equation of the circle, notice that the radius of the circle is given. The only value that is unknown is theta. The value of theta depends on the variable t. Therefore,
   when t = 0, theta = 0
   when t = 1/2, theta = 180 (semicircle)
   and when t = 1, theta = 360 (circle)

2. Enter the following parametric equation of the spiral curve below the dashed line in the notepad:

   IR = 8
   OR = 80
   TURNS = 10
   r = IR + t * (OR - IR)
   theta = t * 360 * TURNS
   z = 0

   In the above equations, the value of r is selected to vary because in the spiral curve, the value of r always increases from center (at center of spiral, r = 0). The internal radius of the spiral IR = 8, outer radius OR = 80, and number of turns TURNS = 10 are given.

3. Choose **File > Save** from the notepad and then exit the notepad.
4. Choose the **Preview** button from the **CURVE: From Equation** dialog box to preview the datum curve.
5. Choose **OK** from the dialog box to exit. The datum curve appears in the graphics window, as shown in Figure 7-41.

Creating Datum Curve by Sketching

 The **Sketch Tool** button is used to sketch a datum curve using the sketcher tools. This is the most commonly used option to create a datum curve. The sketch can be open or closed and is drawn in the sketcher environment.

Intersect Option

The **Intersect** option can be invoked by choosing **Edit > Intersect** from the menu bar. This option is object action tool; this means that you need to select a datum plane and then invoke this option. The **Intersect** option creates a datum curve at the intersection of a face of the model and a datum plane, intersection of a face of the model and a quilt surface, intersection of a quilt surface and a datum plane. Note that you cannot create a datum curve at the intersection of two datum planes, two quilts, or two model faces using this option.

Advanced Modeling Tools-I

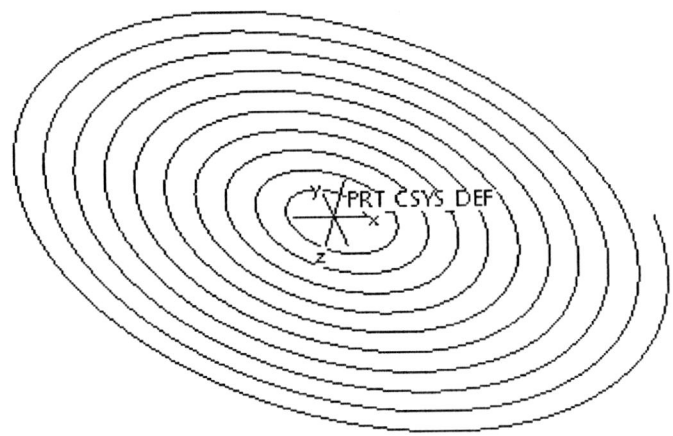

Figure 7-41 Spiral-shaped datum curve

When you choose this option, the **Intersect** dashboard appears, as shown in Figure 7-42. You may need to choose the **Intersect** option twice to invoke the **Intersect** dashboard.

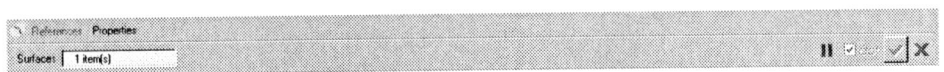

Figure 7-42 The **Intersect** dashboard

To create the ring on the circumference of the cylindrical feature, as shown in Figure 7-43 you need to create a datum curve that is on the circumferential surface of the cylindrical feature. To create the datum curve that lies on the outer surface, follow these steps:

1. Select the datum plane that is intersecting with the circumferential surface of the cylinder.
2. Choose **Edit > Intersect** from the menu bar. The **Intersect** dashboard is displayed.
3. Choose the **References** tab to open the slide-up panel. In the **Surfaces** collector, the selected datum plane is displayed.
4. Use CTRL+left mouse button and select the top and the bottom circumferential surface of the cylindrical feature, as shown in Figure 7-44.
5. Choose the **Build feature** button from the dashboard to exit the feature creation tool. The datum curve is created on the circumferential surface of the cylinder, as shown in Figure 7-45.

Project Option

The **Project** option projects a selected or sketched entity on one or more planar or non-planar surfaces or datum planes. The projected datum curve forms a true projection of the selected or sketched entity on the specified surfaces. The dimensions of the original entity may distort while projecting.

Figure 7-43 Solid model

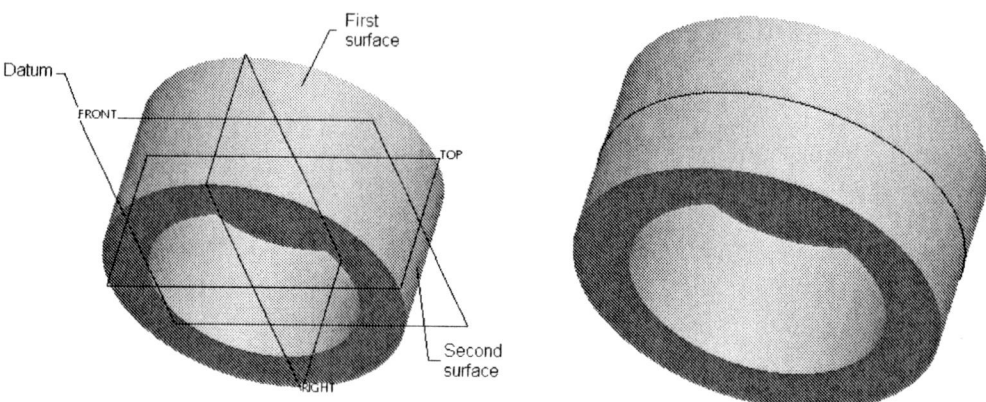

Figure 7-44 Selections to make on the model

Figure 7-45 Datum curve created on the circumferential surface of the cylindrical feature

Choose **Edit > Project** from the menu bar. The **Project** dashboard is displayed, as shown in Figure 7-46. The tools and options available in this dashboard are discussed next.

Figure 7-46 The **Project** dashboard

Surfaces Collector

The **Surfaces** collector is used to select the surface on which you need to project the sketched or selected datum curve.

Direction Drop-down List

The **Direction** drop-down list is used to specify the method of projection of the datum curve on the receiving surface or plane. The two options available in this drop-down list are:
1. Along direction
2. Normal to surface

Along Direction

This option projects the datum curve in a direction shown by the yellow arrow. To specify the direction of projection you can select the default coordinate system, datum plane, edge, or surface. Figure 7-47 shows the datum curves that are overlapping. The datum curve that is selected to project on the receiving surface and the datum curve after projection are of same geometry. This is because, using the **Along direction** option, the true geometry is obtained after projection. Figure 7-48 shows the sketched datum curve after projecting on the receiving surface. The curve is sketched on a datum plane that is parallel to the bottom face of the model.

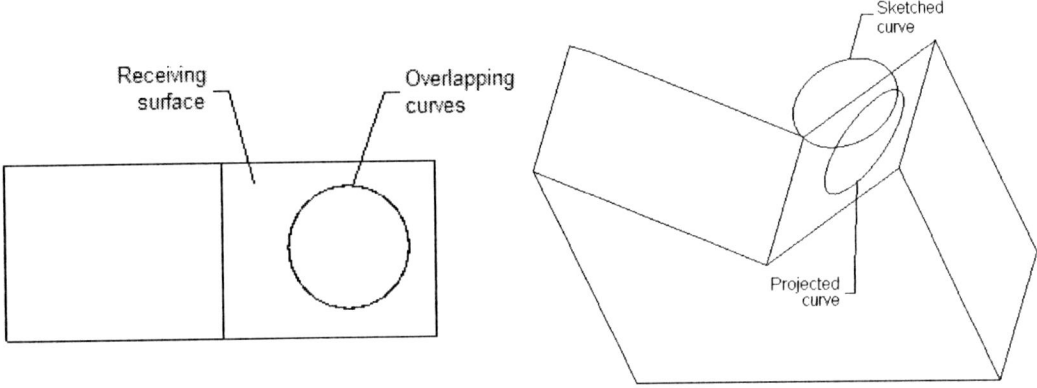

Figure 7-47 Top view of projected datum curve

Figure 7-48 Projecting a datum curve

Normal to Surface

This option of projecting a datum curve projects the datum curve normal to the receiving surface. Figure 7-49 shows the datum curve that is selected to project on the receiving surface and the datum curve after projection. Figure 7-50 shows the sketched datum curve after projecting on the receiving surface.

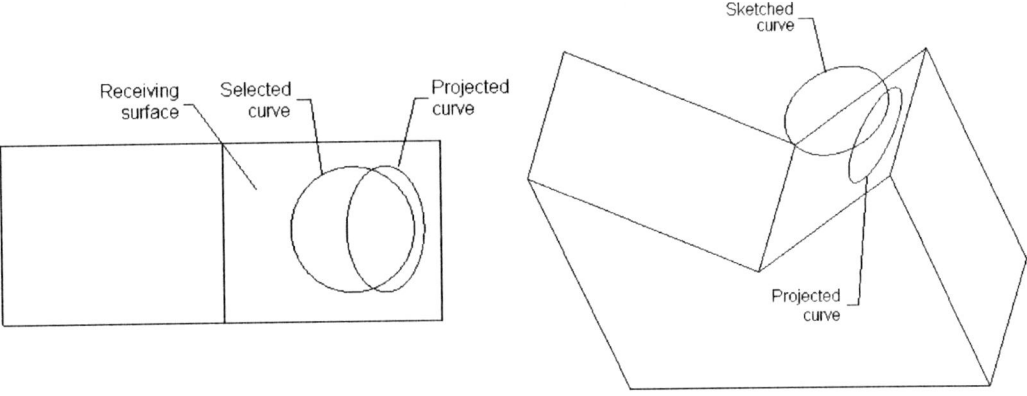

Figure 7-49 Top view of projected datum curve

Figure 7-50 Projecting a datum curve

References Tab

When you choose the **References** tab, the slide-up panel is displayed, as shown in Figure 7-51. Using this slide-up panel you can either select an existing datum curve to project or sketch a datum curve to project. The drop-down list in the slide-up panel lists two options, **Project Chains** and **Project a sketch**. These two options are discussed next.

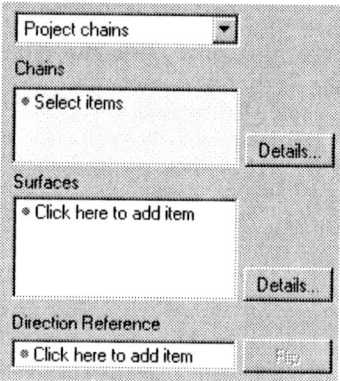

Figure 7-51 References tab slide-up panel

Project chains

The **Project chains** option is used when the datum curve to project exists. Click in the **Chains** collector to make it yellow in color and then select the datum curve that you need to project. After selecting the datum curve, click in the **Surfaces** collector and select the receiving surface (plane or surface to project on to). Then you need to specify the direction of projection. Remember that if you are using the **Normal to surface** option then you do not need to specify the direction of projection. Also in this case, the **Direction Reference** collector displayed in Figure 7-51 does not appear.

Project a sketch

When you select the **Project a sketch** option from the drop-down list in the slide-up panel, the slide-up panel appears, as shown in Figure 7-52.

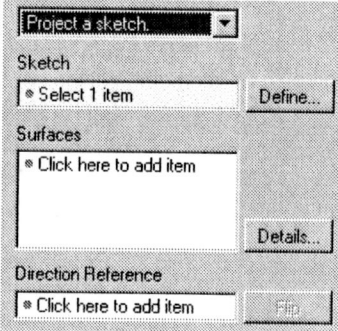

Figure 7-52 References tab slide-up panel

Choose the **Define** button to invoke the **Sketch** dialog box and to select the sketch plane

Advanced Modeling Tools-I

for the curve to be projected. After sketching, exit the sketcher environment and select the surface or plane on which the curve will be projected. If you are using the **Along direction** option to project the curve then you need to click in the **Direction Reference** collector to specify the direction of projection.

Wrap Option

This option is used to create a datum curve by wrapping a sketched entity around a solid or a quilt. Choose **Edit > Wrap** from the menu bar, the **Wrap** dashboard is displayed, as shown in Figure 7-53. The tools and options in this dashboard are discussed next.

Figure 7-53 The **Wrap** dashboard

References Tab

When you choose the **References** tab, the slide-up panel is displayed, as shown in Figure 7-54. The **Define** button is used to invoke the **Sketch** dialog box that you can use to select the sketch plane and enter the sketcher environment to draw the curve.

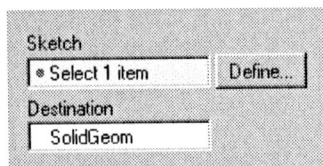

Figure 7-54 **References** tab slide-up panel

The **Destination** collector in the slide-up panel is used to select the object on which you need to wrap the curve. Generally, if there is a single feature in the graphics window then you do not need to select the object to wrap on. Pro/ENGINEER automatically wraps the selected curve or the sketched curve on the object. However, if you want to select a different object to wrap on then you can click in this collector to make it yellow in color and then select the object.

Options Tab

When you choose the **Options** tab, the slide-up panel is displayed, as shown in Figure 7-55.

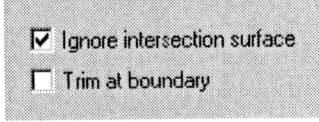

Figure 7-55 **Options** tab slide-up panel

The **Ignore Intersection surface** check box when selected ignores any intersection surface and wraps the selected curve on the destination object.

The **Trim at boundary** check box when selected trims the extra portion of the curve that is beyond the boundary of the destination object.

Follow the following steps to sketch and wrap a curve on the rectangular block, as shown in Figure 7-56. It is assumed that the rectangular block of dimension 20x20x50 exists.

1. Choose **Edit > Wrap** from the menu bar to invoke the **Wrap** dashboard.
2. Choose the **References** tab to invoke the slide-up panel.
3. Choose the **Define** button to invoke the **Sketch** dialog box.
4. Choose the datum plane that is passing through the center of the rectangular block as the sketch plane and then choose a reference for orienting the sketch plane.
5. After entering the sketcher environment, draw a line that starts from the bottom of the rectangular block. Its start point is aligned with the center of the rectangular block and with the bottom edge. The end point of the line is at a distance of **1000** and at a height of **50** from the bottom of the rectangular block, see Figure 7-57.
6. Place a user-defined coordinate system at the start point of the line.
7. Exit the sketcher environment. The sketched curve is automatically wrapped on the circumference of the cylinder.
8. Choose the **Build feature** button to exit the **Wrap** dashboard.

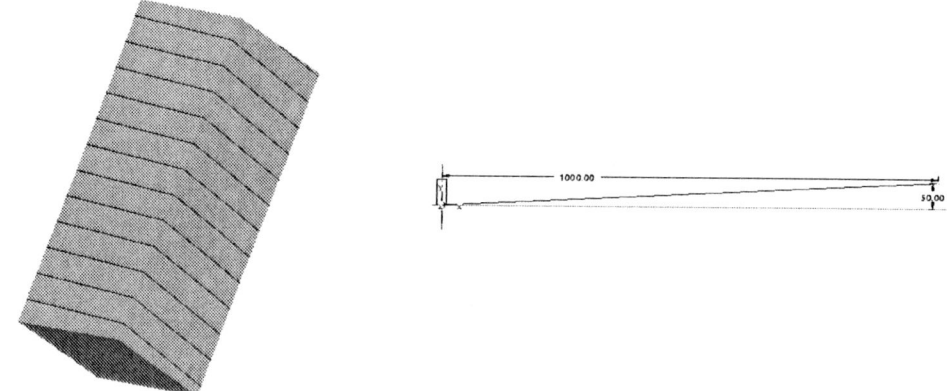

Figure 7-56 Curve wrapped on the cylinder   *Figure 7-57* Sketch of the curve to wrap

CREATING DRAFT FEATURES

In Pro/ENGINEER, drafts are created on existing surfaces. They are created by rotating the selected surface by a certain angle. One of the applications of draft features is in moulds and castings where a taper is required to separate the casting from the mould or vice versa. To create a draft, choose the **Draft Tool** button from the **Engineering Features** toolbar. The **Draft** dashboard is displayed, as shown in Figure 7-58. The options in this dashboard are discussed next.

Advanced Modeling Tools-I

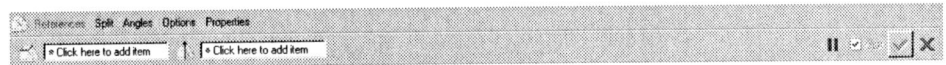

*Figure 7-58 The **Draft** dashboard*

References Tab

When you choose the **References** tab, the slide-up panel is displayed, as shown in Figure 7-59. The options in this slide-up panel are discussed next.

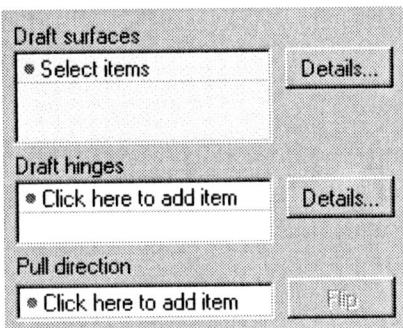

*Figure 7-59 **References** tab slide-up panel*

Draft surfaces Collector

When you invoke the **Draft** dashboard, this collector is selected by default. If this collector is not selected by default, click in this collector and then select the surface to which you need to add a draft angle. The maximum draft angle that can be added to a surface is 30-degrees.

Draft hinges Collector

The draft hinges collector is used to select the hinge of the draft surface. The hinge that you select can be an edge, a surface, an axis, or a datum plane. The draft surface is pivoted on the hinge that you select. In other words, the draft surface is rotated about the hinge. The hinge that you select need not intersect the draft surface. Figure 7-60 shows the draft surface and the hinge selected to create the draft is shown in Figure 7-61.

Pull direction Collector

The **Pull direction** collector is used to specify the direction of rotation of the draft surface. The direction of pull is shown by the direction of the yellow arrow. Generally, when you select the hinge, the pull direction is selected by default. If the pull direction shown by the yellow arrow is not that is desired, click in the **Pull direction** collector and then select the pull direction. You can change the direction of the yellow arrow by choosing the **Reverse pull direction** button.

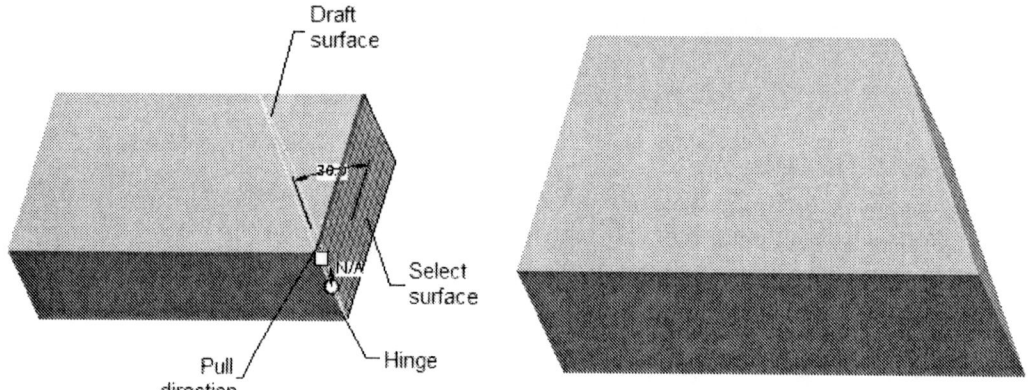

Figure 7-60 Parameters needed to create a basic draft surface

Figure 7-61 Resultant draft surface

Split Tab

When you choose the **Split** tab, the slide-up panel is displayed, as shown in Figure 7-62. The options in this slide-up panel are discussed next.

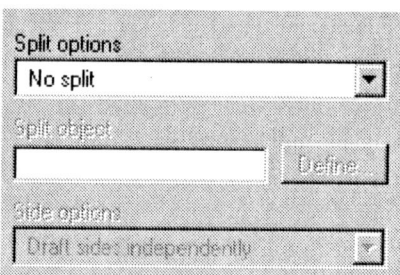

*Figure 7-62 **Split** tab slide-up panel*

Split options Drop-down List

The options in this drop-down list are used to split the surface selected to draft. Using these options the selected surface gets split into two surfaces and different draft angles can be applied to both the surfaces. The options available in the **Split options** drop-down list are discussed next:

No Split

This option is used when you do not want to split the surface selected to give the draft angle.

Split by draft hinge

This option is available only when you have selected the hinge for the draft surface. When you select this option, the selected surface is divided into two surfaces at the location on the surface where the hinge intersects it. Figure 7-63 shows the surface that is split into two surfaces at the hinge.

Advanced Modeling Tools-I

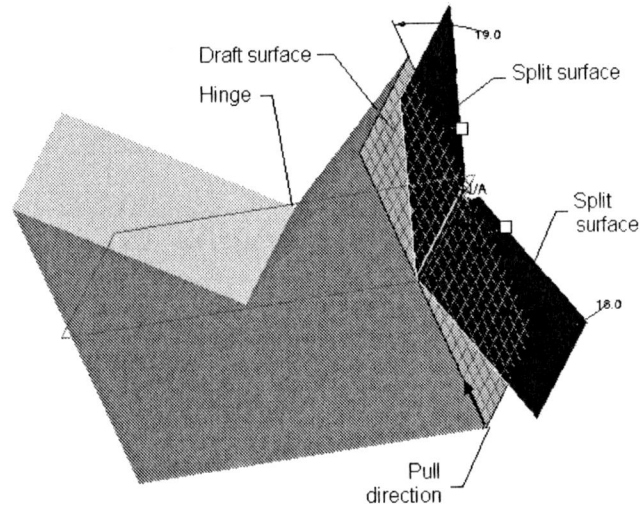

Figure 7-63 Split at draft hinge

Split by split object

This option when chosen activates the **Split object** collector and the **Define** button. This option is used to split the surface selected to draft by drawing a sketch. The surface gets split into two surfaces and the sketch defines the profile of the split. Figure 7-64 shows the parameters that you need to define to create a draft split by sketching. Figure 7-65 shows the draft created on the cylindrical surface.

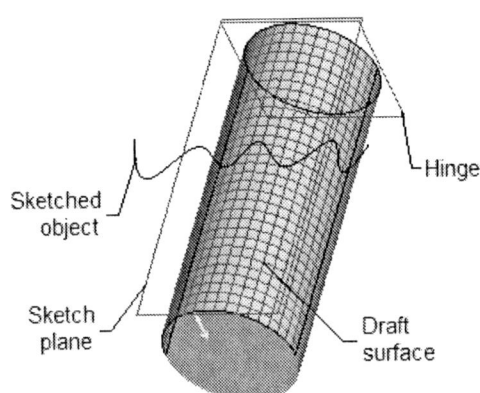 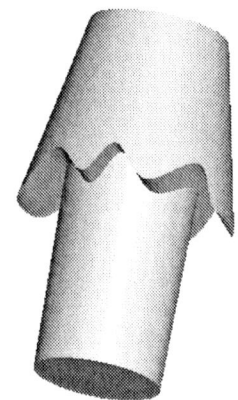

Figure 7-64 Split by split object   *Figure 7-65* Resultant model

Side options Drop-down List

The options in this drop-down list are used to specify how to apply the draft once the surface has been split. The options in this drop-down list are discussed next.

Draft sides independently

The two sides of the draft surface that are formed when the surface is split can be given different draft angles. When you choose this option, the edit boxes appear on the **Draft** dashboard. You can enter the values of the draft angle in these edit boxes.

Draft sides dependently
The two sides of the split surface are given the same draft angle.

Draft first side only
Using this option you can apply the draft angle only to the first side of the split surface.

Draft second side only
Using this option you can apply the draft angle only to the second side of the split surface.

Angles Tab

Choose this tab from the **Draft** dashboard after you have specified some angle for draft surface. When you choose this option, the slide-up panel is displayed, as shown in Figure 7-66. In Pro/ENGINEER, you can create constant and as well as variable angle draft.

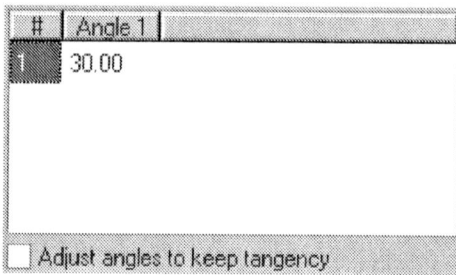

Figure 7-66 Angles option slide-up panel

Constant Angle Drafts
By default, Pro/ENGINEER creates a constant angle draft. This means, the same draft angle is applied to the surface that you have selected.

Variable Angle Drafts
When you apply more than one value of the draft angle to the selected surface, it is called variable angle draft. Apply the variable angle draft after you have selected the hinge, the pull direction, and the draft surface.

To create the draft surface shown in Figure 7-67, follow the steps given next. It is assumed that you have the base feature of the model, as shown in Figure 7-68.

1. Invoke the **Draft** dashboard and click in the first collector from left.
2. Select the top face of the base feature. Notice that in the **Pull direction** collector the direction of pull is selected automatically. Reverse the pull direction by choosing the **Reverse pull direction** button available on the right of the second collector.
3. Hold down the right mouse button to invoke the shortcut menu shown in Figure 7-69.

Figure 7-67 Draft feature on the model

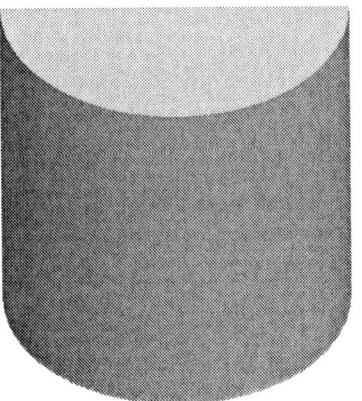

Figure 7-68 Base feature

Figure 7-69 Shortcut menu

4. Choose the **Draft Surfaces** option from the shortcut menu and select the cylindrical surface of the base feature to add the draft angle. The drag handle appears on the base feature; you can use it to vary the angle of the draft. The draft angle value of **1** also appears in the graphics window. Also notice the white ball that appears on the edge of the top face (face selected as hinge).
5. Bring the cursor on the white ball and hold down the right mouse button to invoke the shortcut menu shown in Figure 7-70.

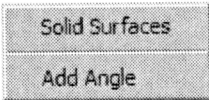

Figure 7-70 Shortcut menu

6. Choose the **Add Angle** option from the shortcut menu. Notice that now there are two white balls on the edge of the top face. On the first ball the value of **0.5** appears. This value varies from one end of the face to the other end. This value represents the location of the point that you need to use for reference in order to apply the draft angles.
7. Again bring the cursor to any one of the two white balls and invoke the shortcut menu.
8. Choose the **Add Angle** option from the shortcut menu. Another point is added with a location value on the edge.
9. Add eleven such points for locations varying from **0** to **1**.

10. After adding eleven points, double-click on any of the location value that is present in the graphics window. The edit box appears, enter the value of **0** in this edit box. Similarly, locate all remaining ten points and increment them by 0.1. Figure 7-71 shows the model after locating all eleven points on the edge.
11. Double-click on the value of the angle that corresponds to the location **0**. The edit box appears, enter a value of **15**.
12. Double-click on the value of the angle that corresponds to the location **0.1** (these values appear above the model and have a default value of 1.0). The edit box appears; enter a value of **5**. Similarly, vary the angle at other locations. Every alternate location should have an angle of **15**. Figure 7-72 shows the preview of the model after modifying the values of the angle at all eleven locations.

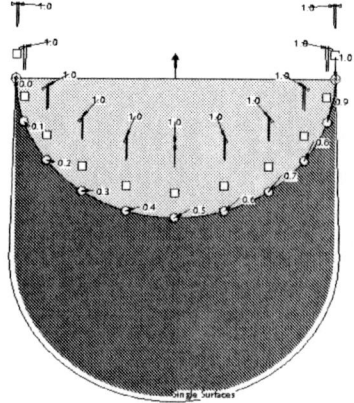

Figure 7-71 Location points on the edge

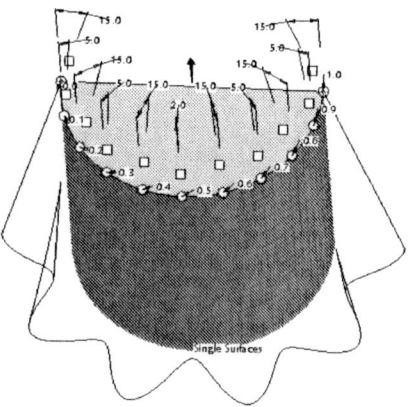

Figure 7-72 Location points with modified angle values

13. After modifying the values of the angle at all locations, choose the **Feature Preview** button from the **Draft** dashboard. The model appears, as shown in Figure 7-73.
14. The model after mirroring the geometry and then shelling the top and bottom faces is shown in Figure 7-74.

Figure 7-73 Model with the draft feature

Figure 7-74 Model of the lamp shade

Advanced Modeling Tools-I

Options Tab

When you choose this tab, the slide-up panel is displayed, as shown in Figure 7-75. The options in this slide-up panel are discussed next.

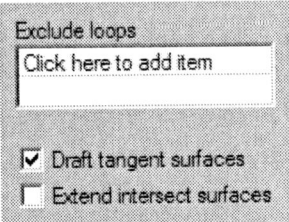

Figure 7-75 **Options** *tab slide-up panel*

Exclude loops Collector

This collector is used to select a surface to which you do not want to add a draft angle. When you select a surface to add a draft angle, all loops that are on the surface are applied the same draft angle. However, using this collector you can select the loop to which you do not want to add a draft angle.

Figure 7-76 shows surface, when selected to add a draft angle also selects the surface of the cylindrical feature. If you continue with the draft feature creation and exit the **Draft** dashboard, the draft is created, as shown in Figure 7-77.

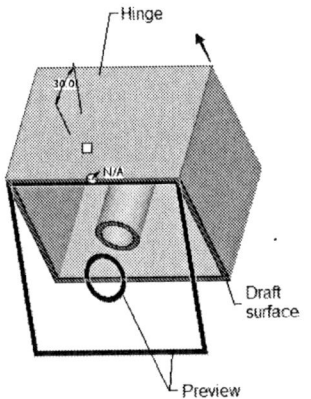
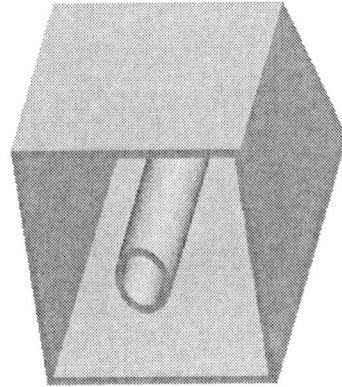

Figure 7-76 Surfaces for draft   *Figure 7-77* After creating draft

To exclude the face of the cylindrical feature from the loop, click in the **Exclude loops** collector to turn it yellow in color. Now, bring the cursor close to the face of the cylindrical feature. The boundary of the face is highlighted in cyan. Select the face to exclude it from the loop. Figure 7-78 shows the draft surface after excluding the face of the cylindrical feature from the loop.

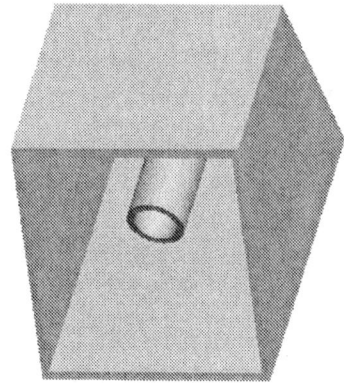

Figure 7-78 Exclude loop draft

Draft tangent surfaces Check Box

When the **Draft tangent surfaces** check box is selected, it applies the draft to the surfaces tangent to the selected surface. In Figure 7-79, the surface shown is selected to add draft angle. Because the **Draft tangent surfaces** check box is selected, all surfaces that are tangent to the selected surface and other surfaces are automatically selected. Figure 7-80 shows the resultant model with the draft feature.

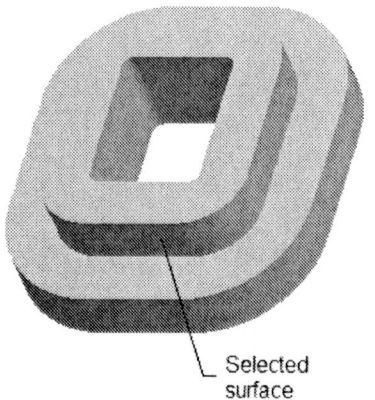 

Figure 7-79 Single surface selected to draft   *Figure 7-80* Draft applied to all tangent surfaces

Extend intersect surfaces Check Box

The **Extend** check box is available only when the **Intersect** radio button is selected. When this check box is selected, the draft surface extends in the direction of the feature it intersects. Figure 7-81 shows the example of the draft that intersects with the adjacent feature and projects outside the edge of the feature at the bottom. Figure 7-82 shows the feature when the draft surface, gets extended.

Advanced Modeling Tools-I

 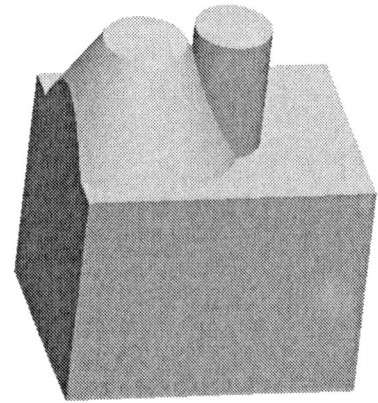

Figure 7-81 Intersect

Figure 7-82 Extend intersect

Figure 7-83 shows another example of extended draft feature. The following steps explain in brief the procedure to create the pencil-shaped model. It is assumed that you have created the model shown in Figure 7-84. The initial model shown is an octagon on the top of which a cylinder is created.

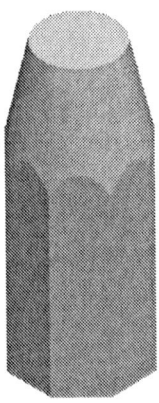 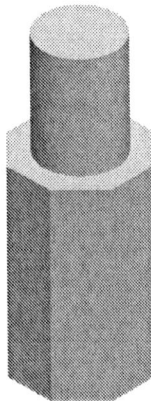

Figure 7-83 Example of extend intersect in a draft surface

Figure 7-84 Model initially created

1. Invoke the **Draft** dashboard and select the cylindrical surface to draft.
2. Select the top face of the cylindrical feature as the hinge. The yellow arrow points in the upward direction.
3. Choose the **Options** tab and from the slide-up panel, select the **Extend intersect surfaces** check box.
4. Use the drag handle and increase the draft angle up to the vertex of the octagon. This can be easily done by viewing the preview of the draft surface as you increase the angle. You may even have to use the edit box present on the dashboard to enter the small increase in angle value. Remember that you will get the desired shape of the intersect only when the draft surface intersects the vertices of the octagonal feature.

Properties Tab

When you choose the **Properties** tab, the slide-up panel displays the feature id of the feature you have creating.

TUTORIALS

Tutorial 1

In this tutorial you will create the model shown in Figure 7-85. This figure also shows the sectioned front view, top view, and the right-side view of the solid model with dimensions. The hidden lines are suppressed for clarity.   **(Expected time: 45 min)**

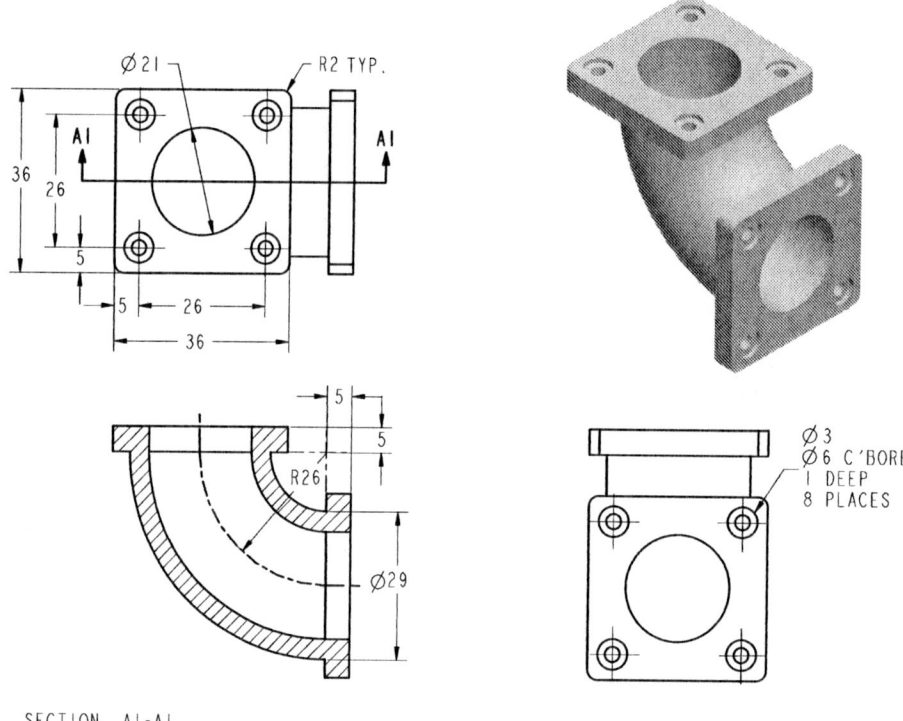

Figure 7-85 Top, front, right-side, and isometric views of the model.

The following steps are required to complete this tutorial:

a. Examine the model and determine the number of features in it. The model is composed of six features, refer to Figure 7-85.
b. Create the base feature, refer to Figures 7-86 through 7-88.
c. Create the shell feature of given thickness, refer to Figure 7-89.
d. Create the third and fourth extrude features on the two ends of the swept feature respectively, refer to Figures 7-90 and 7-91.

Advanced Modeling Tools-I   7-35

e. Create the counterbore hole on the third and forth feature, refer to Figures 7-92 and 7-93.
f. Pattern the counterbore holes, refer to Figure 7-94.

After understanding the procedure for creating the model, you are now ready to create it. When the Pro/ENGINEER session is started, the first task is to set the working directory. Since this is the first tutorial of this chapter, you need to create a folder named *c07* if it does not exist and set it as the **Working Directory**.

Starting a New Object File

1. Open a new part file and name it as *c07tut1*.

The three default datum planes are displayed in the graphics window. The **Model Tree** is also displayed in the graphics window. Close the **Model Tree** by clicking on the sash present on the right edge of the **Model Tree**.

Invoking the Sweep Option

You will use the menu bar present on the top of the screen to invoke the **Sweep** option.

1. Choose **Insert > Sweep > Protrusion** from the menu bar. The **SWEEP TRAJ** menu and the **PROTRUSION: Sweep** dialog box are displayed.

2. Choose the **Sketch Traj** option from the **SWEEP TRAJ** menu. You are prompted to select or create a sketching plane.

Selecting the Sketching Plane

The trajectory of the sweep feature will be sketched on the **FRONT** datum plane.

1. Select the **FRONT** datum plane as the sketching plane. A red arrow points in the direction of viewing the sketch plane.

2. Choose **Okay** from the **DIRECTION** submenu. The **SKET VIEW** submenu is displayed.

3. Choose **Top** from this submenu and select the **TOP** datum plane.

After you select the planes for orientation, the system takes you to the sketcher environment.

Drawing the Trajectory

From the model it is evident that the trajectory for the sweep feature is a quarter circle.

1. Exit the **References** dialog box by choosing the **Close** button from the dialog box.

2. Choose the **Create an arc by picking its center and endpoints** button from the **Sketcher Tools** toolbar. This button is available on the flyout that is displayed when you choose the black arrow on the right of the **Create an arc by 3 points or tangent to an entity at its endpoint** button.

3. Draw an arc such that the center of the arc lies at the intersection of the **TOP** and **RIGHT** datum planes, as shown in Figure 7-86. As you specify the center of the arc, the cursor snaps to the point of intersection of the two planes. Now, draw the arc and exit this tool. The endpoints of the arc are automatically aligned to the **TOP** and **RIGHT** datum planes. You will notice that an arrow is attached at the start point of the trajectory. This arrow points in the direction of sweep.

Tip: *To change the start point on the trajectory, select the point where you want the start point. When the point is highlighted in red color, hold down the right mouse button to invoke a shortcut menu. Choose the **Start Point** option.*

Modifying the Dimensions of the Trajectory

When you were drawing the arc, the arc was dimensioned automatically and a weak radial dimension was assigned to it. You need to modify the dimension as per your requirement.

1. Double-click on the dimension and modify the radial dimension to 26, as shown in Figure 7-87. You will notice the sketch refits on the screen.

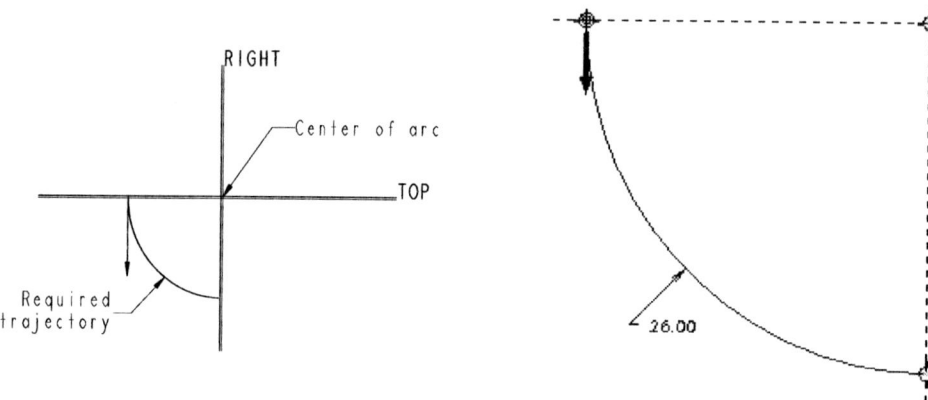

Figure 7-86 The sketch of required trajectory   *Figure 7-87 Dimension for the arc*

2. Choose the **Continue with the current section** button.

Drawing the Section for Sweep

After choosing the **Continue with the current section** button, the direction of viewing is modified such that the start point of the trajectory becomes normal to the screen. A yellow cross of infinite length appears on the screen. This cross consists of two perpendicular lines of infinite length. The intersection point of these lines is the start point of the trajectory. The center of the circle should lie at the intersection of these lines. The **References** dialog box is also displayed. The status displayed in the **Reference status** area of the **References** dialog box is **Fully Placed**. Close the **References** dialog box. You are also prompted to draw the cross-section for the sweep.

Advanced Modeling Tools-I

 **Tip**: *When you enter the sketcher environment to define a section for sweep trajectory, it is often difficult to understand the orientation of the view. For this purpose, a yellow-colored cross that determines the orientation of the section with respect to the trajectory is available.*

1. Choose the **Create circle by picking the center and a point on the circle** button and create a circle such that the center of circle lies at the intersection of the two infinite perpendicular lines.

   When you draw the circle, the cursor snaps to the intersection point of the cross.

Modifying the Dimensions of the Section

1. Choose the **Select items** button and using the left mouse button double-click on the dimension and modify the diameter dimension to 29.

   The sketch accepts the value and refits on the screen.

2. Choose the **Continue with the current section** button.

Previewing the Swept Feature

The sweep feature is completed and now it can be previewed.

1. Choose the **Preview** button from the **PROTRUSION** dialog box that is present on the top right corner of the window.

2. Choose the **Saved view list** button from the **View** toolbar. From the drop-down list, choose the **Default Orientation** option. The model orients in the graphics window, as shown in Figure 7-88.

   You can use the middle mouse button to change the orientation of the model.

3. Now, choose the **OK** button from the **PROTRUSION** dialog box to exit it.

Creating the Shell Feature

The sweep feature is completed and now you can create the next feature. The next feature is shell.

 Note
*Instead of using the **Shell** option, two concentric circles can be drawn while drawing the section for the sweep feature in order to obtain the desired hollow feature. Also, the **Sweep > Thin Protrusion** option can be used to obtain the same hollow feature. However, in this tutorial you will use the **Shell** option.*

1. Choose the **Shell Tool** button from the **Engineering Features** toolbar. The **Shell** dashboard is displayed and you are prompted to select the surfaces to be removed.

2. Select the one end surface of the swept feature and then using CTRL+left mouse button, select the other end surface. The two surfaces selected are highlighted in red.

3. Enter the thickness value of the shell as **4** in the dimension box present on the **Shell** dashboard and press ENTER.

4. Choose the **Build feature** button to exit the **Shell** dashboard.

The default trimetric view of the shell feature is shown in Figure 7-89. You can use the middle mouse button to view the model from various directions.

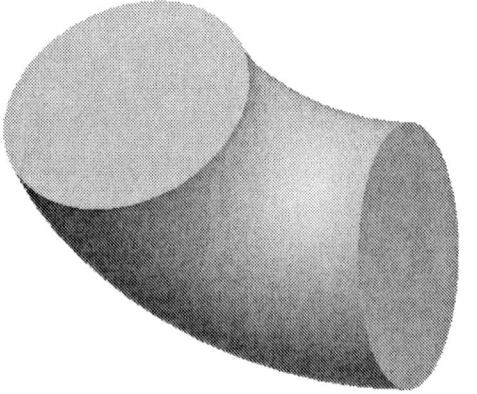
Figure 7-88 Sweep feature without datum planes

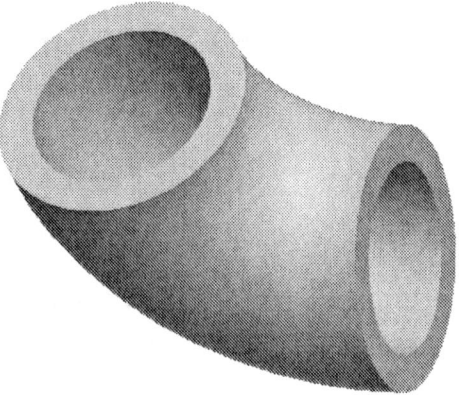
Figure 7-89 Shell feature without datum

Creating the Extrude Features

The next feature is a protrusion feature with a depth of extrusion of 5 and is created at both the ends of the swept feature. While drawing the circle of the sketch for the extrude feature, remember to use the edge of the shell in order to create a hole in the extruded feature also.

1. Choose the **Extrude Tool** button from the **Base Features** toolbar. The **Extrude** dashboard is displayed.

2. Choose the **Placement** tab to display the slide-up panel. Choose the **Define** button. The **Sketch** dialog box is displayed. Select the top face of the swept feature as the sketching plane.

3. Select the **RIGHT** datum plane and then choose the **Right** option from the **Orientation** drop-down list.

4. The **References** dialog box is displayed with the status **Not Placed**. Select the **FRONT** and **RIGHT** datum planes. Now, note that the status displayed under the **Reference Status** area in the **References** dialog box is **Fully Placed**.

5. Draw the sketch of the extrude feature and apply constraints and dimensions, as shown in Figure 7-90.

Advanced Modeling Tools-I

6. Create the extruded feature having a depth of extrusion of 5. Similarly, create the next extruded feature at the other end of the swept feature. Select the sketching plane, draw the sketch similar to the first extruded section, apply the same dimensions and constraints, and extrude the sketch to the given distance.

The protrusion features created at both the ends of the swept feature are shown in Figure 7-91.

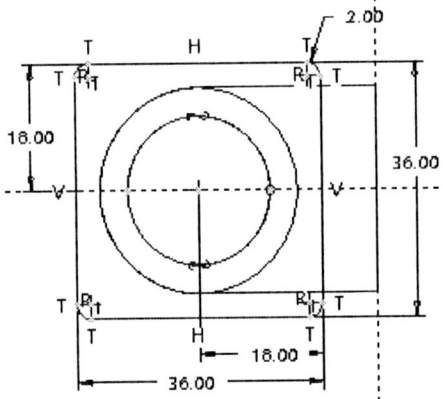

Figure 7-90 Sketch with dimensions and constraints

Figure 7-91 The two extruded features on the ends of the sweep feature

Creating the Hole Feature

After creating extruded features at both the ends of the swept feature, counterbore holes will be created. One hole is to be created on each extruded surface and then they are to be patterned on individual planes separately to create the remaining three instances.

1. Choose the **Hole Tool** button from the **Engineering Features** toolbar. The **Hole** dashboard is displayed with the **Create straight hole** button and the **Simple** option is selected by default.

2. Choose the **Sketched** option from the drop-down list in the **Hole** dashboard.

3. Choose the **Activates Sketcher to create section** button. The system takes you to the sketcher environment; sketch the section for the counterbore hole, as shown in Figure 7-92. Sketch a center line, apply constraints, and diametrically dimension the entities.

4. After completing the sketch, choose the **Continue with the current section** button to exit the sketcher environment.

The system exits the sketcher environment and the **Hole** dashboard is redisplayed. Now, you are prompted to select a surface, axis, or point to place the hole.

5. Select the top face of the third extruded feature for the placement of hole. The preview of the hole appears in the graphics window.

Now, you need to specify the placement parameters for the hole. For this purpose refer to Figure 7-85 and look for the dimensions that can help in placing the hole. The two edges are used to dimension the hole.

6. Choose the **Placement** tab and click in the **Secondary References** collector to turn it yellow.

7. Select the front edge and then use CTRL+left mouse button to select the left edge of the third feature for specifying the linear references. The hole is at a distance of 5 from both the edges. Default dimensions appear on the hole.

   You can also drag the handle and place them at the two edges.

8. Double click on the dimensions and modify the linear distances to **5**.

9. Choose the **Build feature** button from the **Hole** dashboard. The hole is created on the selected face.

10. Create another hole on the fourth feature using the same procedure as discussed above. The default trimetric view of the model that is completed until now is shown in Figure 7-93.

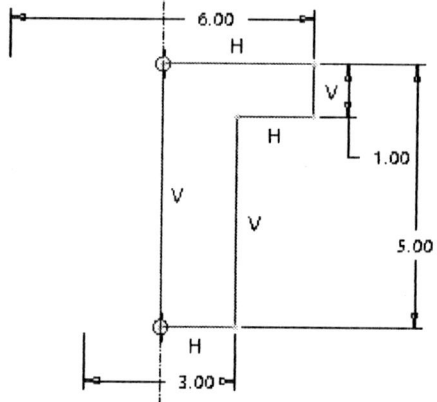

Figure 7-92 Sketch with dimensions and constraints for the counterbore hole

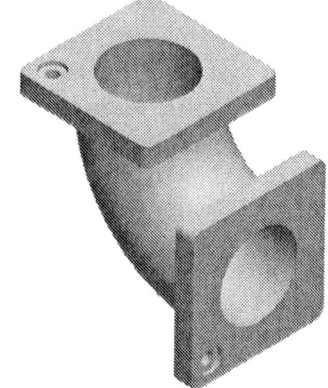

Figure 7-93 One hole each on the two extruded features

Tip: *To draw the sketch of the second hole when you enter the sketcher environment do not again draw the same sketch that was drawn for the first hole. Instead, import the sketch for the hole from the **In Session** folder. The **In Session** folder contains all files that you have created in the current session.*

*To import the sketch of the hole in the sketcher environment, choose **Sketch > Data from File**. The **Open** dialog box appears. Choose the **In Session** folder to display the files created in the current session. Select the third .sec file and open it. The sketch file for the hole will be the third file because you have created three sketches since this tutorial was started.*

Advanced Modeling Tools-I   7-41

Creating the Pattern of the Holes

After one hole is placed on each of the two faces, they are patterned.

1. Select the hole feature from the **Model Tree** or from the graphics window and hold down the right mouse button to invoke the shortcut menu. Choose the **Pattern** option from the shortcut menu. The **Pattern** dashboard is displayed.

2. Create the pattern of the hole.

   Similarly, create pattern of the hole on the other extruded feature also. The default trimetric view of the complete model is shown in Figure 7-94.

Figure 7-94 The default trimetric view of the model

Saving the Model

1. Choose the **Save the active object** button from the **File** toolbar and save the model.

Note

*As discussed in earlier chapters, the model tree is used to get an idea of the order of feature creation. In the **Model Tree**, the id numbers displayed in front of the features may be different when you create the features.*

Tutorial 2

In this tutorial you will create the model of the pencil shown in Figure 7-94 and then change the color of the pencil. Figure 7-95 shows the front view, the section view, and the detail view of the model.   **(Expected time: 30 min)**

The following steps are required to complete this tutorial:

a. Examine the model and determine the number of features in it. The model is composed of six features, refer to Figure 7-95.

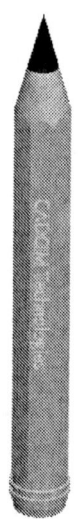

Figure 7-95 Solid model of the pencil

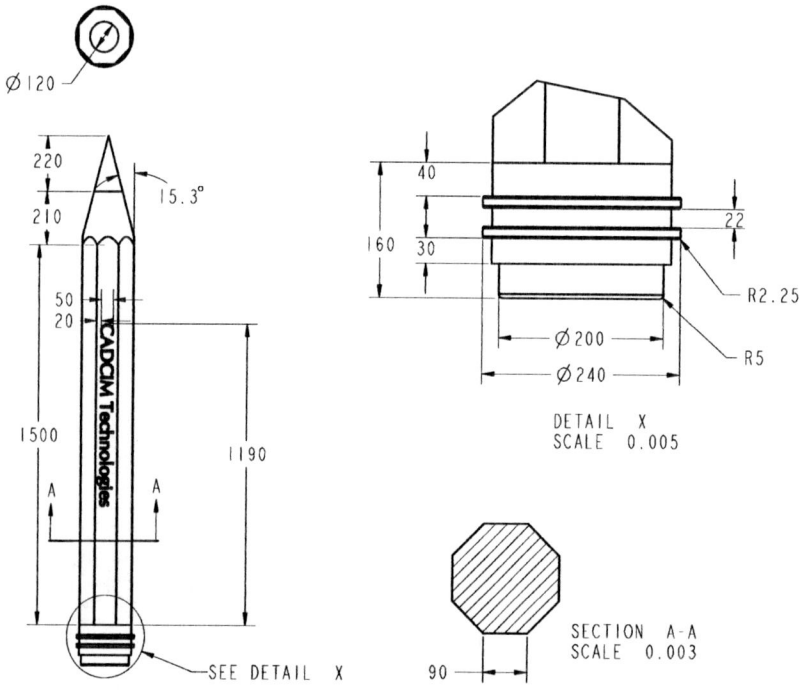

Figure 7-96 Top, front, sectioned, and detail views of the model

b. Create the base feature, refer to Figures 7-97 and 7-98.
c. The second feature is an cylinder on the top face of the base feature, refer to Figure 7-99.

Advanced Modeling Tools-I

d. Create the draft feature, refer to Figure 7-100.
e. Create the revolve feature that will create the tip of the pencil, refer to Figures 7-101 and 7-102.
f. Create the revolve feature that will create the tail-end of the pencil, refer to Figure 7-103 and 7-104.
g. Write the text on the pencil by using the sketched datum curves, refer to Figure 7-105 and 7-105.

After understanding the procedure for creating the model, you are now ready to create it. Make sure that *c07* is the current **Working Directory**.

Starting a New Object File

1. Open a new part file and name it as *c07tut2*.

The three default datum planes are displayed in the graphics window.

Creating the Base Feature

The base feature of the pencil is the protrusion feature in which the octagon is extruded to a depth of 1500.

1. Choose the **Extrude Tool** button from the **Base Features** toolbar. The **Extrude** dashboard is displayed. Choose the **Define** button from the **Placement** slide-up panel.

2. Select the **TOP** datum plane as the sketch plane and the **RIGHT** datum plane for the orientation of the sketch plane.

3. Draw the sketch of the octagon and apply the constraints and dimensions, as shown in Figure 7-97. The construction circle is used to draw the octagon because it becomes easy and less time-consuming.

4. Exit the sketcher environment and extrude the sketch to a depth of 1500. The base feature is, as shown in Figure 7-98.

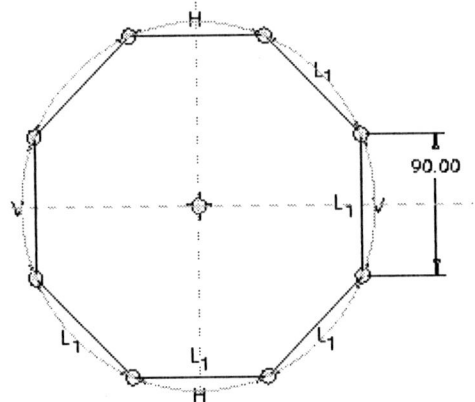

Figure 7-97 Sketch of the octagonal feature   *Figure 7-98 Octagonal sketch after extruding*

Creating the Second Feature

The second feature is a cylindrical feature that is drawn on the top face of the base feature.

1. Choose the **Extrude Tool** button from the **Base Features** toolbar. The **Section** dialog box is displayed.

2. Enter the sketcher environment and draw the sketch for the cylinder. The diameter of the circle is 120.

3. After exiting the sketcher environment, extrude the circle to a depth of 210. The model after creating the second feature is shown in Figure 7-98.

Creating the Draft Feature

The draft feature on the cylindrical surface will be created and the draft surface will be allowed to intersect the base feature. This is because when the draft surface intersects the base feature and is extended, the required shape is obtained at the intersecting edge.

1. Choose the **Draft Tool** button from the **Engineering Features** toolbar. The Draft toolbar is displayed.

2. Select the cylindrical surface of the cylinder to draft.

3. Choose the **References** tab to invoke the slide-up panel. Click the **Draft hinges** collector and then select the top face of the cylindrical feature as the hinge. The yellow arrow points in the upward direction.

4. Choose the **Reverse pull direction** button present on the right of the **Pull direction** collector to change the direction of the arrow so that it points downward.

5. Choose the **Options** tab to invoke the slide-up panel; select the **Extend intersect surfaces** check box.

6. Enter the angle value of **15.3** in the dimension box present on the **Draft** dashboard.

7. Choose the **Build feature** button from the **Draft** dashboard to complete the draft feature, as shown in Figure 7-100.

Creating the Fourth Feature

The fourth feature is a revolve feature that will be used to create the tip of the pencil.

1. Choose the **Revolve Tool** button from the **Base Features** toolbar. The Revolve dashboard is displayed.

2. Invoke the **Sketch** dialog box and select the **FRONT** datum plane as the sketch plane.

Advanced Modeling Tools-I

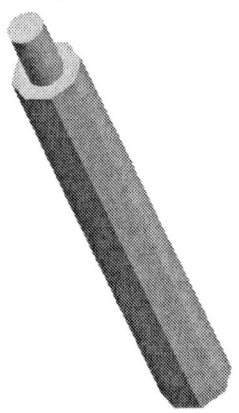

Figure 7-99 Model after creating the cylindrical feature

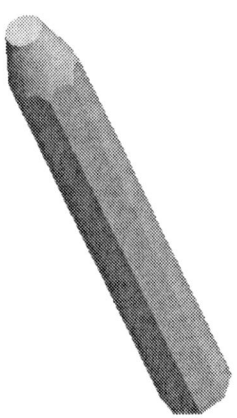

Figure 7-100 Model after creating the draft feature

3. After entering the sketcher environment, draw the sketch, apply constraints, and dimension it, as shown in Figure 7-101.

4. After exiting the sketcher environment, the angle of revolution of 360-degrees is selected by default.

5. Choose the **Build feature** button to complete the feature. The default trimetric view of the model after creating the fourth feature is shown in Figure 7-102.

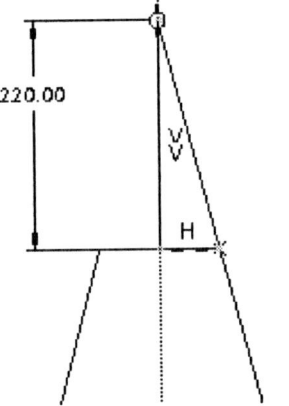

Figure 7-101 Sketch for the tip of the pencil

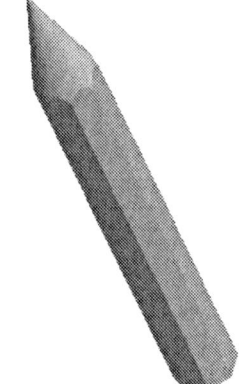

Figure 7-102 Model after creating the tip

Creating the Fifth Feature

The fifth feature is a revolve feature that is used to create the top portion of the pencil.

1. Choose the **Revolve Tool** button from the **Base Features** toolbar. The **Revolve** dashboard is displayed.

2. Invoke the **Sketch** dialog box and select the **FRONT** datum plane as the sketch plane.

3. After entering the sketcher environment, draw the sketch, apply constraints, and dimension it, as shown in Figure 7-103.

4. After exiting the sketcher environment, the angle of revolution of 360-degrees is selected by default.

5. Choose the **Build feature** button to complete the feature. The default trimetric view of the model after creating the fourth feature is shown in Figure 7-104.

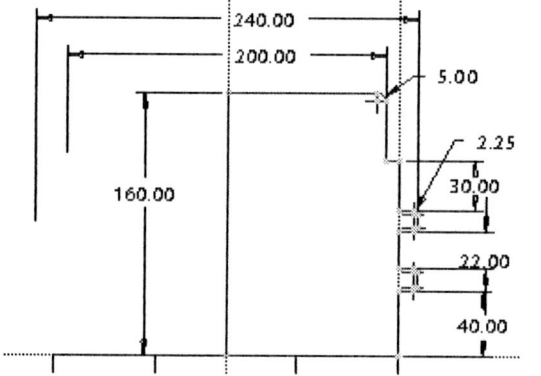

Figure 7-103 Sketch with dimensions and constraints

Figure 7-104 Model after creating the top of the pencil

Writing the Text

Generally, the text on a model should be written using the datum curves.

1. Choose the **Sketched Datum Curve Tool** button from the **Right Toolchest**. The **Sketch** dialog box is displayed.

2. Select the front face of the base feature as the sketch plane.

3. After entering the sketcher environment, create the text using the **Create text as a part of a section** button. Figure 7-105 shows the text written, with the dimensions.

4. Choose the **Continue with the current section** button to exit the sketcher environment.

The model after creating the text is shown in Figure 7-106.

Changing the Colors

The colors in Pro/ENGINEER can be applied on selected surfaces. Remember that the colors you apply are saved with the model and remain on the model until you clear them.

Advanced Modeling Tools-I

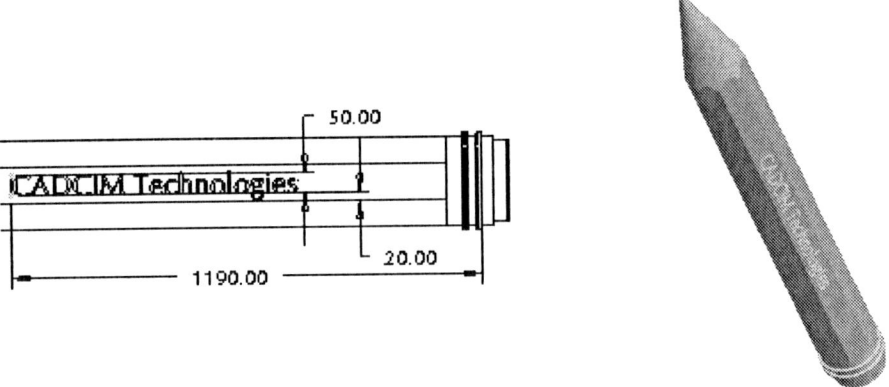

Figure 7-105 Text with dimensions   *Figure 7-106* Model after creating the text

1. Choose **View > Color and Appearance** from the menu bar. The **Appearance Editor** is displayed.

2. Select the +sign. The new color is added.

3. In the **Color** area, choose the **Color** button. The **Color Editor** is displayed. Set the RGB values to 0. The color obtained is black. You need to apply this color to the tip of the pencil.

4. Choose the **Close** button from the **Color Editor**.

5. Choose the **Surfaces** option from the drop-down list present in the **Assignment** rollout. The **SELECT** dialog box is displayed and you are prompted to select the surfaces to alter colors.

6. Select the tip of the pencil and press the middle mouse button. Then choose the **Apply** button to apply the color to the selected surface. The tip of the pencil is changed to black.

7. To add another color to any of the surfaces of the pencil, select the +sign and modify the color to the required color. Follow steps 3 to 6 to add colors to other surfaces of the pencil.

8. After modifying the colors, choose the **Properties** rollout and then choose the **Close** button.

Saving the Model

1. Choose the **Save the active object** button from the **File** toolbar and save the model.

Note
*Colors are not a feature of the model and therefore they will not appear in the **Model Tree**.*

Tutorial 3

In this tutorial you will create a blend feature shown in Figure 7-107. The two views of the blend feature are shown in Figure 7-108 with dimensions. After creating the model, you will redefine it such that the straight blending is changed into a smooth blending.

(Expected time: 45 min)

Figure 7-107 Isometric view of the model

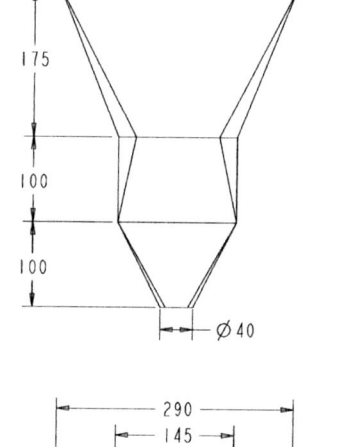

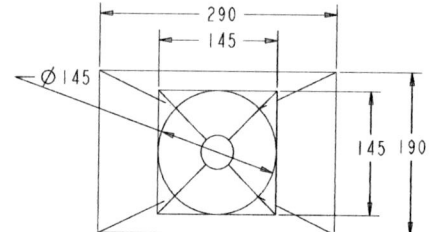

Figure 7-108 Top and front views of the model

The following steps are required to complete this tutorial:

a. Examine the blend feature and determine the number of sections in this feature. The blend consists of four sections, refer to Figure 7-108.
b. Create the sketch for the first section of the blend feature, refer to Figures 7-109.
c. Create the sketch for the second section of the blend feature, refer to Figures 7-110.
d. Create the sketch for the third section of the blend feature.
e. Create the sketch for the fourth section, refer to Figures 7-111. Give the depth between section numbers 1 and 2, 2 and 3, and 3 and 4.
f. Redefine the model to change the straight blending into a smooth blending, refer to Figure 7-113.

After understanding the procedure for creating the model, you are now ready to create it.

Advanced Modeling Tools-I

Starting a New Object File

1. Open a new part file and name it as *c07tut3*.

The three default datum planes are displayed in the graphics window.

Invoking the Blend Option

1. Choose **Insert > Blend > Protrusion** from the menu bar. The **BLEND OPTS** submenu is displayed.

2. Choose **Parallel > Regular Sec > Sketch Sec > Done** options from the **BLEND OPTS** submenu. The **ATTRIBUTES** menu is displayed and you will be prompted to choose the **Straight** or **Smooth** option from this menu.

3. Choose **Straight > Done** option from the **ATTRIBUTES** menu.

A **Smooth** blend will be created during the redefining of the model using the same sections that will be used to create the given model.

Selecting the Sketch Plane

1. Select the **FRONT** datum plane as the sketching plane. A red arrow points in the direction of feature creation and you are prompted to specify the direction of feature creation.

2. Choose **Okay** from the **DIRECTION** submenu. The **SKET VIEW** submenu is displayed.

3. Choose **Top** from this menu and using the left mouse button select the **TOP** datum plane present in the graphics window.

Drawing the First Section

The first section is a rectangle of 290x190 units.

1. Draw the sketch of the rectangular section and then add constraints and dimensions to it, as shown in Figure 7-109.

   After drawing the rectangular section, you need to toggle the section and draw the next section.

   Note
   *While drawing the sections for the blend feature, the start point is very important. The start point should be similar to those shown in the figures. If the start point is not at the desired point then select the point where you need the start point. Hold down the right mouse button to invoke a shortcut menu and choose the **Start Point** option.*

Toggling the Section

Toggling of a section is required in order to sketch the next section. Since in this tutorial four sections are used to create the required model, therefore, whenever you finish drawing one section you need to toggle to the next section.

1. Choose **Sketch > Feature Tools > Toggle Section** from the menu bar. You can also hold down the right mouse button to display the shortcut menu and choose the **Toggle Section** option from the shortcut menu.

When you choose **Toggle Section**, the previous section becomes inactive and appears gray in color.

Tip: *To toggle the section, simply right-click when you are in the selection mode. The selection mode in the sketcher environment can be invoked by choosing the* **Select items** *button.*

Drawing the Next Section

The next section is a circle.

1. Draw the sketch of the circular section, refer to Figure 7-110. Modify the diameter of the circle to 145.

As discussed earlier, the number of entities per section must be equal in a blend feature. Since, a circle is a single entity, it should be divided at four points.

Dividing the Circular Section

The circular section should be divided at four points because the rectangle and square have four entities. When you divide a circle at four points, the number of entities becomes four.

1. Choose the black arrow on the right of the **Dynamically trim section entities** button and then choose the **Divide an entity at the point of selection** button.

2. Select the circle at four points, as shown in Figure 7-110.

As you select points on the circle to divide it, some weak dimensions appear on the circle. Next, you need to apply constraints on the four points that were selected to divide the circle.

Applying Constraints on the Four Points

1. Choose the **Impose sketcher constraints on the section** button. The **Constraints** dialog box is displayed.

2. Choose the **Make line or two vertices vertical** constraint button from the **Constraints** dialog box and select the two division points on the left side of the circle to lie in a vertical line. Similarly, select the two points on the right to apply the constraint.

3. Now, choose the **Make line or two vertices horizontal** constraint button and select the two division points on the upper half and the lower half to lie in a horizontal line.

Advanced Modeling Tools-I

4. Modify the vertical dimension of the upper left division point, as shown in Figure 7-110. After the circular section is completed, the two sections with dimensions should look similar to the sections shown in Figure 7-110.

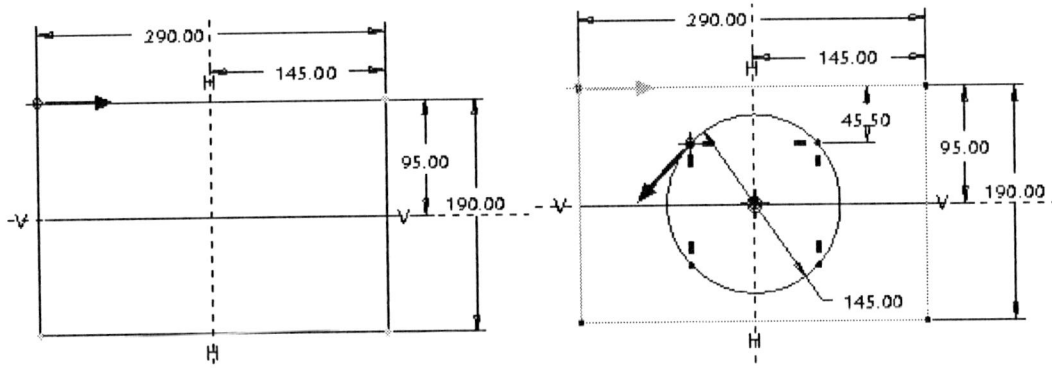

Figure 7-109 First rectangular section with dimensions

Figure 7-110 The two completed sections

5. Now, toggle the section and create the next section. The next section to be drawn is a square. After drawing the square section, draw the circular section. Divide the circular section into four entities similar to section 2 and then constrain and dimension it.

   Figure 7-111 shows all sections completed with dimensions.

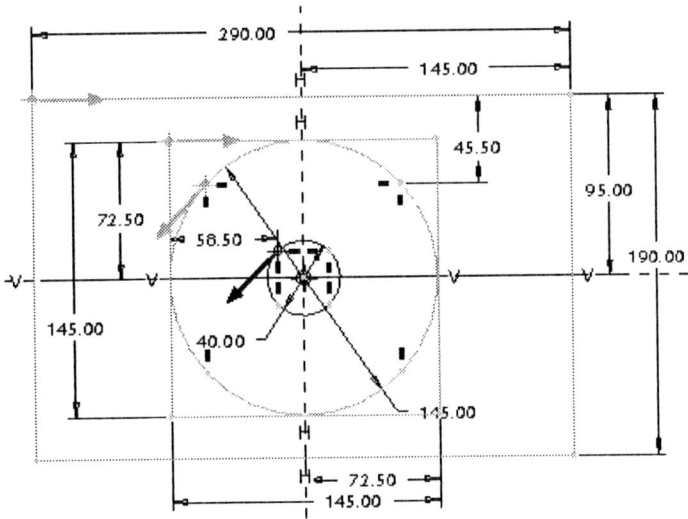

Figure 7-111 All four completed sections before giving depth

Note

In Figure 7-111, note the direction of the start points shown by arrows. These are important to avoid a twisted feature.

Giving Depth to the Sections

After the sketches of all sections are completed, you need to specify the depth between each section. The dimensions for depth between each section can be referred to from Figure 7-108.

1. Choose the **Continue with the current section** button.

   The **Message Input Window** appears and you are prompted to enter the depth for section 2.

2. Enter a value of **175** in the window and press ENTER.

   Similarly, the **Message Input Window** appears again and you are prompted to enter the depth for section 3. Enter 100 for section 3 and 100 for section 4.

Previewing the Blend Feature

The blend feature is completed and it can now be previewed.

1. Choose the **Preview** button from the **PROTRUSION** dialog box that is displayed on the top right corner of the window.

2. Choose the **Saved view list** button from the **View** toolbar. From the drop-down list choose the **Default Orientation** option. The model orients in the graphics window, as shown in Figure 7-112. You can use the middle mouse button to change the orientation of the model.

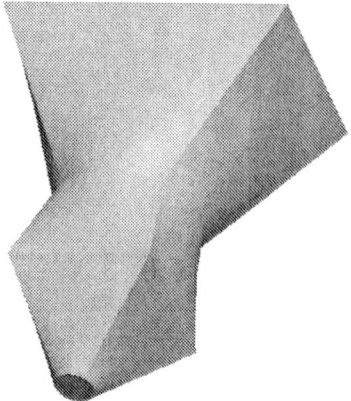

Figure 7-112 Shaded model

3. Now, choose the **OK** button that is present on the **PROTRUSION** dialog box.

Saving the Model

1. Choose the **Save the active object** button from the **File** toolbar and save the model.

Advanced Modeling Tools-I

Redefining the Blend Feature

After saving the straight blend feature, you will redefine this feature so that it is converted into a smooth blend.

1. Select the model in the graphics window. The edges of the model turn red in color.

2. Press and hold down the right mouse button in the graphics window until a shortcut menu appears.

3. Choose the **Edit Definition** option from the shortcut menu. The **PROTRUSION** dialog box is displayed.

4. Select the **Attributes** option under the **Element** tab. The **Attributes** option is highlighted.

5. Choose the **Define** button from the **PROTRUSION** dialog box. The **ATTRIBUTES** menu is displayed.

6. Choose **Smooth > Done** from the **ATTRIBUTES** menu.

7. Choose the **OK** button from the **PROTRUSION** dialog box. The smooth blend is created, as shown in Figure 7-113.

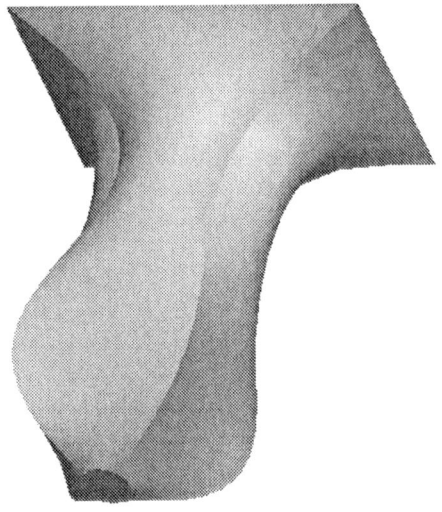

Figure 7-113 Smooth blend feature

Tutorial 4

In this tutorial you will create the model of a tap shown in Figure 7-114. All dimensions for the three sections are shown in the figure. **(Expected time: 30 min)**

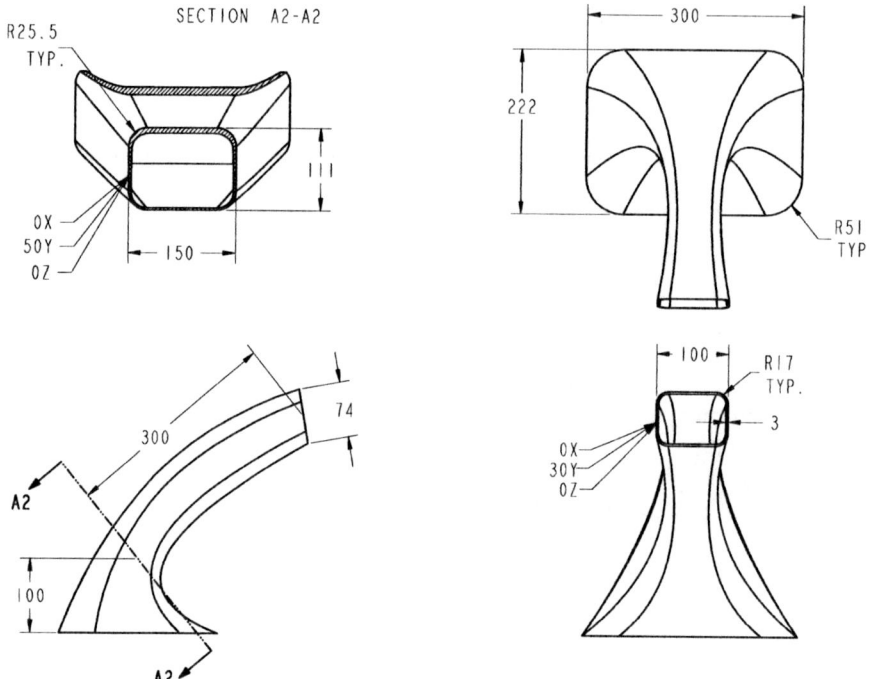

Figure 7-114 Sectioned, left-side, front, and top views of the model

This model is created using the general blend. In general blend, each section should have a coordinate system. The coordinate system helps in the alignment of the sections. Each section will be dimensioned with its coordinate system.

The following steps are required to complete this tutorial:

a. Examine the model and determine the number of sections in the blend feature, refer to Figure 7-114.
b. Create the first section for the blend feature, refer to Figure 7-115.
c. Create the second section for the blend feature, refer to Figure 7-116.
d. Create the third section for the blend feature, refer to Figure 7-117.

After understanding the procedure for creating the model, you are now ready to create it.

Starting a New Object File

1. Start a new part file and name it as *c07tut4*. The three default datum planes appear in the graphics window if they were not turned off in the previous tutorial.

Invoking the Blend Option

The given model is created using the general type of blend.

1. Choose **Insert > Blend > Protrusion** from the menu bar. The **BLEND OPTS** submenu is displayed.

2. Choose **General > Regular Sec > Sketch Sec > Done** from this submenu. The **ATTRIBUTES** menu is displayed.

3. Choose **Smooth > Done**.

   You are prompted to select a sketching plane.

4. Select the **TOP** datum plane and choose **Okay** from the **DIRECTION** submenu. From the **SKET VIEW** submenu, choose the **Right** option and select the **RIGHT** datum plane. The system takes you to the sketcher environment.

Drawing the Sketch for the First Section of the Blend

The blend consists of three sections. The dimensions for the second section are half of the dimensions of the first section. Similarly, the dimensions for the third section are one-third of the dimensions of the first section.

1. Draw the sketch for the first section.

2. Choose **Sketch > Coordinate System** from the menu bar. A reference coordinate system is attached to the cursor. Place the coordinate system at the intersection of the two planes using the left mouse button. It will be automatically aligned with the two datum planes.

3. Apply constraints and modify the dimensions of the sketch, as shown in Figure 7-115.

4. Choose the **Continue with the current section** button. The **Message Input Window** is displayed and you are prompted to enter the x-axis rotation angle for section 2.

5. Enter **0** as the value in this window and press ENTER. You are prompted to enter the y-axis rotation angle for section 2. Enter **50** as the value and press ENTER. Now, you are prompted to enter the z-axis rotation angle for section 2. Enter **0** as the value and press ENTER.

   The system takes you to the sketcher environment and allows you to draw the sketch for the second section. Here, you will notice that the datum planes are not displayed even if they are turned on. The reason for this is that the datum planes are not required now. The use of datum planes is to reference the sketch you draw. But, the second section that you will draw is already referenced to the first section. Hence, the datum planes are not displayed and you can draw the next section anywhere in the graphics window.

Drawing the Sketch for the Second Section

1. Draw the sketch for the second section, insert a reference coordinate system, apply constraints and dimensions to the sketch, and modify the dimensions, as shown in Figure 7-116.

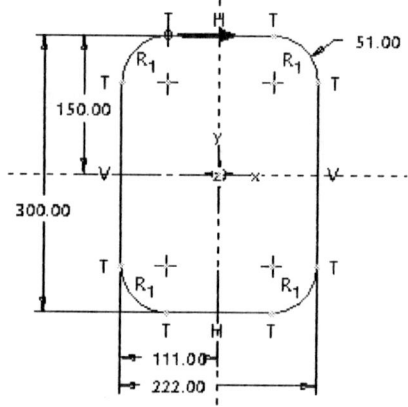

Figure 7-115 Sketch with dimensions and constraints of section 1

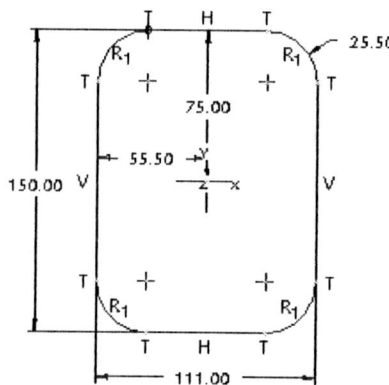

Figure 7-116 Sketch with dimensions and constraints of section 2

2. After completing the sketch, choose the **Continue with the current section** button.

   The **Message Input Window** is displayed and you are prompted to specify if you want to continue to the next section.

3. Enter **Y** in this window and press ENTER. You have entered **Yes** in this window because you have to draw the third section of the blend feature.
   The **Message Input Window** is displayed and you are prompted to enter the x-axis rotation angle for section 3.

4. Enter **0** as the value in this window and press ENTER. You are prompted to enter the y-axis rotation angle for section 3. Enter **30** as the value and press ENTER. Now, you are prompted to enter the z-axis rotation angle for section 3. Enter **0** as the value and press ENTER.

   The system takes you to the sketcher environment and allows you to draw the sketch for the third section.

Drawing the Sketch for the Third Section

1. Draw the sketch for the third section, insert a reference coordinate system, apply constraints and dimension to the sketch, and modify the dimensions, as shown in Figure 7-117.

2. After completing the sketch, choose the **Continue with the current section** button.

   The **Message Input Window** is displayed and you are prompted if you want to continue to the next section.

Advanced Modeling Tools-I

3. Enter **N** in this window and press ENTER. You have entered **No** in this window because all sections that are needed to create the blend feature are completed.

   The **Message Input Window** is displayed and you are prompted to enter the depth for **section 2**.

Specifying the Depths between Sections

1. Enter a value of **100** in the **Message Input Window** and press ENTER. This depth is the distance between the first section and the second section.
   Now, you are prompted to enter the depth for **section 3**.

2. Enter a value of **300** in this window and press ENTER. This depth is the distance between the second section and the third section.

   The blend feature is complete and you can now preview it.

3. Choose **Preview** from the **PROTRUSION** dialog box and choose **OK**. The default trimetric view of the blend feature is shown in Figure 7-118.

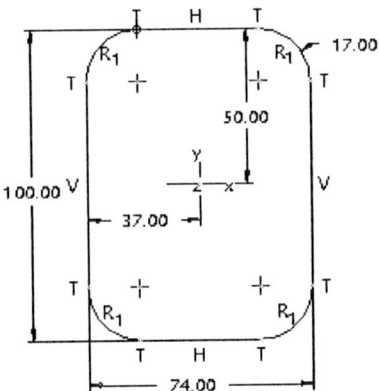
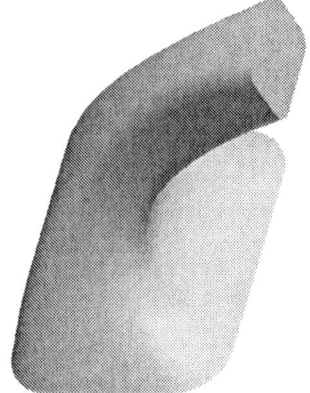

Figure 7-117 Sketch with dimensions and constraints of section 3

Figure 7-118 The default trimetric view of the blend feature

Creating Shell

1. Choose the **Shell Tool** button from the **Engineering Features** toolbar. The **Shell** dashboard is displayed and you are prompted to select the surfaces to be removed from the part.

2. Select the one end surface of the swept feature using the left mouse button and then using CTRL+left mouse button, select the other end surface. The two surfaces selected are highlighted red in color.

3. Enter the thickness value of the shell as **3** in the dimension box present on the **Shell** dashboard and press ENTER.

4. Choose the **Build feature** button to exit the **Shell** dashboard.

The default trimetric view of the shell feature is shown in Figure 7-119. You can use the middle mouse button to change the orientation of the model.

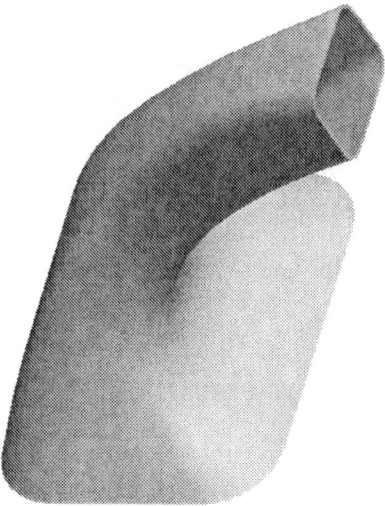

Figure 7-119 The default trimetric view of the model for Tutorial 4

Saving the Model

1. Choose the **Save the active object** button from the **File** toolbar and save the model.

Self-Evaluation Test

Answer the following questions and then compare your answers with those given at the end of this chapter.

1. The types of Protrusion available in Pro/ENGINEER are different from those that are available for Cut. (T/F)

2. **Cut** option is available only when at least a base feature exists. (T/F)

3. If the shell thickness value is negative then the shell thickness is added outside the boundary of the face selected for shelling. (T/F)

4. When trajectory is closed then there are two options that are available: **Add Inn Fcs** and **No Inn Fcs**. (T/F)

5. To create a Sweep Cut feature the procedure to follow is the same as in the case of Sweep Protrusion. (T/F)

6. The **Sweep** option extrudes a section along a _____.

Advanced Modeling Tools-I

7. The cross-section of the swept feature remains _____ throughout the sweep.

8. The sketching plane you select will be _____ to the screen when you draw the trajectory.

9. _____ section and open trajectory are not possible.

10. A quilt is a _____ feature.

Review Questions

Answer the following questions:

1. What is the maximum permissible angle for the rotation of sections in a **Rotational** blend?

    (a) 120 (b) 90
    (c) 180 (d) 45

2. What is the maximum possible draft angle that can be applied in Pro/Engineer?

    (a) 10-degree (b) 30-degree
    (c) 60-degree (d) 90-degree

3. What is the minimum number of sections required for a blend feature?

    (a) one (b) two
    (c) three (d) None of the above

4. Can a trajectory of a sweep feature be modified independent of the geometry of the section?

    (a) No (b) Yes
    (c) In some cases (d) None of the above

5. In which one of the following types of blend, sections are translated and rotated about the x, y, and z-axes?

    (a) **Parallel** (b) **Rotational**
    (c) **General** (d) None of the above

6. You can create a **Cut** feature using the **Sweep** option. (T/F)

7. In creating a Rotational or a General blend you need to create a coordinate system. (T/F)

8. In **General** blend the section is rotated about the y-axis of the coordinate system. (T/F)

9. The **Rotational** blend option is same as the **Parallel** blend option if the rotational blend angle entered between the two sections equals 0-degree. (T/F)

10. There must be an equal number of vertices in each section for blending. (T/F)

Exercises

Exercise 1

Create the foundation bolt shown in Figure 7-120. The shaded model is shown in Figure 7-121.

(Expected time: 30 min)

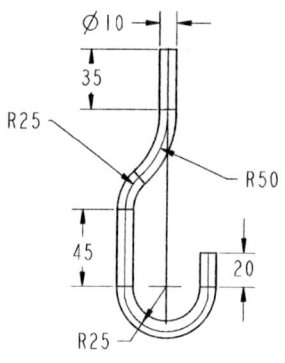

Figure 7-120 Figure for Exercise 1

Figure 7-121 Model for Exercise 1

Exercise 2

In this exercise you will create the model shown in Figure 7-122. The figure 7-123 shows the sectioned front view, top view, and the right-side view of the solid model with dimensions.

(Expected time: 45 min)

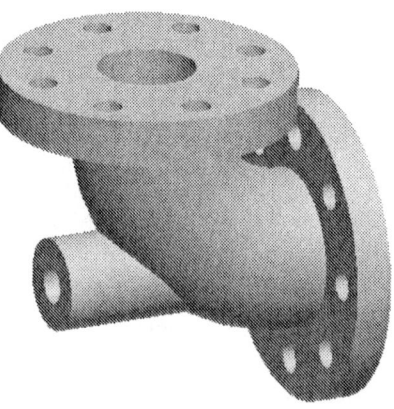

Figure 7-122 Model for Exercise 2

Advanced Modeling Tools-I

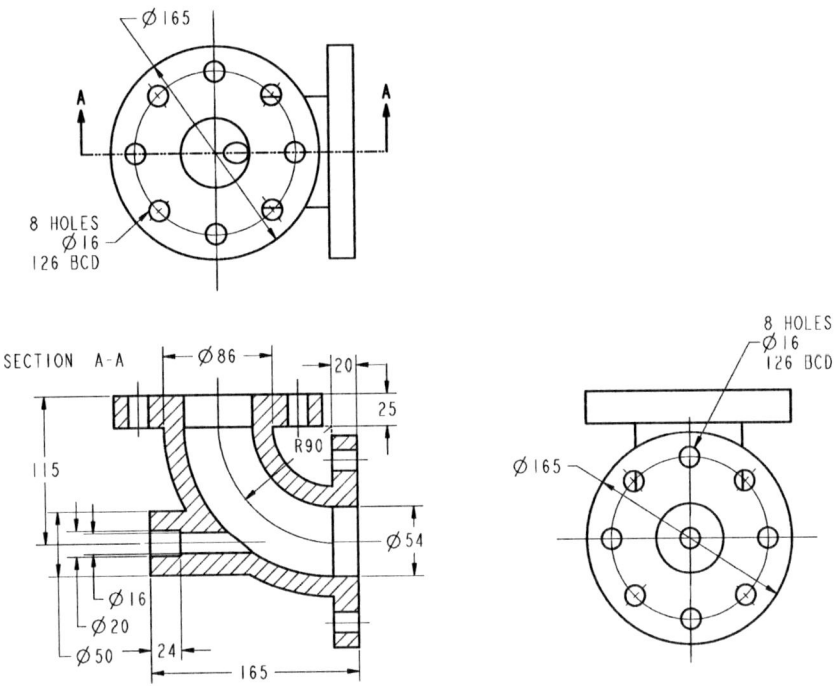

Figure 7-123 Top, front, right-side, and isometric views of the model

Exercise 3

In this exercise you will create the model of a soap case shown in Figure 7-124. Figure 7-125 shows the sectioned front view, top view, right-side view, and the detail view with dimensions.

(Expected time: 45 min)

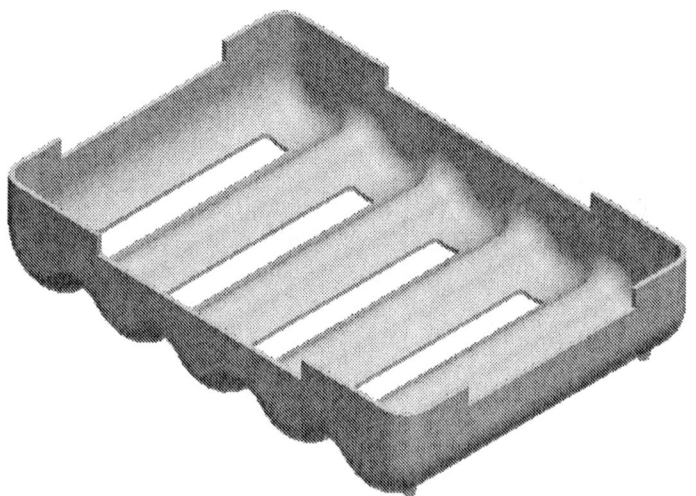

Figure 7-124 *Isometric view of the soap case*

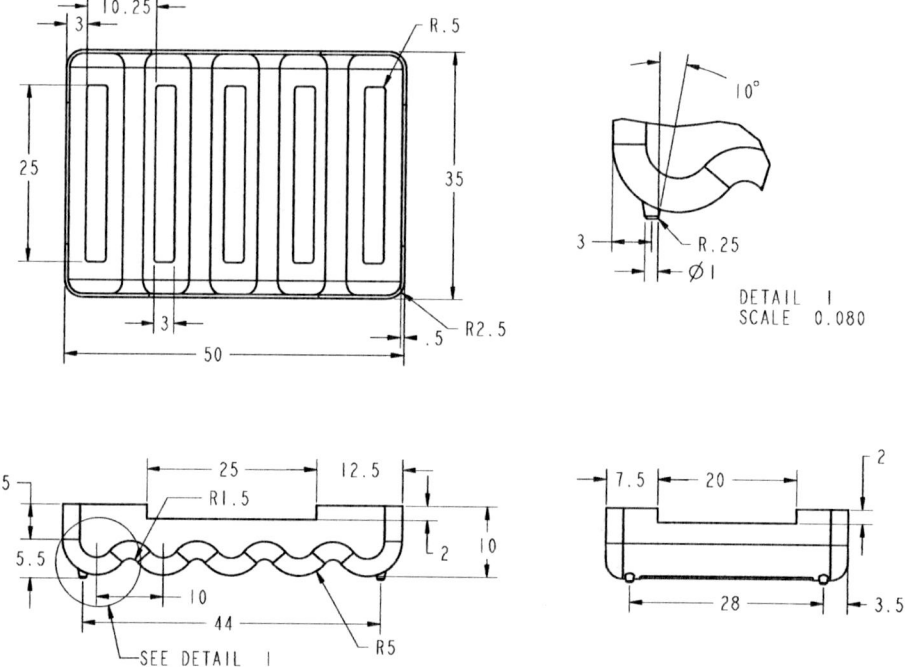

Figure 7-125 *Top, sectioned front, right-side, and detail view of the soap case*

Advanced Modeling Tools-I

Exercise 4

In this exercise you will create the model of a carburettor cover shown in Figure 7-126. Figure 7-127 shows the sectioned top view, sectioned front view, sectioned right-side view, and the sectioned bottom view with dimensions. **(Expected time: 45 min)**

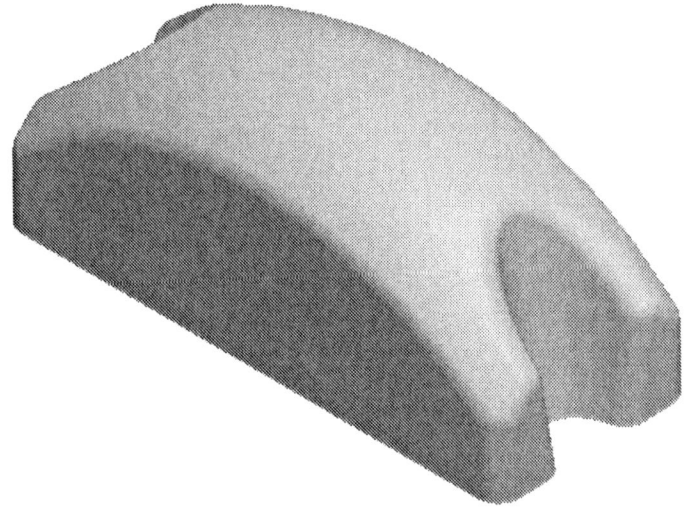

Figure 7-126 Isometric view of the carburettor cover

Hint
1. Create the sketch of the base feature that includes a rectangle of 125x50 and then extrude it.
2. Choose **Insert > Sweep > Cut** from the menu bar.
3. Select the **FRONT** datum plane as the sketching plane for sketching a trajectory.
4. Create a trajectory using the **Create an arc by 3 points or tangent to an entity and its endpoints** button. The start point and the endpoint of the arc should be at a distance of 28 from the bottom of the base feature and the radius of the arc should be 100.
5. Exit the sketcher environment using the **Continue with the current section** button.
6. Choose the **Merge Ends** option from the **ATTRIBUTES** menu.
7. You will again enter the sketching environment to create the section for the sweep feature. Choose the **Create an arc by 3 points or tangent to an entity and its endpoints** button from the **Sketcher Tools** toolbar. The arc created should be tangent to the reference lines, and the endpoints of the arc should be aligned with the edges of the base feature and should have a radius of 35.
8. Choose **OK** from the **SWEEP: Cut** dialog box.
9. Choose **Insert > Blend > Cut** from the menu bar.
10. Choose **Done** from the **BLEND OPTS** submenu.
11. Choose **Done** from the **ATTRIBUTES** menu.
12. Select the bottom face of the base feature as the sketching plane.
13. Choose **Okay** from the **DIRECTION** submenu.
14. Choose the **TOP** option from the **SKET VIEW** submenu and select the **RIGHT** datum plane.
15. Create an ellipse of Rx12 and Ry8 using the **Create full ellipse** button.

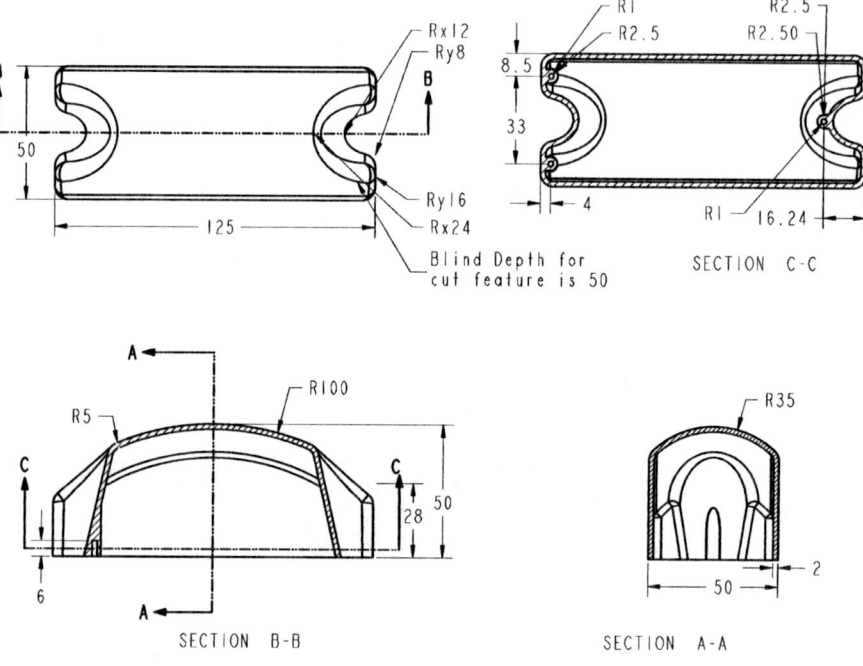

Figure 7-127 *Top view, sectioned front view, sectioned right-side view, and sectioned bottom view of the carburettor cover*

16. Choose **Sketch** > **Feature Tools** > **Toggle Section**.
17. Create another ellipse of Rx24 and Ry16.
18. Exit the sketcher environment using the **Continue with the current section** button.
19. Choose **Okay** from the **DIRECTION** menu.
20. Choose **Done** from the **DEPTH** menu and enter the depth of cut in the **Message Input Window** and press ENTER.
21. Mirror the cut feature about the **RIGHT** datum plane.
22. Create a round feature on all edges except the edges enclosing the bottom planar surface of the base feature.
23. Invoke the **Shell** option from the menu bar. Remove the bottom face of the base feature.
24. Using the bottom face of the base feature as the sketching plane create the three protrusion features that are the supporting structures for the screws. Extrude these features using the **Extrude up to next surface** button.
25. Using the **Hole** dashboard, create the hole in the protrusion features created earlier.

Answers to the Self-Evaluation Test

1 - F, **2** - T, **3** - T, **4** - T, **5** - T, **6** - trajectory, **7** - constant, **8** - parallel, **9** - open, **10** - surface.

Chapter 8

Advanced Modeling Tools-II

Learning Objectives

After completing this chapter you will be able to:
- *Use various advanced options for creating complex models.*
- *Create features using Var Sec Swp option.*
- *Create features using Swept Blend option.*
- *Create features using Helical Sweep option.*
- *Create features using Sect to Srfs option.*
- *Create features using Srfs to Srfs option.*
- *Create features using From File option.*
- *Create features using Toroidal Bend option.*
- *Use Relations.*

ADVANCED FEATURE CREATION TOOLS

Pro/ENGINEER facilitates the design process with some advanced options that you can use to create complex features with greater ease. All these options are available in the **Insert** menu present in the menu bar.

Variable Section Sweep Option

The **Variable Section Sweep** option can be invoked from the **Insert** menu in the menu bar or you can choose the **Variable Section Sweep Tool** button from the **Base Features** toolbar. This option is used to create a sweep feature in which the section of the sweep feature varies along the shape of the trajectories. This feature is created using more than one trajectory. The section is swept along the origin trajectory and the variation of the section is controlled by the X-trajectory and other trajectories. You can sketch the sweep trajectory using various sketcher tools. You can also select a previously created entity as the trajectory for sweep. The **Variable Section Sweep** option is available for both protrusion and cut options.

The following points should be remembered while creating a variable section sweep feature.

1. You should define the origin trajectory to which the section is normal.

2. After creating the origin trajectory you have to create an X-trajectory that defines the horizontal vector of the section.

3. You can also define additional trajectories to facilitate the creation of a complex profile.

4. When you draw the section for the variable section sweep feature, the section should be aligned to the endpoints of the X-trajectory and the other trajectories. If the section is not aligned, then the feature will be created without following the trajectory path and will be similar to a feature created using the **Sweep** option.

When you choose this option the **Variable Section Sweep** dashboard is displayed, as shown in Figure 8-1. Remember that you should have the trajectories present in the graphics window so that they can be selected for the creation of the variable section sweep feature. This is because you cannot sketch the trajectories by using the **Variable Section Sweep** dashboard. However, you can draw the trajectories, that is, datum-on-the-fly using the **Sketched Datum Curve Tool** button. As mentioned earlier, the process of creating a datum feature when a feature creation tool is active is called datum-on-the-fly. The options in the **Variable Section Sweep** dashboard are discussed next.

*Figure 8-1 The **Variable Section Sweep** dashboard*

Advanced Modeling Tools-II

References Tab

When you choose the **References** tab, the slide-up panel is displayed, as shown in Figure 8-2. The options available in this slide-up panel are discussed next.

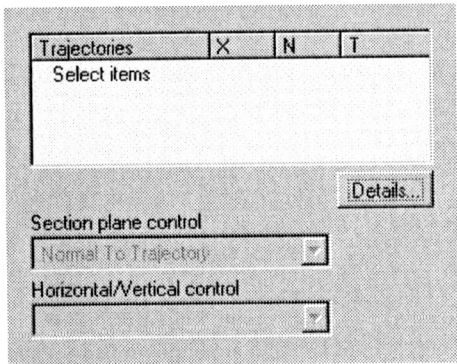

Figure 8-2 References tab slide-up panel

Trajectories collector

This collector is used for selecting the trajectories and after drawing the section. The trajectories that you select are listed in this collector. The first trajectory that you select is by default listed as the origin trajectory.

Note that you need to first select the normal and X trajectories and then draw the sketch of the section. This is because the sketch plane will be oriented only if you have defined the normal trajectory. Also, since the sketch needs to pass through the endpoints of the X-trajectory, it is better to define the X-trajectory before drawing the section. This way you will have the reference point from which the sketch need to pass.

This collector has three columns and the selected trajectories are listed in the rows. The first column is **X**. The check box in this column, when selected, converts the trajectory to the X trajectory. The second column is **N**. The check box in this column, when selected, converts the trajectory in that row to the normal trajectory. Remember that only one trajectory can be a normal trajectory and only one trajectory can be a X trajectory. The third column is **T**. There are two check boxes available in this column and in each row. All these check boxes are gray in color. When you select the check box, the name of the trajectory is filled in that cell and values T=0 appears on the trajectory in the graphics window. This value is displayed on both the ends of the trajectory. Initially, this value is 0. When you enter a value, the point of alignment of the section with the trajectory varies.

> **Section plane control drop-down list**. The options in this list are used to specify the direction of the section in which it start sweeping.

Normal To Trajectory option. When you use the **Normal to trajectory** option to create the variable sweep, the section is normal to the origin trajectory and is swept along the X trajectory. The section should be aligned with the X trajectory to sweep along it. If the section is not aligned with the X trajectory then this option works similar to the Sweep option and the section is swept along the origin trajectory only. Figure 8-3 shows the two trajectories and the section that is used to create the sweep feature shown in Figure 8-4.

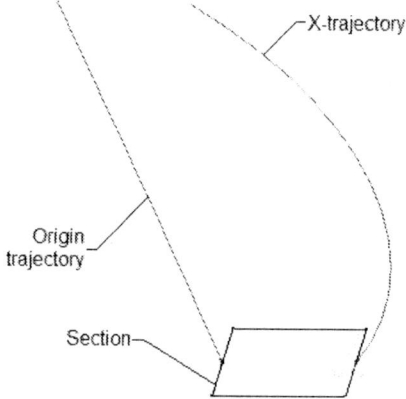
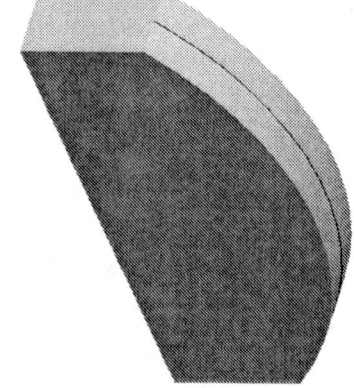

Figure 8-3 The two trajectories   *Figure 8-4 Variable section sweep feature*

Now, in Figure 8-5, the X trajectory and the origin trajectory are interchanged and the sweep feature that is obtained is shown in Figure 8-6. Note the difference in the sweep features. In Figure 8-5, the section is normal to the selected origin trajectory and the curve is selected as the origin trajectory. The resulting sweep feature is shown in Figure 8-6.

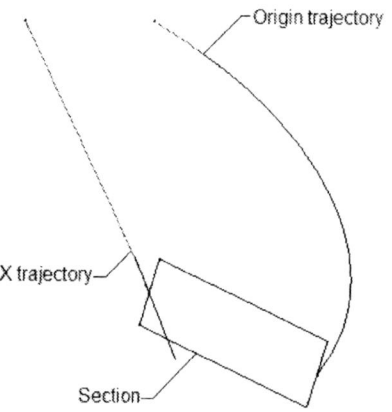
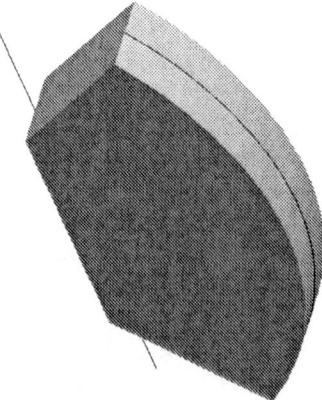

Figure 8-5 The two trajectories and section   *Figure 8-6 Resulting variable section sweep feature*

Advanced Modeling Tools-II

Normal To Projection option. The Normal to projection option is used to create a variable section sweep feature in which the section is normal to the two-dimensional (2D) projection of trajectories. The swept section is normal to the origin trajectory and the variation in the section is controlled by the X-trajectory and the other trajectories. When you use this option, you need to select or sketch an origin trajectory, an X-trajectory, other trajectories if required, and the direction of projection. The direction of projection can be specified by using the default coordinate system.

Figure 8-7 shows the origin trajectory and the X trajectory. The two trajectories are used to create the sweep feature shown in Figure 8-8.

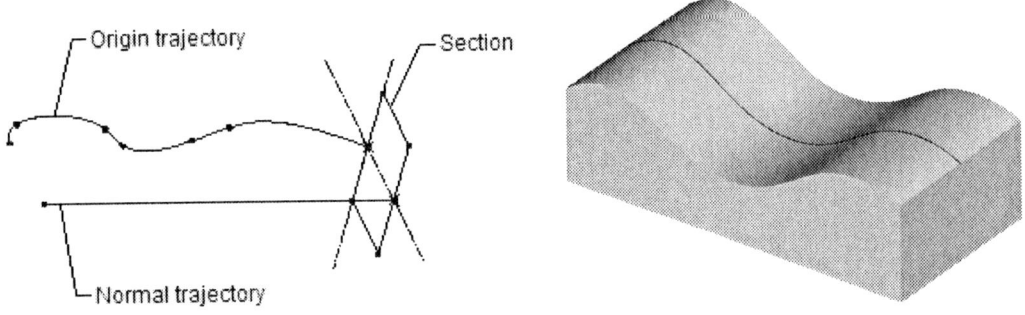

Figure 8-7 Trajectories and section   *Figure 8-8* Resulting swept feature

Constant Normal Direction option. The **Constant normal direction** option is used to create a variable section sweep in which the section to be swept is normal to an edge, normal of the selected plane, or an axis.

When you select the origin trajectory and then select the **Constant normal direction** option, you are required to select or create a plane, an edge, a curve, an axis, or a coordinate system to define the constant normal direction. Select a datum plane and then sketch the section. The system takes you to the sketcher environment and automatically orients the sketch plane such that it is normal to the selected direction. Figure 8-9 shows the origin trajectory, the section, and the normal vector of the datum plane to which the section is perpendicular. Figure 8-10 shows the section that is normal to the origin trajectory. This figure explains the difference between the **Normal to trajectory** option and the **Constant normal direction** option.

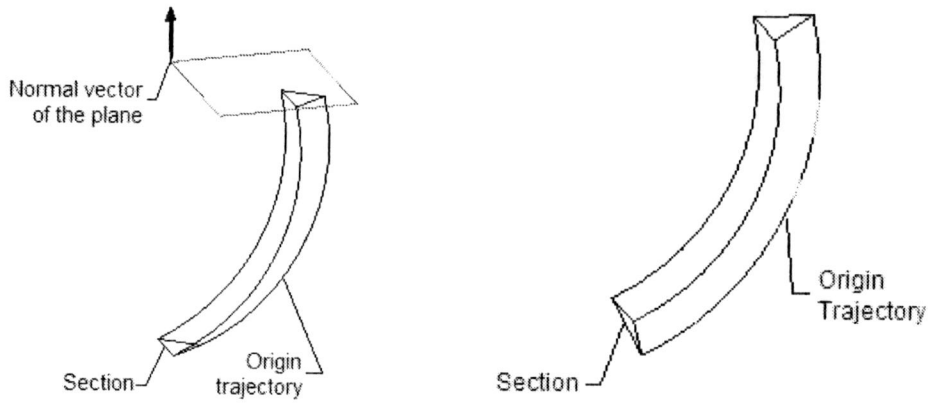

Figure 8-9 Origin trajectory and the pivot plane   *Figure 8-10* Swept feature on the base feature

Options Tab

When you choose the **Options** tab, the slide-up panel is displayed, as shown in Figure 8-11. The options in this slide-up panel are discussed next.

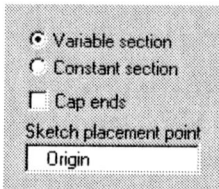

Figure 8-11 **Options** option slide-up panel

Variable section
This option is used to create a sweep feature whose section varies along the trajectory.

Constant section
This option is used to create a sweep feature whose section is constant throughout the trajectory. The feature created using this option is similar to the feature created using the **Sweep** option.

Sketch placement point collector
By default, **Origin** is selected as the start point of the section. This is the reason you draw the section at the start point of the trajectory. However, if you click in this collector to activate it and then select a datum point that lies on the origin trajectory, the section will start from that point. This means that the selected datum point becomes the start point of the trajectory.

Note

The options available on the lower half of the dashboard are the same as those available on the **Extrude** *dashboard. Using these options you can specify the type of feature you want to create. For example, solid, surface, or thin feature.*

Swept Blend

The **Swept Blend** feature is a combination of sweep and blend. To create a sweep feature you need a trajectory along which the section is swept. You have an option of sketching or selecting a previously created feature as the trajectory of the sweep. To create a parallel blend you need to have more than one section to be blended. The combination of both these options is the **Swept Blend** option.

In this option you can blend two or more than two sections having same number of entities and the path of the transition of blending is defined by selecting or sketching a trajectory. You can invoke the **SWEPT BLEND** option from the **Insert** menu in the menu bar.

Choose **Insert > Swept Blend > Protrusion** from the menu bar; the **BLEND OPTS** menu is displayed, as shown in Figure 8-12. The options in this menu are discussed next.

Figure 8-12 BLEND OPTS menu

Select Sec
The **Select Sec** (Select section) option is used to select existing sections that will be used to create a swept blend.

Sketch Sec
The **Sketch Sec** (Sketch section) option is used to sketch the section that will be used to create the swept blend.

NormalToOriginTraj
The **NormalToOriginTraj** option is used to keep the section normal to the origin throughout the trajectory. The sweep option behaves in this manner.

Pivot Dir
The **Pivot Dir** (Pivot direction) option is used to create a swept blend in which the section to be swept is normal to the origin trajectory. You need to select the pivot direction.

Norm To Traj
The **Norm To Traj** (Normal to trajectory) option is used to create a swept blend in which the section is perpendicular to the normal trajectory and is swept along the origin trajectory.

Figure 8-13 shows the three rectangular sections. The three sections are connected through a trajectory by connecting their vertices. The shaded view of the model after shelling it is shown in Figure 8-14.

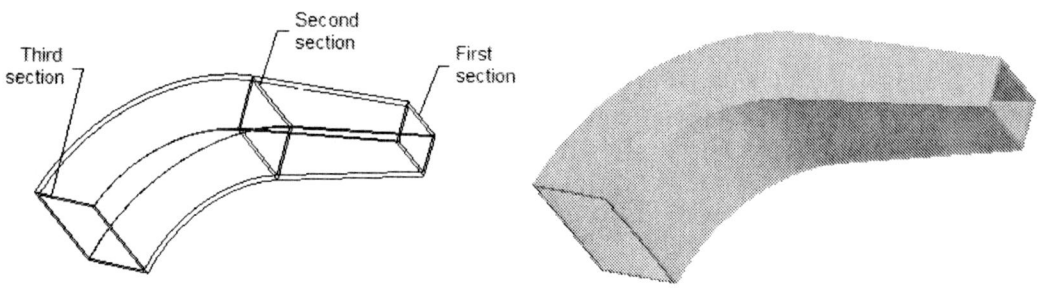

Figure 8-13 Three sections in swept blend   *Figure 8-14 Shaded view of the feature*

In Figure 8-15, the trajectory and the sections that are used to create the feature shown in Figure 8-16 are shown. It should be noted that the number of entities should be same in all sections.

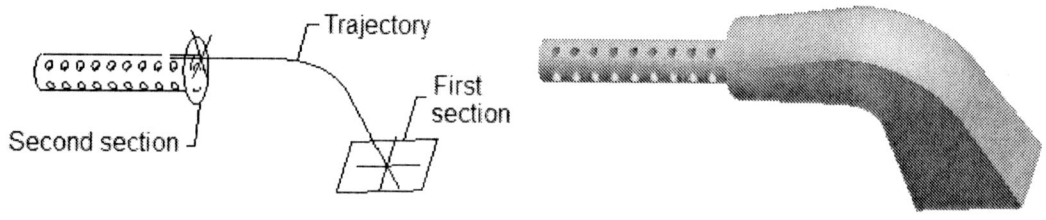

Figure 8-15 Two sections and a trajectory   *Figure 8-16 Resulting feature*

Helical Sweep

The **Helical Sweep** option is used to create helical sweep features. You have to define a trajectory that will specify the shape and height of the helix, a pitch value, and a cross-section to create a helical feature using this option. The distance of the trajectory from the center line defines the radius of the helical path and the length of the trajectory defines the length of the sweep feature. The main use of this option is to create the helical springs and threads.

Advanced Modeling Tools-II

When you choose the **Helical Sweep** option from the **Insert** menu in the menu bar, the **ATTRIBUTES** menu is displayed, as shown in Figure 8-17. The menu options are discussed next.

Figure 8-17 ATTRIBUTES menu

Figure 8-18 shows the parameters to be defined for the helical sweep and Figure 8-19 shows the resulting helical sweep feature.

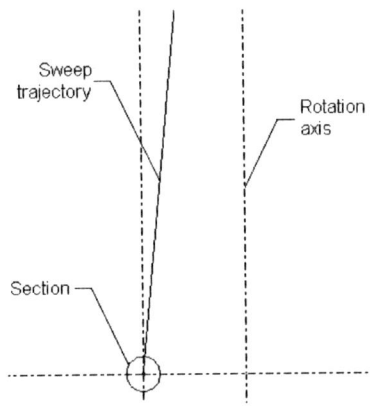

Figure 8-18 Sweep trajectory, sweep section, and axis of rotation to create a helical feature

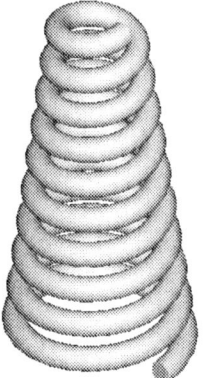

Figure 8-19 The resulting helical feature

Constant

The **Constant** option is used to create a helical feature with constant pitch, as shown in Figure 8-20.

Variable

The **Variable** option is used to create a helical feature of varying pitch. When using this option you have to define the start pitch and the end pitch. You can also define a pitch between the start point and the endpoint of the helix. This pitch value between the endpoint and the start point is specified by creating points on the sweep trajectory. The helical sweep feature with variable pitch is shown in Figure 8-21.

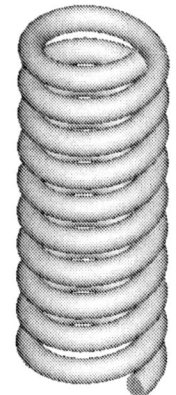 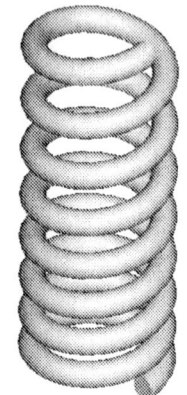

Figure 8-20 Helical sweep feature with constant pitch   *Figure 8-21 Helical sweep feature with variable pitch*

The brief procedure to create a variable pitch spring is discussed next.

1. Select **Insert > Helical Sweep > Protrusion** from the menu bar and then choose **Variable > Thru Axis > Right Handed > Done** from the **ATTRIBUTES** menu.
2. Select the sketch plane and then orient it.
3. Draw the sketch of the sweep trajectory and the center axis.
4. Add points on the sweep trajectory by choosing the **Create points** button. These points will be later used to vary the pitch of the spring.
5. Exit the sketcher environment. You are prompted to enter the pitch at the start point. After entering the pitch value at start you are prompted to specify the pitch value at the end.
6. After specifying the pitch value at the two end, a subwindow is displayed with a pitch graph.
7. Select the points on the trajectory from the main window. The **Message Input Window** is displayed.
8. Enter the pitch value at the selected point. Similarly, specify the pitch values at other points that were defined on the trajectory while sketching it. It should be noted that a specified pitch value can be modified by selecting the **Change Value** option from the **DEFINE GRAPH** submenu.
   As you specify the pitch values to the points, the graph in the subwindow also changes.
9. After all pitch values are entered, choose the **Done** option from the **GRAPH** menu. Now, you enter the sketcher environment to draw the section of the spring.
10. Draw the section of the spring at the intersection of the two infinite yellow dotted lines.
11. Exit the sketcher environment and choose **OK** from the **PROTRUSION** dialog box.

Note
You can use any geometric entity to create the sweep trajectory. For example spline, line, or arcs.

Thru Axis

The **Thru Axis** option is used to create a helical feature around an axis.

Norm To Traj

The **Norm To Traj** option is used to create a helical feature perpendicular to the sketched trajectory.

Right Handed

The **Right Handed** option is used to create a helical feature in which the section is swept in the counterclockwise direction from the start sketch.

Left Handed

The **Left Handed** option is used to create a helical feature in which the section is swept in the clockwise direction from the start sketch.

Blend Section To Surfaces

The **Blend Section to Surfaces** option is used to blend a selected set of tangential surfaces with a sketched contour. You can invoke this option by choosing **Insert > Advanced > Blend Section To Surfaces** from the menu bar. The contour sketched to create the feature using this option must be closed. This option can be used to create a surface or a solid feature. Figure 8-22 shows the section to be blended with the selected set of tangential surfaces and Figure 8-23 shows the resulting blended feature.

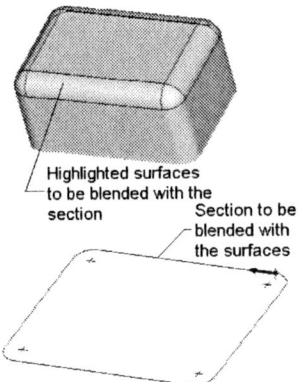
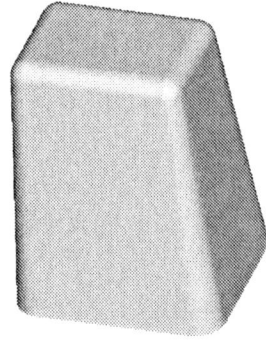

Figure 8-22 Highlighted set of tangential surfaces and the contour sketch.

Figure 8-23 Feature created using the **Sect to Srfs** option

Blend Between Surfaces

The **Blend Between Surfaces** option is used to create a blending between two curved quilts, two curved solid surfaces, or a combination of both. The curved surfaces may be revolved features or spheres. You can invoke this option by choosing **Insert > Advanced > Blend Between Surfaces** from the menu bar. Figure 8-24 shows the two spheres to be blended and Figure 8-25 shows the resulting blended feature created using this option.

Figure 8-26 shows the two curved features to be blended and Figure 8-27 shows the resulting blended feature created using this option.

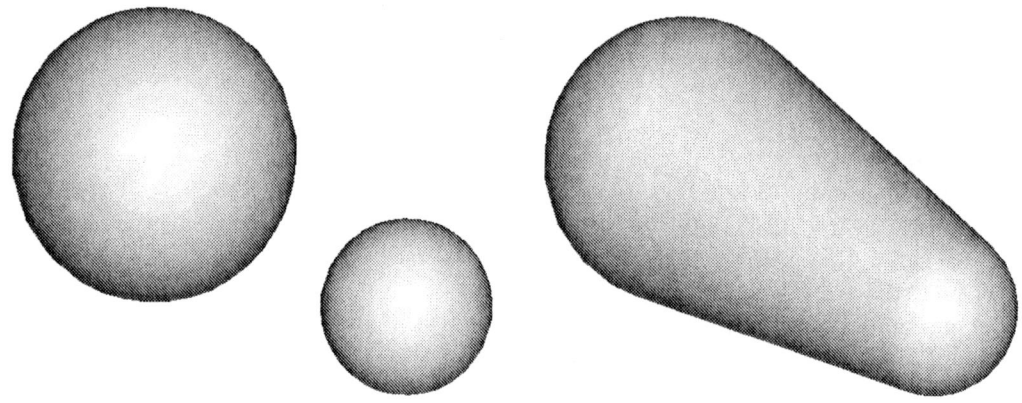

Figure 8-24 Two spheres to be blended   *Figure 8-25* The resulting blended surface.

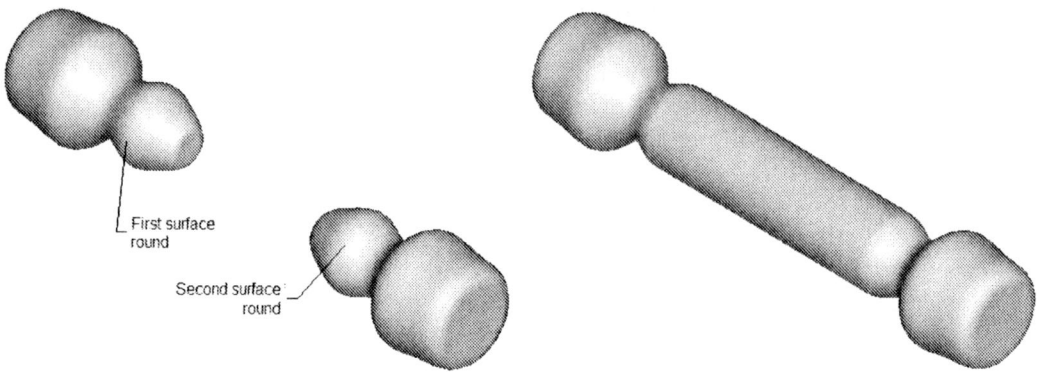

Figure 8-26 Two curved features to be blended   *Figure 8-27* The resulting blended surface.

Toroidal Bend

The **Toroidal Bend** option is used to create features with curved surfaces or the models that generally have a cut profile on the curved surface. For example, if you want to create a feature shown in Figure 8-28, you will first create a rectangular plate and cut the profile on it. Then you will bend the rectangular plate shown in Figure 8-29 to 360-degrees.

To invoke the **Toroidal Bend** option, choose **Insert > Advanced > Toroidal Bend** from the menu bar. The **OPTIONS** menu is displayed, as shown in Figure 8-30. This menu is divided into three parts. The first part of the menu has options to specify the angle of rotation of the bend. The second part has options that specify whether you want to create the feature on one side of the sketch plane or on both sides. The third part of the menu is used to specify the method of creating the toroidal bend.

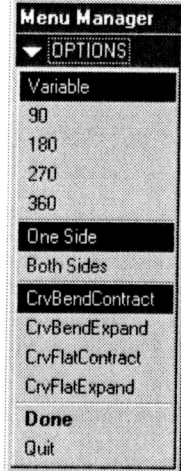

Figure 8-28
OPTIONS menu

Advanced Modeling Tools-II

The following steps explain the procedure to create the model shown in Figure 8-29. It is assumed that you have created the rectangular plate shown in Figure 8-30.

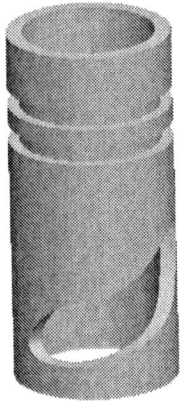

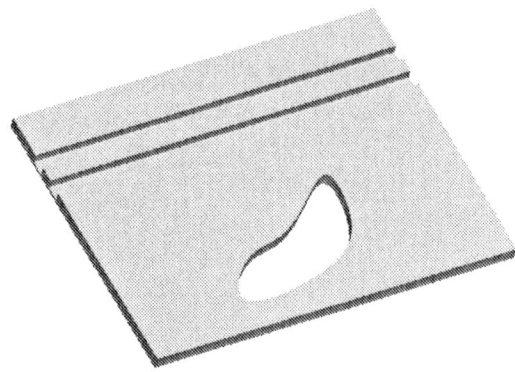

Figure 8-29 Example of toroidal bend

Figure 8-30 Rectangular plate with a profile cut

1. Choose **Insert > Advanced > Toroidal Bend** to invoke the **OPTIONS** menu. Choose **360 > One Side > CrvBendContract > Done**. The **DEFINE BEND** menu is displayed. In this menu, the **Add** option is selected by default and you are prompted to select solids, quilts, or datum curves to bend.
2. Select the top face of the base feature and choose **Done** from the **DEFINE BEND** menu. You are prompted to select the sketch plane.
3. Select the right face of the base feature as the sketch plane and orient the sketch plane.
4. After entering the sketcher environment, draw the sketch, as shown in Figure 8-31. The sketch comprises of a line that is drawn from the left edge to the right edge. A user-defined coordinate system is placed at the midpoint of the line.

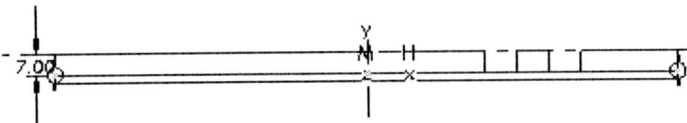

Figure 8-31 Sketch along with user-defined coordinate system

Note
The sketch that is drawn in step 4 controls the side profile of the toroidal bend. If the sketch drawn is a line, then the side walls of the toroidal bend will be vertical but if it is an arc, then the side profile will have the contour of the drawn arc.

5. Exit the sketcher environment. You are prompted to select the two parallel planes to define the length of bend.
6. Select the face that was selected as the sketch plane and the face opposite to it (the left face of the base feature.)
   The face of the base feature that was selected to bend is bent at an angle of 360-degrees and the toroidal bend is created, as shown in Figure 8-29.

TUTORIALS

Tutorial 1

In this tutorial you will create the model shown in Figure 8-32. This figure also shows the sectioned top, front, right-side, and isometric views of the model. **(Expected time: 45 min)**

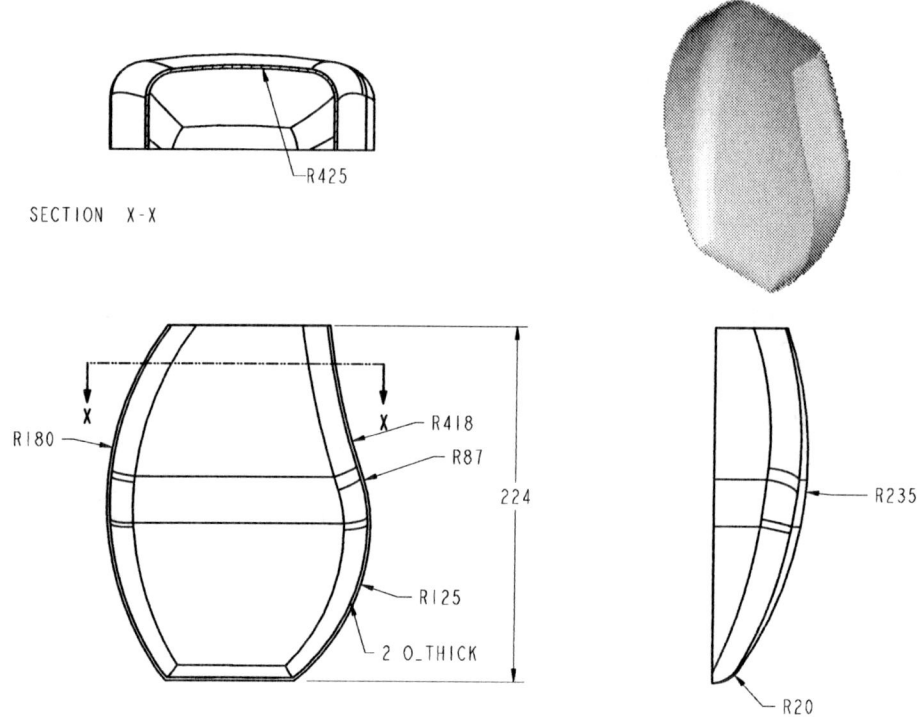

Figure 8-32 Sectioned-top, front, right-side, and isometric views of the model

The following steps are required to complete this tutorial:

a. Examine the model and determine the number of features in it. The model is composed of five features, refer to Figure 8-32.
b. Create the first feature as a set of datum curves that will be drawn on the **FRONT** datum plane, refer to Figures 8-33 through 8-35.
c. The second feature is the datum curve that will be drawn on the **RIGHT** datum plane, refer to Figure 8-36.
d. Create the variable section sweep, refer to Figures 8-37 through 8-39.
e. Create the round feature, refer to Figures 8-41 and 8-41.
f. Create the shell feature of thickness 2, refer to Figures 8-42 and 8-43.

After understanding the procedure for creating the model, you are now ready to create it. When the Pro/ENGINEER session is started, the first task is to set the working directory.

Advanced Modeling Tools-II

Since this is the first tutorial of this chapter, you need to create a directory named *c08*, if it does not exist. Choose the **New Directory** button in the **Select Working Directory** dialog box and create a directory named *c08* at *C:\ProE-WF-2.0*.

Starting a New Object File

1. Open a new part file and name it *c08tut1*. The three default datum planes are displayed in the graphics window.

Creating the Trajectories

You will draw the datum curves that will be selected as trajectories to create the variable section sweep feature. After selecting the sketching plane, draw the sketch for the origin trajectory and dimension it. Then you need to draw the sketch for the X trajectory on the same plane and dimension it. The X trajectory defines the horizontal vector of a section. If the section is aligned with the X trajectory, the section varies along the path of the X trajectory. Similarly, two additional trajectories will be drawn to guide the section throughout the sweep.

1. Choose the **Sketched Datum Curve Tool** button from the **Datum** toolbar. The **Sketched Datum Curve** dialog box is displayed.

2. Select the **FRONT** datum plane from the graphics window.

3. Select the **TOP** datum plane from the graphics window and then select the **Top** option from the **Orientation** drop-down list.

4. Once you enter the sketcher environment, close the **References** dialog box and create the sketch of the origin trajectory. The origin trajectory is a straight line segment aligned to the **RIGHT** datum plane and the start point is aligned to the **TOP** datum plane. The sketch of the origin trajectory is shown in Figure 8-33.

Note
In the sketcher environment you will draw all three datum curves because they are on the same datum plane.

*You are drawing the datum curves as individual features because you cannot sketch the trajectories using the tools available in the **Variable Section Sweep** dashboard.*

5. After the sketch of the origin trajectory is complete, draw the sketch of the X trajectory. The X trajectory is drawn using three arcs. It is recommended that you start the sketch from the top arc. After drawing the first arc, dimension it and then modify its dimension.

6. Draw the second arc and then modify the dimension, as shown in Figure 8-34.

7. Next, draw the third arc. As evident from the sketch of the X trajectory, the center points of the bottom and the second arcs are aligned horizontally.

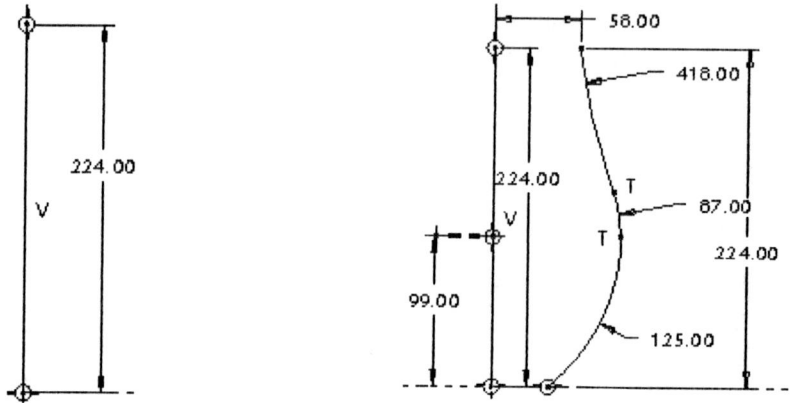

Figure 8-33 Sketch with dimension of the origin trajectory

Figure 8-34 Sketch with dimensions of the X trajectory

8. Create the sketch of the third trajectory, as shown in Figure 8-35. This trajectory is created using a single arc.

9. After the sketch is complete, choose the **Continue with the current section** button to exit the sketcher environment.

10. Similarly, create another trajectory on the **RIGHT** datum plane and keep the **TOP** datum plane at the left. The sketch of the trajectory is shown in Figure 8-36.

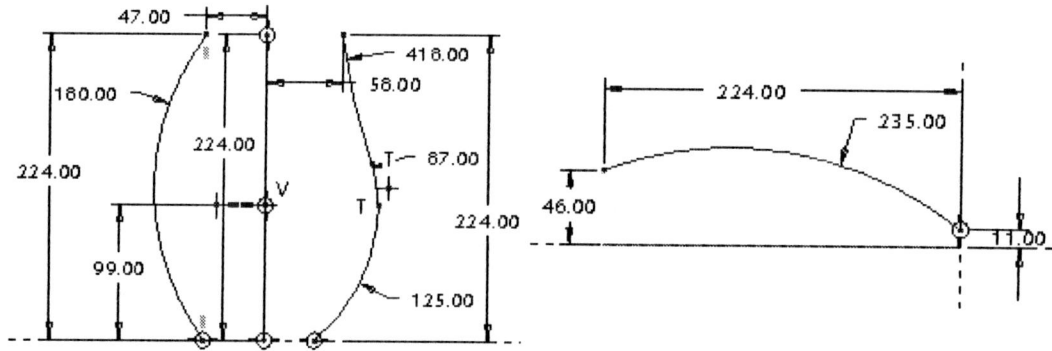

Figure 8-35 Sketch with dimensions of the additional trajectory

Figure 8-36 Sketch with dimensions of the additional trajectory

11. After the sketch is complete, choose the **Continue with the current section** button to exit the sketcher environment.

12. Choose the **Default Orientation** option from the **Saved view list** button in the **Top Toolchest** to view the trajectories you have created, as shown in Figure 8-37. This view gives a better display of trajectories in 3D space.

Creating the Variable Section Sweep feature

The four trajectories that will create the variable section sweep have been created. Now, you need to invoke the feature creation tool and select these trajectories to create the feature.

1. Choose the **Variable Section Sweep Tool** button from the **Base Features** toolbar. The **Variable Section Sweep** dashboard is displayed.

2. Select the **References** tab to invoke the slide-up panel. Select the origin trajectory that was the first datum curve drawn.

3. Use CTRL+left mouse button to select the X trajectory (second datum curve). Note as you are selecting the trajectories, they are being listed in the **Trajectories** collector.

4. Use CTRL+left mouse button to select the remaining two trajectories. Select the check box under the X column for **Chain 1** row. This makes the trajectory that you select second as the X trajectory.

5. Now, choose the **Create or edit sweep section** button from the dashboard to enter the sketcher environment.

6. Draw the closed sketch of the section and apply the dimensions, as shown in Figure 8-38. Note that the section is aligned to all start points of the trajectories that are displayed by crosses. The cursor will automatically snap to these points as you move it close to these points. In order to vary the section with the trajectories, it is necessary to align the section with the trajectories.

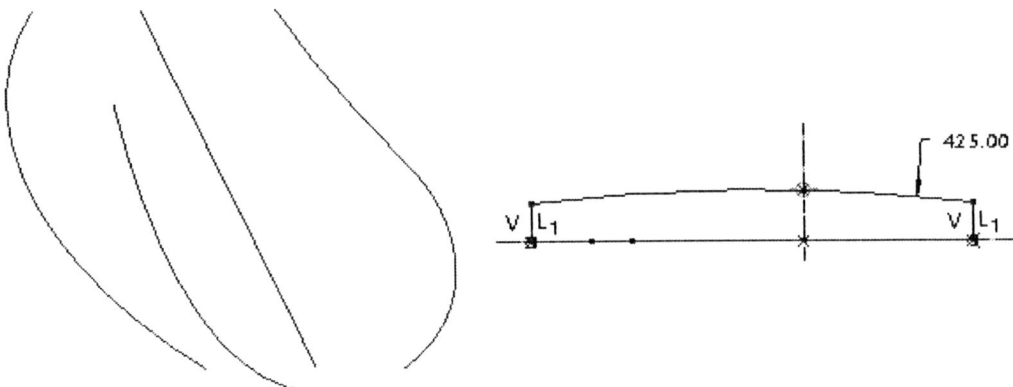

Figure 8-37 Default view of the four trajectories

Figure 8-38 Sketch of the section with dimensions

7. After the sketch is complete, choose the **Continue with the current section** button to exit the sketcher environment.

The preview of the variable section sweep feature is displayed in the graphics window.

8. Choose the **Sweep as solid** button from the dashboard if the preview shows a surface feature.

9. Choose the **Build feature** button from the dashboard. Now you need to hide the datum curves.

10. Choose the **Set layers, layer items and display states** button from the **Top Toolchest** and open the **Model Tree**. In the Navigator, the layers of the current model are displayed. Right-click on the **PRT_ALL_CURVES** to invoke the shortcut menu.

11. From the shortcut menu, choose the **Blank layer** option. Regenerate the model, if needed. All datum curves in the model will disappear from the display.

    The variable section sweep feature is completed and is shown in Figure 8-39. You can use middle mouse button to view the feature, as shown in the figure.

Creating the Round Feature

You will create round on the edges of the base feature. The edges you need to round are shown in Figure 8-41.

1. Choose the **Round Tool** button from the **Engineering Features** toolbar. The **Round** dashboard is displayed.

2. Select the edges shown in Figure 8-40 to round. Remember that you need to use CTRL+left mouse button to select more than one edge.

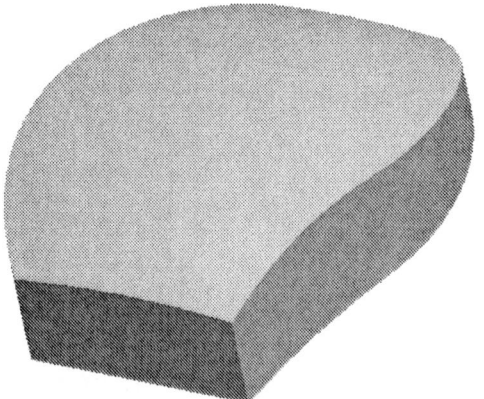
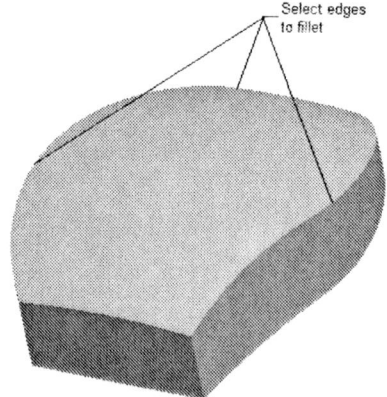

Figure 8-39 Variable section swept feature   *Figure 8-40* Edges to be selected to round

3. Enter **20** in the dimension box present on the dashboard. Choose **Build feature** from the dashboard. The base feature after creating the round is shown in Figure 8-41.

Creating the Shell

Now, you need to create the shell and remove the front and bottom faces of the base feature.

Advanced Modeling Tools-II 8-19

1. Choose the **Shell Tool** button from the **Engineering Features** toolbar. You are prompted to select the surfaces to be removed.

2. Select the surfaces shown in Figure 8-42 to remove. Remember that to select more than one face you need to use CTRL+left mouse button.

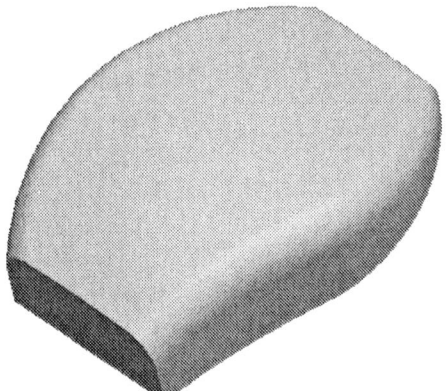 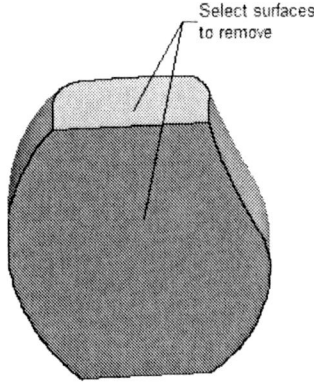

Figure 8-41 Model after creating round   *Figure 8-42* Surfaces to be selected to remove

3. Enter **2** in the dimension edit box. Choose the **Build feature** button from the dashboard. The model after creating the shell feature is shown in Figure 8-43.

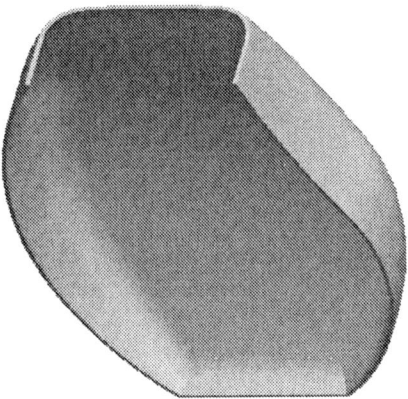

Figure 8-43 Final model after shelling

Saving the Model

You have to save the model because you may need it later.

1. Choose the **Save the active object** button from the **File** toolbar and save the model.

Tutorial 2

In this tutorial you will create the model shown in Figure 8-44. This model is a Upper Housing of a motor blower assembly. Figure 8-45 shows the left-side view of the top view, top view, front view, and the sectioned left-side view of the model.   **(Expected time: 1 hr)**

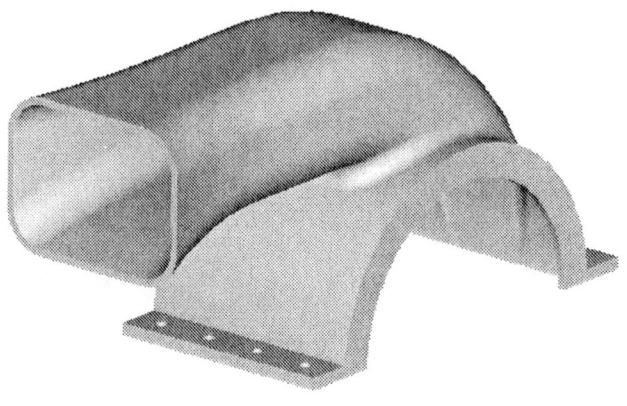

Figure 8-44 Solid model of the Upper Housing

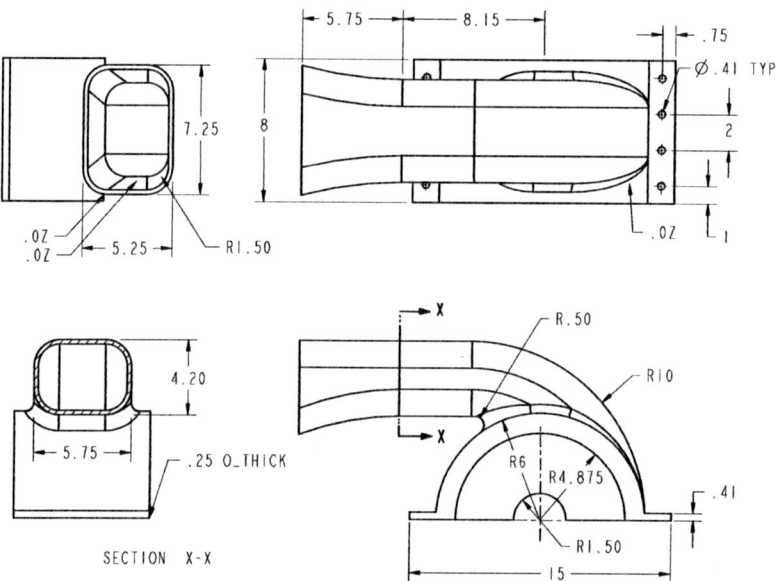

Figure 8-45 Left-side view of the top view, top view, front view, and the sectioned left-side view of the model

Advanced Modeling Tools-II

The following steps are required to complete this tutorial:

a. Examine the model and determine the number of features in it. The model is composed of ten features, refer to Figures 8-44 and 8-45.
b. Create the base feature, refer to Figures 8-46 and 8-47.
c. Create the swept blend feature in which the section is normal to the origin trajectory, refer to Figures 8-48 through 8-53.
d. Create the round feature of radius 1.5, refer to Figures 8-54 and 8-55.
e. Create the round of radius 0.5, refer to Figures 8-56 and 8-57.
f. Create the shell feature of thickness 0.25, refer to Figure 8-58.
g. Create an extruded cut, refer to Figures 8-59 and 8-60.
h. Create another extruded cut, refer to Figures 8-61 and 8-62.
i. Create the eighth feature as an extruded feature, refer to Figures 8-63 and 8-64.
j. Create a mirror copy of the eighth feature, refer to Figure 8-65.
k. Create the hole feature, refer to Figure 8-66, and pattern it, refer to Figure 8-67.

After understanding the procedure for creating the model, you are now ready to create it. If required set the **Working Directory** to *c08*.

Starting a New Object File

1. Open a new part file and name it *c08tut2*. The three default datum planes and the **Model Tree** are displayed in the graphics window. Exit the **Model Tree**. The **Model Tree** will not appear if it was previously closed.

Creating the Base Feature

To create the sketch for the base feature, you first need to select the sketching plane for the base feature. In this model, you need to draw the base feature on the **RIGHT** datum plane.

1. Choose the **Extrude Tool** button from the **Base Features** toolbar.

2. Select the **RIGHT** datum plane as the sketch plane.

3. Select the **TOP** datum plane from the graphics window and then select the **Top** option from the **Orientation** drop-down list.

4. Choose the **Sketch** button to enter the sketcher environment.

5. Once you enter the sketcher environment, create the sketch of the base feature and apply dimensions, as shown in Figure 8-46.

6. After the sketch is complete, choose the **Continue with the current section** button to exit the sketcher environment.

The **Extrude** dashboard reappears below the graphics window. The **Extrude from sketch plane by specified depth value** button is selected by default.

7. Choose the **Extrude on both sides of sketch plane by half the specified depth value in each** button from the depth flyout.

8. Enter a depth of **8** in the dimension box present in the **Extrude** dashboard. Choose the **Build feature** button from the **Extrude** dashboard.

The base feature is completed and the default trimetric view is shown in Figure 8-47.

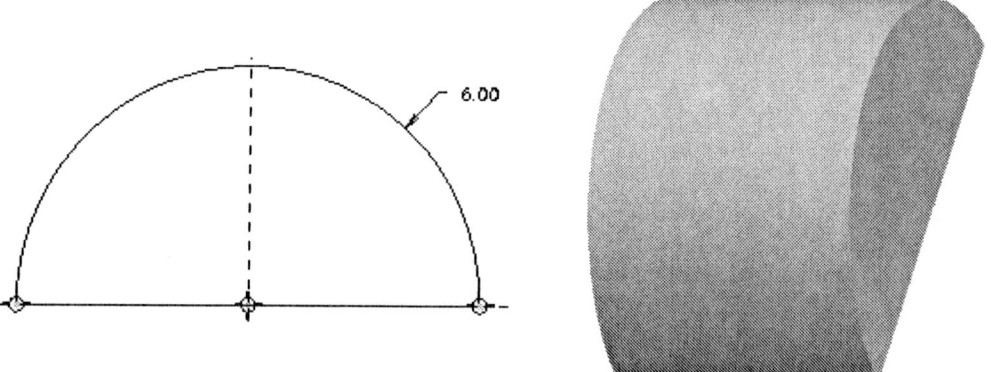

Figure 8-46 Sketch for the base feature   *Figure 8-47* Default trimetric view of the base feature

Creating the Swept Blend

The second feature is a swept blend. You will sketch the trajectory for the blend on the **RIGHT** datum plane. Note that the trajectory is tangent to the curve of the base feature and the start point of the trajectory is aligned with the **TOP** datum plane.

1. Choose **Insert > Swept Blend > Protrusion** from the menu bar. The **BLEND OPTS** menu is displayed. The **Sketch Sec** and the **NrmToOriginTraj** options are selected by default.

2. Choose **Done** from the **BLEND OPTS** menu. The **SWEEP TRAJ** menu is displayed.

3. Choose the **Sketch Traj** option from the **SWEEP TRAJ** menu. You are prompted to select a sketching plane.

4. Select the **RIGHT** datum plane from the graphics window. The **DIRECTION** menu is displayed. Choose **Okay**. The **SKET VIEW** submenu is displayed.

5. Select the **Top** option from the submenu and choose the **TOP** datum plane.

6. Once you enter the sketcher environment, draw the sketch of the trajectory and add constraints and dimensions, as shown in Figure 8-48.

As evident from Figure 8-48, the start point of the trajectory is aligned with the **TOP** datum plane and the trajectory is tangent to the curve of the base feature.

Advanced Modeling Tools-II

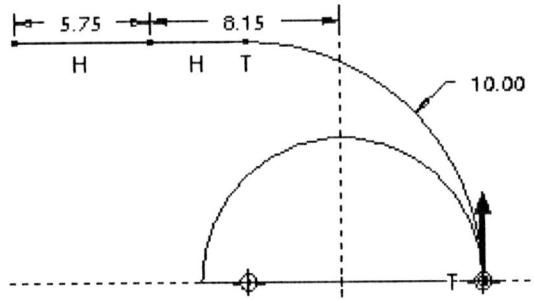

Figure 8-48 Sketch of the origin trajectory

7. After the sketch is complete, choose the **Continue with the current section** button to exit the sketcher environment.

   The **CONFIRM SEL** menu is displayed. Note that on the origin trajectory one red, one white, and two green points are displayed. The green points indicate that these are fixed points for defining the sections. The location for red point can be changed by using the options in the **CONFIRM SEL** menu.

 Note
The two green points are fixed at the start point and at the endpoint of the trajectory, indicating that it is necessary to draw sections at these points in order to complete the swept blend feature. This is because swept blend is created by blending the sections along the trajectory. There can be any number of sections that can be possible between the start point and the endpoint of the trajectory.

8. Choose **Next** from the **CONFIRM SEL** menu to shift the position of the red point. You need to shift the position of the red point because the section of the swept blend feature (SECTION X-X) that is given in Figure 8-48 is at the point where the red point is shifted now. Hence, now you will need to define three sections at three points.

9. Choose the **Accept** option from the **CONFIRM SEL** menu. The **Message Input Window** is displayed and you are prompted to enter the z-axis rotation angle for section 1. Section 1 is at the start point of the origin trajectory.

10. The value **0** is entered by default in the **Message Input Window**; press ENTER.

    The system takes you to the sketcher environment. Now, you need to sketch the first section of the blend at the start point of the trajectory. You will notice a coordinate system is placed at the start point of the trajectory. You need to align the sketch of the section to this coordinate system.

11. Use the middle mouse button to spin the view of the sketch, as shown in Figure 8-49 to have a better understanding of the sketch in 3D space.

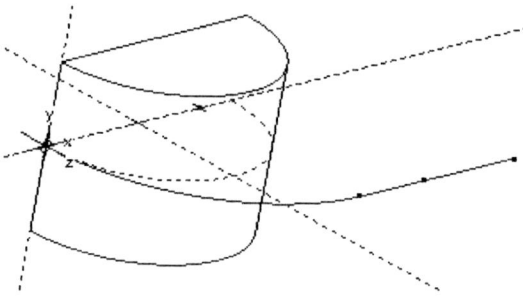

Figure 8-49 Rotated view of the sketch showing the origin trajectory, base feature, and the default coordinate system

12. Choose the **Orient the sketching plane parallel to the screen** button from the **Top Toolchest**. The sketching plane reorients parallel to the graphics window.

13. Draw the sketch of the first section and add dimensions, as shown in Figure 8-50.

14. After the sketch is complete, choose the **Continue with the current section** button to exit the sketcher environment. The **Message Input Window** is displayed and you are prompted to enter the z-axis rotation angle for section 2.

15. The value **0** is entered by default in the **Message Input Window**; press ENTER.

16. Draw the sketch of the second section and add dimensions, as shown in Figure 8-51. This section is at the point between the first and the third sections and in Figure 8-45, this section is named as X-X.

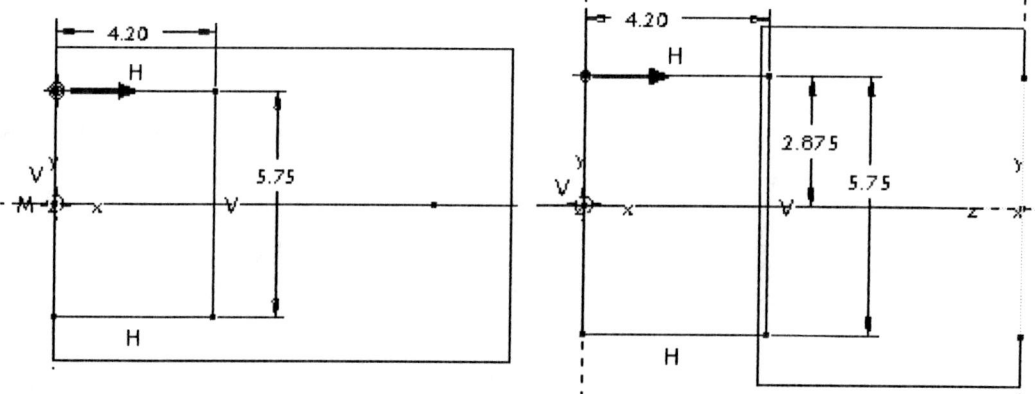

Figure 8-50 Sketch for first section with dimensions

Figure 8-51 Sketch for second section with dimensions

Advanced Modeling Tools-II

17. After the sketch is complete, choose the **Continue with the current section** button to exit the sketcher environment. The **Message Input Window** is displayed and you are prompted to enter the z-axis rotation angle for section 3.

18. The value **0** is entered by default in the **Message Input Window**; press ENTER.

19. Draw the sketch of the third section and add dimensions, as shown in Figure 8-52. This is the last section on the origin trajectory. Remember to align the section to the coordinate system.

20. After the sketch is complete, choose the **Continue with the current section** button to exit the sketcher environment.

21. Choose **OK** from the **PROTRUSION** dialog box. The default trimetric view of the swept blend feature with the base feature is shown in Figure 8-53.

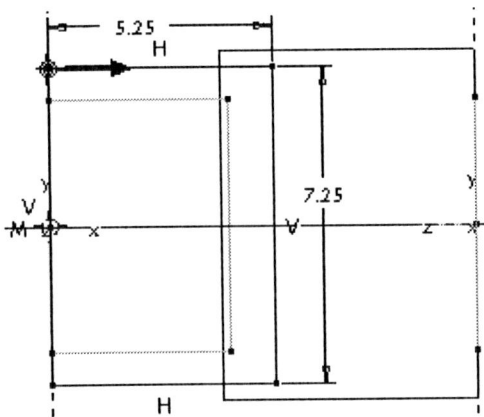

Figure 8-52 Sketch of the third section with dimensions

Figure 8-53 Default view of the swept blend

Creating the Round Features

The two round features are created before shelling because the rounds are also shelled along with the other faces of the model. The third feature in the model is a round of radius 1.5 and the fourth feature is also a round of radius 0.5.

1. Choose the **Round Tool** button from the **Engineering Features** toolbar. The **Round** dashboard is displayed.

2. Select all four longer edges of the swept feature shown in Figure 8-54 to round them. Remember that the first edge is selcted by using the left mouse button and the other remaining edges are selected by using CTRL+left mouse button.

   The preview of the round is displayed in the graphics window along with the default radius value.

3. Double-click on the radius value, enter **1.5** in the edit box that appears, and then press ENTER.

4. Choose **Build feature** from the **Round** dashboard. The default view of the model after creating the round is shown in Figure 8-55.

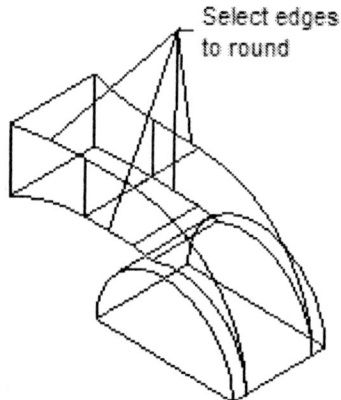 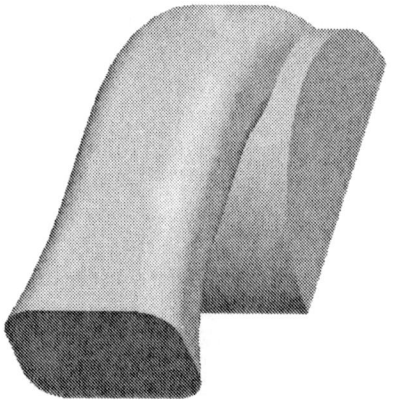

Figure 8-54 Edges to round

Figure 8-55 Default view of the model after creating a round feature of radius 1.5.

Note
The round of radius 0.5 cannot be created by adding a round set. This is because this round is created on the geometry of the previous round. Therefore, until the first round of radius 1.5 is created you cannot create the second round.

The third feature of the model is completed and now you will create the fourth feature, that is, a round of radius 0.5. For this purpose you need to invoke the **Round** dashboard.

5. Choose the **Round Tool** button from the **Engineering Features** toolbar. The **Round** dashboard is displayed.

6. Select the edge where the base feature and the swept feature joins, as shown in Figure 8-56.

   The preview of the round is displayed in the graphics window along with the default radius value.

7. Double-click the radius value, enter **0.5** in the edit box that appears, and then press ENTER.

8. Choose **Build feature** from the **Round** dashboard. The default view of the model after creating the round is shown in Figure 8-57.

Creating the Shell

You need to create a shell of thickness 0.25 on the swept blend feature.

Advanced Modeling Tools-II

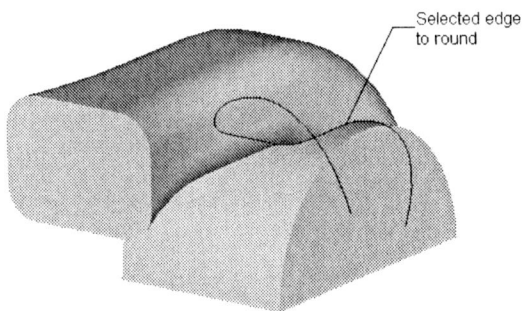

Figure 8-56 *Edge selected to round*

1. Choose the **Shell Tool** button from the **Engineering Features** toolbar. The Shell dashboard is displayed and you are prompted to select the surface to remove.

2. Using the left mouse button, select the front face of the swept blend feature and then use CTRL+left mouse button to select the bottom face of the base feature.

3. Enter a value of **0.25** in the dimension box available on the **Shell** dashboard and press ENTER.

4. Choose the **Build feature** button from the **Shell** dashboard. The shell feature is completed and is shown in Figure 8-58.

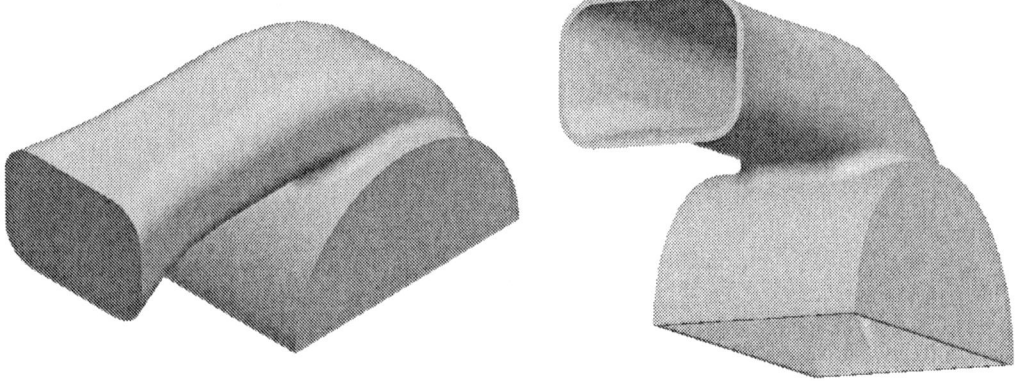

Figure 8-57 *Model after creating a round feature of radius 0.5*

Figure 8-58 *The model after shelling*

Creating the Extruded Cut

The extruded cut is the sixth feature of this model. You need to create the cut on the front face of the base feature.

1. Choose the **Extrude Tool** button from the **Base Features** toolbar. The **Extrude** dashboard is displayed.

2. Choose the **Remove Material** button from the dashboard.

3. Select the front face of the base feature as the sketch plane.

4. Select the **TOP** datum plane from the graphics window and then select the **Top** option from the **Orientation** drop-down list.

5. Choose the **Sketch** button to enter the sketcher environment.

6. Draw the sketch of the cut feature and apply dimension, as shown in Figure 8-59. You can use the **Create concentric arc** button from the **Sketcher Tools** toolbar to draw the sketch.

7. After the sketch is complete, choose the **Continue with the current section** button to exit the sketcher environment.

   The **Extrude** dashboard reappears below the graphics window. The **Extrude from sketch plane by specified depth value** button is selected by default.

8. Choose the **Extrude upto next surface** button from the depth flyout.

9. Choose the **Build feature** button from the **Extrude** dashboard. The cut feature is created and is shown in Figure 8-60.

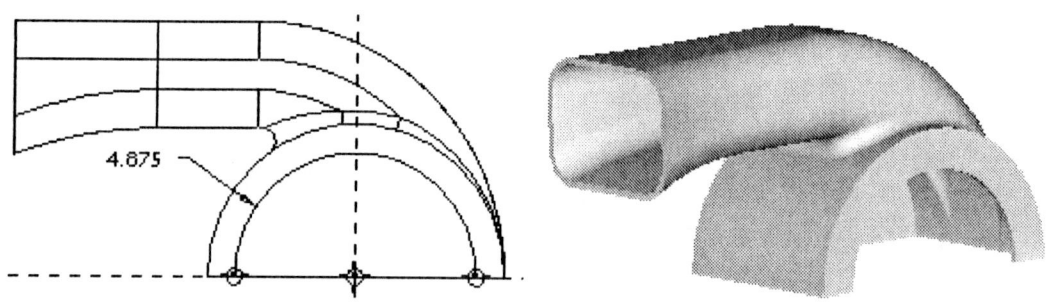

Figure 8-59 Sketch with dimension for cut feature

Figure 8-60 Model with cut feature

Similarly, create the next extruded cut feature at the back face of the base feature. Select the sketch plane, draw the sketch, apply dimension, and extrude the sketch using the **Extrude upto next surface** button. The sketch for the cut feature is shown in Figure 8-62.

The seventh feature of the model, that is, cut feature is shown in Figure 8-62.

Advanced Modeling Tools-II

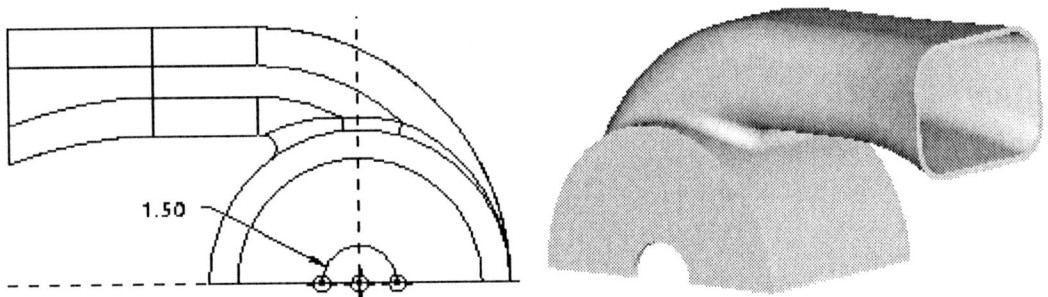

Figure 8-61 Sketch with dimension for cut

Figure 8-62 Model with the cut feature

Creating the Extruded Feature

The eighth feature of the model is an extruded feature and will be created on the front face of the base feature.

1. Select the front face of the base feature as the sketching plane for the extrude feature. Once you enter the sketcher environment, draw the sketch and apply constraints and dimensions, as shown in Figure 8-63.

2. After the sketch is complete, choose the **Continue with the current section** button to exit the sketcher environment.

3. From the depth flyout, select the **Extrude to selected point, curve, plane or surface** button and select the back face of the base feature.

4. Choose the **Build feature** button from the **Extrude** dashboard. The extruded feature is completed and is shown in Figure 8-64.

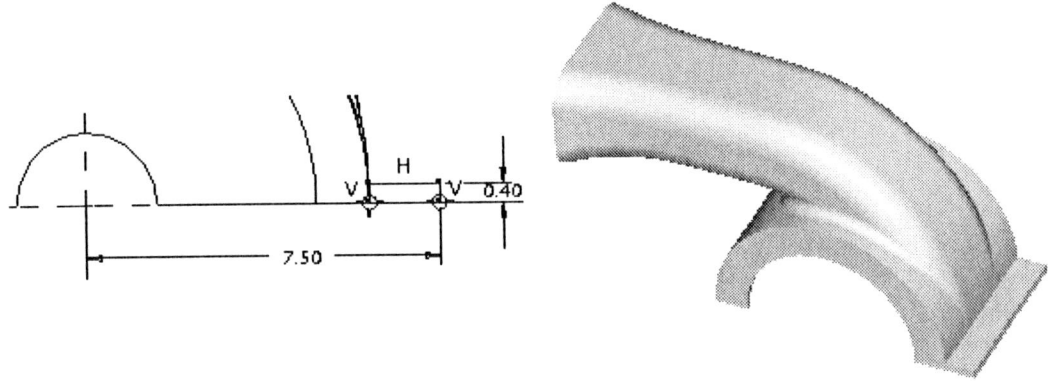

Figure 8-63 Sketch for the extruded feature

Figure 8-64 Model after creating the extrude feature

Creating a Copy of the Eighth Feature

You need to create a mirror copy of the extruded feature, as shown in Figure 8-65 and this will be the ninth feature of the model. The extruded feature will be mirrored about the **FRONT** datum plane.

1. Choose **Edit > Feature Operations** from the menu bar. The **Menu Manager** is displayed.

2. Choose the **Copy** option from the **Menu Manager**. The **COPY FEATURE** submenu is displayed.

3. Choose **Mirror > Select > Dependent > Done** from the **COPY FEATURE** submenu. The **SELECT FEAT** submenu is displayed and you are prompted to select the features to be mirrored.

4. Select the previous extruded feature from the graphics window and choose **Done** from the **SELECT FEAT** submenu. You are prompted to select a plane or create a datum plane to mirror about.

5. Select the **FRONT** datum plane as the mirror plane. The protrusion feature is mirrored about the **FRONT** datum plane, as shown in Figure 8-65.

6. Choose **Done** from the **Menu Manager**.

Creating the Tenth Feature

The tenth feature is a through hole and will be created using the **Hole** dashboard.

1. Choose the **Hole Tool** button from the **Engineering Features** toolbar. The **Hole** dashboard is displayed. The **Create straight hole** button is selected by default.

   Create the hole, as shown in Figure 8-66 by specifying the placement parameters. The placement parameters can be referred from Figure 8-45.

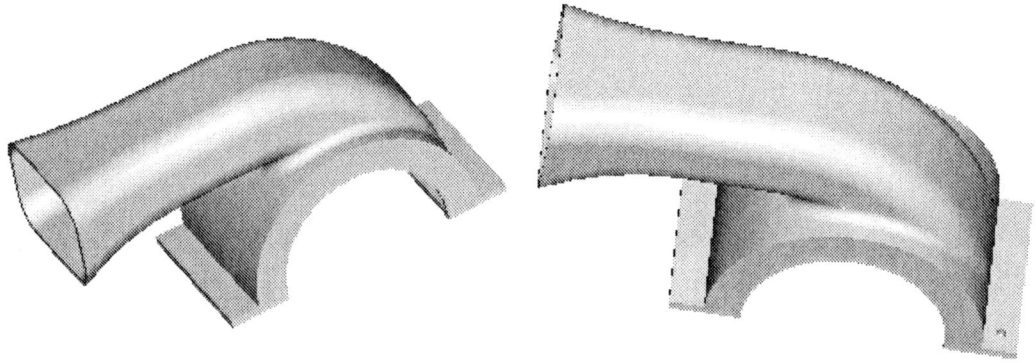

Figure 8-65 Model after mirroring the extruded feature

Figure 8-66 Model after creating the hole

Creating a Pattern of the Hole Feature

As evident from Figure 8-45, you need to create eight instances of the hole. The first instance is created by using the **Hole** dashboard and you can use the **Pattern** option to create the remaining seven instances. You will create a rectangular pattern of the hole feature. You can also create all holes using the **Hole** dashboard and specifying the placement parameters for each of them. But to save time, you will create a pattern of the hole.

1. Select the hole feature from the **Model Tree** or from the model. Hold down the right mouse button to invoke the shortcut menu.

2. Choose the **Pattern** option from the shortcut menu. The **Pattern** dashboard is displayed and the dimensions are displayed on the hole feature.

3. Select the **General** option from the **Options** option slide-up panel.

4. Double-click the dimension value **1** from the graphics window. Enter a value of **2** in the edit box and press ENTER.

5. Enter **4** in the **1** edit box present on the dashboard and press ENTER. Now, you need to enter dimension increment in the second direction.

6. Click the collector that displays **No Items** on the right of the **2** edit box. Double-click the dimension value **0.75** from the graphics window. Enter a value of **13.5** in the edit box and press ENTER.

7. Enter **2** in the **2** dimension box present on the dashboard and press ENTER.

8. Choose the **Build feature** button from the dashboard. You can use the middle mouse button to spin the model and to display the model, as shown in Figure 8-67.

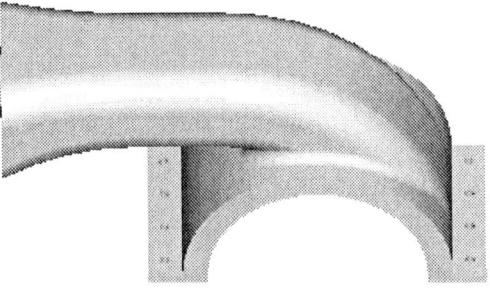

Figure 8-67 Completed model

Saving the Model

1. Choose the **Save the active object** button from the **File** toolbar and save the model.

Tutorial 3

In this tutorial you will create the model shown in Figure 8-68. This model is a wheel of a car. Figure 8-69 shows the top view, sectioned front view, sectioned right view, detail view, and two blend sections with dimensions.   **(Expected time: 45 min)**

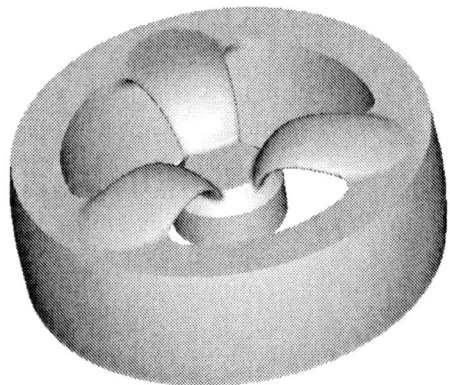

Figure 8-68 Solid model of the wheel

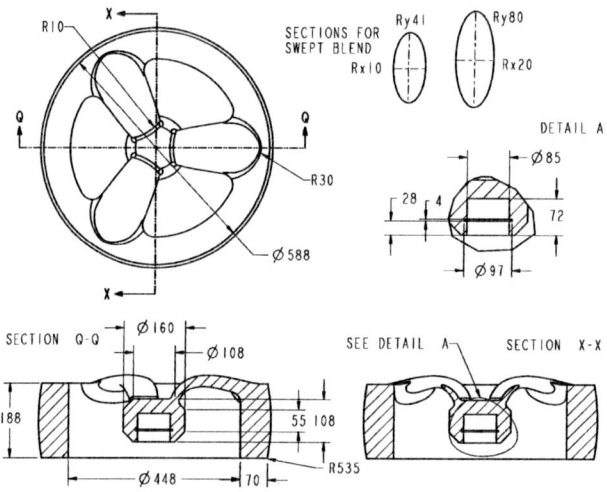

Figure 8-69 *Top view, sectioned front view, sectioned right view, detail view, and two blend sections with dimensions*

The following steps are required to complete this tutorial:

a. Examine the model and determine the number of features in it. The model is composed of four features, refer to Figure 8-69.
b. Create the base feature as a revolved feature, refer to Figures 8-70 and 8-71.

Advanced Modeling Tools-II

c. Create the swept blend feature in which the section is normal to the origin trajectory, refer to Figures 8-72 through 8-75.
d. Pattern the swept blend feature, refer to Figure 8-76.
e. Create the round of radius 30 and radius 10, refer to Figures 8-87 and 8-89 respectively.
f. Create the revolved cut, refer to Figures 8-80 and 8-81.

After understanding the procedure for creating the model, you are now ready to create it. The working directory was already selected in Tutorial 1 and therefore you do not need to select it again.

Starting a New Object File

1. Open a new part file and name it *c08tut3*. The three default datum planes and the **Model Tree** are displayed in the graphics window. Close the **Model Tree**. The **Model Tree** will not appear if it has been previously closed.

Creating the Base Feature

To create the sketch for the base feature, you first need to select the sketch plane. In this model, you need to draw the base feature on the **FRONT** datum plane.

1. Choose the **Revolve Tool** button from the **Base Features** toolbar. The **Revolve** dashboard is displayed.

2. Select the **FRONT** datum plane as the sketch plane.

3. Select the **TOP** datum plane from the graphics window and then select the **Top** option from the **Orientation** drop-down list.

4. Once you enter the sketcher environment, create the sketch of the base feature and apply constraints and dimensions, as shown in Figure 8-70. The sections must be closed.

5. After the sketch is complete, choose the **Continue with the current section** button to exit the sketcher environment. The **Revolve** dashboard is displayed.

6. Choose the **Build feature** button from the **Revolve** dashboard to exit it.

The base feature is completed and the default trimetric view is shown in Figure 8-71.

Creating the Swept Blend

The second feature of the model is a swept blend. You need to create a datum plane that is at an angle of 90-degree from the **RIGHT** datum plane. This is because later in the tutorial, you will create a rotational pattern of the swept blend feature. The 90-degree angle value will be used for the dimensional increment for creating the rotational pattern.

1. Choose **Insert > Swept Blend > Protrusion** from the menu bar. The **BLEND OPTS** submenu is displayed. The **Sketch Sec** and the **NrmToOriginTraj** options are selected by default.

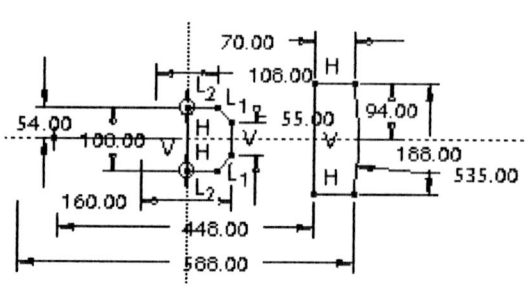

Figure 8-70 Sketch with dimensions for the base feature

Figure 8-71 Default trimetric view of the base feature

2. Choose **Done** from the **BLEND OPTS** submenu. The **SWEEP TRAJ** menu is displayed.

3. Choose the **Sketch Traj** option from the **SWEEP TRAJ** menu. You are prompted to select a sketching plane or create a datum plane.

4. Choose the **Make Datum** option from the **SETUP PLANE** submenu. **The DATUM PLANE** submenu is displayed.

5. Choose the **Through** option from the **DATUM PLANE** submenu and select the axis of revolution of the base feature.

6. Choose the **Angle** option from the **DATUM PLANE** submenu and select the **RIGHT** datum plane.

7. Choose **Done** from the **DATUM PLANE** submenu. The **OFFSET** submenu is displayed.

8. Choose the **Enter Value** option from the **OFFSET** submenu. The **Message Input Window** is displayed.

9. Enter **90** in the **Message Input Window** and press ENTER. The **DIRECTION** menu is displayed. Choose **Okay**. The **SKET VIEW** submenu is displayed.

10. Select the **Top** option from the submenu and choose the **TOP** datum plane.

11. Once you enter the sketcher environment, first you need to specify two references to dimension. Select the **TOP** and **RIGHT** datum planes.

12. Draw the sketch of the trajectory and add constraints and dimensions, as shown in Figure 8-72. In Figure 8-72, the hidden lines are also shown because in this case the display of the sketch is more clear while drawing the sketch with display of hidden lines.

Advanced Modeling Tools-II

The curve is drawn using the spline. The spline has only four control points: one start point, one endpoint, and two intermediate points. The dimensions shown in Figure 8-72 are enough to define the sketch.

Note

The shape of the spline will differ from user to user, and so the shape of the protrusion feature may vary.

13. After the sketch is complete, choose the **Continue with the current section** button to exit the sketcher environment. The **CONFIRM SEL** menu is displayed. Note that two green points are displayed on the origin trajectory. The green points indicate that they are fixed points for defining the sections.

14. Choose the **Next** option twice from the **CONFIRM SEL** menu. The **Message Input Window** is displayed and you are prompted to enter the Z-axis rotation angle for section 1. Section 1 is at the start point of the origin trajectory.

15. The value **0** is entered by default in the **Message Input Window**; press ENTER.

The system takes you to the sketcher environment. Now, you need to sketch the first section of the blend at the start point of the trajectory. You will notice a coordinate system is placed at the start point of the trajectory. You need to align the sketch of the section to this coordinate system.

16. Draw the sketch of the first section and add dimensions, as shown in Figure 8-73.

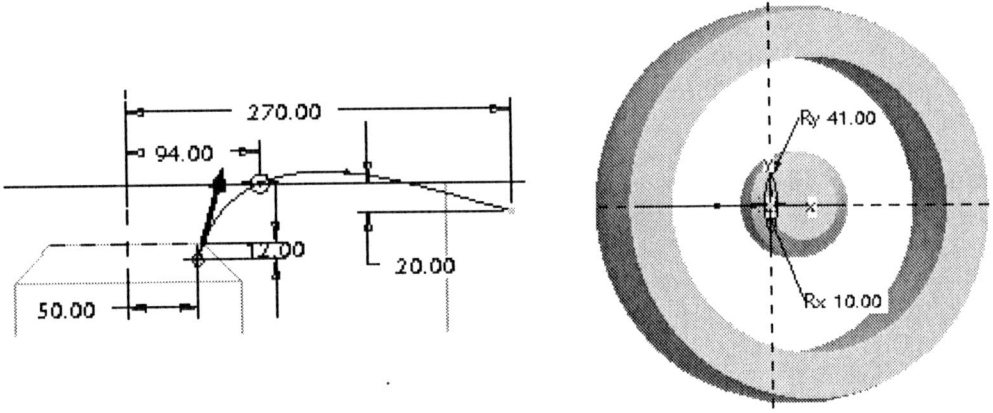

Figure 8-72 Sketch with dimensions for the origin trajectory

Figure 8-73 Sketch for section 1 with dimensions

17. After the sketch is complete, choose the **Continue with the current section** button to exit the sketcher environment. The **Message Input Window** is displayed and you are prompted to enter the Z-axis rotation angle for section 2.

18. The value **0** is displayed by default in the **Message Input Window**; press ENTER.

19. Draw the sketch of the second section and add dimensions, as shown in Figure 8-74. The second section is also elliptical in shape, and therefore an ellipse is drawn in the sketch. The dimensions for this section are shown in Figure 8-69.

20. After the sketch is complete, choose the **Continue with the current section** button to exit the sketcher environment.

21. Choose **OK** from the **PROTRUSION** dialog box. The default trimetric view of the swept blend feature with the base feature is shown in Figure 8-75.

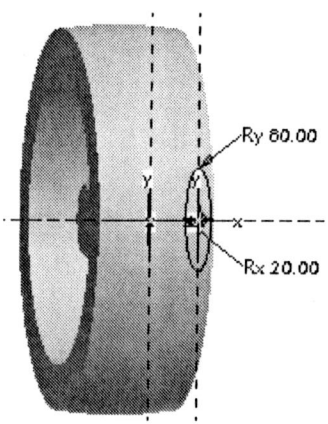

Figure 8-74 Sketch for section 2 with dimensions

Figure 8-75 Model after creating the swept blend

Creating the Rotational Pattern of the Swept Blend

Creating the swept blend features individually on the base feature will consume a lot of time. Therefore, you need to create a rotational pattern of the hole feature. You will use the angular dimension that was defined to create datum on the fly to pattern the feature.

1. Select the swept blend feature from the **Model Tree** or from the model. Hold down the right mouse button to invoke the shortcut menu.

2. Choose the **Pattern** option from the shortcut menu. The **Pattern** dashboard is displayed and the dimensions are displayed on the selected feature.

3. Select the **General** option from the **Options** option slide-up panel.

4. Double-click on the dimension value **90**, which is actually the angular dimension of the datum plane created on the fly. Enter a value of **120** in the edit box.

   You have entered a value of 120 because there are three instances of the swept blend feature on a circular feature.

 Note

*You may need to zoom in the display of the model to select the angular dimension value **90**.*

Advanced Modeling Tools-II

5. Enter **3** in the **1** edit box present on the dashboard and press ENTER.

6. Choose the **Build feature** button from the dashboard. The rotational pattern is created, as shown in Figure 8-76.

Creating the Round Feature

The third feature is a round of radius 30.

1. Choose the **Round Tool** button from the **Engineering Features** toolbar. The dashboard is displayed.

2. Select the edges shown in Figure 8-77 to round. You may need to spin the model to select the inner edges of the swept blend feature. Remember that the first edge is selected by using the left mouse button and the other remaining edges are selected by using CTRL+left mouse button.

   The preview of the round is displayed in the graphics window along with the default radius value.

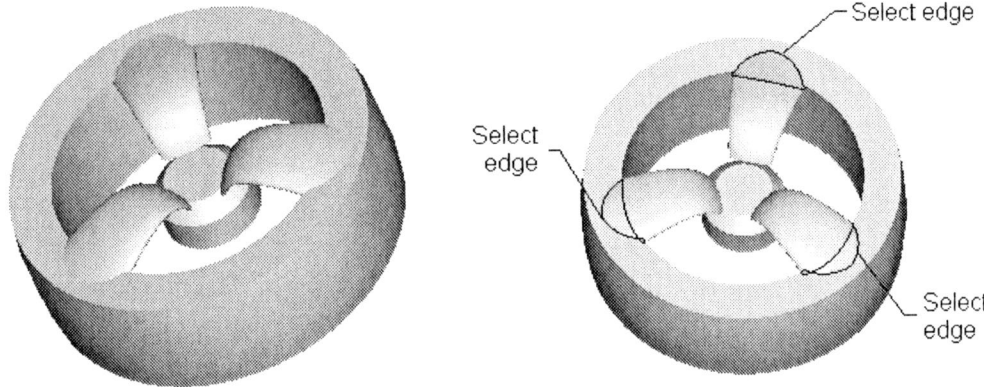

Figure 8-76 Model after creating pattern   *Figure 8-77* Edges to round

3. Double-click the radius value, enter **30** in the edit box that appears, and then press ENTER.

4. Choose **Build feature** from the **Round** dashboard.

   The round feature is completed. The default trimetric view of the round feature is shown in Figure 8-78. Add another set and using the same options create a round feature of radius 10 highlighted in Figure 8-79.

Creating the Revolve Cut Feature

You need to first select a sketching plane to draw the sketch of the cut feature and then specify the angle of revolution.

Figure 8-78 Model after creating a round of radius 30

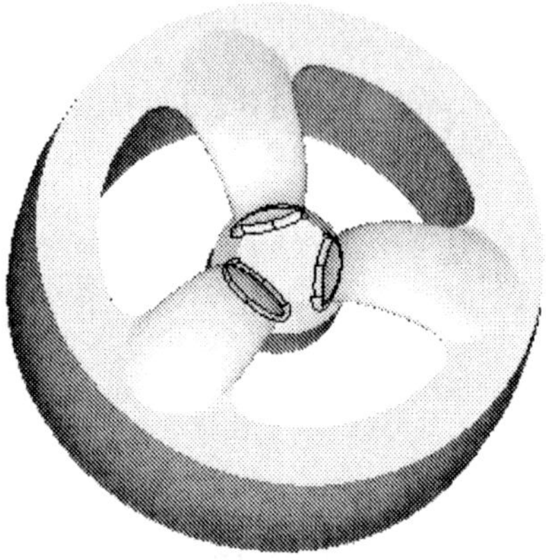

Figure 8-79 Model highlighting the round feature of radius 10

1. Choose the **Revolve Tool** button from the **Base Features** toolbar. The **Revolve** dashboard is displayed.

2. Select the **FRONT** datum plane as the sketch plane.

3. Select the **TOP** datum plane from the graphics window and then select the **Bottom** option from the **Orientation** drop-down list.

4. Once you enter the sketcher environment, draw a center line aligned to the **RIGHT** datum plane. Draw the sketch of the revolved cut feature and add dimensions and constraints, as shown in Figure 8-80.

5. After the sketch is complete, choose the **Continue with the current section** button to exit the sketcher environment. The **Revolve** dashboard is displayed.

6. Choose the **Build feature** button from the **Revolve** dashboard to exit it. The revolve cut feature is shown in Figure 8-81. This completes the model creation.

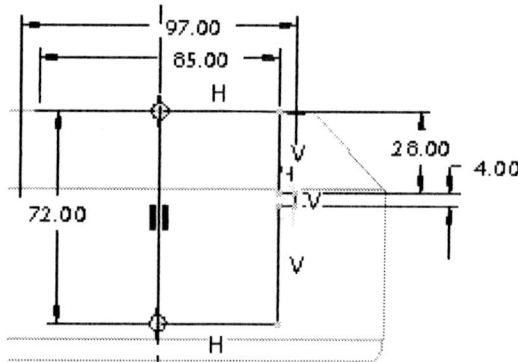

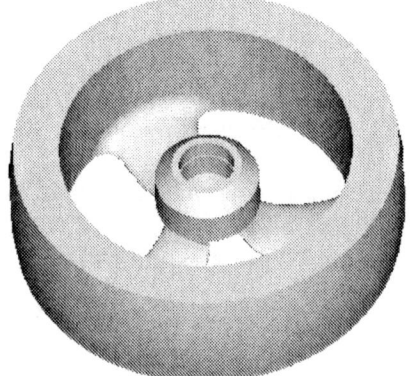

Figure 8-80 Sketch for the revolve cut feature

Figure 8-81 The model with revolve cut feature

Advanced Modeling Tools-II

Saving the Model
You have to save the model because you may need it later.

1. Choose the **Save the active object** button from the **File** toolbar and save the model.

Tutorial 4

In this tutorial you will create the spring shown in Figure 8-82. Figure 8-83 shows the front view with dimensions of the spring.

(Expected time: 15 min)

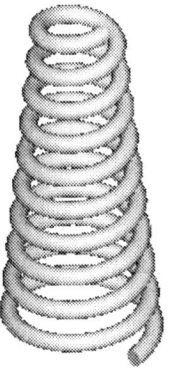

Figure 8-82 Isometric view of the spring

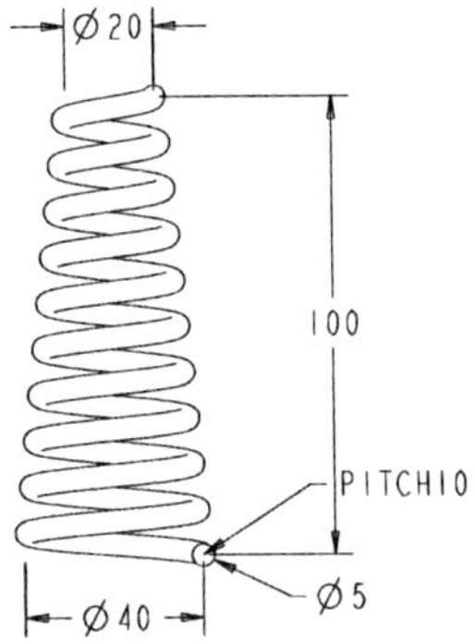

Figure 8-83 Front view of the spring with dimensions

The following steps are required to complete this tutorial:

a. Examine the spring and determine the specifications of the spring, refer to Figure 8-83.
b. Draw the trajectory using the sketcher tools, and apply dimensions, refer to Figures 8-84 and 8-85.
c. Specify the pitch of the spring.
d. Draw the section of the spring, refer to Figure 8-86.

After understanding the procedure for creating the model, you are now ready to draw it. Make sure that the required working directory is selected.

Starting a New Object File
1. Open a new part file and name it *c08tut4*. The three default datum planes and the **Model Tree** are displayed in the graphics window. Close the **Model Tree**. The **Model Tree** will not appear if it has been previously closed.

Creating the Helical Sweep
The spring that you have to create is a right-handed spring of constant pitch.

1. Choose **Insert > Helical Sweep > Protrusion** from the menu bar. The **ATTRIBUTES** menu is displayed. The **Constant > Thru Axis > Right Handed** options are selected by default.

2. Choose **Done** from the **ATTRIBUTES** menu. You are prompted to select a sketching plane.

3. Select the **FRONT** datum plane from the graphics window. The **DIRECTION** submenu is displayed. Choose **Okay**. The **SKET VIEW** submenu is displayed.

4. Select the **Top** option from the **SKET VIEW** submenu and choose the **TOP** datum plane.

5. Once you enter the sketcher environment, draw the sketch of the trajectory and dimension it, as shown in Figure 8-89. As evident from Figure 8-89, the endpoint of the trajectory is aligned to the **TOP** datum plane.

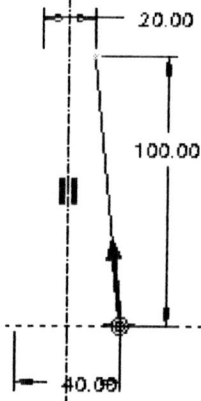

Figure 8-84 *Sketch of the trajectory with dimensions*

You need to draw a center line in the sketch about which the spring will be rotated. This is the axis of the spring and is the first trajectory that is drawn in the sketch.

6. After you complete the sketch of the trajectory, choose the **Continue with the current section** button from the **Right Toolchest**. The **Message Input Window** is displayed and you are prompted to specify the pitch of the spring.

7. Type a value of **10** in the **Message Input Window** and press ENTER. Now, you enter the sketcher environment to draw the section of the spring. You will notice that a yellow cross of infinite length appears on the screen. This cross consists of two perpendicular lines of infinite length. The intersection point of these lines is the start point of the trajectory.

8. Draw the section of the spring such that the center of the circle coincides with the intersection of the two perpendicular lines and dimension it, as shown in Figure 8-90.

9. After completing the sketch, choose the **Continue with the current section** button to exit the sketcher environment.

10. Choose the **OK** button from the **PROTRUSION** dialog box. The spring is created and is shown in Figure 8-91.

Advanced Modeling Tools-II

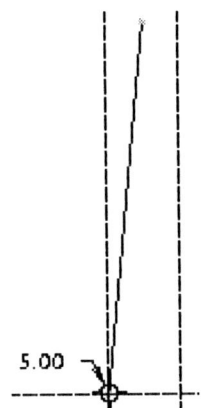
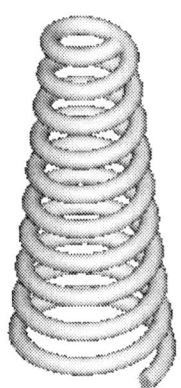

Figure 8-85 Sketch of the section with dimensions   *Figure 8-86* Isometric view of the spring

Saving the Model

1. Choose the **Save the active object** button from the **File** toolbar and save the model.

Self Evaluation Test

Answer the following questions and then compare your answers with those given at the end of this chapter.

1. You can create Constant pitch as well as Variable pitch helical sweep features. (T/F)

2. The **Helical Sweep** option is used to create cut as well as protrusion features. (T/F)

3. You can create a helical feature having 40 units pitch and 50 units as the diameter of the circular cross-section. (T/F)

4. You can add as well as delete points while specifying the pitch values during the creation of variable pitch helical sweep. (T/F)

5. The **Var Sec Sweep** option is available for both protrusion and cut options. (T/F)

6. While using the **Var Sec Sweep** option, the contour of the sweep feature depends on the shape of the _____.

7. While using the **Var Sec Sweep** option, when you draw the section for the sweep feature, the section should be _____ to the endpoints of the trajectories.

8. When you choose the **Done** option from the **BLEND OPTS** menu, the _____ menu is displayed.

9. When you use the **Normal to trajectory** option to create the variable sweep, the section is _____ to the origin trajectory and is swept along the X trajectory.

10. The **Swept Blend** option is used to create a model that is a combination of _____ and _____.

Review Questions

Answer the following questions:

1. Which of the following menus is displayed when you choose the **Helical Sweep** option from the **Insert** menu in the menu bar?

   (a) **ADV FEAT OPT**
   (c) **SWEEP OPTS**
   (b) **ATTRIBUTES**
   (d) None

2. Which of the following options is used to create a helical feature with constant pitch throughout the sweep?

   (a) **Right Handed**
   (c) **Thru Axis**
   (b) **Constant**
   (d) None

3. Which of the following options in the **BLEND OPTS** is used to create a sweep in which the swept section is normal to the trajectory defined as the origin trajectory?

   (a) **Pivot Dir**
   (c) **Norm To Traj**
   (b) **NrmToOriginTraj**
   (d) None

4. Which of the following options in the **Relations** dialog box is used to add dimensions that appear on the screen?

   (a) **Insert > From Screen**
   (c) **Insert > From List**
   (b) **Insert > Function**
   (d) None

5. Which of the following options is used to create a swept blend in which you need to select a plane, edge, curve, or axis to which the section to be swept is perpendicular?

   (a) **Pivot Dir**
   (c) **NrmToOriginTraj**
   (b) **Nrm To Traj**
   (d) None

6. The **Var Sec Swp** option can be used to create a sweep in which the section of the sweep is constant throughout the sweep. (T/F)

7. The trajectory called X trajectory is used to guide the section of the sweep feature along it. (T/F)

8. When you choose the **Edit Rel** option, the relations are not checked for errors. (T/F)

9. The **Switch Dim** option of the **RELATIONS** submenu is used to toggle the dimension value mode and the symbols. (T/F)

Advanced Modeling Tools-II

10. The **Helical Sweep** option is also used to create a cut. (T/F)

Exercises

Exercise 1

Create the model shown in Figure 8-78. The front view is shown in Figure 8-88.

(Expected time: 15 min)

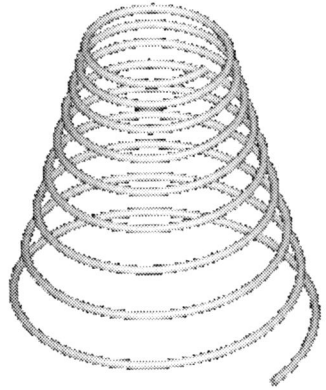

Figure 8-87 Isometric view of the spring

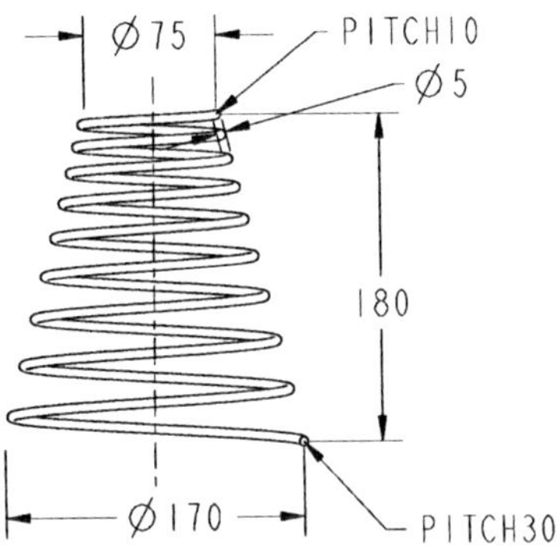

Figure 8-88 Front view of the spring

Exercise 2

Create the model shown in Figure 8-89. The front view and the top view of the model is shown in Figure 8-90.

(Expected time: 30 min)

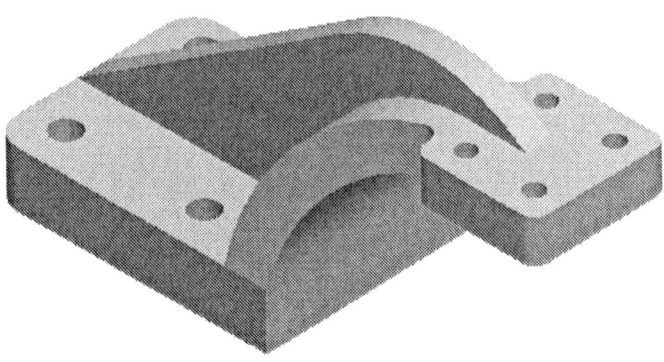

Figure 8-89 Solid model for Exercise 2

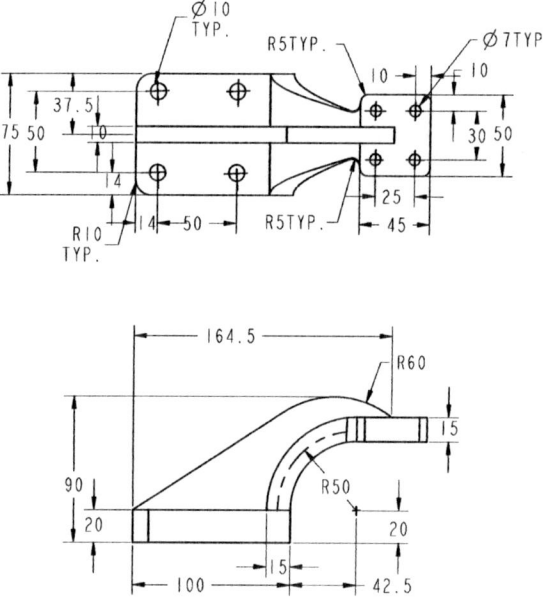

Figure 8-90 Front view and the top view of the model

Hints to create the swept blend feature:
1. Choose **Insert > Swept Blend > Protrusion** from the menu bar. The **BLEND OPTS** menu is displayed.
2. Accept the options that are selected by default and choose **Done**. The **SWEEP TRAJ** menu is displayed.
3. Choose the **Sketch Traj** option. You are prompted to select a sketch plane.
4. Select the **FRONT** datum plane as the sketch plane and select the top face of the base feature to be at the top while sketching.
5. Draw the arc of radius 50 and locate its center from the base feature.
6. Exit the sketcher environment by choosing the **Continue with the current section** button. You are prompted to enter the z-axis rotation angle for section 1.
7. Press ENTER.
8. Draw the rectangle of dimensions 15x50. Exit the sketcher environment.
9. You are prompted to enter the z-axis rotation angle for section 2.
10. Draw a rectangle of dimensions 15x75. Exit the sketcher environment and choose **OK** from the **PROTRUSION** dialog box.

Answers to the Self-Evaluation Test
1 - T, **2** - T, **3** - F, **4** - T, **5** - T, **6** - X trajectory, **7**- aligned, **8** - **SWEEP TRAJ** **9** - normal, **10** - sweep, blend.

Chapter 9

Assembly Modeling

Learning Objectives

After completing this chapter you will be able to:
- *Understand the top-down assembly approach.*
- *Understand the bottom-up assembly approach.*
- *Assemble the components of the assembly using the assembly constraints.*
- *Understand packaging of components.*
- *Create the simplified representation for viewing various components of assembly.*
- *Use View Manager.*
- *Edit the assembly constraints after assembling.*
- *Modify the components of the assembly within the assembly.*
- *Create the exploded state of the assembly.*
- *Add the offset lines to the exploded components.*
- *Understand the Bill Of Material in the assemblies.*

ASSEMBLY MODELING

An assembly is defined as a design consisting of two or more components bonded together at their respective working positions using the assembly constraints. These assembly designs are created in the **Assembly** mode of Pro/ENGINEER. To proceed to the **Assembly** mode, choose the **Create a new object** button from the **Top Toolchest**. The **New** dialog box is displayed; select the **Assembly** radio button from the **Type** area and then select the **Design** radio button from the **Sub-type** area, as shown in Figure 9-1. Specify the name of the assembly in the **Name** edit box and choose **OK**.

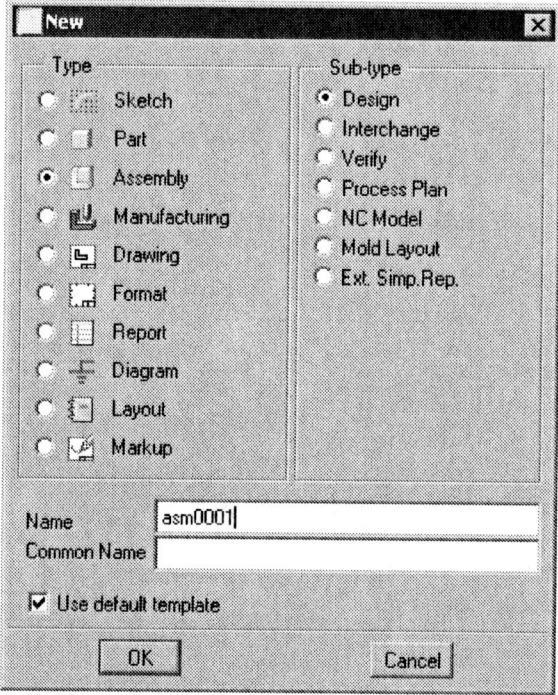

Figure 9-1 Selecting the Assembly mode from the New dialog box

When you choose the **OK** button from the **New** dialog box, the **Assembly** mode is activated. The initial screen appearance of the **Assembly** mode in Pro/ENGINEER Wildfire 2.0 is shown in Figure 9-2.

IMPORTANT TERMS RELATED TO ASSEMBLY MODE

Before proceeding with this chapter, it is very important for you to understand the following terms.

Top-Down Approach

This is the method of assembling the components in which the components of the assembly are created in the same assembly file. In this type of assembly modeling approach,

Assembly Modeling 9-3

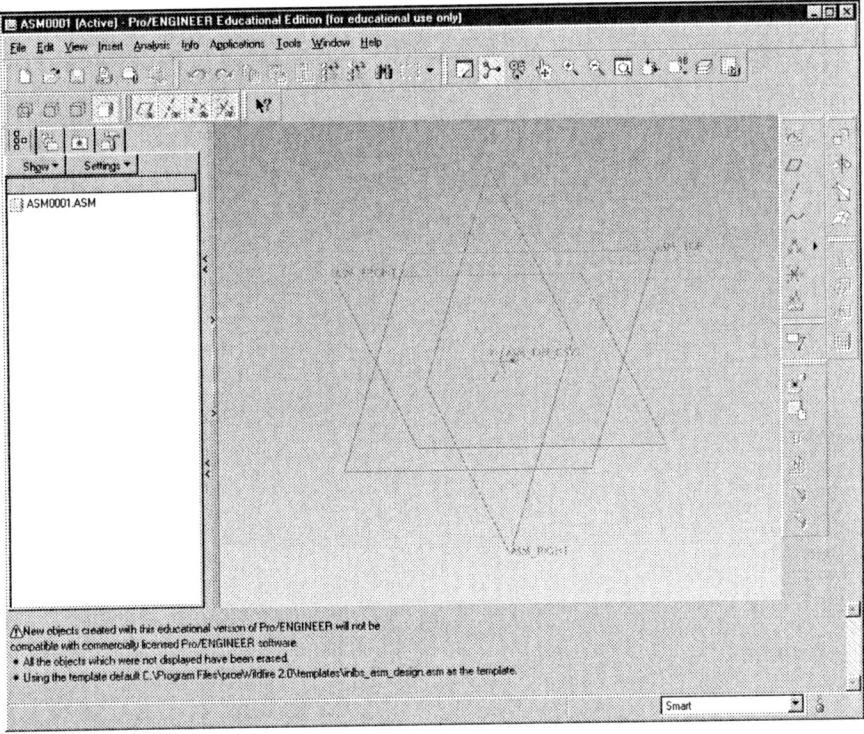

*Figure 9-2 Initial screen appearance of the **Assembly** mode*

the components are created in the assembly file and then assembled using the assembly constraints. Note that the parts you create in the **Assembly** mode are saved as separate *.prt* files.

Bottom-Up Approach

This is the method of assembling the components that are created as separate parts in the **Part** mode and are saved as *.prt* files. Once all parts of an assembly have been created, you will create a new assembly file (*.asm*) and then assemble the parts using the assembly constraints available in the **Assembly** mode. Since the assembly file has information related only to the assembling of components, the file size is small and therefore requires less hard disk space.

Remember that if any of the assembly component is moved from its original location on the hard disk, the assembly will not open the next time. This is because Pro/ENGINEER searches for the assembly component in the folder in which it was originally saved. If it does not find the name of the component in that folder, it gives an error while opening the assembly file and the **RESOLVE FEAT** menu is displayed, as shown in Figure 9-3. Choose the **Quick Fix** option from this menu; the **QUICK FIX** submenu is displayed, as shown in Figure 9-4. You can use the options in this submenu to locate the part file.

Figure 9-3 RESOLVE FEAT menu

Figure 9-4 QUICK FIX menu

Tip: *It is advised that you create separate folders for all assemblies that you create in Pro/ENGINEER.*

The other important thing that you need to remember while opening an existing assembly is that for example, Nut is a component that is common and used in most of the assemblies. Therefore, if the current Pro/ENGINEER session already has a component with the name Nut, it will use this component only when you open the assembly and not the original Nut used in the assembly.

Placement Constraints

The placement constraints are the constraints that are used to rigidly bind the components of the assembly to their respective positions in the assembly. These constraints are also called the assembly constraints. Generally, these constraints are used in combinations and you can constrain upto six degrees of freedom of a component.

Package

It is the state in which the component that is being assembled is not fully constrained and thus is not rigidly placed at its actual location.

CREATING TOP-DOWN ASSEMBLIES

As mentioned earlier, top-down assemblies are those in which all components are created in the **Assembly** mode and in the same assembly file. The components that you create in the **Assembly** mode are saved as separate *.prt* files when you choose to save the assembly file in which they are created. As you know, to create the components you need the sketcher environment and to invoke the sketcher environment you need the **Part** mode. You can invoke the **Part** mode from within the **Assembly** mode and create components using the tools available in the **Part** mode. The method to create the component from within the **Assembly** mode is discussed next.

Creating Components in the Assembly Mode

To create a component in the **Assembly** mode, choose the **Create a component in assembly mode** button from the **Right Toolchest**. The **Component Create** dialog

Assembly Modeling

box is displayed, as shown in Figure 9-5. Select the **Part** radio button from the **Type** area and the **Solid** radio button from the **Sub-type** area to create a *.prt* file. Enter the name of the component in the **Name** edit box and then choose the **OK** button. The **Creation Options** dialog box is displayed. From this dialog box, select the method for creating the component.

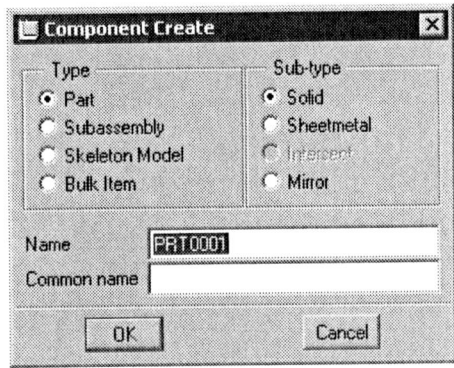

*Figure 9-5 The **Component Create** dialog box*

CREATING BOTTOM-UP ASSEMBLIES

As discussed earlier, the bottom-up assemblies are created by inserting the part files one by one in the assembly file and then assembling them using the assembly constraints. When the first component is inserted in the **Assembly** mode, its three default datum planes are placed in the same orientation as that of the default datum planes of the **Assembly** mode. The method of inserting the component in the assembly is discussed next.

Inserting Components in the Assembly

To insert the component, you need to choose the **Add component to the assembly** button, the **Open** dialog box is displayed. This dialog box is the same as the dialog box that is used to open an existing file. Select the part that you need to insert in the assembly. You can also preview the component before adding it to the assembly. After the component is displayed in the graphics window in the **Assembly** mode, you need to specify constraints in order to assemble them. Even if you are placing the first component, you need to constrain it using the assembly constrains. Remember that no component is placed automatically; you need to manually specify the position of the component.

PLACEMENT CONSTRAINTS

The assembly constraints are also called placement constraints and are available in the **Component Placement** dialog box, which is discussed later in this chapter. The placement constraints that are available in Pro/ENGINEER are discussed next.

Automatic

When you choose this constraint, Pro/ENGINEER assumes the constraint and applies it according to the type of entity selected. For example, if you select axes of two components to assemble, then Pro/ENGINEER will understand that you want to apply the **Align** constraint and the **Align** constraint will be applied to the two components.

Mate

The **Mate** placement constraint allows you to make two selected planes, datum planes, faces, or a combination of a datum plane and a face coplanar with respect to each other. The faces or datums selected may or may not be in contact with each other.

The **Mate** constraint is used in combination with three options: **Offset**, **Coincident**, and **Oriented**.

Mate Offset

If you need to keep the two faces or planes at some distance apart, then you can use the **Mate** constraint with some offset value. The offset distance between the two coplanar faces can have a positive or a negative value. Figure 9-6 shows the faces that are selected to apply the **Mate Offset** constraint. Figure 9-7 shows the component assembled at an offset distance from the selected face.

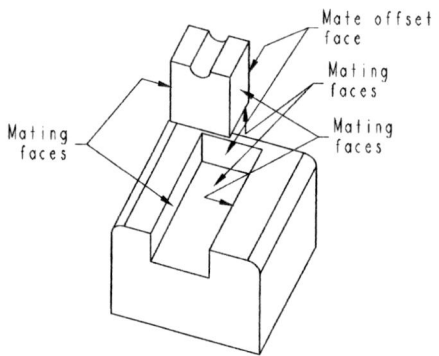

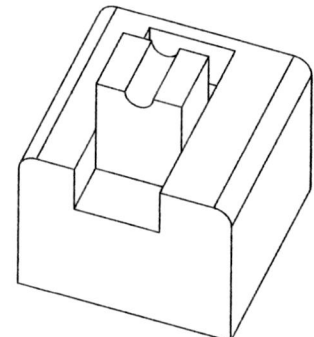

Figure 9-6 Components before assembling   *Figure 9-7 Components after assembling*

Mate Coincident

The **Mate Coincident** combination of constraints allows you to make two selected planes or faces coplanar to each other. Figure 9-8 shows the faces that are selected to mate. Figure 9-9 shows the components after assembling.

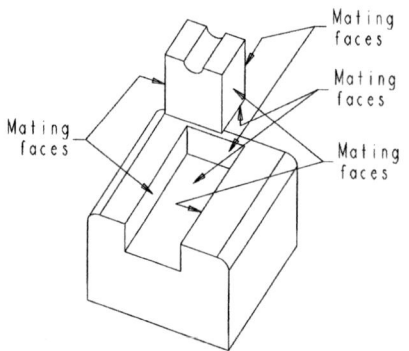

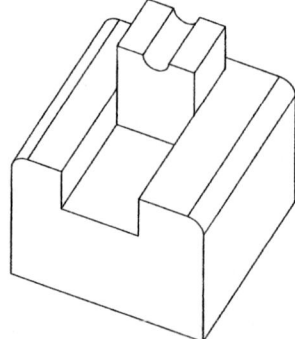

Figure 9-8 Components before assembling   *Figure 9-9 Components after assembling*

Assembly Modeling

Mate Oriented

The **Mate Oriented** combination of constraints is used when you want to mate two faces or planes or a combination of the two and at the same time you want to orient some other faces or planes of the two components to face in the same direction. Figure 9-10 shows the faces that are selected to mate and Figure 9-11 shows the components after assembling.

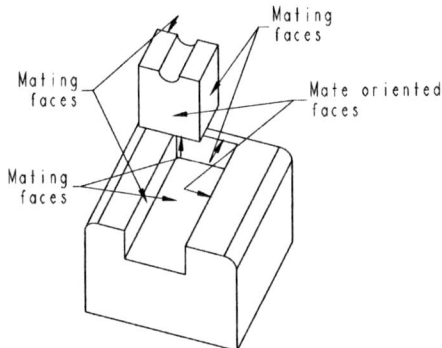
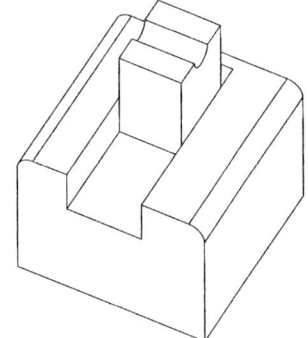

Figure 9-10 Components before assembling   *Figure 9-11* Components after assembling

Align

The **Align** constraint is used to assemble two components by making the selected faces or planes coplanar such that the aligned faces or planes are facing in the same direction. Apart from planes and faces, you can also select datum axes, datum points, edges, or vertices for applying the **Align** constraint.

The **Align** constraint is used in combination with three options: **Offset**, **Coincident**, and **Oriented**.

Align Offset

The **Align Offset** combination of constraints is used to align two entities with some offset distance between the aligning entities. The offset value can be positive or negative. Figure 9-12 shows the faces to align offset and Figure 9-13 shows the components after applying the constraints.

Align Coincident

The **Align Coincident** combination of constraints is used to assemble two components by making the selected faces or planes coplanar such that the aligned faces or planes are facing in the same direction. Figure 9-14 shows the faces that are selected to align. Figure 9-15 shows the components after assembling.

Align Oriented

The **Align Oriented** combination of constraints makes the two selected faces or planes or the combination of both, face in the same direction. Figure 9-16 shows the constraints applied on the two components and Figure 9-17 shows the two components after applying the constraints.

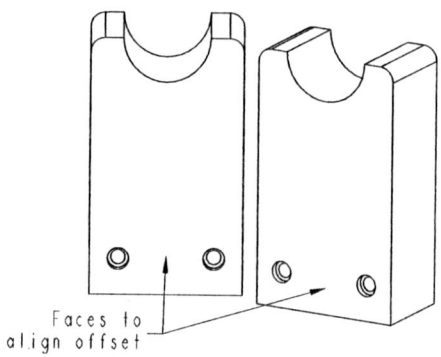

Figure 9-12 Components before applying Align Offset constraint

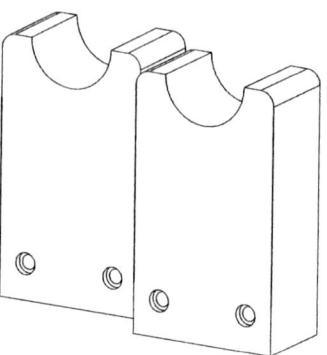

Figure 9-13 Components after applying Align Offset constraint

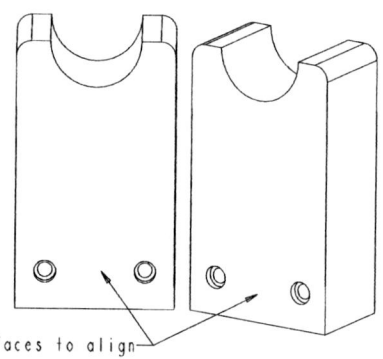

Figure 9-14 Components before applying Align Coincident constraint

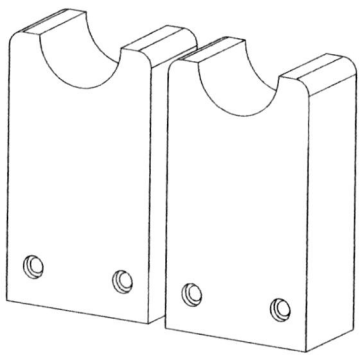

Figure 9-15 Components after applying Align Coincident constraint

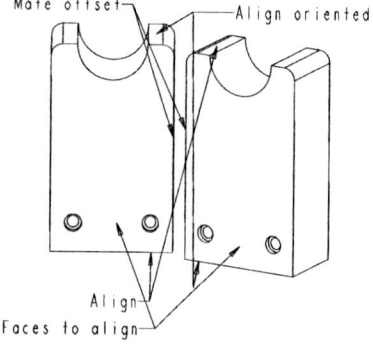

Figure 9-16 Components before applying Align Oriented constraint

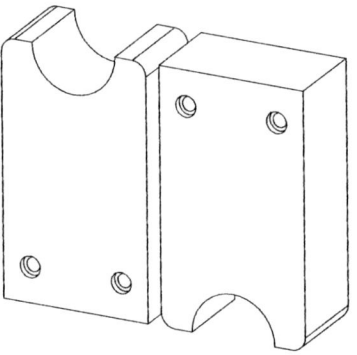

Figure 9-17 Components after applying Align Oriented constraint

Insert

The **Insert** constraint is used to assemble the revolved components. Applying this constraint allows the revolved component, holes, or a combination of the both to share the same orientation of the central axis. Figure 9-18 shows the faces to insert and Figure 9-19 shows the components after assembling.

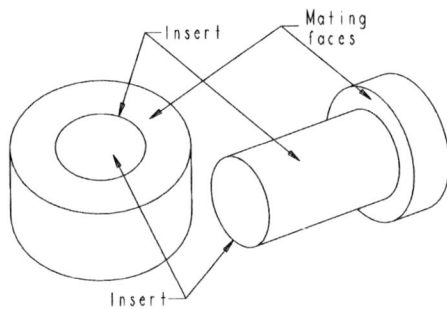
Figure 9-18 Components before assembling

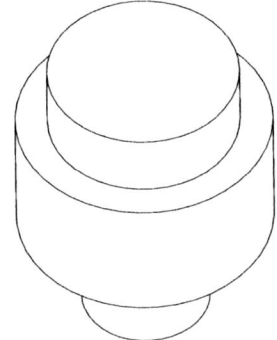
Figure 9-19 Components after assembling

Coord Sys

The **Coord Sys** constraint aligns the coordinate system of the first component with the coordinate system of the second component.

Tangent

The **Tangent** constraint is used to make the selected circular face tangent to the other selected face or plane and at the same time make them coplanar.

Pnt On Line

The **Pnt On Line** constraint is used to align the selected datum point or the vertex on the first part with the selected edge, datum axis, or datum curve on the second part.

Pnt On Srf

The **Pnt On Srf** constraint is used to align the selected datum point or vertex on the first part with the selected surface or datum plane on the second part.

Edge On Srf

The **Edge On Srf** constraint is used to align the selected edge of the first part with the selected surface or datum plane on the second part.

Default

 The **Default** constraint is used to assemble the component in the assembly by aligning the default coordinate system of the component with the default coordinate system of the assembly. Also, the part datum planes are aligned with the assembly datum planes.

Fix

The **Fix component to current position** button packs the two components and displays that the two components are fully constrained.

Convert

The **Convert** button converts the applied constrains to mechanism connections and vice-versa.

ASSEMBLY DATUM PLANES

When you enter the assembly environment, the three default assembly datum planes are displayed in the graphics window. If you do not want the default assembly datum planes in the assembly file, then while opening a new file you need to clear the **Use default template** check box that is available in the **New** dialog box. It is recommended that you use the assembly datum planes as the first feature of the assembly and assemble the components of the assembly taking the reference of these datum planes. Using the datum planes as the first feature will help you during the modification of the assembly.

Some of the advantages of using the assembly datum planes are:

1. The components of the assembly can be redefined in such a way that the component placed later can be made the first component.

2. The first component can be replaced by some other component.

3. The placement constraint of the first component can be modified.

ASSEMBLING THE COMPONENTS

The components in an assembly can be placed parametrically or nonparametrically. If the components are placed using the placement constraints, it is called parametric assembly. On the other hand if the components are packaged, it is called nonparametric assembly.

To assemble the components parametrically, choose the **Add component to the assembly** button. The **Open** dialog box is displayed. Once you select the part to assemble from the **Open** dialog box, the **Component Placement** dialog box is displayed, as shown in Figure 9-20. The **Component Placement** dialog box has multiple functions. This dialog box is used for both constrained and packaged placement of the components. There are three tabs in this dialog box: **Place, Move, and Connect**. In this chapter you will learn about **Place** and **Move** tabs. The various options available in this dialog box are discussed next.

Displaying the Components in a Separate Window

The **Show component in a separate window while specifying constraints** button is used to display the component to be assembled in a separate window. When you choose this button, a new window will open displaying the component to be assembled. Note that viewing the component to be assembled, and the assembly, in separate windows, prevents cluttering of the components in the assembly window.

Assembly Modeling

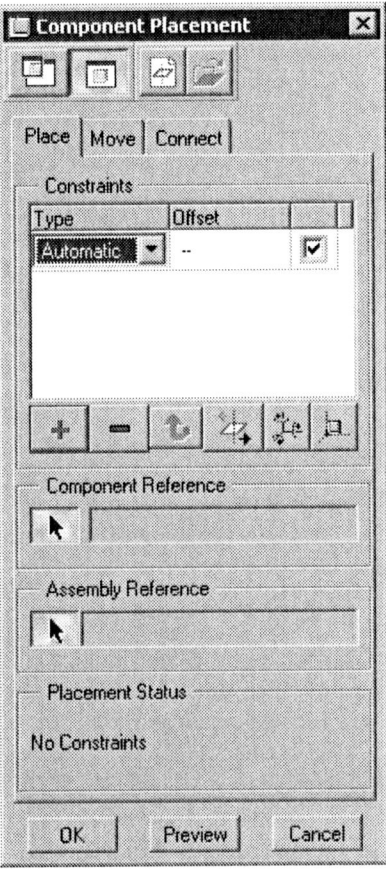

Figure 9-20 The **Component Placement** *dialog box*

However, there is a limitation when the assembly and the component to be assembled are displayed in separate windows. If the components are in the same window, they change their orientation as you apply the constraints, thus giving you an idea of the next constraint to be applied. In other words, you can easily find out which degree of freedom of the component is constrained and which needs to be constrained.

Displaying the Components in the Same Window

The **Show component in the assembly window while specifying constraints** button is used to display the component to be assembled in the assembly window itself. This button is selected by default when you invoke the **Component Placement** dialog box. As you apply constraint, the component in the assembly changes its position according to the constraints.

Place Tab

The options that are available in the **Place** tab of the **Component Placement** dialog box are discussed next.

Constraints Area

This area displays the constraints applied to the component, along with the offset value (if any), in the order in which they were applied. It is recommended that you check this area after applying a constraint. Make sure that the constraint you intended to apply is listed in this area.

You will notice a check box on the extreme right in the **Constraints** area. After you apply a constraint, the check box is automatically selected. As a result, you cannot move the component in the degree of freedom that is locked by the constraint applied. However, if the check box is cleared, the constrained degree of freedom is unlocked.

The other options in this area are discussed next.

Type Drop-down List

This drop-down list displays the placement constraints that are available to assemble the component. You can select the constraints from this drop-down list to assemble components in an assembly. The different constraints that are available in this drop-down list were explained earlier.

Offset Drop-down List

This drop-down list is used to specify the offset value (if any) for the placement constraint. This drop-down list has two more options, **Coincident** and **Orient**. The **Coincident** and **Orient** options are used in combination with **Mate** and **Align** options.

Specify a new constraint Button

This button is chosen to add a constraint to assemble the components. When you choose this button, the **SELECT** dialog box is displayed. To exit this dialog box, you need to choose the **Cancel** button.

Remove the selected constraint Button

This button is chosen to delete the constraint that has already been applied. When you choose this button, the constraint that is selected in the **Constraints** area is deleted.

Change orientation of constraint Button

This constraint button is used to reverse the orientation of the constraint applied. This button is used only after a constraint has been applied to the components. This button is not available initially when the **Component Placement** dialog box is invoked.

Component Reference Area

The button under this area is chosen to select the reference for specifying the placement constraint on the component to be assembled. When you choose this button, you can select the reference from the assembly or from the component.

Assembly Modeling

Assembly Reference Area

The button under this area is chosen to select the reference for specifying the placement constraint on the assembly. This button is automatically chosen when you have finished specifying the reference on the component to be placed.

Placement Status Area

This area displays the placement status of the component in the assembly. The status displayed in this area is usually partially constrained or fully constrained. If you choose **OK** from the **Component Placement** dialog box when the status displayed is **Partially Placed**, then in the **Model Tree** the component assembled is displayed with a mark on it. This mark indicates that the component is packaged.

Note
Pro/ENGINEER Wildfire 2.0 assumes a few placement constraints after you have added some constraints. This is the reason sometimes the allow assumptions check box is displayed in the placement status area.

Move Tab

The options under the **Move** tab can be used only if you have selected the **Show component in the assembly window while specifying constraints** button. This tab is used to move or rotate a component along a degree of freedom that is not restricted. The **Move** tab is shown in Figure 9-21 and the options under this tab are discussed next.

Motion Type Area

The options under this area are discussed next.

Orient Mode

This radio button is selected to orient the part in the assembly about the spin center of the part.

Translate

This radio button is selected to move the component from its current location in the assembly. However, remember that the component can be moved only along the degrees of freedom that are not restricted.

Tip: *It is recommended that you use CTRL+ALT+left mouse button to translate a component and CTRL+ALT+middle mouse button to rotate a component. It is always easier to use the above combinations to move the component rather than switching to a separate tab in the **Component Placement** dialog box and using its options to move a component. Direct manipulation of the component can speed up the process of locating the component correctly in the assembly and establishing constraints.*

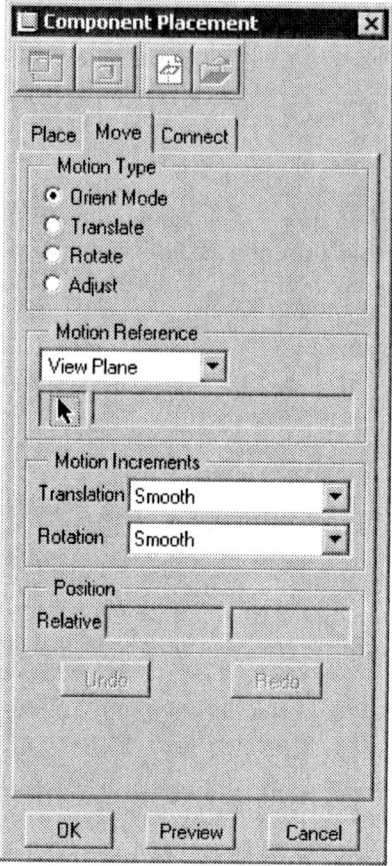

*Figure 9-21 Move tab of the **Component Placement** dialog box*

Rotate
This radio button is selected to rotate the component in the assembly along its available degrees of freedom.

Adjust
This radio button is selected to pack a component with reference to the assembly. The message "**Select surface on packaged component to adjust to ref plane**" is displayed in the **Message Area** when you select this radio button.

Motion Reference Area
The drop-down list under this area displays the references that can be used for the motion of the component. The options provided under this drop-down list are discussed next.

View Plane
This option uses the view plane as the reference for the motion of the component.

Sel Plane

This option allows you to select a planar surface or a datum plane that you want to use as reference for the motion of the component.

Entity/Edge

This option allows you to select a straight edge, an axis, or a datum curve as the reference for the motion of the component.

Plane Normal

This option allows you to select a plane that will allow the motion of the component in a direction that is normal the selected plane.

2 Points

This option is used to select two datum points that will specify the direction for the movement of the component.

Csys

This option is used to select the coordinate system to specify the motion of the component. When you select this option, three buttons appear besides this drop-down list. These buttons are **X**, **Y**, and **Z**. They are used to specify the axis of the selected coordinate system for specifying the direction of motion.

Motion Increments Area

The options available under this area are discussed next.

Translation

This drop-down list is used to select the predefined increment values by which the component will be moved along the specified direction when using the options in the **Motion Reference** drop-down list.

Rotation

This drop-down list is used to select the predefined increment angles by which the component will be rotated when using the options in the **Motion Reference** drop-down list.

Position Area

Relative

These boxes display the coordinates of the current position of the component.

Undo

This button is chosen to undo the changes made when using the options in the **Move** tab.

Redo

This button is chosen to reinstate the changes that were removed using the **Undo** button.

Preview

This button is used to preview the component in the assembly before it is actually assembled.

PACKAGING THE COMPONENTS

Sometimes, you do not know the exact location of a component in the assembly or you want to specify the exact location later. In such cases, you can nonparametrically place the component in the assembly. This method of nonparametrically placing the components in the assembly is called packaging. There are two methods to package a component in an assembly. The first method is to use the **Insert** menu in the menu bar. The second method is to assemble the component using the **Component Placement** dialog box and then exit the dialog box before fully constraining it.

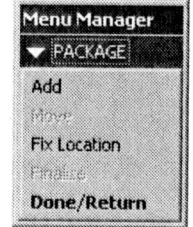

To package the component, choose **Insert > Component > Package** from the menu bar. When you choose this option, the **PACKAGE** menu will be displayed, as shown in Figure 9-22. This submenu provides the following options.

Figure 9-22 The PACKAGE menu

Add

The **Add** option is used to add the component to be packaged. You can open a new component using the **Open** dialog box, select it from the assembly, or select the last component from the assembly to assemble again.

When you select the component from the **Open** dialog box, the **Move** dialog box is displayed, as shown in Figure 9-23. The options under this dialog box are same as those under the **Move** tab of the **Component Placement** dialog box. After you have specified the location of the component in the assembly file, choose **OK** to place the component and close the **Move** dialog box. Remember that you can move only the packaged component using this option.

Move

The **Move** option is used to move the packaged component to a new location. When you choose this option, the **Move** dialog box will be displayed, as shown in Figure 9-23.

Fix Location

The **Fix Location** option is used to fix the location of an existing packaged components. When you choose this option, you are prompted to select the component to fix.

Finalize

As mentioned earlier, the packaged components are not assembled parametrically. Hence, packaging does not relate the component with its neighboring components in the assembly. Once the location of the component is decided, then it can be finalized using the **Finalize** option. When you choose this option, you will be prompted to select the packaged component. The **Component Placement** dialog box will be displayed when you select the packaged component. You can parametrically assemble the component by adding the placement constraints to the component by using this dialog box.

Assembly Modeling

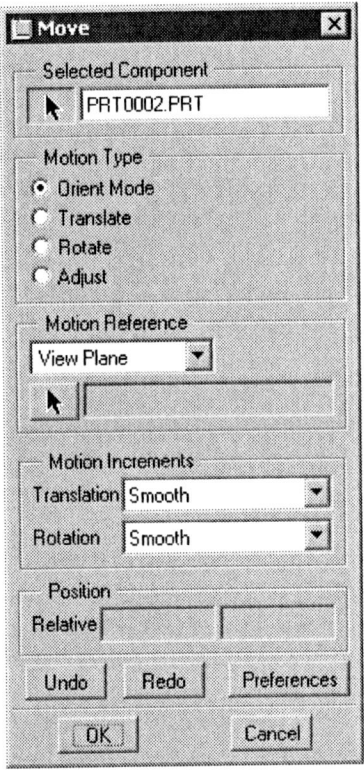

*Figure 9-23 The **Move** dialog box*

Tip: *The components that are packaged in an assembly are displayed in the **Model Tree** with a mark on them.*

CREATING SIMPLIFIED REPRESENTATIONS

As you know, the assembly designs consist of a number of parts and subassemblies. In case of some complicated assemblies, such as a computer CPU, once the cover is assembled, the components inside it are not visible. So there have to be some means to temporarily remove certain components that are not desired at some point of time from the current display or force them to be displayed in the wireframe so that you can see through them. This process of removing some components or changing their type of display from the current display is called **Simplified Representation**.

Simplified representations can be created using the cascading menu or by using the **View Manager**. The only difference is that by using the **View Manager** you can give a name to the representation. To create a simplified representation using the cascading menu, select a component from the **Model Tree** and then choose **View > Representation** from the menu bar. The cascading menu is displayed, as shown in Figure 9-24.

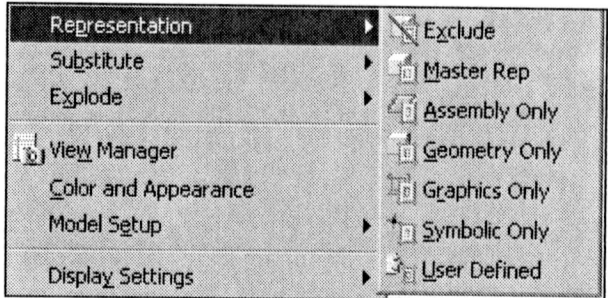

Figure 9-24 Cascading menu

Using the Cascading menu

The options available in the cascading menu are used to create a simplified representation without saving it. The simplified representation reduces the regeneration time of the assembly, especially the complex assemblies. The options in this menu are discussed next.

Exclude

When you choose this option, the selected component is removed from the display. You can even select more than one component from the **Model Tree** or from the assembly to remove them from the display. However, the component remains in the memory of Pro/ENGINEER and is even displayed in the **Model Tree**. This component can be, at any time, forced to be displayed again. To redisplay the components, choose the **Master Rep** option from the cascading menu. Figure 9-25 shows a butterfly valve assembly in which the Body is removed from the current display. This kind of representation helps in assembling the components inside the body, like the plate and the screw in this case.

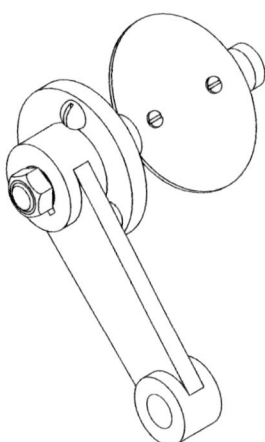

Figure 9-25 Components excluded from the assembly

Master Rep

This option is used to restore the original representation of the assembly, thus displaying all

Assembly Modeling

components in the assembly. When you use this option, the components that were removed from the current display after using the **Exclude** option, are also redisplayed, see Figure 9-26.

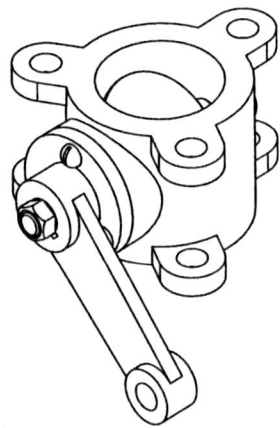

Figure 9-26 Master representation of the assembly

 **Tip**: *When you create a simplified representation, some information regarding the type of representation is displayed in the graphics window. To remove this information, choose **View** > **Shade** from the menu bar.*

Assembly Only

The **Assembly Only** option is used to hide all components of an assembly except the components of the subassembly.

Geometry Only

In this type of simplified representation the complete geometry of the components is displayed in the current model display style. This type of simplified representation takes a long time for regeneration. The components that are not selected in this type of representation cannot be modified. However, these components can be redefined.

Graphics Only

When you choose the **Graphics Only** option, the selected components will be displayed in wireframe, as shown in Figure 9-27. This type of representation is generally used for assembling the components inside some other component. This type of simplified representation also reduces the regeneration time for large assemblies. However, note that the components that are displayed as the wireframe model in the graphic representation cannot be modified, redefined, or referenced.

Using the View Manager

 The **View Manager** is used to create, modify, and switch between simplified representations, exploded views, and orientation of the assembly. To

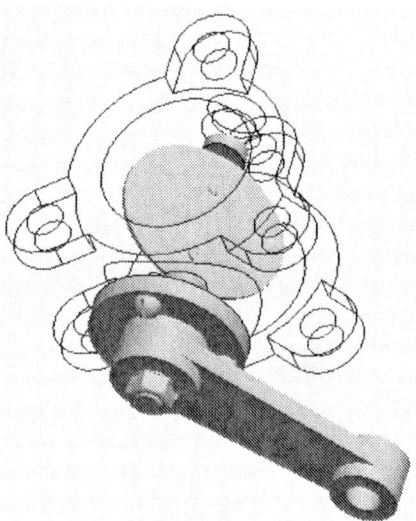

Figure 9-27 The graphics representation of an assembly

invoke the **View Manager**, choose the **Start the view manager** button from the **Top Toolchest**. The **View Manager** dialog box is displayed, as shown in Figure 9-28. The various uses of this dialog box and the operations that you can perform on an assembly using this dialog box are discussed next.

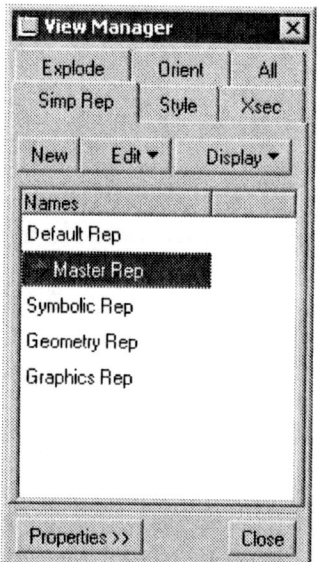

Figure 9-28 View Manager dialog box

Creating Simplified Representation Using the View Manager

When the **View Manager** dialog box is displayed, the **Simp Rep** tab is selected by default. The options available under this tab are similar to those that were available in the cascading

Assembly Modeling

menu. The red arrow in front of **Master Rep** indicates that this representation is set currently. To set any of the listed representation types, select the representation and right-click to invoke the shortcut menu. Choose the **Set Active** option from the shortcut menu to set the display of the assembly to that type of representation or double click on one of the representation types.

To create a user-defined representation, right-click in the dialog box and choose the **New** option. You can specify a name to the representation.

To redefine a representation type, choose the representation from the list and right-click to invoke the shortcut menu. Choose the **Redefine** option from the shortcut menu; the **EDIT** dialog box is displayed, as shown in Figure 9-29. Using this dialog box, you can select the components that you want to include, exclude, and substitute from the display.

Creating Display Style Using the View Manager

To create a new display style from the **View Manager** dialog box, choose the **Style** tab. Two styles are available in the **Names** list of the dialog box. You can create a new style or redefine an existing style. The options to perform these operations are available in the shortcut menu that is displayed when you right-click in this dialog box.

Figure 9-29 The **EDIT** *dialog box*

Creating Explode States Using the View Manager

The exploded view of the assembly created by using the **View Manager** can be saved and then later retrieved in the **Drawing** mode. Choose the **Explode** tab to view the list of explode states available. When you choose the **Explode** tab, the **Default Explode** state is listed in the **Names** list. You can create a new explode state by choosing the **New** option from the shortcut menu. Enter the name of the explode state and then redefine it to display the **MOD EXPLODE** menu, as shown in Figure 9-30. Using this dialog box, you can create an exploded state of the assembly. The creation of exploded state is discussed later in the chapter.

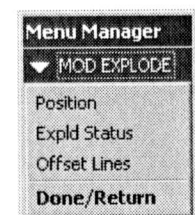

Figure 9-30 **MOD EXPLODE** *menu*

Setting the Orientation of the Assembly Using the View Manager

The various orientations that are given in the **Saved view** list (from the **Top Toolchest**) are used to display the assembly as viewed from different directions. Using the **View Manager** you can invoke the **Orientation** dialog box to set a user-defined orientation. To invoke the **Orientation** dialog box, choose the **Orient** tab and create a new orientation by choosing the **New** option from the shortcut menu. Name the view and then redefine it to display the **Orientation** dialog box. Use this dialog box to set the orientation of the assembly and then you can save the orientation by giving it a name. This name is listed in the **Saved view list**.

Creating X-section of the Assembly Using the View Manager

Using the **View Manager,** you can create the section view of the assembly. Choose the **Xsec** tab from the **View Manager** dialog box and use shortcut menu to create a **New** section. Name the section and press the middle mouse button to display the **X-SEC OPTS** menu. Use this menu to select the options to create the required section.

REDEFINING THE COMPONENTS OF THE ASSEMBLY

Sometimes, after you have assembled the components of the assembly, you may need to modify or redefine the location or the placement constraints of the components. To do this, select the component from the **Model Tree** and right-click to invoke the shortcut menu. Choose the **Edit Definition** option from the shortcut menu. You can also right-click on the component to be redefined in the graphics window and choose the **Edit Definition** option from the shortcut menu. All components assembled after the selected component will become invisible and the **Component Placement** dialog box will be displayed. You can modify the existing placement constraints or add new constraints using this dialog box.

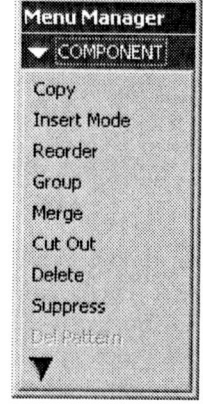

REORDERING THE COMPONENTS

The order of assembling the components can be changed by choosing **Edit > Component Operations** from the menu bar. The **COMPONENT** menu is displayed, as shown in Figure 9-31. Choose the **Reorder** option from this menu. When you choose this option, you will be prompted to select the component to be reordered. When you select a component to reorder from the **Model Tree** or from the graphics window and choose **Done**, a message will be displayed in the **Message Area** that specifies the components before which you can insert the selected component.

Figure 9-31 The COMPONENT menu

*Tip: You can also reorder the components in an assembly by using the **Model Tree**. This can be done by selecting the component in the **Model Tree** and then dragging it to the position where it can be placed. Note that this method of reordering does not inform you where the component can be placed and where it cannot be placed. However, the parent component of the selected child component gets highlighted when you start dragging the child component. This means that if a component has a parent, then the parent is also reordered with the child. Remember that in the **Model Tree**, the parent component will always be placed before its child component.*

SUPPRESSING/RESUMING THE COMPONENTS

If you do not want certain components of the assembly to appear in the current display or in the drawing views; you can suppress them. To suppress the components, choose **Edit > Component Operations** from the menu bar; the **COMPONENT** menu is displayed. Choose the **Suppress** option from this menu.

Similarly, the suppressed components can be resumed by choosing the **Resume** option from the same menu. The components can also be suppressed/resumed using the shortcut menu that is displayed when you right-click on the component in the **Model Tree** or in the graphics window. In the shortcut menu, choose **Suppress** to suppress and **Resume** to resume the components.

REPLACING COMPONENTS

The existing components of the assembly that are assembled using the assembly constraints can be replaced with some other component, if necessary. To replace the component, choose **Edit > Replace** from the menu bar. The **Replace** dialog box is displayed, as shown in Figure 9-32 and you will be prompted to select the model for the replacement of the component. The new component can be selected using the **Browse** button of the **Replace** dialog box. The new component selected will be displayed in the **Browse** edit box. The new component can be assembled using a family table, an interchange assembly, by layout, or manually. After you have selected one of these options, choose **OK** from the **Replace** dialog box to display the **Component Placement** dialog box for adding the placement constraints.

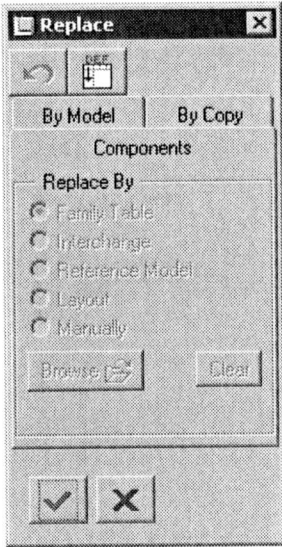

Figure 9-32 The **Replace** *dialog box*

ASSEMBLING REPEATED COPIES OF A COMPONENT

Sometimes, the assembly design demands the assembling of a particular component more than once. One option is that you should assemble the component every time and add the placement constraints all over again. However, this is a very tedious and time-consuming process, especially in the assemblies that have a large number of similar components. To solve this problem, Pro/ENGINEER provides you with an option of assembling multiple copies of the components.

To repeat a component in an assembly, select the component to repeat from the **Model Tree** or from the graphics window and then choose **Edit > Repeat** from the menu bar. When you choose this option, the **Repeat Component** dialog box will be displayed, as shown in Figure 9-33. The options in the **Repeat Component** dialog box are discussed next.

Component Area

The button in this area is chosen to select the component whose multiple copies are to be assembled. The name of the selected component is displayed in the display box provided in this area.

Variable Assembly Refs Area

The **Variable Assembly Refs** area displays the placement constraints that are applied on the selected component. You can keep the placement constraints that need to be present for the new component and modify the remaining constraints by selecting them from this area. The field in this area consists of the following columns.

Type

The **Type** column displays the type of placement constraints that are applied to the selected component.

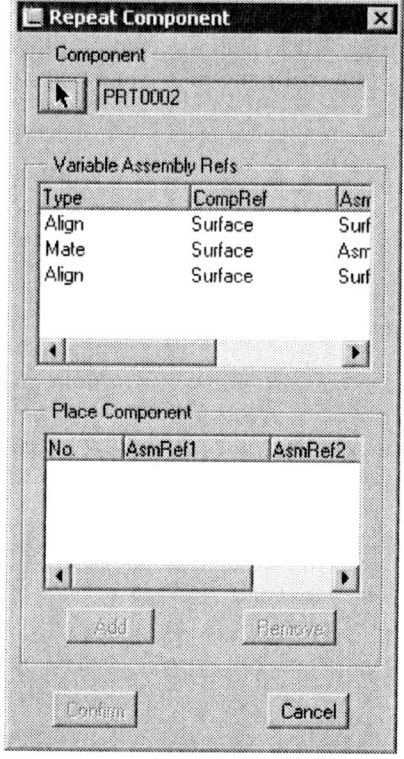

*Figure 9-33 The **Repeat Component** dialog box*

CompRef

The **CompRef** column displays the references that were used to assemble the component. In case you want to change this for the new component, select this constraint and then choose the **Add** button. Now, select a reference on the assembly. A copy of the selected component will be added to the assembly at a new location. Similarly, to add another copy, repeat this procedure.

AsmRef

The **AsmRef** column displays the reference on the assembly that was used to assemble the selected component.

Place Component Area

The **Placement Component** area displays the new references added for the newly placed copies of the selected component.

Add

The **Add** button is chosen to add the required references for the new copy of the selected component. This button will be available only when you select one of the constraints from the **Variable Assembly Refs** area.

Remove

The **Remove** button is chosen to remove the selected reference from the new component. This button is available only when you select one of the references from the **Place Component** area.

MODIFYING THE COMPONENTS OF THE ASSEMBLY

Sometimes, during or after the assembly of a components, you may need to modify the dimensions of the components. The modification in the dimensions of the components can be directly made in the **Assembly** mode and you do not need to open the part file to make the modifications. You can even redefine the selected feature in the **Assembly** mode itself. The methods used to modify and redefine the features are discussed below.

Modifying the Dimensions of a Feature of a Component

The dimensions of a selected feature of the component can be modified by selecting the component from the **Model Tree** or from the graphics window. When the outline of the model turns red in color, right-click to invoke the shortcut menu. Choose the **Activate** option, the name of the component appears on the graphics window. Also a small green button appears on the active component in the model tree. Now select the feature of the component and then hold down the right mouse button to invoke the shortcut menu. Choose the **Edit** option from the shortcut menu. The dimensions of the selected feature of the component are displayed in the graphics window. After modifying the dimensions, remember to regenerate the component by choosing the **Regenerate Model** button available in the **Edit** toolbar. After regenerating the component, you can view the effect of the modification.

Redefining a Feature of a Component

Redefining the features in the **Assembly** mode is available. This saves time in opening the part in the **Part** mode and then redefining the features. To redefine the feature of a component, you need to select the component from the **Model Tree** and make it active. After the component is activated, select the feature that you need to redefine from the graphics window. Hold down the right mouse button to invoke the shortcut menu. Choose the **Edit Definition** option. After choosing the option, the feature creation tool is activated and now you can redefine the feature.

CREATING THE EXPLODED STATE

In an assembly design, the components that are assembled inside some other components may not be visible. This could be misleading and may confuse the viewer as he cannot view all components. To eliminate this confusion, generally an exploded view is provided along with the assembled view. The exploded view is a state in which all components are moved from their actual location so that they are visible, as shown in Figure 9-34.

The exploded state can be created by choosing **View > Explode > Edit Position** from the menu bar. When you choose this option, the **Explode Position** dialog box is displayed, as shown in Figure 9-35. This dialog box provides you the following options to move the components from their actual location.

Figure 9-34 The exploded state of an assembly

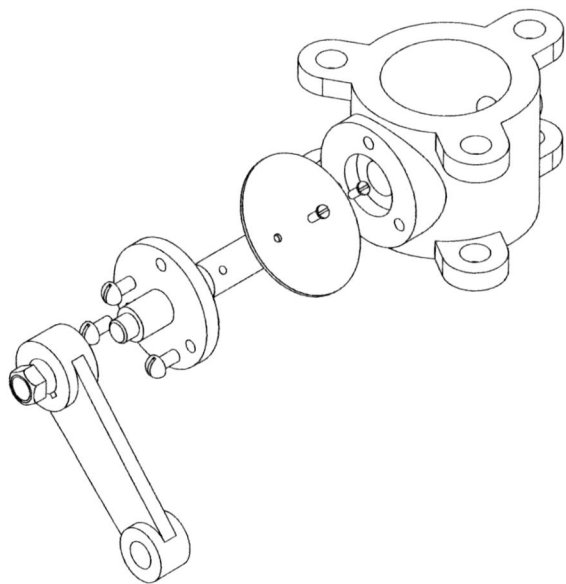

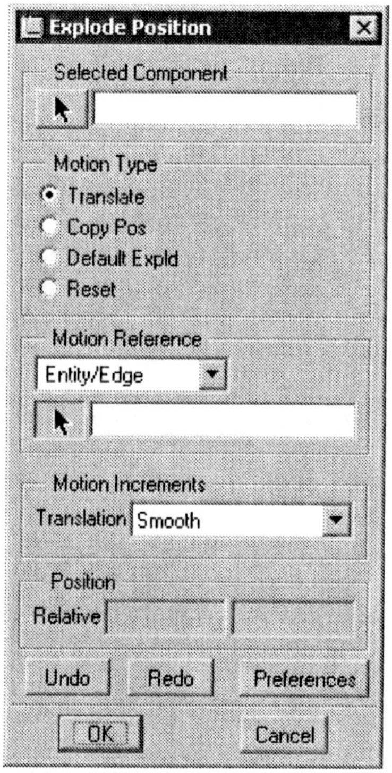

Figure 9-35 The **Explode Position** dialog box

Motion Type Area

The **Motion Type Area** has following four radio buttons, which are discussed next.

Translate

This radio button allows you to move a component to a new location in the graphics window from its current location in the assembly.

Copy Pos

This option is used to restore the exploded component to its original location. This radio button when selected, prompts you to select a component whose position you want to copy. For example, a component is exploded and is moved away from the assembly in the graphics window and you want to assemble it back to its original position. Select the **Copy Pos** radio button; you will be prompted to select a component to copy the position from. Select a component from the assembly that is at a similar location in the assembly. Now, you are prompted to select a component to move. Select the exploded component that is moved away from the assembly. The selected component will be assembled at its original position in the assembly.

Default Expld

This radio button when selected, prompts you to select a component. When you select a component from the assembly, the component is exploded from the assembly to a default position in the graphics window.

Reset

The **Reset** radio button is used to set the selected component to its original location in the assembly.

Motion Reference Area

The options available in the drop-down list available in this area discussed next.

View Plane

The **View Plane** option is used to specify the current view plane (that is, the screen of the monitor) as the reference for exploding the selected component.

Sel Plane

The **Sel Plane** option is used to select a datum plane or a planar face for specifying the reference for the motion of the component. When you choose this option, you will be prompted to select the datum plane or a planar face and then you will be prompted to select the component to move.

Entity/Edge

The **Entity/Edge** option is selected by default and is used to select a linear edge or an axis for specifying the reference direction for exploding the component.

Plane Normal

The **Plane Normal** option is used to select a planar face or a datum plane, to which the reference for exploding the components will be normal.

2 Points

The **2 Points** option is used to select two datum points or two vertices to specify the reference direction for exploding the components.

Csys

The **Csys** option is used to select one of the axes of the selected coordinate system to specify the reference direction for exploding the components.

Note

*If you have not created an exploded state of an assembly, a default explode state is available when you choose **View** > **Explode** > **Explode View** from the menu bar. To redisplay the unexploded state, choose **View** > **Explode** > **Unexplode** from the **View** menu in the menu bar.*

OFFSET LINES

The offset lines are the lines that are used to display the actual path and direction of the mating components. These lines are generally used for easy visualization of the exploded components of the assemblies. The exploded state of an assembly using the offset lines is shown in Figure 9-36.

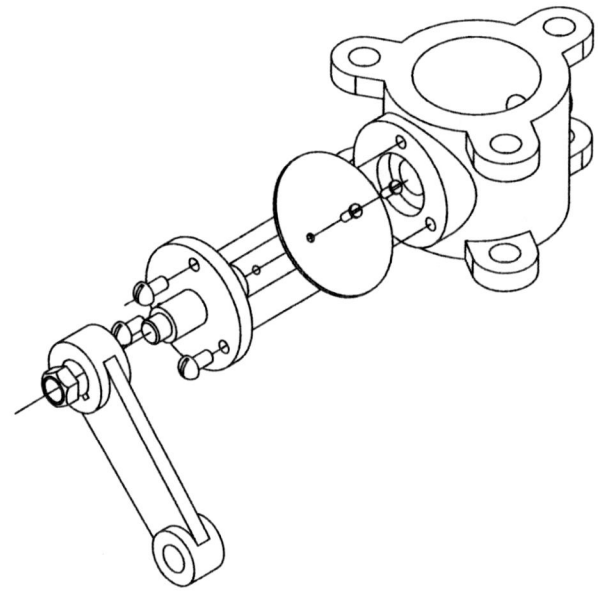

Figure 9-36 The exploded state of an assembly with offset lines

Creating the Offset Lines

To create offset lines, an exploded state should exist in which the offset lines will be created. To create the offset lines, choose **View > Explode > Offset Lines > Create** from the menu bar. The **Entity Select** menu is displayed, as shown in Figure 9-37. This menu provides you with the following options for selecting the objects for creating the offset lines.

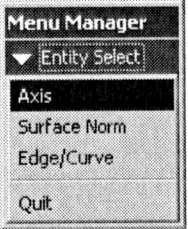

Figure 9-37 Entity Select menu

Axis

The **Axis** option is used to select the axis of the components for creating the offset lines.

Surface Norm

The **Surface Norm** option is used to create the offset lines normal to the selected surface at the specified point.

The **Edge/Curve** option is used to select an edge or a curve for creating the offset lines.

Note
The display of the offset lines is automatically turned off once you display the unexploded state of the assembly.

Modifying the Offset Lines

To modify the offset lines that are already created, choose **View > Explode > Offset Lines > Modify** from the menu bar. The **EXPL LINES MODIFY** menu is displayed, as shown in Figure 9-38. The options in this menu are discussed next.

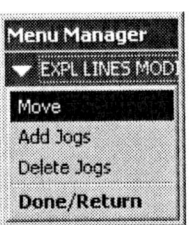

Figure 9-38 EXPL LINES MODIFY menu

Move

This option is used to move the offset line that has already been created. When you choose this option, you are prompted to select the offset line segment to modify. Select the end of the offset line and move the cursor to extend or shorten the line. When you move the cursor the offset line that is being extended is not visible in the graphics window. To pause the extension of line press the right mouse button. To end the modification of the line press the left mouse button. To abort the modification process, press the middle mouse button.

Add Jogs

This option is used to add a jog to an existing offset line segment. A jog is a break in the continuation of a line segment.

Delete Jogs

As the name of the option suggests, this option is used to delete a jog, if any, from a line segment.

Note
*To delete an offset line, choose **View** > **Explode** > **Offset Lines** > **Delete** from the menu bar. You can also set the line style for the offset lines you create in an exploded view. To set a line style for offset lines, choose **View** > **Explode** > **Offset Lines** > **Mod Line Style** from the menu bar.*

THE BILL OF MATERIAL

The Bill Of Material or the BOM is the tabular representation of all components of the assembly, along with the information associated with them. The information can be the material of the components, the additional note with the components, and so on. Pro/ENGINEER Wildfire 2.0 provides you with a ready-made BOM that can be directly utilized. This BOM is automatically updated during the assembly of the components. The BOM can be viewed using the **BOM** dialog box shown in Figure 9-39, which is displayed by choosing **Info >Bill of Materials** in the menu bar. The options in this dialog box are discussed next.

Select Model Area

The options provided under the **Select Model** area are used to select the type of assembly whose BOM has to be displayed. These options are discussed next.

Figure 9-39 The BOM dialog box

Top Level

The **Top Level** radio button is used to select the top level assembly for displaying the BOM. In this case, the subassemblies inserted in the current assembly are considered as a component.

Subassembly

The **Subassembly** radio button is used to list the components of the selected subassemblies, as individual components, in the **BOM**. When you select this option, the button available under this area will be available and you will be prompted to select the subassembly.

Include Area
Skeletons

The **Skeletons** check box is selected so that the skeletons, if any, in the assembly could also be included in the BOM.

Unplaced

The **Unplaced** check box is selected to include the unplaced components in the BOM.

Assembly Modeling

Designated Objects

The **Designated Objects** check box is selected to include the bulk items, if any, that are used in the assembly. Bulk items are items whose solid models are not created but they are used in the assembly. In Pro/ENGINEER, you can include the weight and other properties of these bulk items to calculate certain parameters related to the assembly like total weight of assembly. Examples of bulk items are paint, glue, cotton lace, and so on.

After you have selected the options in the **BOM** dialog box, choose **OK** to open the browser that displays the BOM for the selected assembly. The file of the bill of material is automatically saved with the name of the assembly. The extension of this file is *.bom*. You can later retrieve it in the **Drawing** mode to display the BOM in the drawing views of the assembly.

TUTORIALS

Tutorial 1

In this tutorial you will create all components of the Shock assembly and then assemble them, as shown in Figure 9-40. Also create an exploded state of the assembly displaying the offset lines, as shown in Figure 9-41. The BOM is shown in Figure 9-42. The dimensions of the components are shown in Figures 9-43 through 9-50.   **(Estimated time: 2 hrs)**

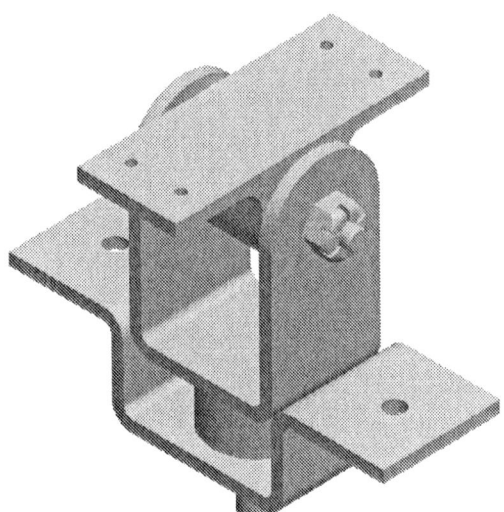

Figure 9-40 *The Shock assembly*

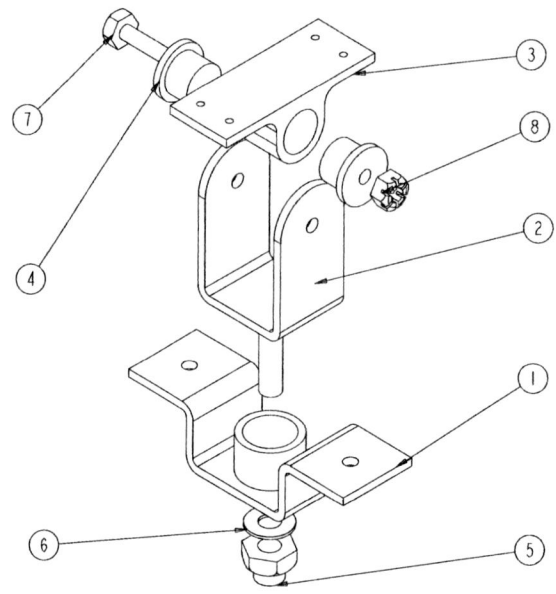

Figure 9-41 *The exploded state of the Shock assembly*

ITEM	QTY.	NAME	MATERIAL
1	1	BRACKET	STEEL
2	1	U-SUPPORT	STEEL
3	1	PIVOT	STEEL
4	2	BUSHING	BRONZE
5	1	SELF LOCKING NUT	0.625-11UNC
6	1	WASHER	USER DEFINED
7	1	HEXAGONAL BOLT	STEEL
8	1	CASTLE NUT	STEEL

Figure 9-42 *The Bill Of Material for the Shock assembly*

Assembly Modeling

9-33

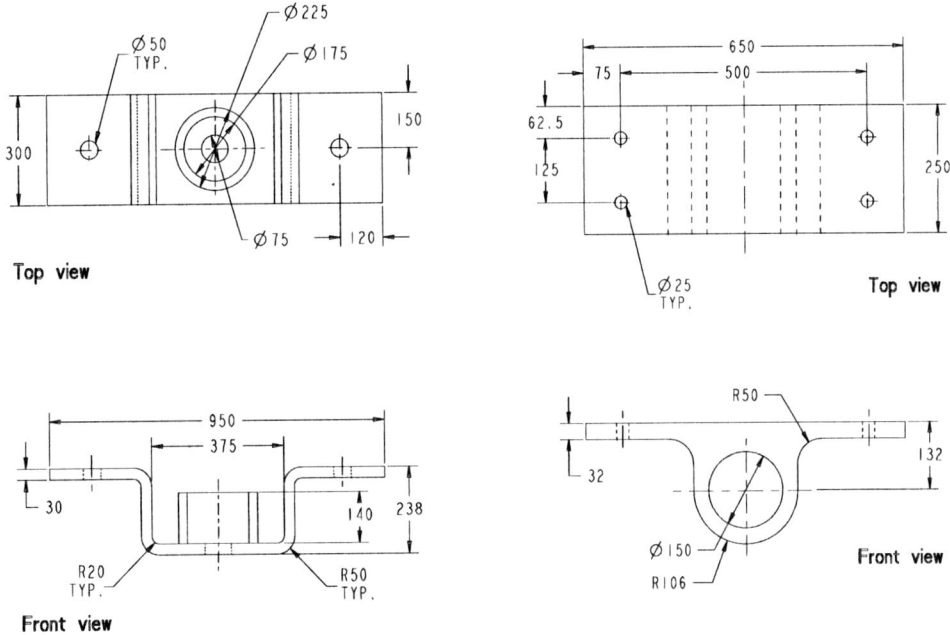

Figure 9-43 Dimensions for Bracket

Figure 9-44 Dimensions for Pivot

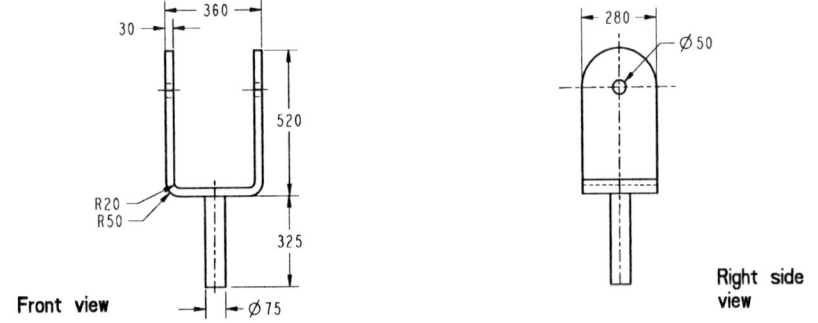

Figure 9-45 Dimensions for U-Support

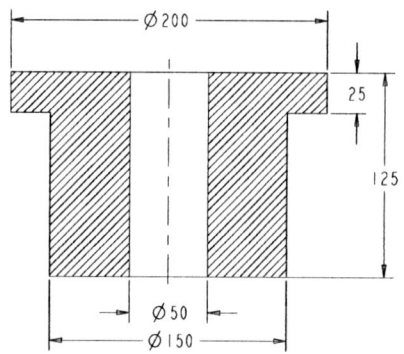

Figure 9-46 Dimensions for Bushing

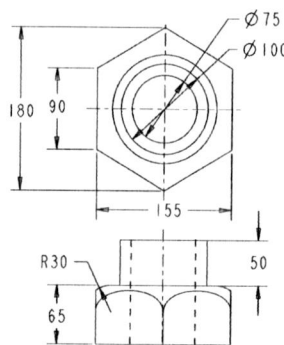

Figure 9-47 Dimensions for Self-locking nut

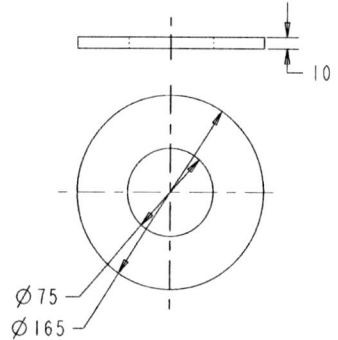

Figure 9-48 Dimensions for Washer

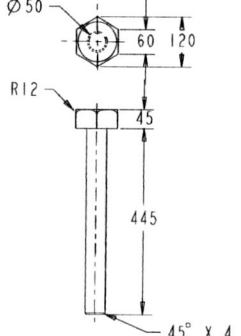

Figure 9-49 Dimensions for Hexagonal bolt

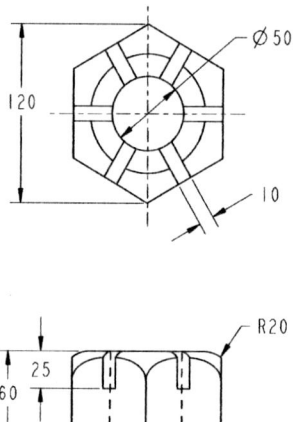

Figure 9-50 Dimensions for Castle nut

Assembly Modeling

The following steps are required to complete this tutorial:

a. Create all components of the assembly as separate part files in the **Part** mode.
b. Create a new file in the **Assembly** mode and then assemble the U-Support with the default datum planes.
c. Assemble the Bushing with the U-Support and then repeat the bushing in the assembly, refer to Figures 9-51 through 9-53.
d. Assemble the Pivot with the Bushing, refer to Figures 9-54 and 9-55.
e. Suppress the two Bushings and the Pivot.
f. Assemble the Bracket with the assembly, refer to Figures 9-56 and 9-57.
g. Assemble the Washer, Bolt, Castle nut, and Self-locking nut in the assembly, refer to Figures 9-58 through 9-60.
h. Unsuppress the suppressed components.
i. After assembling all components, create the exploded state of the assembly, refer to Figures 9-61 and 9-62.
j. Generate offset lines in the exploded state, refer to Figure 9-63.
k. Save the assembly and then exit the **Assembly** mode.

Before you start creating the components, set the working directory to *C:\ProE-WF-2.0\c09\Shockassembly*. Also, make sure that the *c09* directory exists inside the *ProE-WF-2.0* directory. If these directories do not exist, create them and set the working directory.

Starting a the Components of the Assembly

The given assembly is created by using the bottom-up approach. As mentioned earlier, in bottom-up approach assemblies, the components are created as separate files and are placed in the assembly file. Therefore, you first need to create the components of the assembly.

1. Create all components of the assembly as separate part files and save them in the current working directory.

2. Close the part files, if opened.

Creating New Assembly File

All components that you have created are assembled in an assembly file having an extension *.asm*. You need to open a new assembly file to assemble the components.

1. Choose the **Create a new object** button from the **File** toolbar to display the **New** dialog box.

2. Select the **Assembly** radio button in the **Type** area of the **New** dialog box. In the **Sub-type** area of the **New** dialog box, the **Design** radio button is selected by default. Enter the name of the assembly in the **Name** edit box as **SHOCKASSEMBLY**.

Note
*Make sure that in the **New** dialog box, the **Use default template** check box is selected. If this check box is not selected, you will not be able to use the template provided by Pro/ENGINEER. Hence, in that case you will need to create the default datum planes by choosing the **Insert a datum plane** button from the **Right Toolchest**.*

3. Choose **OK** from the **New** dialog box to proceed to the assembly environment.

Assembling the U-Support with the Default Datum Planes

When you choose **OK** from the **New** dialog box, the three default assembly datum planes are displayed in the graphics window and the **Model Tree** is displayed on the left of the graphics window. You can close the **Model Tree** by clicking on the sash available on the right edge of the **Model Tree**. Now, you can start assembling the components.

1. Choose the **Add component to the assembly** button. The **Open** dialog box is displayed.

2. Select **u-support** from the **Open** dialog box and choose the **Open** button. The **Component Placement** dialog box is displayed with the **Show Component in the assembly window while specifying constraints** button selected by default.

   Since the **Show component in the assembly window while specifying constraints** button is selected by default, the U-Support is placed with the default datum planes by assuming some assembly constraints.

   As discussed earlier, when you open a component to assemble, there are two display modes in which the component can be displayed. These two display modes are controlled by two buttons at the top in the **Component Placement** dialog box. You will use the display mode in which the component is displayed in a separate window called the **Component Window**.

   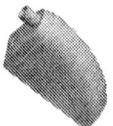 **Tip**: *Viewing the components in separate windows helps the user to locate the references for the assembly constraints easily and also the user has independent viewing control over the two windows. You can spin, pan, and zoom the components in their respective windows.*

3. Choose the **Show component in a separate window while specifying constraints** button.

   The subwindow is displayed on the left of the graphics window with the U-Support. You will notice that the component is still displayed in the graphics window. This is because the **Show component in the assembly window while specifying constraints** button is still selected in the **Component Placement** dialog box.

4. Choose the **Show component in the assembly window while specifying constraints** button again from the **Component Placement** dialog box to clear it. Now, the component is not displayed in the graphics window.

   You can drag the subwindow to a desired position so that you can view the prompt messages of both the windows. You can also use the keyboard and mouse button combinations to zoom in and zoom out the component in the two windows and to spin the models individually.

5. Choose **Assemble component at a default location** button from the **Component Placement** dialog box.

Assembly Modeling

As discussed earlier, the **Assemble component at a default location** button aligns the datum planes of the component with the assembly datum planes. The advantages of assembling the first component of the assembly with the default datum planes were discussed earlier.

6. The **Placement Status** area shows **Fully Constrained**. Choose **OK**. The **Component Placement** dialog box is closed and the U-Support is assembled with the default assembly datum planes.

Assembling the Bushing with the U-Support

The next component you need to assemble is the Bushing.

1. Choose **Add component to the assembly** button to display the **Open** dialog box.

2. Select **bushing** and choose **Open** to display the **Component Placement** dialog box. The component is displayed in the **Component Window**. You may need to drag the **Component Window** so that you can view the prompts in the **Message Area**.

3. Select the **Mate** option from the **Type** drop-down list in the **Constraints** area of the **Component Placement** dialog box to add the mate constraint. You are prompted to select a mating surface.

4. Select the planar surface of the Bushing, as shown in Figure 9-51.

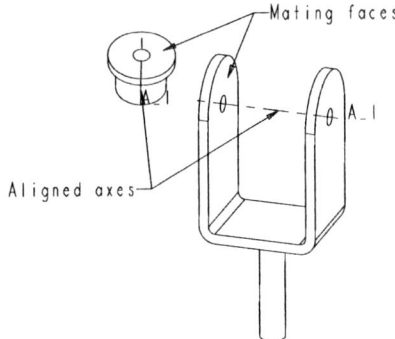

Figure 9-51 Constraints and the location to apply them

5. Select the left inner face of the U-Support, as shown in Figure 9-51, as the other face to add the mate constraint.

6. Select the **Align** option from the **Type** drop-down list in the second row to add the align constraint.

7. Select the axis of the Bushing, shown in Figure 9-51 to add the **Align** constraint.

8. Select the axis of the hole on the vertical face of the U-Support, as shown in Figure 9-51. The **Placement Status** area shows **Fully Constrained**.

   Note that the **Allow Assumptions** check box is selected. This means that Pro/ENGINEER has assumed one or more constraints to fully constrain the assembly. If you clear the **Allow Assumptions** check box, the **Partially Constrained** message is displayed in the **Placement Status** area. Therefore, let Pro/ENGINEER assume the constraints and let the **Allow Assumptions** check box be selected.

   > **Tip**: *The arrow in the **Component Placement** dialog box under the **Component Reference** area is used to make selections. If you have made a wrong selection then you can choose this arrow to display the **SELECT** dialog box and make the selection again.*

9. You can preview the assembly of the U-Support and the Bushing by choosing the **Preview** button from the **Component Placement** dialog box. The bushing is displayed as a wireframe model in the assembly. Choose **OK**. The two components are assembled, as shown in Figure 9-52.

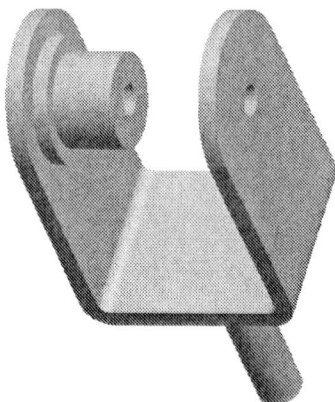

Figure 9-52 The Bushing assembled with the U-Support

10. The next instance of the Bushing can be placed directly using the **Repeat** option. To repeat Bushing in the assembly, select it from the **Model Tree** or from the graphics window and then choose **Edit > Repeat** from the menu bar. When you choose this option, the **Repeat Component** dialog box is displayed.

    In the dialog box, the **Mate** and **Align** constraints are displayed in the **Variable Assembly Refs** area.

11. Select the **Mate** constraint from this area and then choose the **Add** button.

12. Select the inner right face of the U-Support as the mating face. The copy of the Bushing will be assembled, as shown in Figure 9-53. Choose **Confirm** from the **Repeat Component** dialog box.

Assembly Modeling

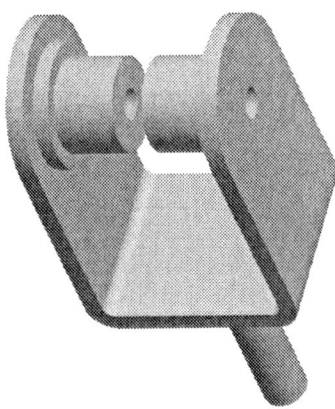

Figure 9-53 Assembled Bushings and the U-Support

Assembling the Pivot with the Bushing

The next component that you need to assemble is the Pivot.

1. Choose the **Add component to the assembly** button to display the **Open** dialog box.

2. Select **pivot** and choose **Open** to display the **Component Placement** dialog box.

3. Select the **Mate** option from the **Type** drop-down list and then select coincident from the drop-down list that is displayed when you click in the field of the offset column. This will add the mate coincident constraint.

4. Choose the select component reference button from the component reference area and then select the face of the Pivot, as shown in Figure 9-54 to add the mate constraint.

5. Select the front face of the Bushing, as shown in Figure 9-54 to add the mate constraint.

6. Select the **Align** option from the **Type** drop-down list in the second row to add the align constraint.

 **Tip**: *If the display of datum axis is turned on, the datum axis will not be displayed in the subwindow until you choose **View > Repaint** from menu bar.*

7. Select the axis of the Bushing, as shown in Figure 9-54.

8. Select the central axis of the Pivot, as shown in Figure 9-54. The **Fully Constrained** message is displayed in the **Placement Status** area of the **Component Placement** dialog box. But note that the **Allow Assumptions** check box is selected. This suggests that Pro/ENGINEER has assumed some constraints to constrain the Pivot with the Bushing.

9. Choose the **Preview** button from the **Component Placement** dialog box.

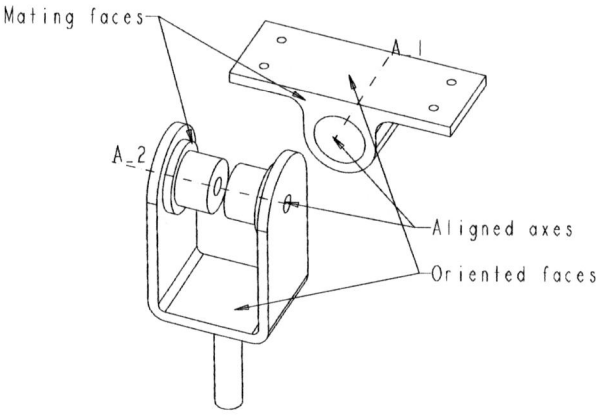

Figure 9-54 Constraints and the location to apply them

The orientation of the Pivot with the Bushing is not what is required. Therefore, you need to clear the **Allow Assumptions** check box and define one more constraint.

10. Clear the **Allow Assumptions** check box in the **Component Placement** dialog box. Now, the **Partially Constrained** message is displayed in the **Placement Status** area.

11. To add a constraint, choose the **Specify a new constraint** button from the **Constraints** area in the **Component Placement** dialog box. The **SELECT** dialog box is displayed.

12. Select the field in the **Offset** column. The field will be changed into a drop-down list. Select the **Oriented** option from the drop-down list and select the top face of the Pivot, as shown in Figure 9-54.

13. Now, select the upper horizontal face of the U-Support, as shown in Figure 9-54. The message displayed in the **Placement Status** area is **Fully Constrained**. Choose the **Preview** button from the **Component Placement** dialog box.

14. If the surface is oriented toward the bottom side, choose the **Change orientation of the constraint** button from the **Component Placement** dialog box. Choose **OK**.

The assembly after placing the Pivot should look similar to the one shown in Figure 9-55.

Suppressing the components

Next, you need to suppress both the Bushings and the Pivot. This is because when the components in the assembly increase, some component are hidden behind the others and it gets difficult to select them. Also, more the number of components, more are the datum planes. Thus, it becomes difficult to make selections on the components. However, when you suppress a component, its datum planes are also suppressed and the complications in the graphics window are reduced.

Assembly Modeling 9-41

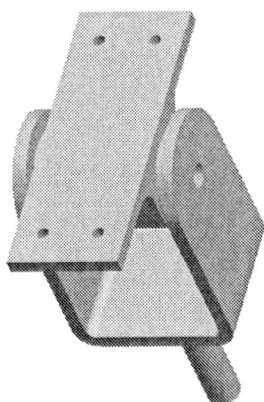

Figure 9-55 *Assembled view of the Pivot, U-Support, and the Bushing*

1. Spin the model and select the first bushing from the graphics window or from the **Model Tree** and right-click to display the shortcut menu. Choose **Suppress** from the shortcut menu to suppress the component.

   The **Suppress** window is displayed. Choose **OK**. The Pivot will also get suppressed because it was dependent on the first bushing.

2. Similarly, suppress the other Bushing to make the assembly of the other components easy.

Note
*Very often you will need to use the datum planes and the datum axes to assemble the components. Therefore, you will need to turn their display on or off from the **Datum Display** toolbar whenever required.*

Assembling the Bracket with the Assembly

The next component you need to assemble is the Bracket.

1. Choose the **Add component to the assembly** button to display the **Open** dialog box.

2. Open the **bracket** to display the **Component Placement** dialog box.

3. Select the **Mate** option from the **Type** drop-down list to add the mate constraint.

4. Select the upper face of the cylindrical feature on the Bracket as the first face to add the mate constraint and then select the bottom central horizontal face, of the U-Support as the other mating face, as shown in Figure 9-56.

5. Select the **Align** option from the **Type** drop-down list in the second row to add the align constraint.

6. Select the axis of the Bracket to add the align constraint and then select the axis of the U-Support, as shown in Figure 9-56.

7. Select the field in the **Offset** column in the third row. The field will be changed into a drop-down list. Select the **Oriented** option from the drop-down list and select the faces, as shown in Figure 9-56.

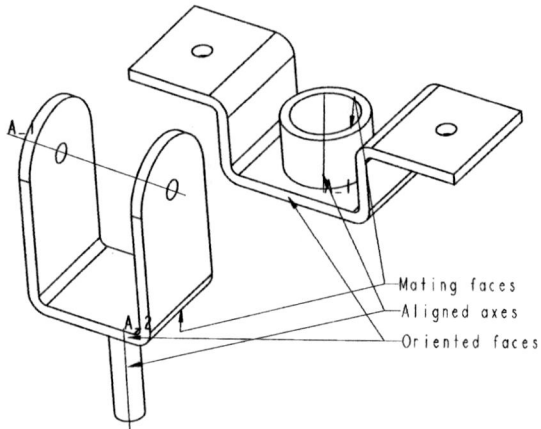

Figure 9-56 Constraints and the location to apply them

8. Choose **OK** from the **Component Placement** dialog box. The assembled Bracket is shown in Figure 9-57.

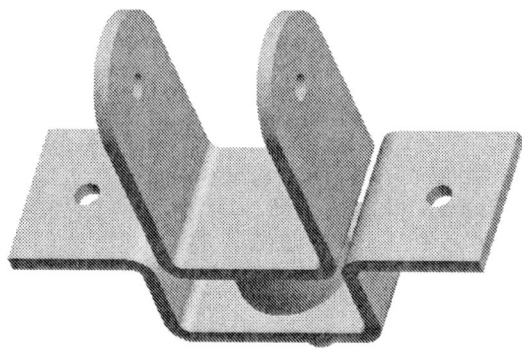

Figure 9-57 Assembled view of the Bracket with the U-Support

Note
*If another row of constraints is not displayed automatically, choose the **Specify a new constraint** button from the **Component Placement** dialog box. This will allow you to add a new constraint.*

Assembling the Washer with the Assembly

The next component that you will assemble is the Washer.

Assembly Modeling

1. Choose the **Add component to the assembly** button to display the **Open** dialog box.

2. Open the **washer** to display the **Component Placement** dialog box.

3. Select the **Mate** option from the **Type** drop-down list to add the mate constraint.

4. Select the top face of the Washer as the first face to add the mate constraint and then select the bottom face of the Bracket as the other mating face.

Note
*Sometimes when you select a reference from a component to apply a constraint, the **Message Input Window** is displayed and you are prompted to specify the offset value for the constraint applied. Enter a value of 0 in the **Message Input Window** if the offset distance is not required. Now, if the constraint changes from **Mate** to **Align**, change it back to **Mate** by selecting it from the **Type** drop-down list.*

5. Select the **Align** option from the **Type** drop-down list in the second row.

6. Select the axis of the Washer and then select the axis of the cylindrical feature on the Bracket.

7. Choose the **Cancel** button from the **SELECT** dialog box.

8. Select the **Allow Assumptions** check box. The message displayed in the **Placement Status** area is **Fully Placed**. Choose OK from the component placement dialog box.

You have defined two constraints for the Washer and the Bracket, but you need to define one more constraint that will constrain the rotatory degree of freedom of the Washer. That is why you selected the **Allow Assumptions** check box. You can also restrict the rotatory movement of the Washer by selecting a datum plane that is perpendicular to the direction of rotation of the Washer and orient it with a plane or a face parallel to the selected datum plane.

The washer is assembled to the Bracket, as shown in Figure 9-58.

Inserting the Hexagonal Bolt in the Assembly

The next component that you will assemble is the Hexagonal Bolt.

1. Choose the **Add component to the assembly** button to display the **Open** dialog box.

2. Open the **hexagonal bolt** to display the **Component Placement** dialog box.

3. Select the **Mate** option from the **Type** drop-down list to add the mate constraint.

4. Select the inner face of the head of the Hexagonal bolt and then select the outer left face of the U-Support as the other mating face. Enter 0 if you are prompted to specify offset.

Figure 9-58 Assembled view of the Washer with the Bracket

5. Select the **Align** option from the **Type** drop-down list in the second row.

6. Select the axis of the Hexagonal bolt and then select the axis of the hole feature on the U-Support.

7. Clear the **Allow Assumptions** check box and then choose the **Specify a new constraint** button. Select the field in the **Offset** column. The field will be changed into a drop-down list. Select the **Oriented** option from the drop-down list.

8. Out of six faces of the head of the Hexagonal bolt, select any one face and then select the top face of the Bracket. The **Placement Status** area shows **Fully Constrained**. Choose **OK**. The Hexagonal bolt is inserted in the assembly, as shown in Figure 9-59.

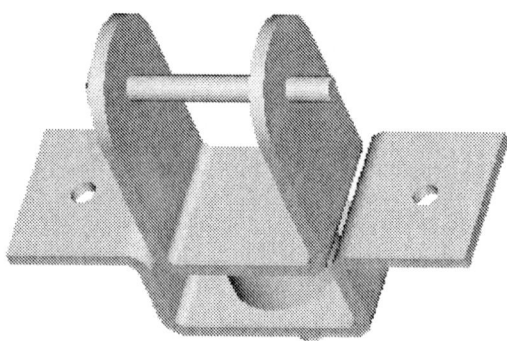

Figure 9-59 Assembly after assembling the hexagonal bolt

Assembling Other Components

Similarly, assemble the Castle nut and the Self-locking nut.

Assembly Modeling

Figure 9-60 Completed assembly with all components

Unsuppressing the Components

All components are assembled at their required positions. Now, you need to unsuppress the components that were suppressed.

1. Choose **Edit** > **Resume** > **All** from the menu bar. The suppressed components appear in the assembly where they were assembled. The completed assembly is shown in Figure 9-60.

Creating the Exploded State of the Assembly

To clearly view all components in an assembly, you need to create an exploded state of the assembly.

 **Tip**: *You can also view the default exploded state of an assembly by choosing* **View** > **Explode** > **Explode View** *from the* **View** *menu in the menu bar.*

1. Choose the **Start the view manager** button from the **Top Toolchest**; the **View Manager** dialog box is displayed. Choose the **Explode** tab and right-click in the Names list area of dialog box to invoke the shortcut menu. Choose the **New** option, name it **EXP1**, and press the middle mouse button.

2. Right-click on the **EXP1** in the dialog box and choose the **Redefine** option. The **MOD EXPLODE** menu is displayed. Choose the **Position** option from the menu; the **Explode Position** dialog box is displayed. The **SELECT** dialog box is displayed and you are prompted to select an axis or a straight edge as the motion reference.

   Note that the **Translate** radio button is selected by default in the **Motion Type** area of the **Explode Position** dialog box. The **Entity/Edge** option is selected by default in the **Motion Reference** drop-down list. The **Smooth** option is selected by default in the **Translation** drop-down list. These are the default settings and you will use these settings to create the exploded state.

3. In the **Filter** drop-down list available in the status bar that is below the **Message Area**, select the **Axis** filter and then select the axis of the cylindrical feature on the Bracket to specify the direction for the motion of all components.

4. After selecting the axis, select the Self-locking nut and move the cursor down to move the component. Note that the Self-locking nut is moving along its axis only. Place the Self locking nut at an appropriate position by pressing the left mouse button, refer to Figure 9-61. You are prompted to select another component to move.

5. One by one select the Washer and Bracket to move in the direction, as shown in Figure 9-61.

   Next, you need to move the Castle nut, the two Bushings, the Hexagonal bolt, and the Pivot together as one component along the vertical axis selected earlier.

6. Choose the **Preferences** button from the **Explode Position** dialog box. The **Preferences** dialog box is displayed. Select the **Move Many** radio button and choose **Close**.

7. Press and hold CTRL key down and select the Castle nut, the two Bushings, the Hexagonal bolt, and the Pivot. The selected components turn red in color. You may need to spin the assembly using the middle mouse button to make selections.

8. After selecting the above-mentioned components, choose the **Default Orientation** option from the **Saved view list** button. The assembly orients to its default orientation.

9. Now press the middle mouse button to exit the selection mode. Press the left mouse button on the graphics window and move the cursor upward to move the selected components, as shown in Figure 9-61. After moving the components to the required location, press the left mouse button to place them at that location.

Figure 9-61 Creating the exploded state

Assembly Modeling 9-47

10. After moving all components to the desired positions, choose **OK** from the **Explode Position** dialog box. The **MOD EXPLODE** menu is redisplayed.

11. Choose the **Position** option from the submenu. The **Explode Position** dialog box is displayed. Select the axis of the Hexagonal bolt and move Bushings, Hexagonal bolt, and Castle nut, as shown in Figure 9-62.

Figure 9-62 *Exploded state of the assembly*

12. After moving the components, choose **OK** from the **Explode Position** dialog box. The **MOD EXPLODE** submenu is redisplayed.

13. Choose the **Offset Lines** option from the **MOD EXPLODE** menu. The **OFFSET LINES** submenu is displayed. Choose the **Create** option from the **OFFSET LINES** submenu. The **Entity Select** submenu is displayed and the **Axis** option is selected by default.

14. Select the axis of the Self-locking nut and then select the axis of the shaft feature on the U-Support. The cyan-colored offset line is created.

15. Similarly, select the axes of the other components to create the offset lines displaying the path and direction of the mating components, as shown in Figure 9-63.

16. After creating all offset lines, choose **Done/Return** from the **OFFSET LINES** submenu and then choose **Done/Return** from the **MOD EXPLODE** menu. Choose the **Close** button to exit the **View Manager** dialog box.

Tip: *To unexplode the assembly, choose the View > Explode > Unexplode View from the View menu in the menu bar.*

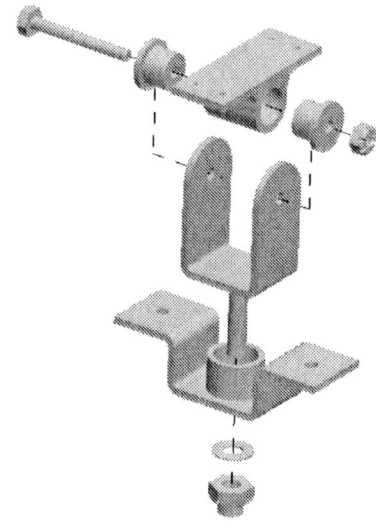

Figure 9-63 *Displaying the offset lines in the exploded state*

Saving the Assembly

1. Choose the **Save the active object** button from the **File** toolbar and save the model.

Closing the Window

Now, you have saved the assembly and the window can be closed.

1. Choose the **Close** option from the **Window** menu in the menu bar.

Tutorial 2

In this tutorial you will create all components of the Pedestal Bearing assembly and then assemble them, as shown in Figure 9-64. You will also create the exploded state, shown in Figure 9-65, displaying the offset lines. The dimensions of the components are shown in Figures 9-66 through 9-69. The BOM is shown in Figure 9-70.   **(Estimated time: 2 hrs)**

Figure 9-64 *Assembly of the Pedestal Bearing*

Assembly Modeling

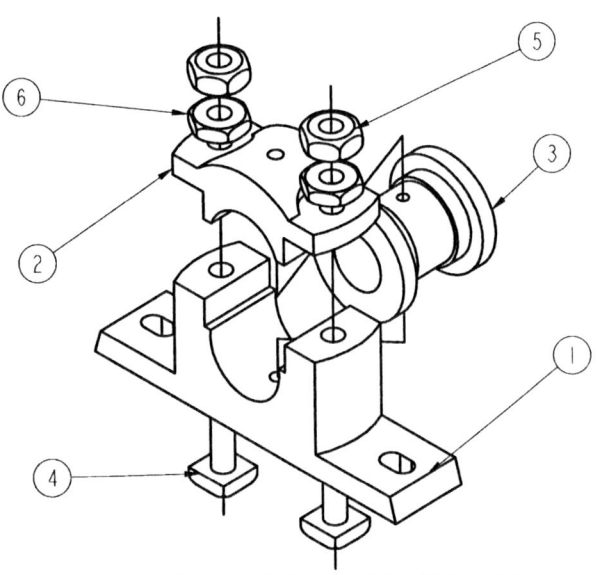

Figure 9-65 The exploded state of the Pedestal Bearing

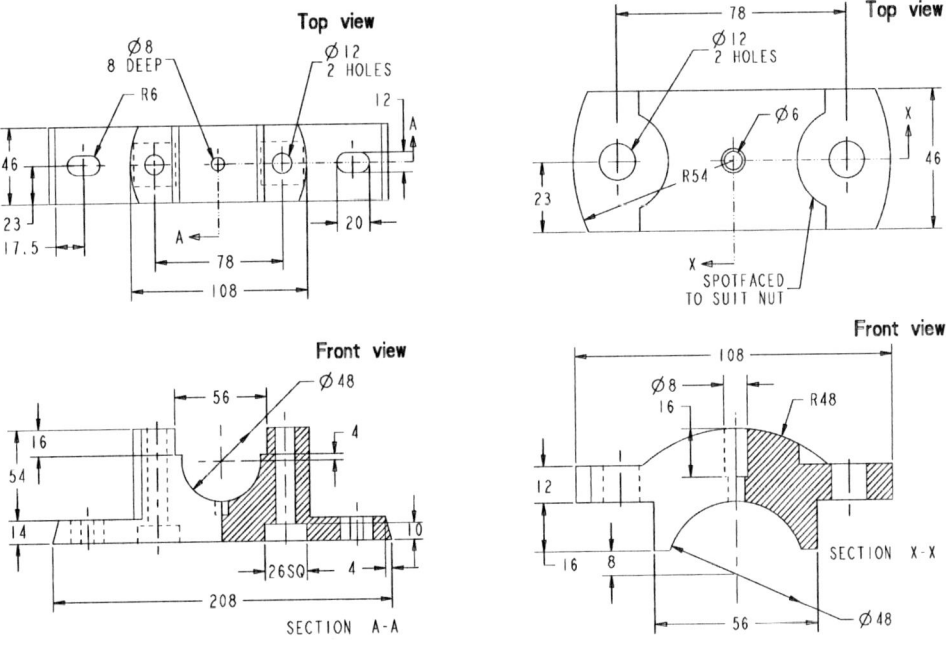

Figure 9-66 Dimensions for Casting

Figure 9-67 Dimensions for Cap

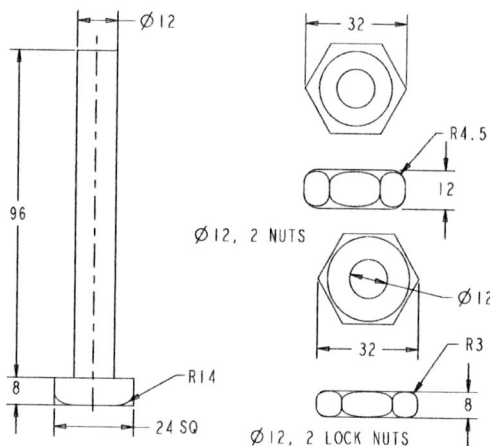

Figure 9-68 *Dimensions for the components of the assembly*

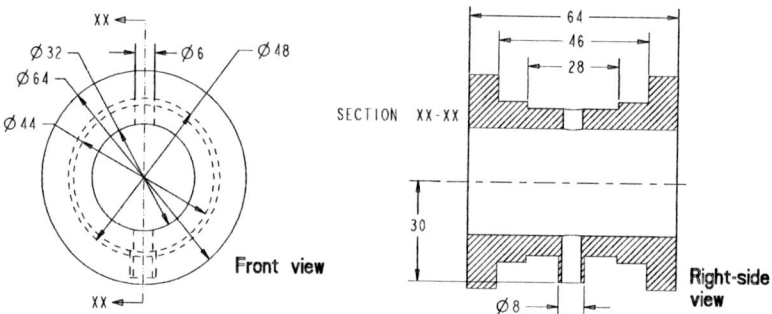

Figure 9-69 *Dimensions for the Brasses*

S. NO.	NO. OFF	PART'S NAME	MATERIAL
1	1	BODY	CAST IRON
2	1	CAP	CAST IRON
3	1	BRASSES	GUN METAL
4	2	SQUARE HEADED BOLT	MILD STEEL
5	2	NUT	MILD STEEL
6	2	LOCK NUT	MILD STEEL

Figure 9-70 *The Bill Of Material for Pedestal Bearing*

Assembly Modeling

The following steps are required to complete this tutorial:

a. Create all components of the assembly as separate part files in the **Part** mode.
b. Create a new file in the **Assembly** mode and then assemble the Casting with the default assembly datum planes.
c. Assemble the Cap with the Casting, refer to Figures 9-71 and 9-72.
d. Suppress the Cap from the assembly.
e. Assemble the Brasses with the Casting, refer to Figures 9-74 and 9-74.
f. Insert Square headed bolt, nut, and lock nut in the assembly and assemble them, refer to Figure 9-75.
g. Resume the suppressed components.
h. Create the exploded state of the assembly, refer to Figure 9-76.
i. Save the assembly and close the file.

Before you start creating the components, set the working directory to *C:\ProE-WF-2.0\c09\Pedestalbearing*.

Starting a the Components for Assembly

To create the assembly, all components must be created first. All components will be created in the **Part** mode. You will use the bottom-up approach of assembly creation.

1. Create all components of the assembly as separate part files and save them in the current working directory.

2. Close the part files, if it is open.

Creating New Assembly File

As mentioned earlier, all components that you have created are assembled in an assembly file that has an extension *.asm*. Therefore you need to open a new *.asm* file.

1. Choose the **Create a new object** button from the **File** toolbar to display the **New** dialog box.

2. Select the **Assembly** radio button in the **Type** area of the **New** dialog box. In the **Sub-type** area of the **New** dialog box the **Design** radio button is selected by default. Enter the name of the assembly in the **Name** edit box as **PEDESTALBEARING**.

3. Choose **OK** to proceed to the assembly modeling environment.

Assembling the Casting with the Default Datum Planes

In the new assembly file, the three default assembly datum planes are displayed in the graphics window and the **Model Tree** is displayed on the left of the graphics window. If the display of the **Model Tree** was turned off in the previous tutorial, it will not appear. Now, you can start assembling the components.

1. Choose the **Add component to the assembly** button to display the **Open** dialog box.

2. Select **casting** from the **Open** dialog box and choose the **Open** button. The **Component Placement** dialog box is displayed with the **Show component in a separate window while specifying constraints** button chosen by default.

Since the **Show component in a separate window while specifying constraints** button was selected in the previous tutorial, the Casting is placed in the **Component Window**. In this tutorial, you will use the display mode in which the component is displayed with the assembly in the same window.

Note

*If the datum planes are displayed in the graphics window, you need to turn their display off by choosing the **Datum planes on/off** button from the **Datum Display** toolbar. This is because the datum planes clutter the graphics window and then the selection of the references for applying assembly constraints becomes difficult. You can turn their display on when the datum planes are required.*

3. Choose the **Show component in the assembly window while specifying constraints** button. Now, choose the **Show component in a separate window while specifying constraints** button to clear it.

4. Select the **Assemble component at a default location** button from the **Component Placement** dialog box.

5. The **Placement Status** area shows **Fully Constrained**. Choose **OK**. The **Component Placement** dialog box is closed and the Casting is assembled with the default datum planes.

Assembling the Cap with the Casting

First the Cap will be assembled with the Casting.

1. Choose the **Add component to the assembly** button to display the **Open** dialog box.

2. Select the **cap** and choose **Open** to display the **Component Placement** dialog box.

3. Select the **Mate** option from the **Type** drop-down list. You are prompted to select a mating surface.

4. Select the top face of the Casting as the first mating face. Now, use the CTRL+ALT+middle mouse button to spin the Cap and then select the lower flat face of the Cap as the other mating face, as shown in Figure 9-71.

The CTRL+ALT+middle mouse button is used to spin the component to be assembled with the assembly and the CTRL+ALT+left mouse button is used to move the component. However, to spin the assembly you can use the middle mouse button.

5. Enter a value of **4** in the **Message Input Window**. If this window is not displayed, click on the field in the **Offset** column list and select **0.0**. Now, enter offset distance of **4** and press ENTER.

Assembly Modeling

6. Choose the **Specify a new constraint** button and then select the **Align** option from the **Type** drop-down list.

7. Select the axis of the left hole on the Casting as the first aligning axis, and then select the axis of the left hole on the Cap as the second aligning axis, as shown in Figure 9-71.

   Note that in the **Component Placement** dialog box, the **Allow Assumptions** check box is selected and the Cap is assembled with the Casting. The status displayed in the **Placement Status** area is **Fully Constrained**. This means that Pro/ENGINEER assumed some constraints and has assembled the two components. However, while assembling the previous component if you cleared the **Allow Assumptions** check box, then it will not be selected and the status will be **Partially Constrained**. You will not accept the default constraints. You will define one more constraint to assemble the two components.

8. Clear the **Allow Assumptions** check box, is selected. The status displayed in the **Placement Status** area is **Partially Constrained**.

   The Cap is still assembled with the Casting. To find out what degree of freedoms of the Cap is not constrained with the Casting, perform the next step.

9. Use CTRL+ALT+middle mouse button to move the Cap. You will notice that the cap rotate around the central axis of the left hole. Therefore the second hole of both the Cap and the Casting needs to be aligned.

10. If the **Automatic** option is not displayed in the **Type** drop-down list, choose **Specify a new constraint** button from the **Constraints** area in the **Component Placement** dialog box. The **SELECT** dialog box is displayed. Select the **Align** option from the **Type** drop-down list.

11. Align the axis of the right hole on the Casting with the axis of the right hole on the Cap by selecting them, as shown in Figure 9-71. The **Placement Status** area shows **Fully Constrained**. Choose **OK** to exit the **Component Placement** dialog box.

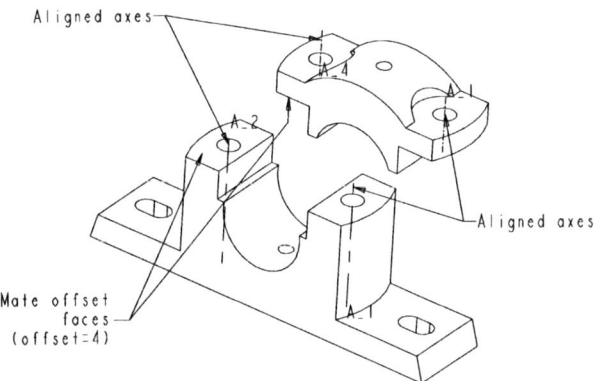

Figure 9-71 Constraints used for assembling the components

12. After assembling the Cap, the assembly should look similar to the one shown in Figure 9-72.

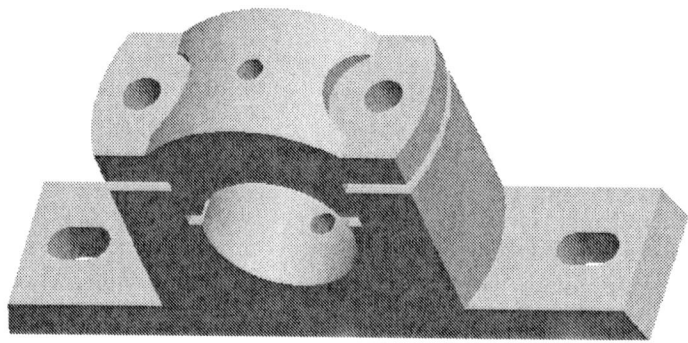

Figure 9-72 After assembling the Cap with the Casting

Suppressing the Cap from the Assembly

You need to suppress the Cap because the next component has to be assembled with the Casting. When you suppress the Cap from the assembly, it becomes easier to assemble the new component in the assembly. The Cap will be unsuppressed later.

1. Select the Cap from the assembly in the graphics window and press and hold the right mouse button to display the shortcut menu.

2. Choose the **Suppress** option. You are prompted to confirm the suppression. Choose **OK**.

Assembling the Brasses with the Casting

Next, you need to assemble the Brasses with the Casting.

1. Choose the **Add component to the assembly** button to display the **Open** dialog box.

2. Select **brasses** and choose **Open** to display the **Component Placement** dialog box.

3. Select the **Mate** option from the **Type** drop-down list and mate the face of the Brasses with that of the Casting, as shown in Figure 9-73.

4. Select the **Insert** option from the **Type** drop-down list and insert the oil hole of the Brasses into that of the Casting, as shown in Figure 9-73.

5. Use CTRL+ALT+left mouse button to move the Brasses, if it is inside the casting. Select the **Align** option from the **Type** drop-down list.

6. Select the **TOP** datum plane from the Brasses and then select the face of the Casting shown in Figure 9-73. The datum of the Brasses is the datum that goes through the center of Bushings. The red arrow is displayed and you are prompted to enter the offset distance in the direction of the arrow. If you are not prompted to enter the offset distance, select **0.0** from the drop-down list that is displayed when you click on the field in the offset column.

Assembly Modeling

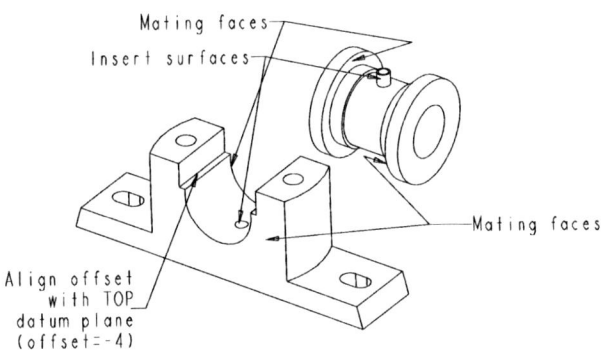

Figure 9-73 Constraints used for assembling the components

7. Enter a distance of **-4** as the offset distance and press ENTER. The **Placement Status** area shows **Fully Constrained**.

8. If the join feature of the oil hole is at the top then choose the **Change orientation of the constraint** button from the **Component Placement** dialog box. The oil hole is now oriented correctly. Choose **OK** to exit the **Component Placement** dialog box.

   After assembling the Brasses, the assembly should look similar to the one shown in Figure 9-74.

Figure 9-74 Inserting the Brasses in the assembly

9. Similarly, assemble the remaining components and then resume the Cap. The final assembly is shown in Figure 9-75.

Creating the Exploded State of the Assembly

To view all components in an assembly clearly, you need to create an exploded state of the assembly.

Figure 9-75 The final Pedestal Bearing assembly

1. Choose the **Start the view manager** button from the **Top Toolchest**. The **View Manager** dialog box is displayed. Choose the **Explode** tab and right-click in the Names list box to invoke the shortcut menu. Choose the **New** option, name it **EXP1**, and press the middle mouse button.

2. Right-click on **EXP1** in the dialog box and choose the **Redefine** option. The **MOD EXPLODE** menu is displayed. Choose the **Position** option from the menu; the **Explode Position** dialog box is displayed. The **SELECT** dialog box is displayed and you are prompted to select an axis or edge as the motion reference.

   Note that the **Translate** radio button is selected by default in the **Motion Type** area of the **Explode Position** dialog box. The **Entity/Edge** option is selected by default in the **Motion Reference** drop-down list. The **Smooth** option is selected by default in the **Translation** drop-down list. You will use these default settings to create the exploded state.

3. In the **Filter** drop-down list available in the status bar that is below the **Message Area**, select the **Axis** filter and then select the central axis of the Square headed bolt to specify the direction of motion of the components and then move the Square headed bolts, Nuts, Cap, Lock Nuts, and Casting, as shown in Figure 9-76.

4. After moving the above components along the axis of the Square headed bolt, choose the arrow button in the **Motion Reference** area.

5. Select the axis of the Brasses and then move it, as shown in Figure 9-76.

6. After moving the components, choose **OK** from the **Explode Position** dialog box. The **MOD EXPLODE** menu is displayed.

7. Choose the **Offset Lines** option from the **MOD EXPLODE** menu. The **OFFSET LINES** submenu is displayed.

Assembly Modeling

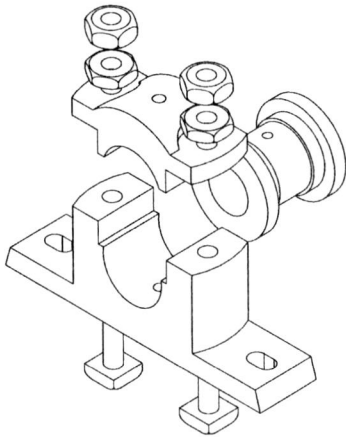

Figure 9-76 The Exploded Pedestal Bearing assembly

8. Choose the **Create** option from the **OFFSET LINES** submenu. The **Entity Select** submenu is displayed and the **Axis** option is selected by default.

9. Select the axes of the two Lock Nuts and the Square headed bolts to create the offset lines, as shown in Figure 9-77.

10. Now, spin the model and select the axis of the oil hole on the Cap from the lower curved surface and then select the axis of the upper oil hole on the Brasses to create the offset lines between them. Similarly, create the offset lines between the lower oil hole on the Brasses and that on the Casting, as shown in Figure 9-77.

11. The final assembly after creating the offset lines in the exploded state should look similar to the one shown in Figure 9-77.

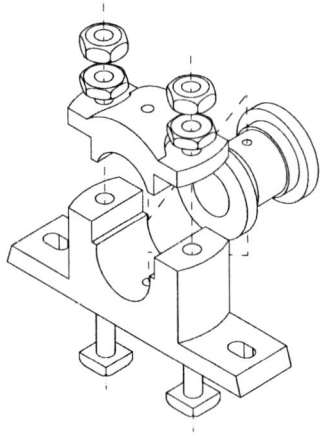

Figure 9-77 Exploded state of the assembly displaying the offset lines

12. After creating all offset lines, choose **Done/Return** from the **OFFSET LINES** submenu and then choose **Done/Return** from the **MOD EXPLODE** menu. Choose the **Close** button to exit the **View Manager** dialog box.

Note
You can delete or modify an offset line by using the Delete and Modify options respectively from the OFFSET LINES submenu.

Saving the Assembly
1. Choose the **Save the active object** button from the **File** toolbar and save the assembly.

Closing the Window
Now, you have saved the assembly and the window can be closed.

1. Choose **Window >Close** option from the menu bar to close the file.

Self-Evaluation Test

Answer the following questions and then compare your answers with those given at the end of this chapter.

1. When you select a datum plane for applying the mate constraint then you will have to specify whether you want to select the cream side or the brown side of the datum plane to be coplanar with the other face or datum plane. (T/F)

2. The exploded view is a state in which all components are moved from their actual location so that they are visible. (T/F)

3. The dimensions of a feature cannot be modified in the **Assembly** mode. (T/F)

4. You can display the unexploded state of the assembly by choosing **Unexplode** from the **View** menu. (T/F)

5. The offset lines are automatically made invisible once you display the unexploded state of the assembly. (T/F)

6. In the _____ type of simplified representation the selected component will be displayed as shaded.

7. The _____ type of simplified representation reduces the regeneration time in case of large assemblies.

8. The _____ constraint is similar to the **Mate** constraint, with the only difference being that this constraint allows you to specify some offset distance between the two coplanar faces.

9. The _____ constraint is used to assemble the circular components.

Assembly Modeling   9-59

10. The _____ is the tabular representation of all components of the assembly along with the information associated with them.

Review Questions

Answer the following questions:

1. Which constraint is used to align the selected datum point or vertex on the first part with the selected surface or datum plane on the second part?

    (a) **Orient**
    (c) **Edge on Srf**
    (b) **Pnt on Srf**
    (d) **Tangent**

2. When the assembly is constrained fully the message displayed in the **Placement Status** area of the **Component Placement** dialog box is

    (a) **Partially constrained**
    (c) **Completely constrained**
    (b) **Fully constrained**
    (d) None of the above

3. In which of the following types of simplified representation the complete geometry of the components is displayed in the current model display style?

    (a) Geometry representation
    (c) Master representation
    (b) Graphics representation
    (d) None of the above

4. Which constraint is used to assemble two components by making the selected faces or planes coplanar such that the mating faces or planes are facing in the same direction?

    (a) **Insert**
    (c) **Tangent**
    (b) **Orient**
    (d) **Align**

5. The unexploded state of an assembly can be displayed by choosing **Unexplode** from which of the following menu in the menu bar?

    (a) **View**
    (c) **Info**
    (b) **File**
    (d) **Window**

6. A component of an assembly that is assembled using the assembly constraints cannot be replaced with some other component. (T/F)

7. A Part of an assembly can be made hidden from the graphics window. (T/F)

8. Offset lines can only be created in the exploded state of an assembly. (T/F)

9. The order of assembling the components in an assembly can be changed. (T/F)

10. Pro/ENGINEER provides you with a ready-made BOM that can be directly used. (T/F)

Exercise

Exercise 1

In this exercise you will create all components of the Crosshead assembly and then assemble them, as shown in Figure 9-78. Also you will create an exploded state, shown in Figure 9-79, displaying the offset lines. The dimensions of the components are shown in Figures 9-80 through 9-84.

(Estimated time: 2 hrs)

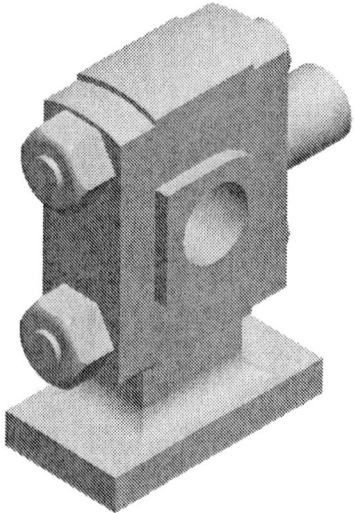

Figure 9-78 The Crosshead assembly

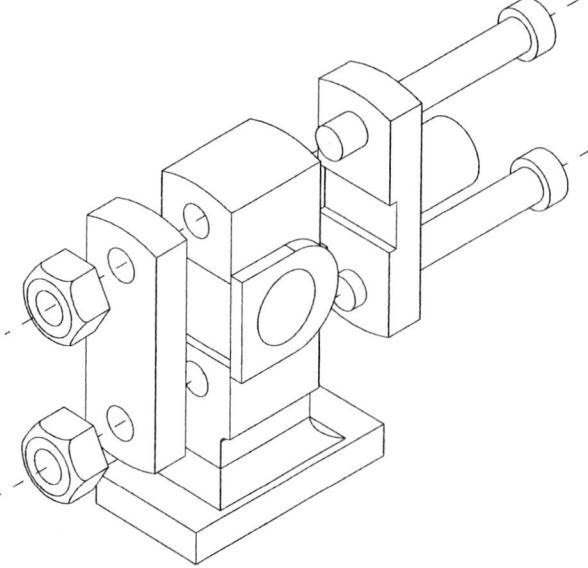

Figure 9-79 The exploded state of Crosshead assembly

Assembly Modeling

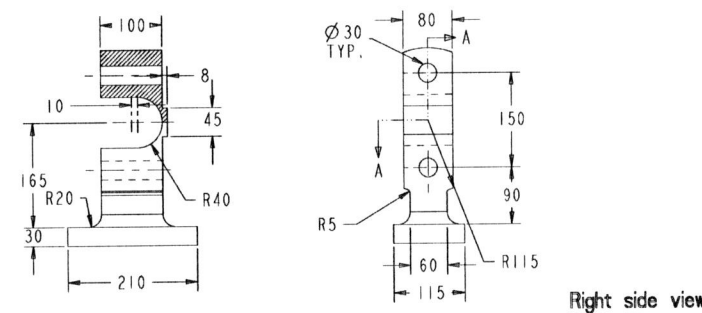

Figure 9-80 Front view and the right-side view of the Body

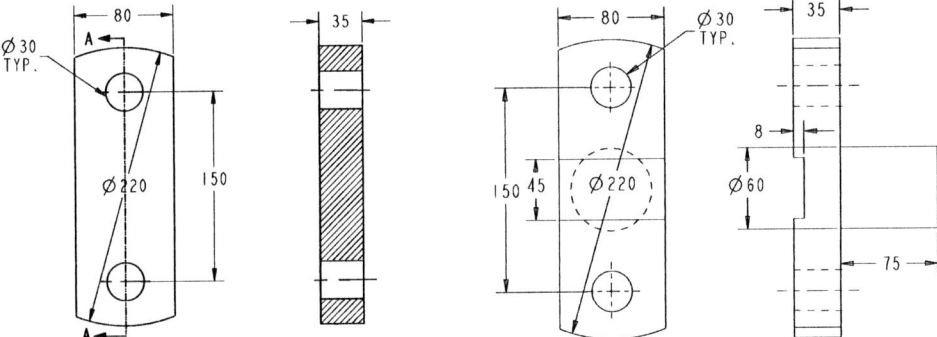

Figure 9-81 Dimensions of Keep Plate

Figure 9-82 Dimensions of Piston Rod

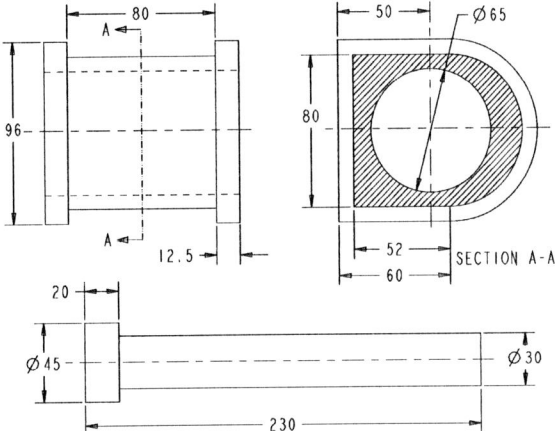

Figure 9-83 Dimensions of the Brasses and Bolt

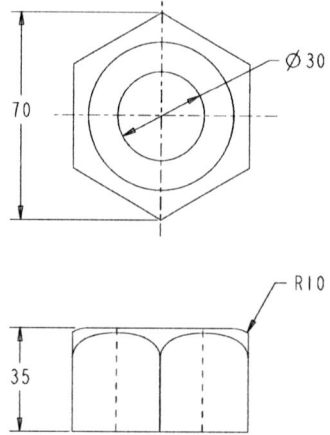

Figure 9-84 Dimensions of Nut

Answers to the Self-Evaluation Test
1 - F, **2** - T, **3** - F, **4** - T, **5** - T, **6** - Master representation, **7** - Graphics representation, **8** - Mate offset, **9** - Insert, **10** - BOM.

Chapter 10

Generating, Editing, and Modifying the Drawing Views

Learning Objectives

After completing this chapter you will be able to:
• *Create and retrieve the drawing sheet formats.*
• *Generate different drawing views of an existing part.*
• *Edit the existing drawing views and parameters associated with the views.*
• *Modify the existing drawing views.*

THE DRAWING MODE

In the previous chapters you have learned about creating the parts in the **Part** mode and assembling different parts in the **Assembly** mode. In this chapter you will learn to generate the drawing views of the parts and assemblies created earlier. Drawing views are generated in the **Drawing** mode. One of the major advantages of working with this software package is its bidirectional associative nature. This is the property that ensures that if any modifications are made in the model in the **Part** mode, its drawing views are updated automatically and if any modifications are made in the dimensions of the model in the drawing views, the model gets updated automatically in the **Part** and the **Assembly** mode. In Pro/ENGINEER Wildfire 2.0 there are two types of drafting methods: Interactive drafting and Generative drafting. In this chapter you will learn Generative drafting. To generate the drawing views first you need to start a new file in the **Drawing** mode of Pro/ENGINEER Wildfire 2.0.

Choose the **Create a new object** button from the **File** toolbar to display the **New** dialog box. Select the **Drawing** radio button in the **New** dialog box, as shown in Figure 10-1.

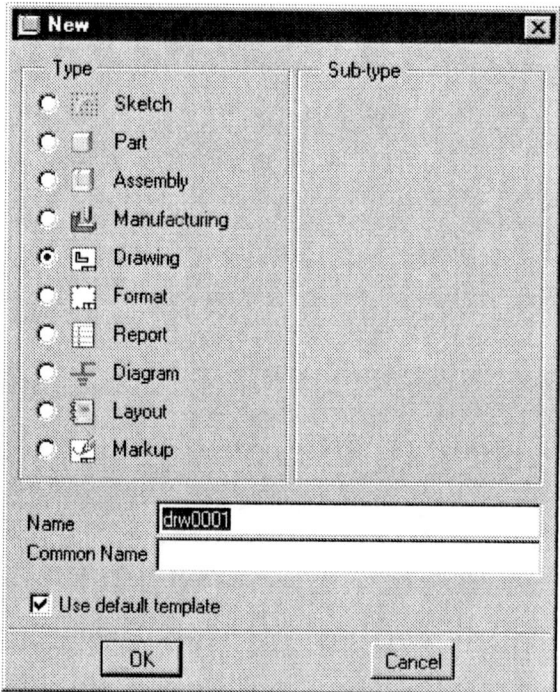

Figure 10-1 The New dialog box

Specify the name of the drawing in the **Name** edit box and then choose **OK** to display the **New Drawing** dialog box, as shown in Figure 10-2.

New Drawing Dialog Box

The **New Drawing** dialog box is used to specify the template that will be used while starting a new file in the **Drawing** mode. The option available in this dialog box are discussed next.

Default Model Area

The **Default Model** area is used to specify the name of the model whose drawing views you want to generate. You can specify the name of the model in the **Name** edit box or select the model using the **Open** dialog box that is displayed when you choose the **Browse** button. If a part or assembly file is already opened in the current session then the name of that model is displayed by default in the **Name** edit box of the **Default Model** area.

Specify Template Area

The **Specify Template** area is used to specify whether you want to use the default templates available in Pro/ENGINEER, use predefined formats, or use an empty sheet. There are three radio buttons in this area. The **Use template** radio button is selected by default. When you select the **Empty with format** radio button, the dialog box changes, as shown in Figure 10-3.

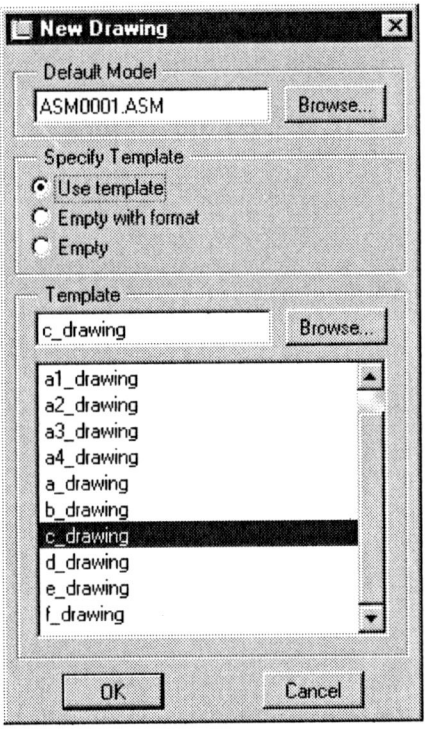
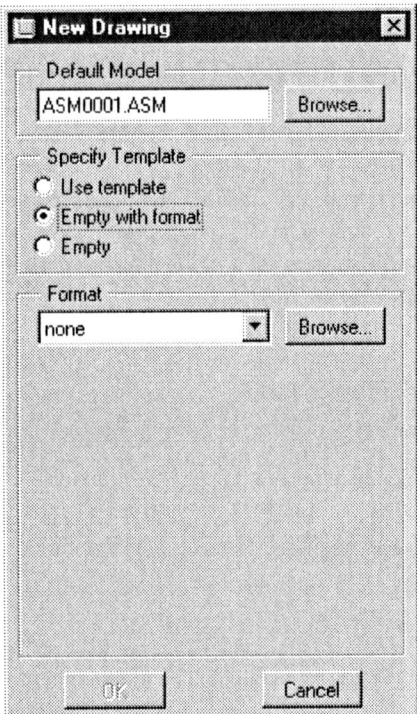

Figure 10-2 The **New Drawing** dialog box with the **Use template** radio button selected

Figure 10-3 The **New Drawing** dialog box with the **Empty with format** radio button selected

When you select the **Empty** radio button, the **New Drawing** dialog box is modified, as shown in Figure 10-4.

Orientation Area

The **Orientation** area is available only when you select the **Empty** radio button from the **Specify Template** area. The buttons provided in this area are used to specify the orientation of the sheet. You can select a standard size sheet with a portrait or a landscape orientation using the **Portrait** or the **Landscape** button. You can also specify a sheet with the user-defined size by choosing the **Variable** button.

Size Area

The options available in the **Size** area are used to set the size and the units of the sheet. The **Size** area is available only when you select the **Empty** radio button from the **Specify Template** area. The options available in this area are discussed next.

Standard Size

This drop-down list is used to select a drawing sheet of standard size. This drop-down list is available only when you select the **Portrait** or **Landscape** button from the **Orientation** area.

Inches/Millimeters

These radio buttons are selected to set the standards for the user-defined sheets. You can set the size of the sheet in inches or in millimeters. These buttons are available only when you select the **Variable** button from the **Orientation** area.

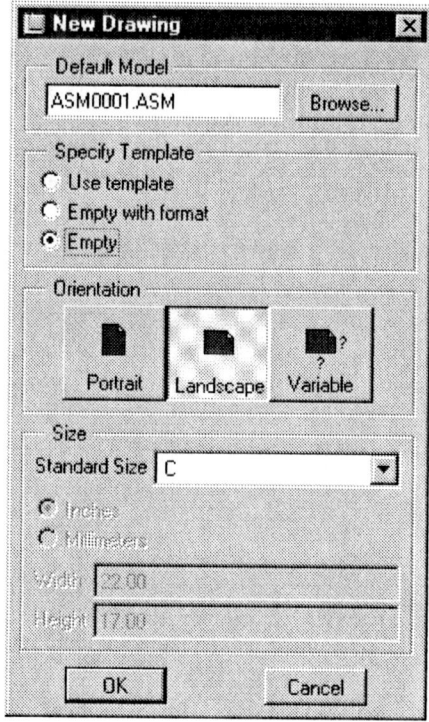

*Figure 10-4 The **New Drawing** dialog box with the **Empty** radio button selected*

Width/Height

These edit boxes are used to specify the width and the height of the user-defined drawing sheets. These edit boxes are available only when you select the **Variable** button from the **Orientation** area.

Format Area

The **Format** area is available only when you select the **Empty with format** radio button from the **Specify Template** area.

Format

The **Format** drop-down list is used to select the available formats.

Browse

The **Browse** button is chosen to display the **Open** dialog box for retrieving the drawing formats. By default, there are only eight standard system formats that can be retrieved. However, you can create your own user-defined formats that can be retrieved later.

Choose **OK** from the **New Drawing** dialog box to proceed to the **Drawing** mode. A drawing sheet of the specified size and orientation will be placed on which you can now generate the drawing views.

Tip: *If you have not specified any model in the **Default Model** area of the **New Drawing** dialog box and you choose **OK** to enter the Drawing mode. When you choose the **Create a general view** button from the **Top Toolchest**, the **Open** dialog box is displayed and you can select the model from the **Open** dialog box.*

*Pro/ENGINEER automatically selects the model in the **Default Area** of the **New Drawing** dialog box. This model is selected from the in session memory of Pro/ENGINEER. If the model that is selected automatically is not the desired model then change the model by selecting the **Browse** button in the **New Drawing** dialog box.*

GENERATING DRAWING VIEWS

In Pro/ENGINEER Wildfire 2.0, the first view that you need to generate is the general view. This view mostly acts as the parent view for the remaining views. The method of generating various views in Pro/E Wildfire 2.0 is discussed next.

Generating the General View

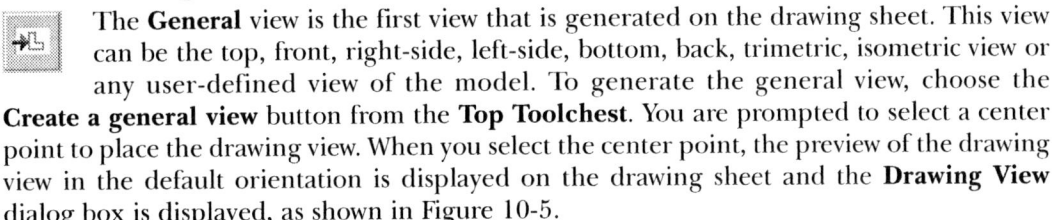

The **General** view is the first view that is generated on the drawing sheet. This view can be the top, front, right-side, left-side, bottom, back, trimetric, isometric view or any user-defined view of the model. To generate the general view, choose the **Create a general view** button from the **Top Toolchest**. You are prompted to select a center point to place the drawing view. When you select the center point, the preview of the drawing view in the default orientation is displayed on the drawing sheet and the **Drawing View** dialog box is displayed, as shown in Figure 10-5.

This dialog box contains the options that are required to generate a general view. The **Categories** list box in the dialog box lists all parameters that are required to generate a drawing view in Pro/ENGINEER Wildfire 2.0. The **View Type** option is selected by default in the **Categories** list box and its related options are displayed on the right side of the **Categories** list box. These parameters/options are discussed later in this chapter.

The procedure to generate a general view is given below.

1. Choose the **Create a general view** button from the **Top Toolchest** or choose **Insert > Drawing View > General** from the menu bar. You are prompted to select a center point.

2. As soon as you select the center point, the **Drawing View** dialog box is displayed on the screen and the default view of the model is displayed on the drawing sheet. In the **Type** drop-down list, the **General** option is selected by default and in the **View orientation** area, the **Views names from the model** radio button is selected by default. This radio button allows you to select the standard orientations of the model from the **Model view names** list box. If you select the **Geometry references** radio button, you can select the planar faces or the datum planes to orient in a particular orientation.

3. Select the view that you need to generate as the general view from the list provided below the **Model view names** list box.

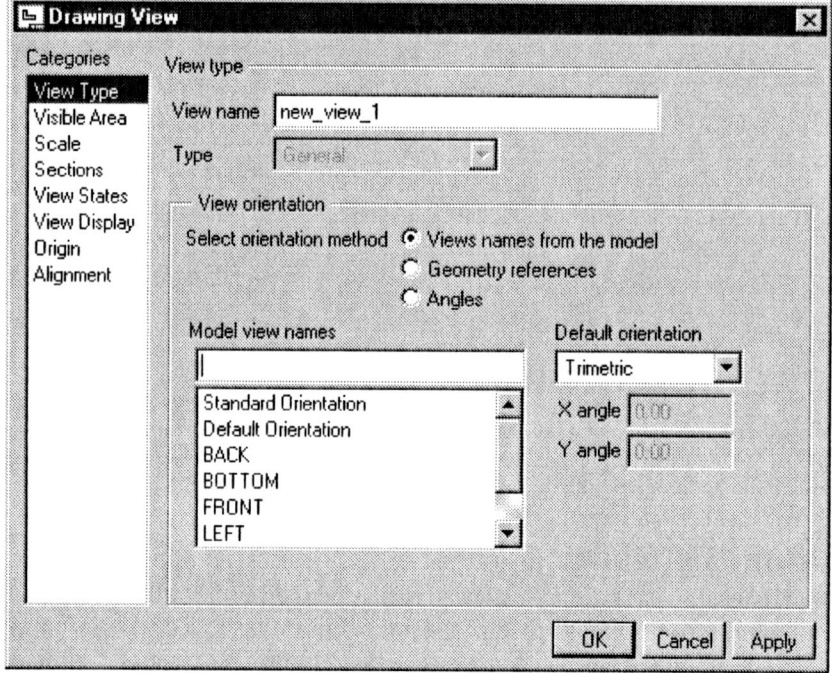

*Figure 10-5 The **Drawing View** dialog box*

4. After selecting the required options choose the **Apply** button and then choose the **Close** button from the **Drawing View** dialog box.

> **Tip**: *To set the model orientation using the **Geometry references** radio button, select an option from the **Reference 1** drop-down list and then from the model select the face or the datum plane that you need to see from that selected view. Similarly, select an option from the **Reference 2** drop-down list and then select a datum plane or a face from the model. For example, if you select **Top** from the **Reference 1** drop-down list then from the model you need to select the face or the datum plane that you need to keep in the top orientation in the drawing view.*

Generating the Projection View

The projection views are the orthographic views generated by projecting lines normal from an existing view, as shown in Figure 10-6. Before generating a projected view, you need to make sure that at least one parent view is already present on the drawing sheet. To generate a projected view, choose **Insert > Drawing View > Projection** from the menu bar. Now, move the cursor to a location where you need to place the view and specify a point on this location. Depending on the point selected to place the view, the resulting view will be top, front, right-side, or left-side view.

The scale factor of these views will be the same as that of the parent view from which they are projected. If there exists more than one view that can be the parent view of the projection

Generating, Editing, and Modifying the Drawing Views

view, then you will be prompted to select the parent view for the new view.

Generating the Detailed View

Detailed views are used to provide the enlarged view of a particular portion of an existing view. To generate a detailed view, choose **Insert > Drawing View > Detailed** from the menu bar. You are prompted to specify the center point for detail on an existing view. Define a center point on the view whose detail needs to be generated. Next, you are prompted to sketch a spline, without intersecting other splines, to define an outline. Draw a closed spline to define the outline of the detail view. Next, you need to specify the placement point for placing the detail view. You can change the name, scale, or the reference point on the parent view and can also select the boundary type of the detailed view using the options from the **Drawing View** dialog box, which is displayed when you double-click on the detailed view.

Figure 10-6 shows a drawing sheet with various drawing views.

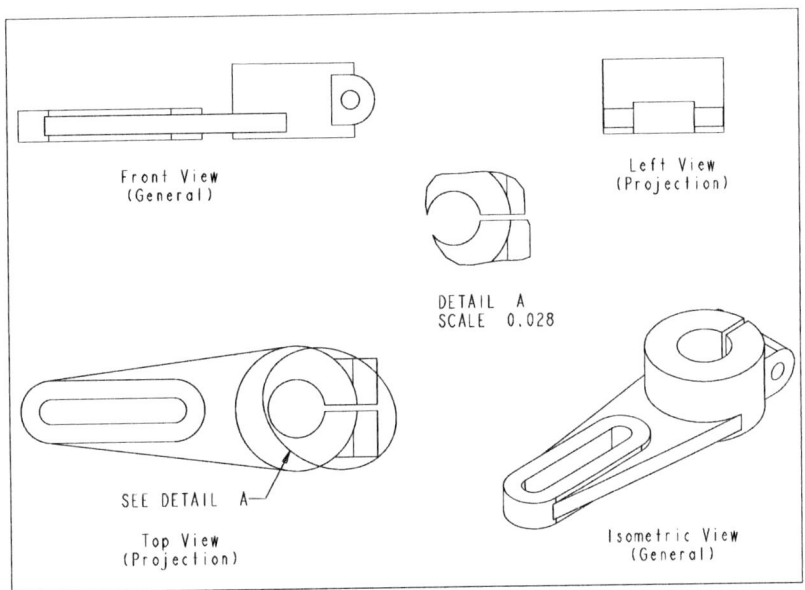

Figure 10-6 Sheet with various drawing views

Note

*After generating a view you need to choose the **Update the display of all views in the active sheet** button from the **Top Toolchest**. This is because sometimes the scale or the other information related to generated view is not displayed along with the it.*

Generating the Auxiliary View

The auxiliary views are generated by projecting normal lines from a specified edge, an axis, or a datum plane of an existing view. The view scale will be the same as that of the parent view. To generate an auxiliary view, choose **Insert > Drawing View > Auxiliary** from the menu bar. You are prompted to select edge of or axis through, or datum plane as, front surface on main view. Select the edge or surface normal to which you need to place the generated view. The preview of the drawing view is attached to the cursor. Select a point on the drawing sheet to place the view on the drawing sheet. Figure 10-7 shows the edge to be selected as reference to generate the drawing view and the resulting auxiliary view.

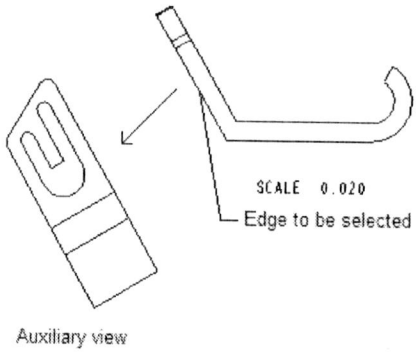

Figure 10-7 Auxiliary view

Generating the Revolved Section View

A revolved section view is the view that is generated from an existing view by revolving the section through an angle of 90-degree about the cutting plane and then projecting it along the length. Remember that the cutting plane of the revolved section view is normal to the parent view. The procedure to generate a revolved section view is discussed next.

1. Choose **Insert > Drawing View > Revolved** from the menu bar. You are prompted to select a parent view for the revolved section.

2. Select the view from the drawing sheet that will be defined as the parent view for generating the revolved view.

3. Next, you need to select a center point on the drawing sheet to place the view. Select a point anywhere on the drawing sheet but the resulting view will be placed in-line with the section plane. The **Drawing View** dialog box is displayed. The **Create New** option is selected by default in the **Revolved view properties** area of the **Drawing View** dialog box. The **XSEC CREATE** menu is also displayed on the lower right corner of the screen.

4. Choose **Planar > Single > Done** from the **XSEC CREATE** menu. You need to specify the name of the cross-section in the **Message Input Window** that is displayed. The **SETUP PLANE** menu is displayed and you are prompted to select a planar surface or datum plane.

Generating, Editing, and Modifying the Drawing Views

5. Select the plane along which the view will be sectioned as shown in Figure 10-8. The resulting section view will be placed on the point specified earlier.

6. Choose the **Apply** button and then choose the **Close** button to exit the **Drawing View** dialog box. The resulting sectioned view is shown in Figure 10-9.

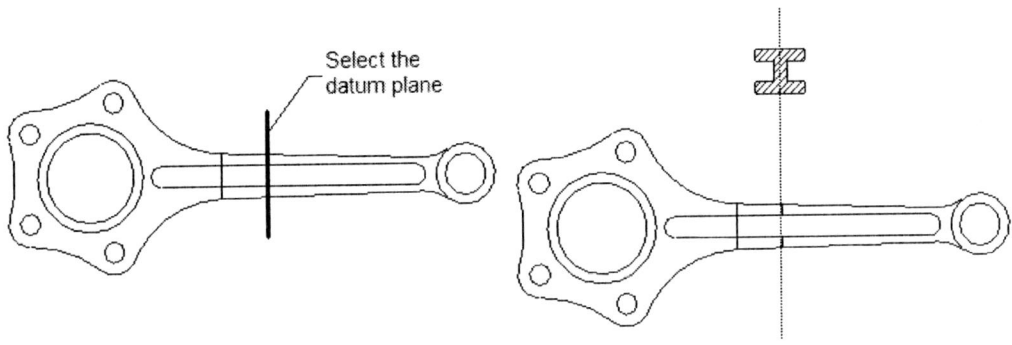

Figure 10-8 Datum plane for the revolved section view

Figure 10-9 Revolved section view

Note
The copy and align view will be discussed later in this chapter.

Drawing View Dialog Box Options

The remaining types of drawing views that you can generate in Pro/ENGINEER Wildfire 2.0 require the options available in the **Drawing View** dialog box. Therefore, it is important for you to understand these options before proceeding further.

View Type Option

When you invoke the **Drawing View** dialog box, the **View Type** option is automatically selected. The **View name** edit box is used to enter the name of the drawing view. The **View orientation** area lists the options to orient the drawing view. These options are the standard options of orienting the model, which in this case are used to orient the model.

Visible Area Option

The **Visible Area** option is used to set the displayed of the view. You can display a drawing view as a full view, partial view, a half view, and so on. The **Drawing View** dialog box after selecting the **Visible Area** option is displayed in Figure 10-10. The options available in this dialog box when the **Visible Area** option is selected are discussed next.

Full View

The **Full View** option is by default selected in the **View visibility** drop-down list. This option can be combined with any of the view type to generate a drawing view displaying the complete view.

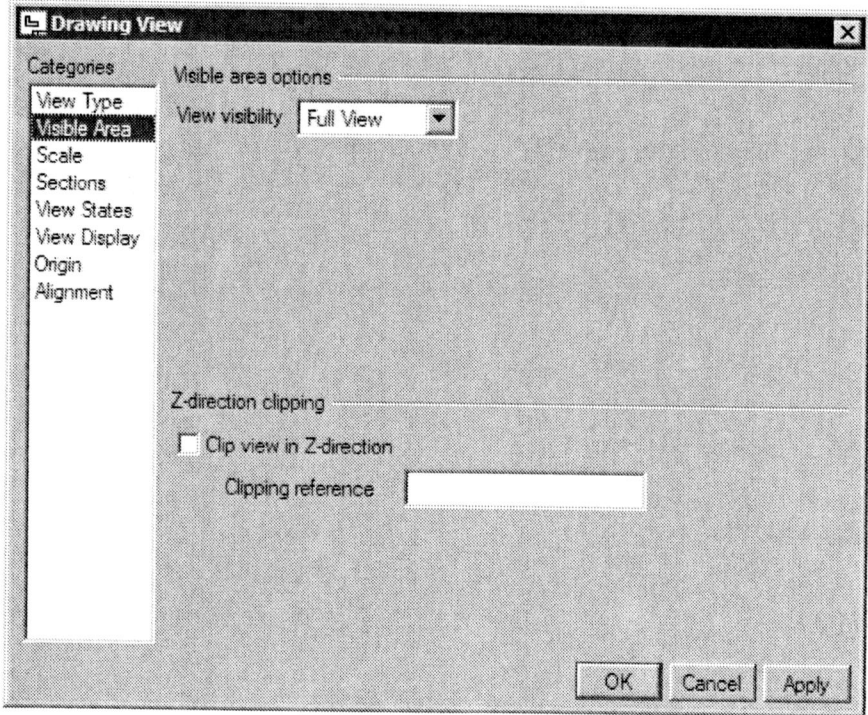

*Figure 10-10 The **Drawing View** dialog box with the **Visible Area** option selected*

Half View

The **Half View** option can be used on the projection, auxiliary, or general views to generate a drawing view that displays only half view of the part. Generally, an existing view is selected to display the half view. For generating a half view, double click on an existing view to display the **Drawing View** dialog box. From the **Categories** list box, choose the **Visible Area** option in the **Categories** to invoke the **Visible Area** options and then select **Half View** from **View visibility** drop-down list. You are prompted to select a reference plane that will be used to remove half of the drawing view. The reference plane can be a datum plane or a planar surface, and must be normal to the screen in the selected view. Select the reference plane, an arrow will be displayed attached to the selected plane. This arrow indicates the portion of the view to be removed. You can also flip the direction of the arrow using the **Side to keep** button available below the **Half view reference plane** collector in the dialog box. Generally, this type of view is generated for symmetric parts. Therefore, you can specify the type of symmetry line using the **Symmetry line standard** drop-down list. Figure 10-11 shows the reference plane to be selected and Figure 10-12 shows the resulting half view.

Partial View

The **Partial View** option can be used on the projection, auxiliary, revolved, or general views to generate a view that displays a specified portion of the view. To convert an existing view to a partial view, double-click on that view. Choose the **Visible Area** option from the **Categories** list box and select the **Partial View** option from the **View visibility** drop-down list; you are prompted to draw a spline that will be the boundary of the

Generating, Editing, and Modifying the Drawing Views

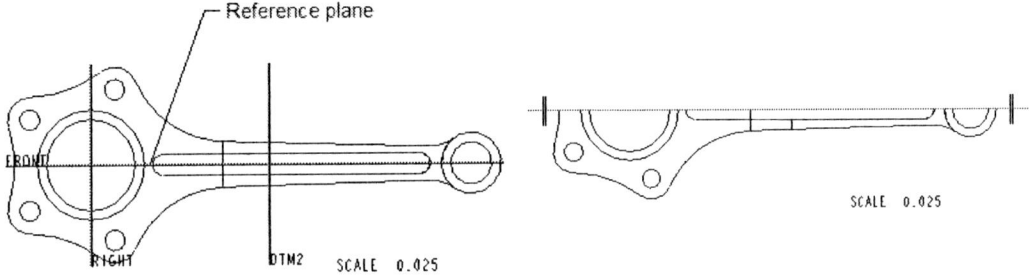

Figure 10-11 Reference plane to be selected to generate a half view

Figure 10-12 Half view

portion of the view you want to display, see Figure 10-13. Select the center point on the view and then draw the spline. Now, choose the **Apply** button from the **Drawing View** dialog box. You will notice that only the area of the view inside the spline is retained and the rest of potion of drawing view is cropped, see Figure 10-14.

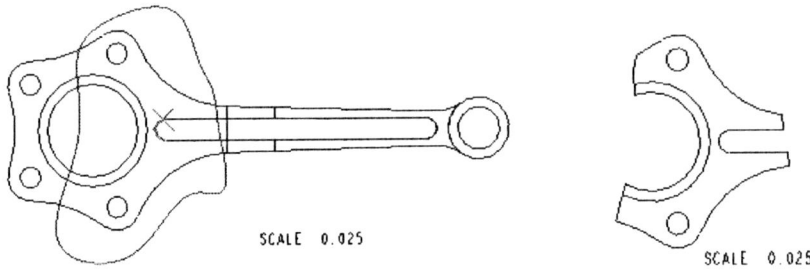

Figure 10-13 Spline drawn on the view

Figure 10-14 Partial view

Broken View

This type of view is used for the parts having high length to width ratio. The **Broken View** option can be used on the projection or the general views to generate a view that is broken along the horizontal or the vertical direction using horizontal or vertical lines. To generate a broken view, select the **Broken View** option from the **View visibility** drop-down list. Select the **Add break** collector available in the **Drawing View** dialog box to add break lines. Select any one edge from the view and move the cursor vertically and select a point on the screen to draw the first vertical break line. Move the cursor and select a point up to which you want to break the view, see Figure 10-15. You can also specify the style of break line using the **Break Line Style** drop-down list. Now, choose the **Apply** button from the Drawing View dialog box to create the resulting broken view and then choose the **Close** button to exit it. Figure 10-16 shows the resulting broken view.

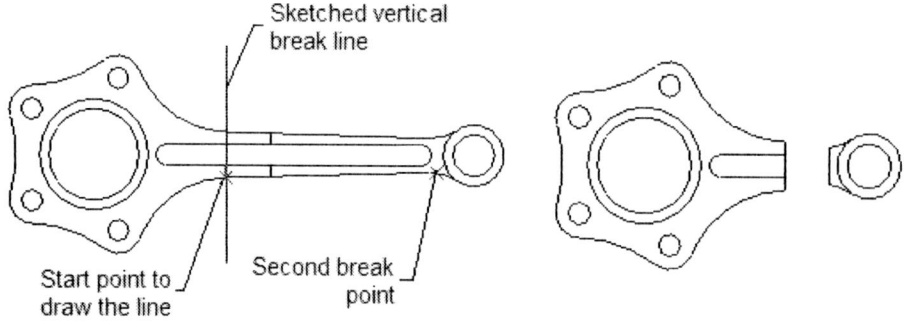

Figure 10-15 References to break the view   *Figure 10-16 Broken view*

Figure 10-19 shows the different views that are already discussed.

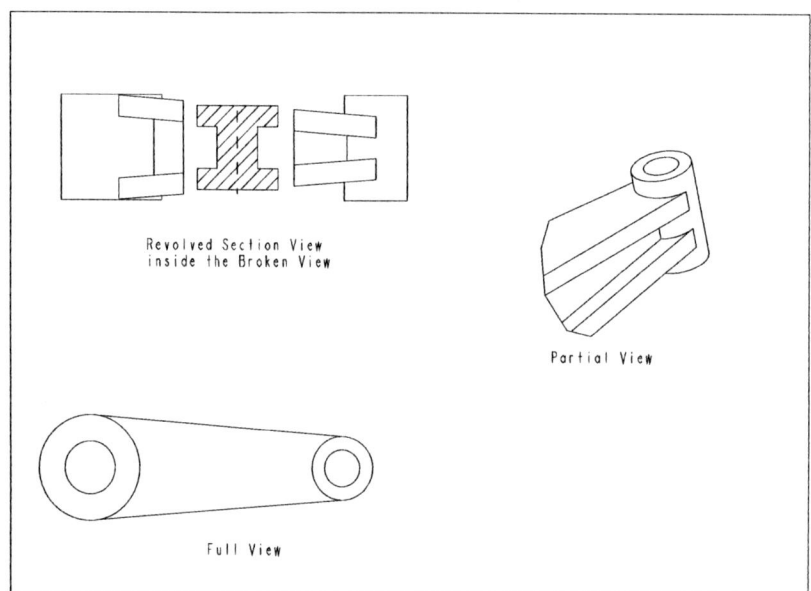

Figure 10-17 The different drawing views

Scale Option

When you select the **Scale** option, the **Scale and perspective options** area is displayed on the right of the **Categories** list box. The **Default scale for sheet** radio button is selected by default. Therefore, all views that are generated are scaled with the default sheet scale. The second radio button is **Custom scale**. This radio button when selected allows you to enter a scale for the drawing view. When you modify the scale factor of a view, the view associated with this parent view are also scaled.

The third radio button is **Perspective**. This radio button is selected to generate a perspective view. After selecting the **Perspective** radio button you can specify the distance from the eye and the diameter of the view circle. Figure 10-18 shows a perspective view.

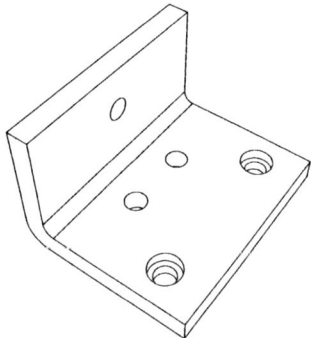

Figure 10-18 *A general perspective view*

Sections Option

The section views are generally used for the views that are complicated from inside. As it is not possible to display the inside of the part using the conventional views, therefore, these views are cut (sectioned) using a datum plane or a planar surface and the resulting section view is displayed. When you choose **Sections** option from the **Categories** list box in the **Drawing View** dialog box, **Section options** area is displayed. The **No section** radio button is selected by default. Select the **2D cross-section** radio button and then select plus button to activate the options to create a section. The options available in the Section options area and the methods of section plane creation are discussed next.

Full section view

Consider a part that is cut throughout its length, width, or height and the cut portion is removed from the display. The remaining portion when projected normal to the cutting plane displays the full section view. The **Full** option in the **Sectioned area** drop-down list allows you to create a full section view.

To generate a full section view, you first need to generate a projected view from one of the views existing on the drawing sheet. After generating the projected view, double-click on it to invoke the **Drawing View** dialog box. Choose the **Sections** option from the **Categories** list box. The options available for generating the section view are displayed on the right of the **Drawing View** dialog box. Select the **2D cross-section** radio button to define a section plane. Now, choose the plus button to add a section plane. Some options in the drop-down lists are invoked in the area provided below the plus button.

The **XSEC CREATE** menu is displayed. Choose **Planar > Single > Done** from the **XSEC CREATE** menu; you are prompted to specify the name of the cross-section. Specify the name of the cross-section in the **Message Input Window**. The **SETUP PLANE** menu is displayed and you are prompted to select a planar surface or a datum plane. Select the section plane from the other view; the name of the section view is displayed in the **Name** drop-down list. Scroll to the right in the dialog box and click once in the **Arrow Display** collector to invoke the selection mode and select the view in which you need to display the section view arrows. You can also flip the material side of the section view using the **Flip** button. Choose the **Apply** button from the **Drawing View** dialog box. Figure 10-19 shows the section view.

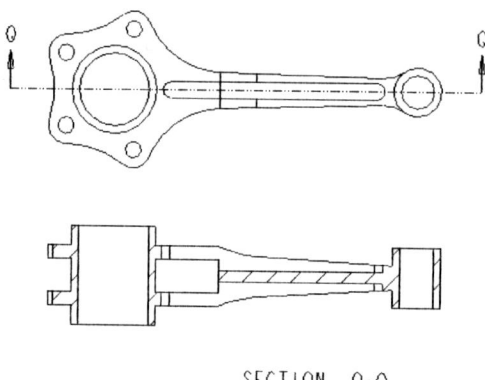

Figure 10-19 Figure showing top and section front view

In the above section, you have learned to generate a full section view by defining a planar section plane. Now, you will learn to generate a full section view by defining an offset section plane. You need to make sure that before generating a section view by defining an offset section plane, the part file of the current model should be closed.

Invoke the **Drawing View** dialog box by double-clicking on an already generated projected views. Now, choose the **Sections** option from the **Categories** list box available on the left of this dialog box. Select the **2D cross-section** radio button and choose the plus button to define the section plane. If there are some existing section planes, select the **Create New** option from the **Name** drop-down list; the **XSEC CREATE** menu is displayed. Choose **Offset > Both Sides > Single > Done** from the **XSEC CREATE** menu; the **Message Input Window** is displayed. Specify the name of the cross-section.

The model is displayed in a subwindow on the left of the screen. The **SETUP SK PLN** menu is displayed and you need to select a sketch plane for drawing the sketch that represents an offset section plane. Select the sketch plane from the subwindow. Choose **Okay** from the **DIRECTION** submenu. The **SKET VIEW** submenu is displayed; choose **Default** from this submenu. Draw the sketch of the offset section using the sketch tools available in the **Sketch** menu of the subwindow, refer to Figure 10-20. After drawing the sketch, choose **Sketch > Done** from the menu bar to exit the sketching environment.

Click once in the **Arrow Display** collector to invoke the selection mode and then select the view in which the section arrows will be displayed. Choose the **Apply** button from the **Drawing View** dialog box. Figure 10-20 shows the top view, offset sectioned front view, and the sectioned general view of a part.

Generating, Editing, and Modifying the Drawing Views

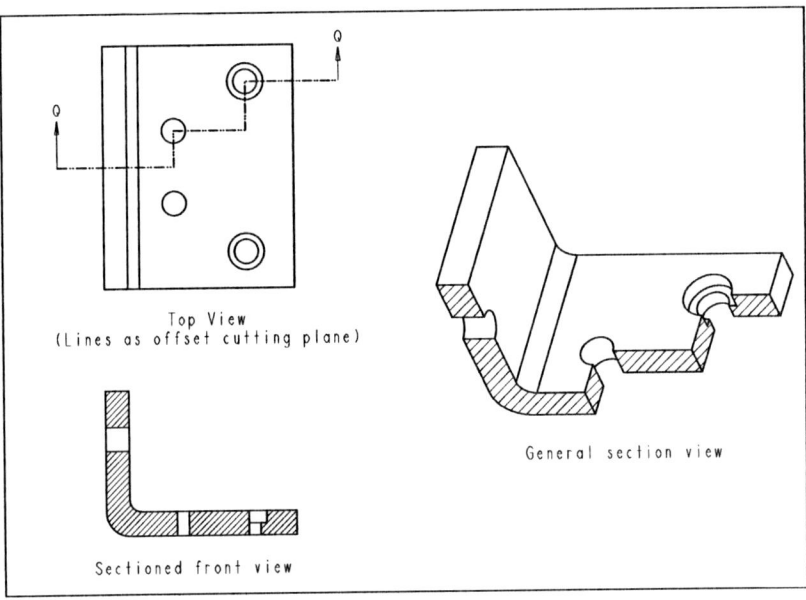

Figure 10-20 The section views created using the offset section plane

To generate an isometric section view, generate the general view and switch the orientation of the general view to isometric. Now, select the **Sections** option from the **Categories** list box and sketch a section plane using the same subwindow method as discussed earlier. Note that you cannot select a datum plane to define the section plane for an isometric section view. Figure 10-21 shows the full section front view and the full section isometric view.

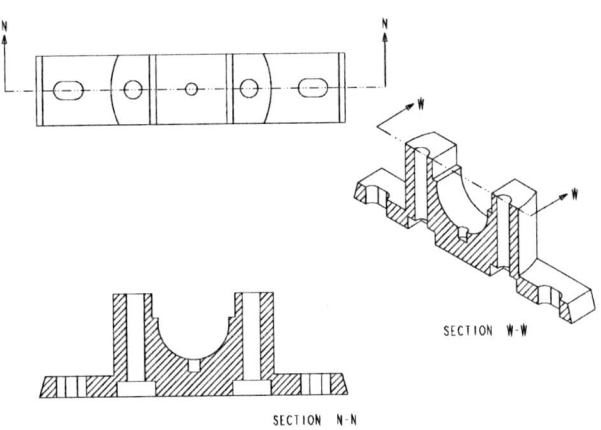

Figure 10-21 Top view, full section front view, and full section isometric view

Half section view

Consider a part that is cut half way through the length, width, or height and the front cut portion (front quarter) is removed from the display. This part when projected is called the half section view. In this projected view, only half of the part is displayed sectioned and the other half of the part is displayed as it is. To generate the half section view, you need to select the **Half** option from the **Sectioned Area** drop-down list while specifying the parameters for sectioning the view. After selecting this option, you are prompted to select reference plane for half section creation. Select the plane from the drawing view; you are prompted to pick the side to keep. Select a point on the side of the section view that you need to keep. Select the parent view where the section arrow will be placed and choose the **Apply** button from the **Drawing View** dialog box. Figure 10-22 shows the resulting half section view.

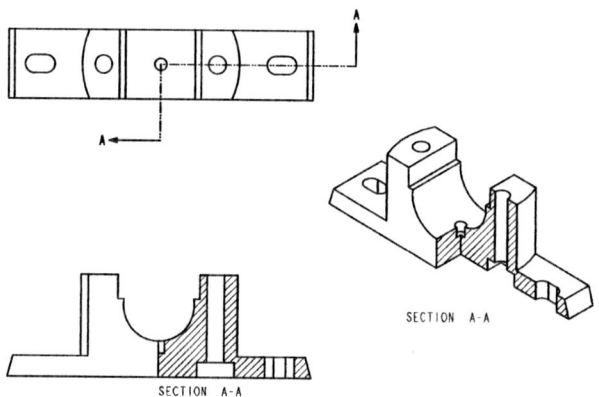

Figure 10-22 Top view, half section front view, and half section isometric view

Local section view

The **Local** section view is used when you want to show a particular portion of the view in section and at the same time not section the remaining view. The local section area is specified by drawing a spline around it. To generate the local section view, you first generate the full section view as discussed earlier. After generating the full section view, choose the **Local** option from the **Sectioned Area** drop-down list. Next you need to draw a spline that will define the area of the drawing that needs to be sectioned.

To draw a local section isometric view, you need to draw a section plane and then choose the **Local** option. Define a center point for the local section and sketch the spline. This method of generating the local section isometric view is similar to generating the other section isometric views. Figure 10-23 shows the local section front view and the local section isometric view.

Model edge visibility

Pro/ENGINEER Wildfire 2.0 provides you with two options of displaying the edges of the

Generating, Editing, and Modifying the Drawing Views

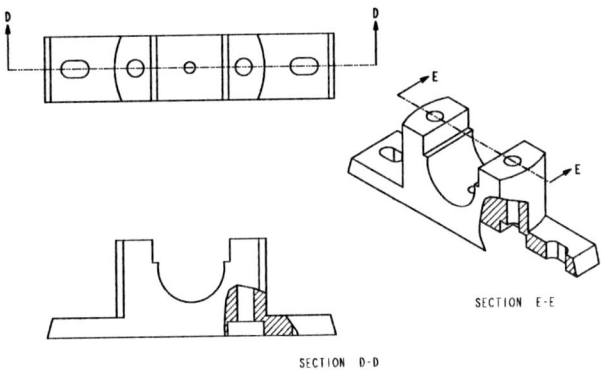

Figure 10-23 Top view, Local section front view, and Local section isometric view

section drawing views. The **Total** radio button when selected in the **Section options** area of the **Drawing View** dialog box creates a section view that displays all visible edges of the section view in addition to the section area. This option can be combined with the full, half, local, full (unfold) and full (aligned). Figure 10-24 shows the front and total sectioned left view.

The **Area** radio button when selected displays only that area of the section view that is sectioned. No other edges of the view are displayed in the area cross section view, as shown in Figure 10-25. This option can be combined with the full, half, local, full (unfold) and full (aligned).

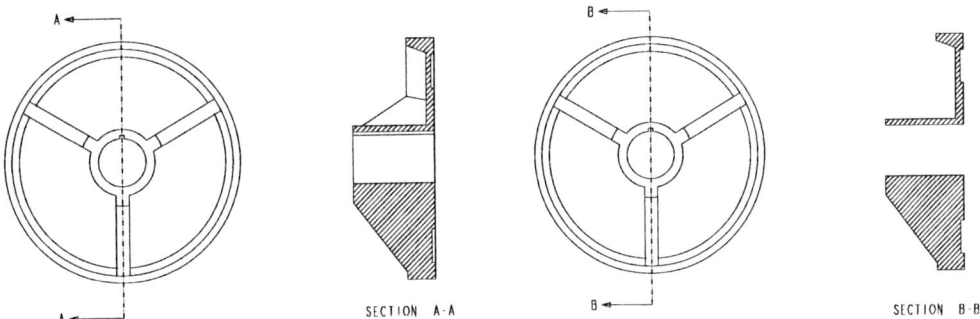

Figure 10-24 The front view and the total cross section view

Figure 10-25 The front view and the area cross section view

Tip: *You can also create the cross sections in the **Part** mode and later retrieve those sections in the **Drawing** mode and use them for generating the section views.*

Full (Aligned)

This view is used to section those features that are created at a certain angle to the main section planes. Align sections straighten these features by revolving them about an axis that is normal to the parent view. Generally for this section view, the section plane is sketched. Remember that the axis about which the feature is straightened should lie on all cutting planes. This means that the lines that are used to sketch the section plane should pass through the axis about which the feature will be revolved. Figures 10-26 and 10-27 show these views when the **Area** radio button is selected. Figures 10-28 and 10-29 shows these views when the **Total** radio button is selected.

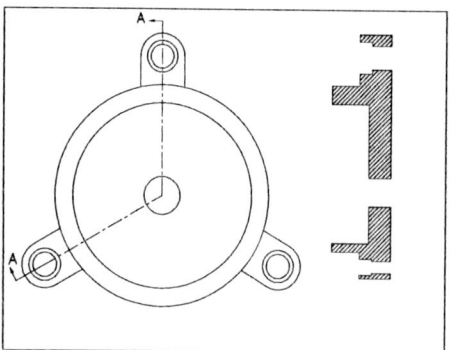

Figure 10-26 The area cross section view with normal lines of projection

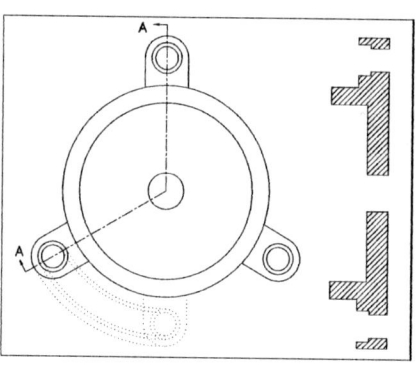

Figure 10-27 The area cross section view with aligned lines of projection

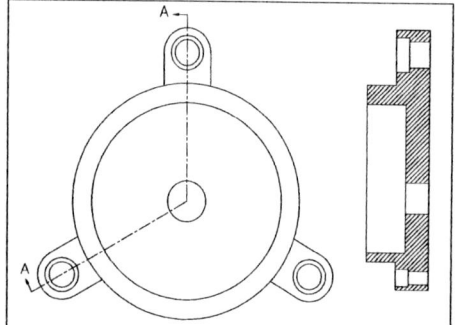

Figure 10-28 The total cross section view with normal lines of projection

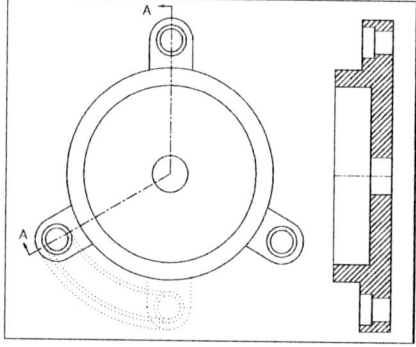

Figure 10-29 The total cross section view with aligned lines of projection

Full(Unfold)

The **Full(Unfold)** option is used to generate the section view by unfolding the section surface of an offset section view. This type of view is only generated using a general view. To generate this type of view, first you need to generate a general view in the required orientation. You can place this general view anywhere on the drawing sheet. Now, choose the **Sections** option from the **Categories** on the left of the **Drawing View** dialog box. Now, define the section plane using the offset option and specify the view in which you

Generating, Editing, and Modifying the Drawing Views 10-19

need to display the arrows. Now, choose the **Full(Unfold)** option from the **Sectioned Area** drop-down and complete the creation of full unfold section view. Figure 10-30 shows the model and the section plane. Figure 10-31 shows an offset section view and a full unfold section view.

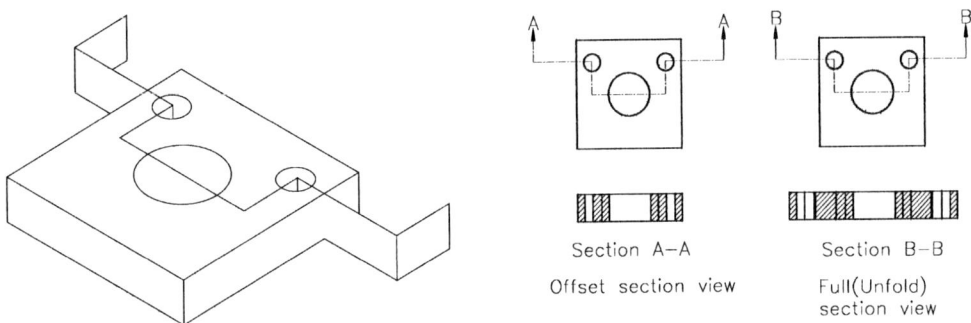

Figure 10-30 Model and the section plane

Figure 10-31 Offset section view and the full unfold section view

Once you have understood the main options of the **Drawing View** dialog box, you can generate the copy and align view. This view is discussed next.

Generating the Copy and Align View

These are the aligned views that are generated from an existing partial view. Therefore, to generate this type of drawing view, it is necessary that first a partial view should be generated. You will learn more about the partial views later in this chapter.

The procedure to generate a **Copy and Align** view is discussed next.

1. To generate the **Copy and Align** view, choose **Insert > Drawing > Copy and Align** from the menu bar. You will be prompted to select an existing partial view to be aligned with. Once you have selected a partial view, you are prompted to select center point for drawing view.

2. Select center point anywhere on the drawing sheet to place the view. The view is placed at the selected location. You are prompted to specify a center point for detail on the current view and to sketch a spline to define the outline of the **Copy and Align** view.

3. Specify the center point of the spline and draw the spline. Press middle mouse button to finish sketching the spline. The drawing view will be cropped along the spline drawn. You are prompted to select a straight line (axis, segment, datum curve) alignment on the current view.

4. Select an edge from the current view, as shown in Figure 10-32 to align it with the partial view. The copied view is aligned with its parent partial view, as shown in Figure 10-33.

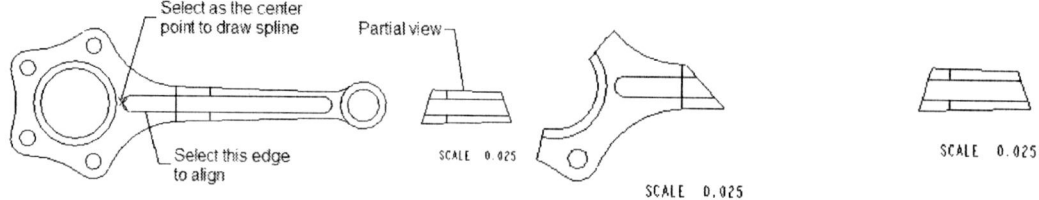

Figure 10-32 *Center point to draw spline and the edge to align and the partial view*

Figure 10-33 *Resulting copy and align view*

EDITING THE DRAWING VIEWS

With this release of Pro/ENGINEER, it has become very easy to edit a drawing view and the items in the drawing views. All options of modifying the drawing views or items related to it can be chosen from the shortcut menu that is displayed when you right-click after selecting the view or the item. However, these options can also be invoked from the menus available in the menu bar. Pro/ENGINEER allows you to perform the following types of editing operations on the drawing views.

Moving the Drawing View

When a view is generated, its movement is locked by default. To unlock the drawing view, choose the **Disallow the movement of drawing views with the mouse** button from the **Top Toolchest**. When you select the drawing view, it is displayed inside a boundary. To move the drawing view, select it and then drag it to the required location on the drawing sheet. Remember that if you select the view that has some child views, the child views will also move along with the parent view in order to maintain their alignment with the parent view. Also, the projected views can be moved only in the direction of projection.

Note
The **General** *and* **Detailed** *views can be moved to any new location because they are not the projections of any view.*

Erasing the Drawing View

Choose **View > Drawing Display > Drawing View Visibility** from the menu bar to invoke the **VIEWS** menu. The **Erase View** option in the **VIEWS** menu is used to temporarily remove the selected drawing view from the sheet. However, the view still remains in the memory of the drawing and can be resumed at any point of time. As the view is not completely removed from the memory, you can also erase a view that has some child views associated to it and it will not affect the child view. Once a view is erased, a box is displayed in place of the view displaying the name of the view. To resume the view, choose the **Resume View** option from the **VIEWS** menu.

Note
When you erase a view, the leaders and dimensions that were attached with the view are also erased. When you resume an erased view, the leaders and dimension values are redisplayed.

Deleting the Drawing Views

To delete a drawing view from the drawing sheet, select the view and hold down the right mouse button to invoke a shortcut menu. Choose the **Delete** option from the shortcut menu. You can also use the **Delete selected items** button available in the **Top Toolchest** after you select the drawing view to delete the selected drawing view. Once the view is deleted, no information related to the deleted view remains in the memory of the drawing. Remember that if a view that has some child views associated with it is deleted, then before deletion if informs you that views associate with this view will also be deleted. You can use the **Undo** button from the **Top Toolchest** for restoring the deleted views.

Adding New Parts or Assemblies to the Current Drawing

You can also add more parts or assemblies in addition to the default part or assembly for generating the drawing views. This is done by choosing **File > Properties** from the menu bar. The **FILE PROPERTIES** menu is displayed, as shown in Figure 10-34. Choose the **Drawing Models** option, the **DWG MODELS** menu is displayed, as shown in Figure 10-35.

Figure 10-34 FILE PROPERTIES menu

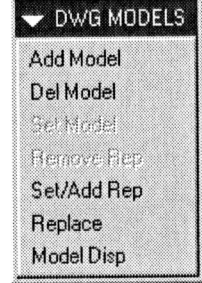

Figure 10-35 DWG MODELS submenu

Choose the **Add Model** option to add a model to the current drawing. When you choose this option, the **Open** dialog box is displayed. You can select the new model to be added using this dialog box. Remember that the latest model added will be the current model and the drawing views that will be generated now will be of this model. You can change the current model by choosing the **Set Model** option from the **DWG MODELS** menu. Similarly, you can also delete a model from the current drawing. However, only the model that does not have any view generated from it can be deleted.

MODIFYING THE DRAWING VIEWS

You can also make the following modifications in the existing drawing views.

Changing the View Type

Select the drawing view that needs to be modified and hold down the right mouse button to invoke the shortcut menu. Choose **Properties** from the shortcut menu; the **Drawing View** dialog box is displayed, as shown in Figure 10-36. The **Drawing View** dialog box can also be invoked by double-clicking the view to be modified. The **View Type** option in the **Categories** is selected by default. You can modify the view type by selecting one of the types from the drop-down list. Remember that only the general, projection, and auxiliary view types can be modified using this option.

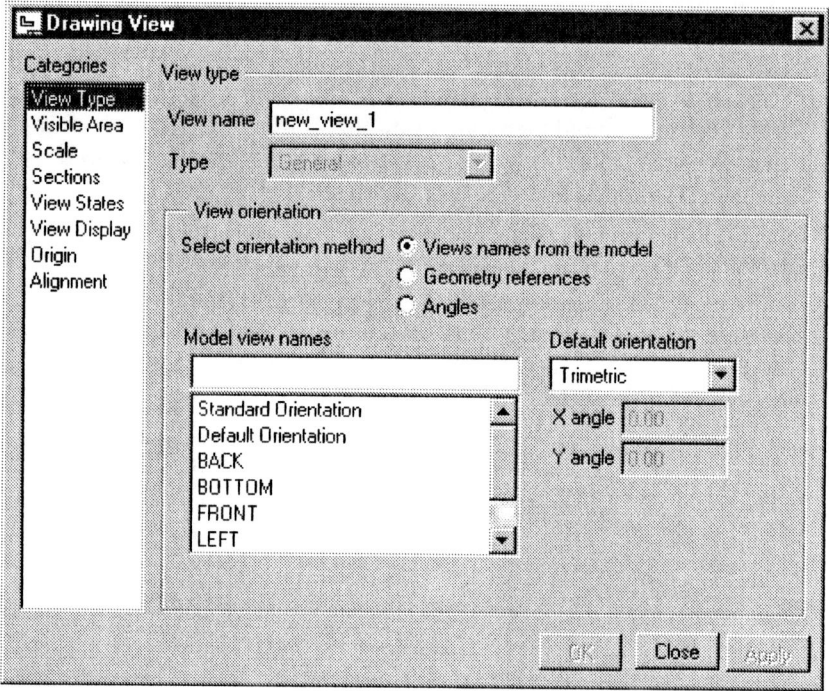

*Figure 10-36 The **Drawing View** dialog box*

Changing the View Scale

The scale of a view can be modified using the **Scale** option from the **Categories** list box of the **Drawing View** dialog box. You can modify the scale factor of views those are generated using the general view and detailed view options.

Note
*The scale factor represents the drawing to model scaling. For example, if the scale set is **0.25**, Pro/ENGINEER scales the drawing views to one-quarter (1/4) of the actual size of the model.*

*You can also modify the scale factor by selecting it from the drawing sheet and holding down the right mouse button. From the shortcut menu that is displayed, choose the **Edit Value** option. Specify the scale factor in the **Message Input Window** that is displayed.*

Reorienting the Views

Double click on the general view to be reoriented to display the **Drawing View** dialog box.

Modifying the Cross-sections

You can flip the side of sectioning and replace, delete, or rename the sections in the views using the **Section** option from the Categories in the **Drawing View** dialog box.

Modifying Boundaries of Views

You can modify the boundaries of the detailed or partial views using the option available in the **Drawing View** dialog box. To modify the boundary of a detailed view, double-click on it to invoke the **Drawing View** dialog box. The selection mode is invoked by default in the **Reference point on parent view** selection area. Therefore, you can specify the center point for the boundary of the detailed view. To sketch the boundary of the detailed view again, click once in the **Spline boundary on parent view** selection area and draw the boundary on the parent view. To modify the boundary of a partial view, invoke the **Drawing View** dialog box and choose the **Visible Area** option from the **Categories** list box. Using the options available on the right of this dialog box you can modify the boundary of the detailed view.

Adding or Removing the Cross-section Arrows

If you have not specified the cross-section arrows while generating the section views, you can specify them by selecting the section view and holding down the right mouse button to invoke the shortcut menu. Choose the **Add Arrows** option from the shortcut menu, you will be prompted to select the view where the arrows should be displayed. Select the view in which you need to display the arrows.

To remove arrows, select the arrows and choose the **Delete selected items** button from the **Top Toolchest**.

Modifying the Perspective Views

Double-click on the perspective view to display the **Drawing View** dialog box and choose **Scale** option from the **Categories** list box. The options related to the perspective are available. You can modify the eye point distance or the view diameter using these options.

MODIFYING OTHER PARAMETERS

Apart from modifying the drawing views, you can also modify other parameters related to the drawing views. For example, you can modify the size and the style of the text, scale factor of all drawing views, cross-section hatching, and so on. All this is done by selecting the item. When the item is highlighted in red color, hold down the right mouse button to invoke the shortcut menu. You can select the options from the shortcut menu to modify that item.

You can modify any parameter associated with the drawing views. Depending on the item selected to modify, the options related to it vary. The options in the shortcut menu that are displayed vary from item to item.

Editing the Cross-section Hatching

Select the hatching from a drawing view. When the hatching turns red in color, hold down the right mouse button to invoke the shortcut menu. Choose the **Properties** option from the shortcut menu; the **MOD XHATCH** menu is displayed, as shown in Figure 10-37.

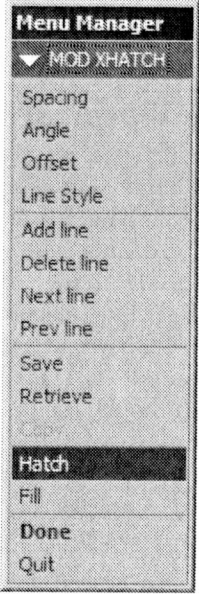

Figure 10-37 MOD XHATCH menu

The parameters related to cross-section hatching that can be modified are the spacing of the hatching, angle of the hatching lines, offset value, and the line style of the hatching lines. There are also some standard hatch patterns that are available in Pro/ENGINEER. You can retrieve these standard patterns by using the **Open** dialog box displayed upon choosing **Retrieve** from the **MOD XHATCH** menu.

TUTORIALS

Tutorial 1

In this tutorial you will generate the drawing views of the model created in Exercise 4 of Chapter 6 shown in Figure 10-38. Select the A4 size drawing sheet and generate all drawing views shown in Figure 10-39. **(Estimated time: 45 min)**

Tip: *If you delete the part file from which the drawing views are generated, the drawing file will not open. Also, if the part file is removed from the folder where its drawing file is placed or if you rename the part file after generating the drawing views, the drawing file will not open.*

Generating, Editing, and Modifying the Drawing Views

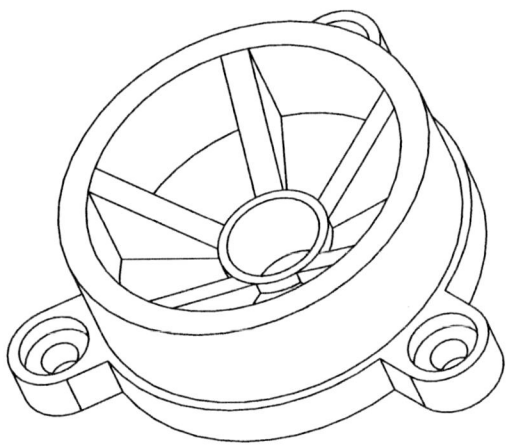

Figure 10-38 Part for generating the drawing views

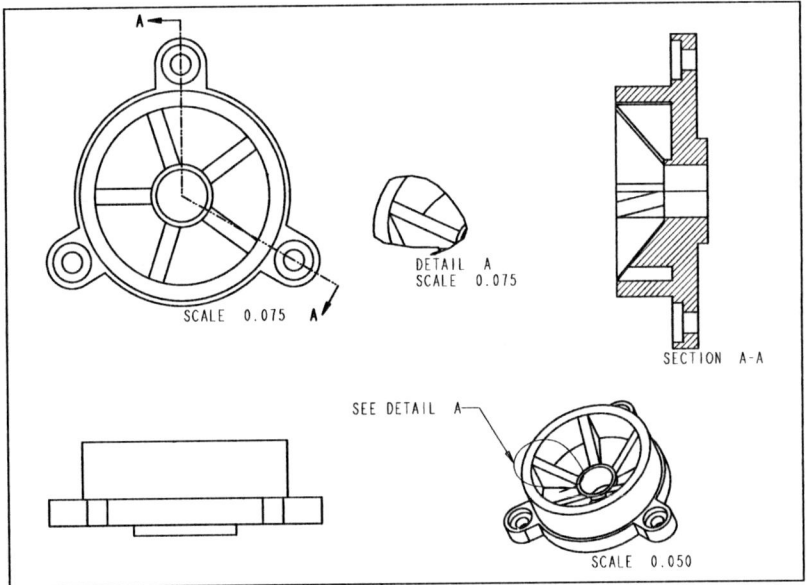

Figure 10-39 The drawing views to be generated

The following steps are required to complete this tutorial:

a. Start a new drawing file and select the required drawing sheet.
b. Generate the top view, refer to Figure 10-41.
c. Generate the front view taking the top view as the parent view, refer to Figure 10-41.
d. Generate the sectioned right view by defining a plane on the top view of the model, refer to Figures 10-42 and 10-43.

e. Generate the default 3D view of the model, see Figure 10-43.
f. Generate the detail view from the 3D view, see Figure 10-44.

Before you start generating the drawing views, set the working directory to *C:\ProE-WF-2.0\c10*. Copy and paste the model *c06exr4.prt* file in the folder named *c10* from *c06* folder. The drawing that you will generate will be saved in the *c10* directory with an extension *.drw*. The part file *(.prt)* and the drawing file *(.drw)* should lie in the same directory or folder.

Starting a New Drawing File

To generate drawing views in Pro/ENGINEER Wildfire 2.0, you first need you start a new file in the **Drawing** mode.

1. Choose the **Create a new object** button from the **File** toolbar to display the **New** dialog box.

2. Select the **Drawing** radio button and then enter the name of the file as *c10tut1*.

3. Choose **OK** from the **New** dialog box to display the **New Drawing** dialog box.

4. Choose the **Browse** button to select *c06exr4.prt* for generating the drawing views.

5. Select the **Empty** radio button from the **Specify Template** area.

6. Choose the **Landscape** button from the **Orientation** area, if it is not chosen by default, as shown in Figure 10-40.

7. Select **A4** from the **Standard Size** drop-down list. Choose **OK** to proceed to the **Drawing** mode.

   An A4 size sheet with landscape orientation is displayed in the graphics window.

Generating the Top View

Generally, you can generate any view as the first view on the drawing sheet. However, in this tutorial, you will generate the top view as the first view.

1. Choose the **Create a general view** button from the **Top Toolchest**.

   You will be prompted to specify the center point for drawing view.

2. Specify the placement point for the first view close to the top left corner of the sheet, as shown

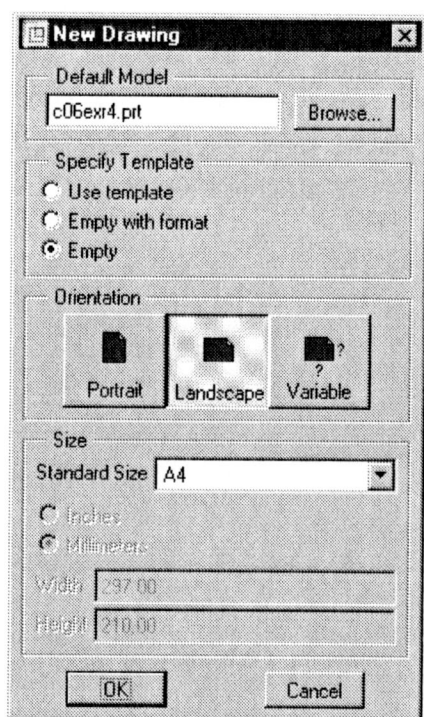

Figure 10-40 New Drawing dialog box with the Empty radio button selected

Generating, Editing, and Modifying the Drawing Views 10-27

in Figure 10-34. As soon as you specify the point, the **Drawing View** dialog box is displayed along with the preview of the drawing view in the default orientation. From the **Model view names** list box, select the **TOP** option and choose the **Apply** button. The default view is oriented as top view.

Note
Although the views once placed can be moved on the sheet, yet while specifying the placement point, try to place the drawing view inside the sheet and also leave space for the other views to be generated.

3. Choose the **Scale** option from the **Categories** list box in the **Drawing View** dialog box; the **Scale and perspective options** area is displayed. Select the **Custom scale** radio button and enter the scale factor for the view as **0.075**. Choose the **Apply** button; the scale of the view will be modified.

4. Choose the **Close** button from the **Drawing View** dialog box to complete placing the top view.

5. Turn off the display of datum planes, axes, points, and coordinate system by choosing their respective buttons from the **Datum Display** toolbar.

6. Also, change the model display to no hidden by choosing the **No Hidden** button from the **Model Display** toolbar. Now, choose the **Redraw the current view** button from the **Top Toolchest** to repaint the screen.

Tip: *If the view that you have placed on the drawing sheet is not at the proper location on the drawing sheet then you need to move the drawing view. To unlock the movement, choose the **Disallow the movement of drawing views with the mouse** button from the **Top Toolchest**. Select the drawing view; the selected drawing view is enclosed in an red box. Now, press the left mouse button inside the box and drag it to place it at the desired location.*

Generating the Front View

The front view of the model is generated from the top view, which is already placed on the drawing sheet.

1. Choose **Insert > Drawing View > Projection** from the menu bar; you will be prompted to specify a center point for drawing view.

2. Specify the center point for the front view below the top view, as shown in Figure 10-41.

3. Click anywhere on the drawing sheet to clear the current selection set.

Tip: *It is recommended that you use the **Redraw the current view** button available in the **View** toolbar to remove any temporary information in the graphics window and to refresh the screen.*

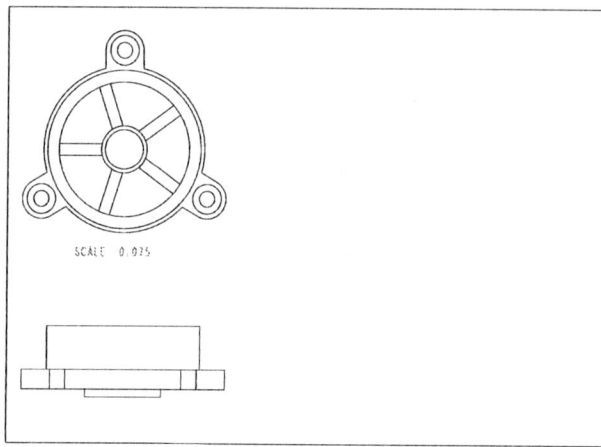

Figure 10-41 The top and front views

Generating the Section View

To generate section view, a section must be defined on the model. You will sketch this section to generate the section view for this tutorial.

1. Choose **Insert > Drawing View > Projection** from the menu bar; you are prompted to select a projection parent view.

2. Select top view as projection parent view. You are again prompted to specify a center point for drawing view.

3. Specify the center point for placement of the view on the right side of the top view.

4. Once the view is placed, double-click on it to display the **Drawing View** dialog box.

5. Select **Sections** from the **Categories** list box in the **Drawing View** dialog box; the section options are invoked in **Drawing View** dialog box.

6. Select the **2D cross-section** radio button and choose the plus (+) button; the window area below is activated. Select the **Create New** from the **Name** drop-down list; the **XSEC CREATE** menu is displayed.

7. Choose **Offset > Both Sides > Single > Done** from the **XSEC CREATE** menu.

8. Specify the name of the cross section as **X** in the **Message Input Window** and press ENTER. Once you have specified the name of the cross section, a separate window will appear displaying the part. You are prompted to select a sketching plane.

Generating, Editing, and Modifying the Drawing Views 10-29

Note
*If the model in the **Part** mode is opened in another window, the subwindow will not appear and you have to manually change the window. Choose the **Window** menu, and select the part file that is opened. Now, you can continue with the sketcher environment.*

9. Choose the **Datum planes on/off** button from the **Datum Display** toolbar in the original window to display the datum planes. You need to repaint the screen in the subwindow by choosing **View > Repaint** from the menu bar to view the datum planes.

10. Select the **TOP** datum plane from the subwindow. Choose **Okay** from the **DIRECTION** submenu. The **SKET VIEW** submenu is displayed.

11. Choose the **Bottom** option from the **SKET VIEW** submenu and select the **FRONT** datum plane from the subwindow. The **References** dialog box is displayed. Choose the **Close** button to exit the **References** dialog box.

12. Choose the **Sketch >Line >Line** from the menu bar and sketch the line, as shown in Figure 10-42. These lines define the section plane.

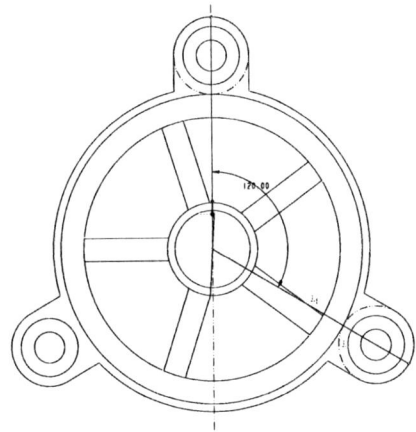

Figure 10-42 Sketch for the total align section

13. Align the lines and modify the angular dimension to **120**, as shown in Figure 10-42.

14. Choose **Sketch > Done** from the menu bar.

15. Select the **Full (Aligned)** option from the **Sectioned Area** drop-down list; you are prompted to select an axis.

16. Select the central axis of the model from the front drawing view. You may need to turn on the display of the central axis if its display is turned off.

17. Scroll the bar in the **Drawing View** dialog box to the right edge. Click once in the field

 **Tip**: *If you do not want to display the section arrows in any view, you can use the middle mouse button (in case of three button mouse) to abort the creation of arrows.*

below the **Arrow Display** column; you will be prompted to pick a view for arrows where the section is perpendicular. Select the top view to display the arrows.

18. Choose the **Apply** button from the **Drawing View** dialog box and then exit the dialog box. The section view is displayed.

19. Turn off the display of the datum planes and the datum axes.

Modifying the Hatching

You need to modify the spacing between the hatch lines in order to make the hatching dense. In this tutorial, you will select the filter that is available to select the hatch lines. This filter is available from the **Filter** drop-down list.

1. Select the **X-Section** filter from the **Filter** drop-down list that is present in the status bar below the **Message Area**. Select the hatching from the drawing sheet; the hatching lines turn red in color.

2. Press and hold the right mouse button to invoke the shortcut menu. Choose the **Properties** option from the shortcut menu; the **MOD XHATCH** menu is displayed.

3. From the **MOD XHATCH** menu, choose the **Spacing** option to display the **MODIFY MODE** submenu.

4. The spacing between the hatching lines has to be reduced. Choose the **Half** option twice and then choose **Done** in the **MOD XHATCH** menu. Now, the hatching appears to be more dense, as shown in Figure 10-43.

5. Set the filter back to the **Drawing Item and View** option using the **Filter** drop-down list.

Generating the General View

The **General** view is generated to show a 3D view of the model, which is the trimetric view.

1. Choose the **Create a general view** button from the **Top Toolchest**.

2. Specify the center point for the placement of the general view below the section view, as shown in Figure 10-43. The **Drawing View** dialog box is displayed.

3. Set the value of the scale for the new view to **0.05** using the **Scale** option from the **Drawing View** dialog box.

4. Choose **Apply** and then exit the dialog box.

Generating, Editing, and Modifying the Drawing Views 10-31

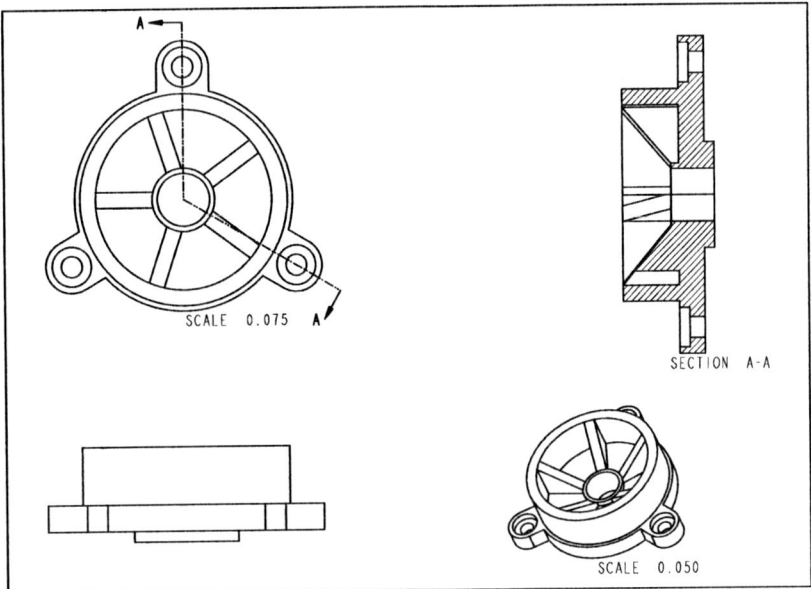

Figure 10-43 Different drawing views

Generating the Detail View

As mentioned earlier, the detail view is required to provide details of a particular portion of the drawing view. In this tutorial you need to give the details of one of the ribs of the model.

1. If the general view is selected (highlighted in red box), then you need to click once on the drawing sheet to exit the current selection set. Choose the **Insert > Drawing View > Detailed** from the menu bar.

   You will be prompted to select a center point for detail on an existing view. This center point will be the center of the detailed view.

2. Select the center point for the detail on one of the ribs in the trimetric view, refer to Figure 10-44.

   You will be prompted to sketch a spline about the center point that you selected.

3. Draw the spline. After the spline is drawn, press the middle mouse button to quit the spline tool.

4. Select the center point for the placement of the drawing view. The detailed view is generated and is selected by default.

5. Press and hold down the right mouse button on the detail view to invoke the shortcut menu and choose the **Properties** option; the **Drawing View** dialog box is displayed.

6. Choose **Spline** from the **Boundary type on parent view** drop down list.

7. Choose the **Scale** option from the **Categories** list box. Set the value of the custom scale to 0.075. Choose the **Apply** button and exit the dialog box.

A note is also displayed with an arrow attached to the trimeric view. You may need to move the note to a suitable position by selecting it and then dragging the cursor. The final sheet after generating all views is shown in Figure 10-44.

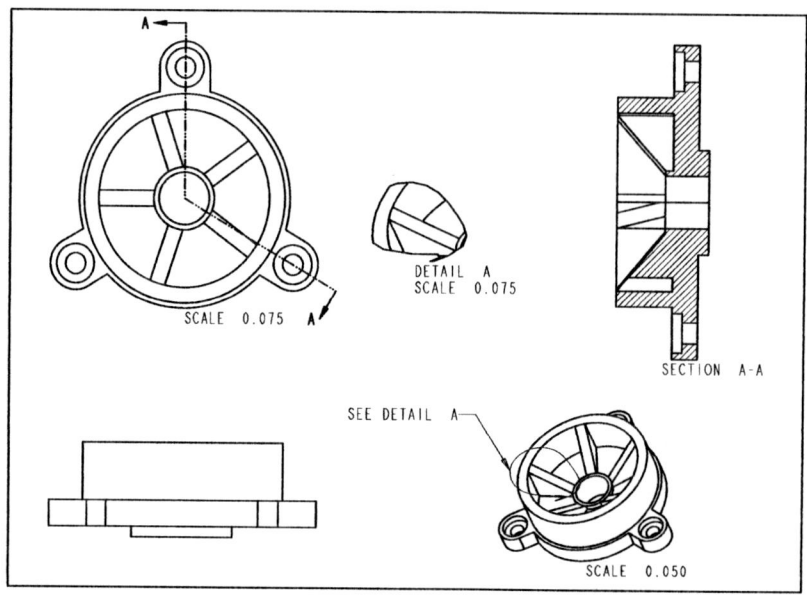

Figure 10-44 The drawing views for Tutorial 1

Note
*To select any item or drawing view from the drawing sheet, you have to invoke the selection mode. The selection mode is invoked by choosing the **Select items** button from the **Right Toolchest**.*

Saving the Drawing File

You need to save the drawing file that you have created as you may need it later.

1. Choose the **Save the active object** button from the **Top Toolchest**. The **Save Object** dialog box is displayed with the name of the drawing file that you entered earlier.

2. Press ENTER to confirm the saving of the file.

 **Tip**: *If any of the views or the text on the drawing is overlapping or is not at the desired place on the sheet, select it and then move it by dragging the mouse.*

Generating, Editing, and Modifying the Drawing Views

Closing the Drawing File

After you have saved the drawing file that you have created, you need to close the drawing file.

1. Choose **Window > Close** from the menu bar to close the drawing window.

Tutorial 2

In this tutorial, you will generate drawing views of the part created in Tutorial 1 of Chapter 7. The part is shown in Figure 10-45. The drawing views that need to be generated are shown in Figure 10-46. **(Estimated time: 45 min)**

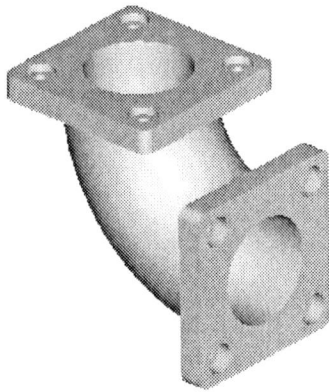

Figure 10-45 Model for generating the drawing views

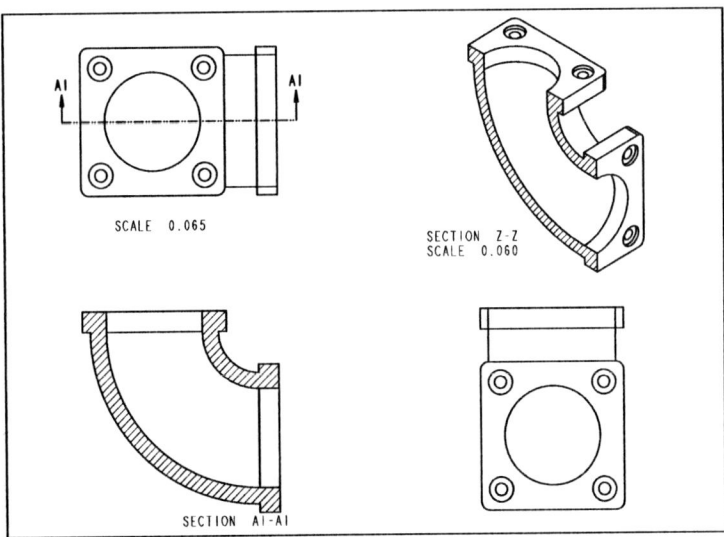

Figure 10-46 Drawing views to be generated

Before you start generating the drawing views, copy the file *c07tut1.prt* from *c07* folder in the current directory.

The following steps are required to complete this tutorial:

a. Start a new drawing file and select the size of the drawing sheet.
b. Generate the top view, refer to Figure 10-47.
c. Generate the sectioned front view by defining the **FRONT** datum plane as the section plane, refer to Figure 10-47.
d. Generate the right-side view of the sectioned front view, refer to Figure 10-48.
e. Generate the isometric sectioned view. The section will be defined by drawing a line on the **TOP** datum plane, refer to Figures 10-49 and 10-50.

Before starting to generate the drawing views, set the working directory to *C:\ProE-WF-2.0\c10*. The part *(.prt)* file and the drawing *(.drw)* file should lie in the same directory or folder.

Starting New Drawing File

To generate the drawing views, you first need to start a new drawing file.

1. Choose the **Create a new object** button from the **File** toolbar to display the **New** dialog box.

2. Select the **Drawing** radio button and then enter the name of the file as *c10tut2*.

3. Choose **OK** from the **New** dialog box to display the **New Drawing** dialog box.

4. Choose the **Browse** button to select *c07tut1.prt* from *c10* for generating the drawing views.

5. Select the **Empty** radio button from the **Specify Template** area.

6. Choose the **Landscape** button from the **Orientation** area.

7. Select **A4** from the **Standard Size** drop-down list. Choose **OK** from the **New Drawing** dialog box to proceed to the **Drawing** mode.

Generating the Top View

First the top view will be generated. All other views, except the sectioned isometric view, will be the child views of the top view. You need to generate the top view first because the required sectioned front view can be generated only from the top view. The right-side view can also be generated independently, but then this view will not help to generate any other required view.

1. Choose the **Create a general view** button from the **Top Toolchest**.

2. Specify the center point for the placement of the top view close to the upper left corner of the drawing sheet; the **Drawing View** dialog is displayed.

Generating, Editing, and Modifying the Drawing Views

3. Select the **TOP** option from the **Model view names** list box and choose the **Apply** button.

4. Choose **Scale** option from the **Categories** list box. Select the **Custom Scale** radio button and enter **0.065** in the edit box.

6. Choose **OK** to exit the **Drawing View** dialog box.

   If necessary, move the view, as shown in Figure 10-47, and choose **No Hidden** from the **Model Display** toolbar. You may also need to repaint the screen.

Generating the Sectioned Front View

The sectioned front view of the model is generated from the top view. Before proceeding further, use the **Datum planes on/off** button from the **Datum Display** toolbar to turn on the display of datum planes and repaint the screen.

1. Choose the **Insert > Drawing View > Projection** from the menu bar.

2. Specify the center point for the placement of the front view below the top view, as shown in Figure 10-47.

3. Select the newly generated view and invoke the shortcut menu. Choose the **Properties** option to display the **Drawing View** dialog box.

4. Select the **Sections** option from the **Categories** list box to display the section related options in the dialog box.

5. Select the **2D cross-section** radio button and choose the plus (+) button, the window area below it gets activated. Select the **Create New** from the **Name** drop-down list, if it is not selected. The **XSEC CREATE** menu is displayed.

6. Choose **Planar > Single > Done** from the **XSEC CREATE** menu.

7. Enter the name of the cross section in the **Message Input Window** as **A1** and press ENTER to display the **SETUP PLANE** submenu. You are prompted to select a planar surface or a datum plane.

8. Select the **FRONT** datum plane (the plane that cuts the part horizontally from the center of the cylindrical feature in the top view) from the graphics window.

9. Scroll the bar in the **Drawing View** dialog box to the right edge. Click on the field below the **Arrow Display** column. You will be prompted to pick a view for arrows where the section is perpendicular. Select the top view to display the arrows.

10. Choose the **Apply** button and then exit the dialog box.

Modifying the Hatching

The offset distance between the hatching lines in the front sectioned view is large. You need to reduce the distance between the hatching lines.

1. Select the **X-Section** filter from the **Filter** drop-down list present in the status bar. Select the hatching from the sectioned front view in the drawing sheet; the hatching lines turn red in color. Hold down the right mouse button to invoke the shortcut menu. Choose the **Properties** option from the shortcut menu.

   The **MOD XHATCH** menu is displayed.

2. From the **MOD XHATCH** menu, choose the **Spacing** option to display the **MODIFY MODE** submenu.

3. The spacing between the hatching lines has to be reduced. Choose the **Half** option twice and then choose **Done** in the **MOD XHATCH** menu. Now, the hatching appears to be more dense.

4. Click once in the graphics window to remove the X-hatch from the current selection set. The sheet after placing these two views should look similar to the one shown in Figure 10-47.

5. Set the selection filter back to **Drawing Item and View**.

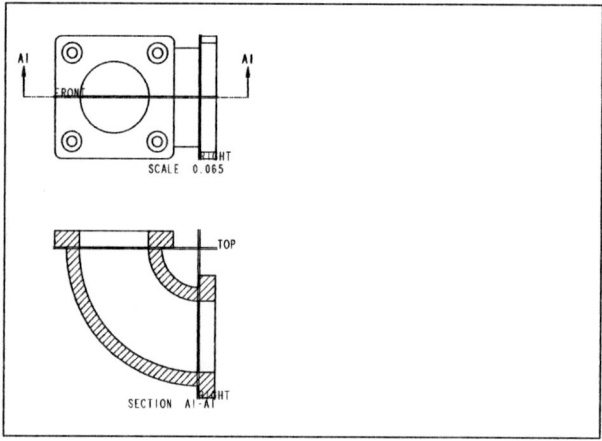

Figure 10-47 Drawing sheet after generating the top view and the sectioned front view with the display of datum planes

Generating the Right-Side View

The right-side view is the projection of the sectioned front view. Before proceeding further, turn off the display of datum planes.

Generating, Editing, and Modifying the Drawing Views　　　　　　　　　　　　　　　　　　**10-37**

1. Choose **Insert > Drawing View > Projection** from the menu bar; you will be prompted to select projection parent view.

2. Select sectioned front view as the parent view.

3. Specify the center point for the placement of the drawing view on the right side of the sectioned front view. The right-side view of the model is placed, as shown in Figure 10-48.

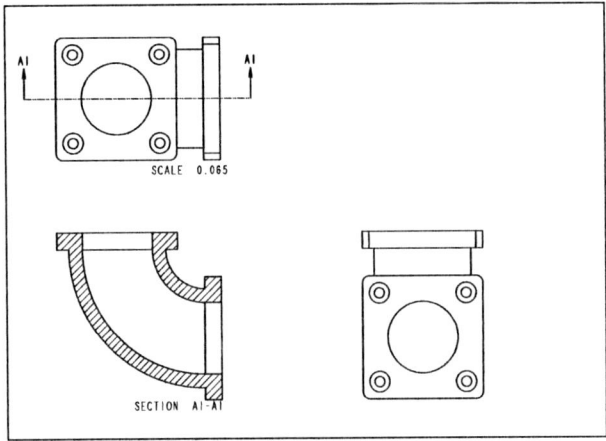

Figure 10-48 Drawing sheet after generating the top, sectioned front, and right-side views

Generating the Isometric Section View

The isometric section view is an independent view and will be generated by using the **Create a general view** option.

1. Choose the **Create a general view** button from the **Top Toolchest**.

2. Specify the center point for the placement of the view close to the upper right corner of the drawing sheet; the **Drawing View** dialog is displayed.

   The default view is a trimetric view but you need the isometric view to be displayed. Therefore, you need to change the orientation.

3. Select the **Isometric** option from the **Default orientation** drop-down list; the isometric view of the model is displayed on the drawing sheet.

4. Choose the **Scale** option from the **Categories** list box. Enter scale factor value as **0.06** by selecting **Custom scale** radio button and click the **Apply** button.

5. Select the **Sections** option from the **Categories** list box to display the section options in the dialog box.

6. Select the **2D cross-section** radio button and then choose the plus (+) button. The collector below it gets activated. Select **Create New** from the name drop-down list; the **XSEC CREATE** menu manager is displayed.

7. Choose **Offset > Both Sides > Single > Done** from the **XSEC CREATE** menu.

8. Enter the name of the section as **Z** in the **Message Input Window** that appears and press ENTER. Once you have specified the name of the section, a separate window will appear displaying the model.

9. Select the **TOP** datum plane from the subwindow. You may need to turn on the display of the datum planes. Choose **Okay** from the **DIRECTION** submenu. The **SKET VIEW** submenu is displayed.

10. Choose the **Right** option from the **SKET VIEW** submenu and select the **RIGHT** datum plane from the subwindow; the **References** dialog box is displayed. Choose the **Close** button to exit the **References** dialog box.

11. Choose the **Sketch > Line > Line** from the menu bar and draw the line, as shown in Figure 10-49. This line creates a section plane.

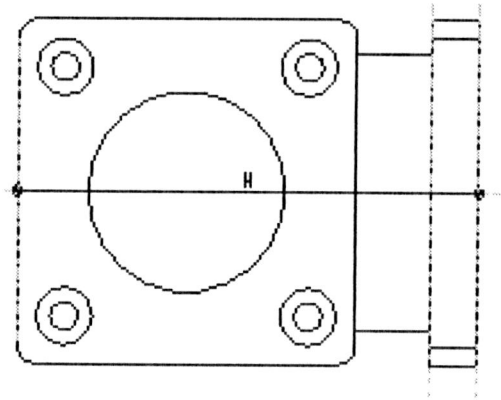

Figure 10-49 Sketch for the section line with the constraint

12. Choose the **Constrain** option from the **Sketch** menu; the **Constraints** dialog box is displayed.

13. Choose the **Create same points, points on entity or collinear constraint** button from the **Constraints** dialog box. Align both the ends of the line to the edges.

14. Choose **Sketch > Done** from the menu bar.

15. Choose the **Apply** button to place the view and then exit the dialog box.

Note
If the drawing view is placed on the sheet such that it overlaps the boundary of the drawing sheet then you can move the drawing view. Select the drawing view; the selected drawing view is enclosed in an red box. Now, press the left mouse button inside the box and drag it to place it at the desired location.

Modifying the Hatching

The spacing between the hatching lines in the sectioned isometric view is large. Therefore, you need to reduce the distance between the hatching lines. Use the same procedure

Generating, Editing, and Modifying the Drawing Views

that was discussed earlier in this tutorial to modify the spacing between the hatch lines. The sheet after placing all views should look similar to the one shown in Figure 10-50.

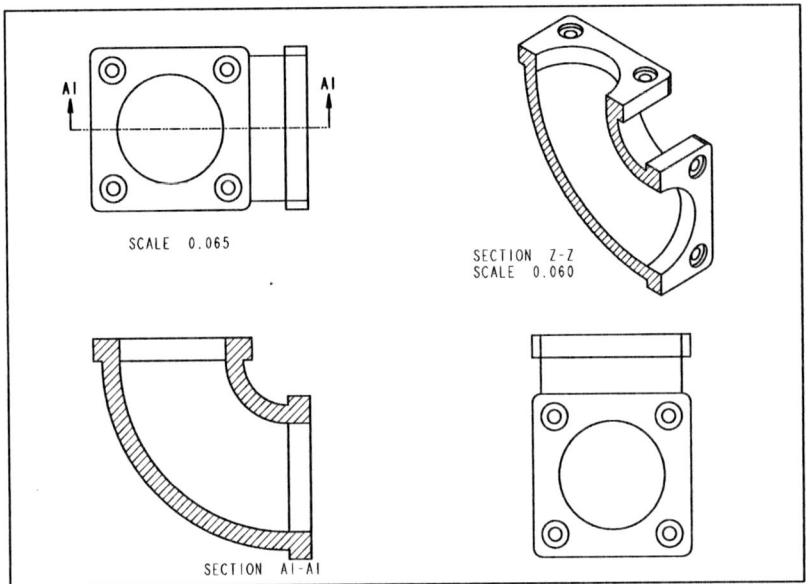

Figure 10-50 The different drawing views generated

Saving the Drawing File

You need to save the drawing file that you have created as you may need it later.

1. Choose the **Save the active object** button from the **Top Toolchest**. The **Save Object** dialog box is displayed with the name of the drawing file that you had entered earlier.

2. Press ENTER to confirm the saving of the file.

Closing the Drawing File

After you have saved the drawing file that you have created, you need to close the drawing file.

1. Choose **Window > Close** from the menu bar; the drawing window is closed.

Self-Evaluation Test

Answer the following questions and then compare your answers to the answers given at the end of this chapter.

1. The bidirectional associative nature of a software package means that when one file related to the part model is modified the corresponding modification can be seen in other related files. (T/F)

2. General view is the first view that is generated in the sheet. (T/F)

3. The full Section view is the most widely used type of section view. (T/F)

4. The section views are generally created for the models having features that are not clearly visible in the standard view. (T/F)

5. Broken views are used for parts that have a high length to width ratio. (T/F)

6. The _____ view is used when you want to show a particular portion of the view in section and at the same time not section the remaining view.

7. The _____ option allows you to temporarily remove the selected drawing view from the sheet.

8. The _____ option is used to redisplay the drawing views that are erased using the **Erase View** option.

9. Using the _____ dialog box, you can reorient the general view.

10. The cross-section hatching on the sectioned portion can be modified using the _____ menu.

Review Questions

Answer the following questions:

1. Which of the following options when selected, displays all erased drawing views on the sheet?

    (a) **Delete** (b) **Resume View**
    (c) **Move** (d) **Erase**

2. Which of the following buttons on the **View** toolbar is used to refresh the screen?

    (a) **Zoom Out** (b) **Zoom In**
    (c) **Redraw the current view** (d) **None**

Generating, Editing, and Modifying the Drawing Views

3. Which of the following options when selected, displays only that area of the section view that is sectioned?

   (a) **Total**   (b) **Area**
   (c) **Align**   (d) None

4. Which of the following options is used to permanently remove a drawing view from the sheet?

   (a) **Resume** (b) **Erase**
   (c) **Delete** (d) **Move**

5. Which of the following options is used to modify any numeric value associated with the drawing views?

   (a) **Xhatching** (b) **Any Item**
   (c) **Value**   (d) None

6. The view type of an existing view can be changed. (T/F)

7. You can reorient only the general views. (T/F)

8. You can flip the side of the cross section views. (T/F)

9. The **Orientation** area in the **New Drawing** dialog box is available only when you select the **Empty** radio button from the **Specify Template** area. (T/F)

10. The orientation of a model saved in the **Part** mode can be used to orient the drawing view in the **Drawing** mode. (T/F)

Exercise

Exercise 1

Generate the drawing views of the model created in Tutorial 3 of Chapter 8 on an A4 size sheet. The model is shown in Figure 10-51. The drawing views to be generated are shown in Figure 10-52.   **(Estimated time: 45 min)**

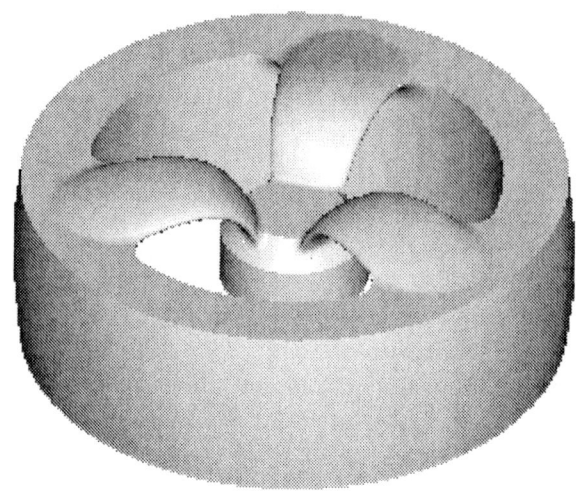

Figure 10-51 Part for generating the drawing views

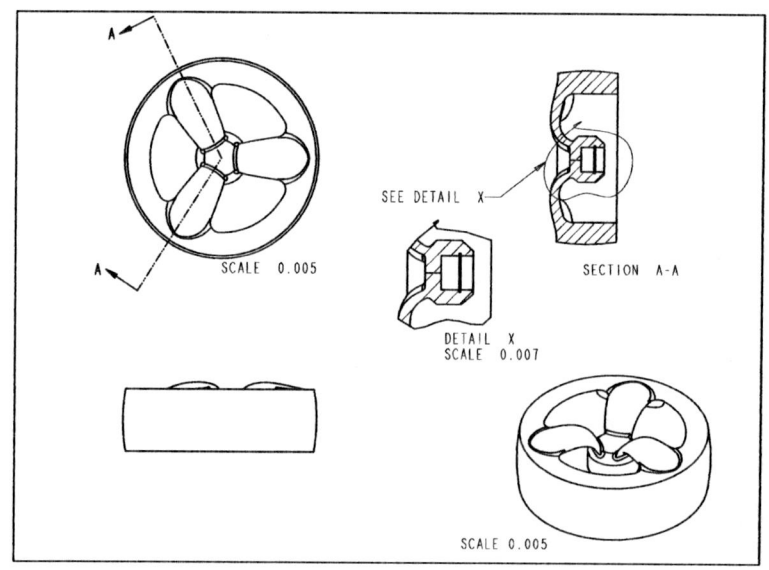

Figure 10-52 Drawing views to be generated in Exercise 1

Answers to the Self-Evaluation Test

1 - T, 2 - T, 3 - T, 4 - T, 5 - T, 6 - Local Section, 7 - Erase, 8 - Resume View, 9 - Drawing View, 10 - MOD XHATCH

Chapter 11

Dimensioning the Drawing Views

Learning Objectives
After completing this chapter you will be able to:
- *Show or erase the dimensions in the drawing views.*
- *Modify and edit the dimensions.*
- *Add reference datums to the drawing views.*
- *Add dimensional and geometric tolerance to the drawing views.*
- *Edit the geometric tolerance.*
- *Add notes to the drawing.*
- *Add balloons to the exploded assembly views.*

DIMENSIONING THE DRAWING VIEWS

Once you have generated the drawing views, you need to generate the dimensions, add notes, symbols, balloons, and so on in the drawing views. These dimensions are assigned to each entity of the sketches in the model or to the features associated with the model. These dimensions are associative in nature and hence can be used to modify or drive the dimensions of a part. Therefore, these dimensions are also called driving dimensions. Dimensions are generated using the **Show / Erase** dialog box shown in Figure 11-1.

Note
*You can also create dimensions in the **Drawing** mode, but this type of dimension cannot drive the dimensions of the part. Therefore, in Pro/ENGINEER Wildfire 2.0 the dimensions you create are called driven dimensions. The creation of dimensions is discussed in Chapter 12.*

Tip: *You must have noticed that some options in the menus available in the menu bar are followed by dots. The dots indicate that after you choose any of these options, a menu, a dialog box, or a dashboard will be displayed.*

The **Show / Erase** dialog box is displayed when you choose the **Open the Show \ Erase dialog box** button from the **Top Toolchest**. It can be also invoked by choosing View > **Show and Erase** from the menu bar.

Show / Erase Dialog Box Options
Show
The **Show** button is chosen to select the options to generate the various parameters related to the model in the drawing views. The options that are displayed when you choose this button are discussed next.

Type Area
This area contains the buttons for various parameters related to the drawing views. The options to display the various parameters will be available only when you select at least one button from this area.

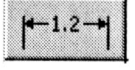
Dimension. The **Dimension** button is chosen to generate the parametric dimensions in the drawing views. The bidirectional associative nature of this software package is valid when you change the dimension value generated using this option.

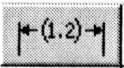
Reference Dimension. The **Reference Dimension** button is chosen to generate the reference dimensions in the drawing views.

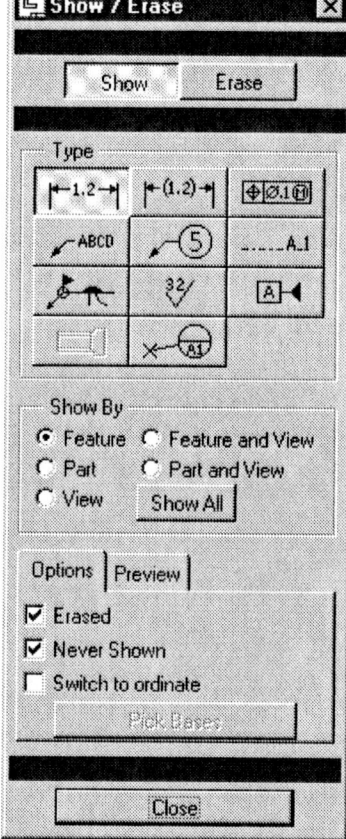

*Figure 11-1 The **Show / Erase** dialog box with the **Show** button*

Dimensioning the Drawing Views   11-3

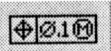 **Geometric Tolerance**. The **Geometric Tolerance** button is chosen to display the geometric tolerances in the dimensions.

 **Tip**: *The dimensional tolerance will be displayed only when the **tol_display** is set to **yes** in the configuration file. To invoke this file, choose **File** > **Properties** from the menu bar; the **FILE PROPERTIES** menu is displayed. Choose the **Drawing Options** option to open the configuration file. Now, scroll down to **tol_display** option in the **These options contrl dimensions** heading in the left pane of the dialog box. By default, this option is set to **no**. Select this option and then select **yes** from the **Value** drop-down list available below the two panes.*

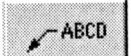 **Note**. The **Note** button is chosen to display the notes that were created in the **Part** mode. If any change is made to a note in the **Drawing** mode, the note in the **Part** mode is automatically modified. Similarly, if you modify this note in the **Part** mode, the modifications are reflected automatically in the **Drawing** mode.

 **Balloon**. The **Balloon** button is chosen to display the balloons associated with an assembly in the drawing views.

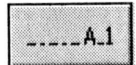 **Axis**. The **Axis** button is chosen to display the axis of the selected hole or cylindrical feature in the drawing views.

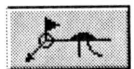 **Symbol**. The **Symbol** button is chosen to display any symbol associated with the model.

 **Surface Finish**. The **Surface Finish** button is chosen to display the surface finish symbols associated with the part in the drawing views.

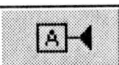 **Datum Plane**. The **Datum Plane** button is chosen to display the reference datum planes in the drawing views.

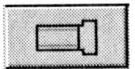 **Cosmetic Feature**. The **Cosmetic Feature** button is chosen to display the cosmetic features in the drawing views that are associated with the model.

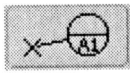 **Datum Target**. The **Datum Target** button is used to show the critical measurement locations on the drawing view.

Show By Area
The options in this area are discussed next.

Feature. The **Feature** radio button is selected to display the required annotation of the selected feature. If the required annotation of the selected feature cannot be displayed in the view in which the feature was selected, the required annotation will be displayed in the appropriate view.

Part. The **Part** radio button is selected to display all required annotations of the selected part. The required annotations are displayed in all views irrespective of the view in which the part was selected.

View. The **View** radio button is selected to display all required annotations that can be displayed in the selected view.

Feature and View. The **Feature and View** radio button is selected to display the required annotations of a particular selected feature that can be displayed in the selected view.

Part and View. The **Part and View** radio button is selected to display all required annotations of the selected part that can be displayed in the selected view.

Show All. The **Show All** radio button is selected to show all required annotations in the drawing views. When you choose this button all required annotations that were used in creating the model are displayed in the drawing views.

Options Tab
The options under this tab of the **Show / Erase** dialog box are discussed next.

Erased. The **Erased** check box is selected to display the annotations that were once erased. If this check box is cleared, the annotations that were once erased using the **Show / Erase** dialog box will not be displayed the next time you wish to display them.

Never Shown. The **Never Shown** check box is selected to display the annotations that are never displayed in the drawing views. At least one of the **Erased** or the **Never Shown** check box remains selected while specifying the annotation.

Switch to ordinate. The **Switch to ordinate** check box is selected to switch to the ordinate mode of dimensioning. When you select this check box, you will be prompted to select a dimension that will be the baseline dimension for the other ordinate dimensions. Ordinate dimensioning is discussed in Chapter 2.

Preview Tab
The options under this tab of the **Show / Erase** dialog box are discussed next.

With Preview. The **With Preview** check box is selected to allow you to confirm the placement of the annotation. When you show the annotations with this check box selected, you will be prompted to confirm the annotations that you want to keep or delete. You can select the annotations to keep, to delete, accept all, or delete all using the buttons provided under this area. The remaining options under this area will be available only after you place the annotations. If this check box is cleared, you will not be prompted to confirm the placement of the annotation. All displayed annotations will be automatically placed in the drawing views.

Erase

The options under the **Erase** button are shown in Figure 11-2. These options are used to erase the annotations placed using the options under the **Show** button. The annotations once placed in the drawing view can also be erased by selecting it. When the selected annotation turns red in color, hold down the right mouse button and choose the **Erase** option from the shortcut menu that appears. The selected annotation gets erased from the drawing view.

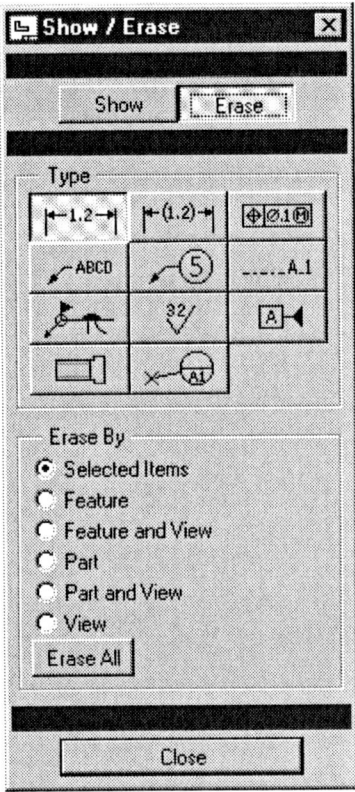

*Figure 11-2 The **Show / Erase** dialog box with the **Erase** tab*

Note
*The annotations erased using the **Erase** options are not deleted from the model.*

Type Area

The buttons provided under this area are similar to those under the **Type** area displayed when you choose the **Show** button. The only difference is that here these buttons are selected to choose the type of items to be erased.

Erase By Area

The options in this area are discussed next.

Selected Items. If the **Selected Items** radio button is selected then you can select the annotations to be erased using the left mouse button.

Feature. If the **Feature** radio button is selected then all annotations of the feature that you select will be erased from all views.

Feature and View. If the **Feature and View** radio button is selected then all annotations related to the selected feature will be erased from the selected view.

Part. If the **Part** radio button is selected then all annotations related to the selected part will be erased from all drawing views.

Part and View. If the **Part and View** radio button is selected then all annotations related to the selected part will be erased from the selected view.

View. The **View** radio button is selected to erase all annotations displayed in the selected view.

Erase All. The **Erase All** button is chosen to erase all annotations displayed in all views of the current sheet. When you choose this button, you will be prompted to confirm that all annotations can be erased.

Figure 11-3 shows the dimensioned drawing views.

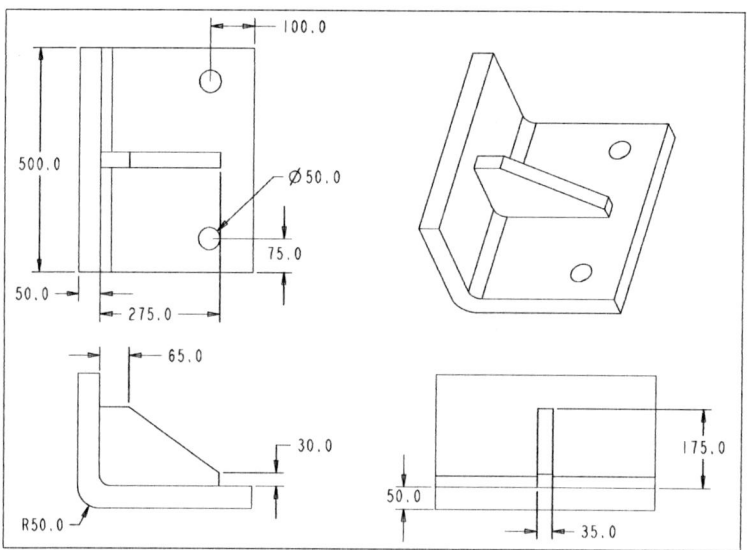

Figure 11-3 Dimensioned drawing views

MODIFYING AND EDITING DIMENSIONS

Pro/ENGINEER allows you to edit and modify the dimensions assigned to the drawing views. There are various methods of editing and modifying the driving dimensions. Before these methods are discussed, it is important for you to learn the terminology used in these methods. Figure 11-4 shows the terminology that is used in the **Drawing** mode of Pro/ENGINEER Wildfire 2.0.

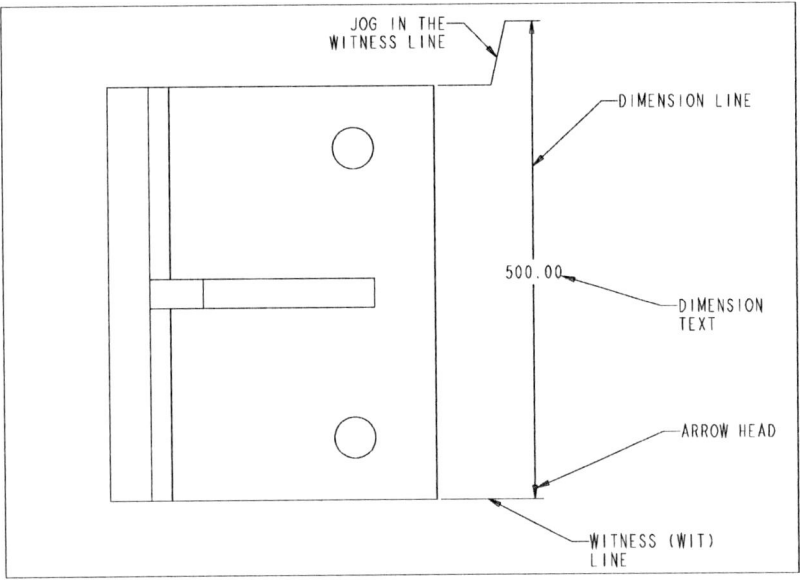

Figure 11-4 *Parameters associated with the dimensions*

 **Tip**: *To display the dimensions in a 3D view, you will have to first set the value of **allow_3d_dimensions** to **yes** in the configuration file. If this value is set to **no** then the dimensions will not be displayed in 3D views. To open the configuration file, choose **File > Properties** from the menu bar; the **FILE PROPERTIES** menu is displayed. Choose the **Drawing Options** option to open the **Options** dialog box. This dialog box lists all variables that can be varied for the current session.*

Modifying the Dimensions Using the Dimension Properties Dialog Box

To modify the dimensions using this option, select the dimension and then choose **Edit > Properties** from the menu bar; the **Dimension Properties** dialog box is displayed. The second method of invoking the **Dimension Properties** dialog box is to select the dimension and hold down the right mouse button to invoke the shortcut menu. Choose the **Properties** option from the shortcut menu. The **Dimension Properties** dialog box is displayed, as shown in Figure 11-5. The third method of invoking this dialog box is to double-click on the dimension value. The options in this dialog box are discussed next.

Properties Tab
Value and tolerance Area
The options provided under this area are related to the tolerance in the drawing.

Tolerance mode. This drop-down list will be available only when the value of **tol_display** is set to **yes** in the configuration file. The options under this drop-down list are used to select the type of tolerance mode. Depending on the option selected from this drop-down list, the other options in this area change.

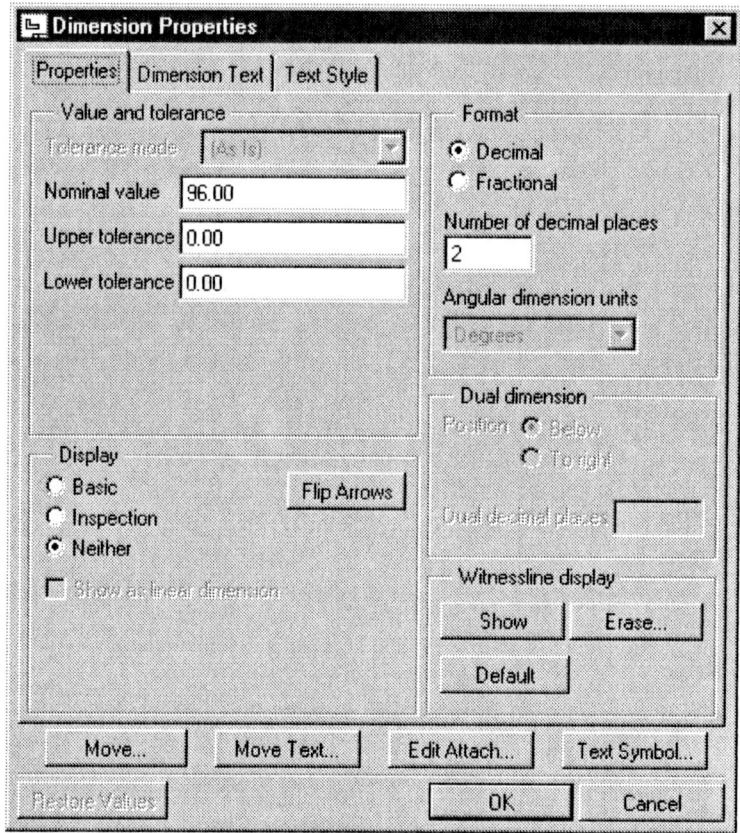

*Figure 11-5 The **Properties** tab of the **Dimension Properties** dialog box*

Format Area

Decimal. The **Decimal** radio button is selected to display the dimensions in the decimal format. The number of decimal values can be set in the **Number of decimal places** edit box.

Fractional. The **Fractional** radio button is selected to display the dimensions in the fractional format. The largest denominator can be set using the **Largest denominator** edit box that is displayed when you select the **Fractional** radio button.

Dual dimension Area

The options under the **Dual dimension** area will be available only when the value of **dual_dimensioning** is not set to **no** in the configuration file. The options under this area are used to specify the placement point for the dual dimensioning and the number of decimal places.

Display Area

Basic. If the **Basic** check box is selected then the selected dimension values will be displayed inside a rectangular box.

Dimensioning the Drawing Views   11-9

Tip: *If you modify the nominal value of a dimension in the **Value and tolerance** area, you will have to regenerate the model using the **Update the display of all views in the active sheet** button to view the effect of modifications. The modified dimension appears blue in color.*

Inspection. If the **Inspection** check box is selected then the selected dimension will be displayed inside an elliptical box.

Neither. This radio button is used to use neither of the display types from this area.

Dimension Text Tab

The **Dimension Properties** dialog box with the options under the **Dimension Text** tab is shown in Figure 11-6.

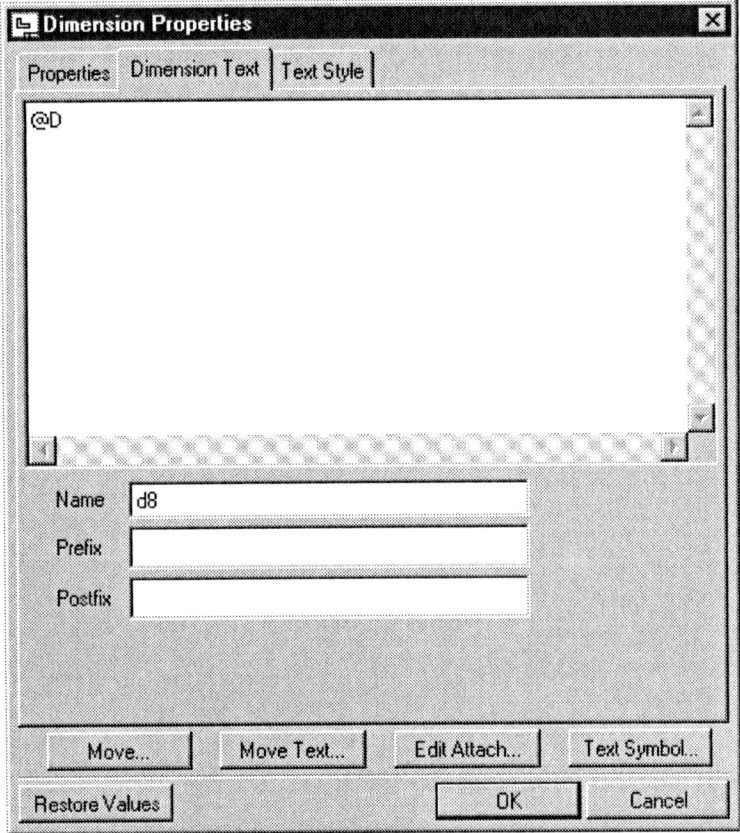

*Figure 11-6 The **Dimension Text** tab of the **Dimension Properties** dialog box*

Dimension Text Area

The **Dimension text** box displays the text of the dimension. You can edit the text using this box.

Name

The **Name** edit box display the dimension symbol. You can change the dimension symbol using this box.

Prefix

The **Prefix** edit box is used to add some additional text as prefix to the default dimension text.

Postfix

The **Postfix** edit box is used to add some additional text as suffix to the dimension text.

Tip: *The modifications made using the* **Dimension Properties** *dialog box are applied dynamically on the dimensions. You can view this in the background by moving the* **Dimension Properties** *dialog box on the side of the screen.*

Text Style Tab

The **Dimension Properties** dialog box with the options under the **Text Style** tab is shown in Figure 11-7.

*Figure 11-7 The **Text Style** tab of the **Dimension Properties** dialog box*

Dimensioning the Drawing Views

Copy from Area
The **Copy from** area displays the options to modify the text style of the dimension text. You can edit the text using this area.

Character Area
The **Character** area displays the options to modify the font style of the dimension text. Various parameters like height, font name, width of font, font style, and so on can be modified using the options available in this area.

Note/Dimension Area
The **Note/Dimension** area displays the options to modify the dimension or the text in the form of a note. The parameters that can be modified using this area are text justification, color, line spacing, angle, and so on.

Preview button
The **Preview** button is used to preview the changes that are made using the **Dimension Properties** dialog box.

Reset button
The **Reset** button is used to reset the default values.

In addition to the options available in various tabs, the following four buttons are available in the lower portion of this dialog box.

Move
The **Move** button is chosen to move the dimension to a new location on the current sheet. You can also move the dimension by selecting it and then dragging it by moving the mouse.

Move Text
The **Move Text** button is chosen to move the dimension text to a new location on the screen. Remember that only the text will be moved. The witness lines and the dimension lines do not move from their original location.

Modify Attach
The **Modify Attach** button is used to modify the attachment point of dimensions of some features like draft, round, and chamfer.

Text Symbol
The **Text Symbol** button is chosen to display the **Text Symbol** dialog box using which you can add special symbols to the default dimension text.

Modifying the Drawing Items Using the Shortcut Menu
When you select an item from a drawing view and hold down the right mouse button down, the shortcut menu with various options are displayed. These options depend on the item selected to modify. These options vary from item to item. Some of the items related to the drawing that can be modified using the shortcut menu are discussed next.

Value

You can modify the dimension value by selecting a dimension value. When the dimension value and the arrows turn red, click on the dimension value. Notice that now only the value is highlighted in red. Hold down the right mouse button to invoke the shortcut menu. From the shortcut menu choose the **Edit Value** option. The dimension value is displayed in the edit box. Modify the value in the edit box and press ENTER. To view the effect of the modification, you will have to regenerate the drawing view using **Update the model and redraw all views as specified by the latest modification** button from the **Top Toolchest**. This option is also used to modify the scale of the drawing view and the tolerance values.

Text

The parameters that can be modified are the selected text line, the entire text material, text height, text style, style library for the text, and the current style.

Miscellaneous Modifications

The modifications that can be made using the shortcut menu are: move a dimension, flip the arrows of the dimensions, switch the selected dimension from one view to the other, make a jog in the witness line, modify the values or text of the dimension, modify the arrow style, break the witness line, remove the break from the witness line, erase the witness line, show the erased witness line, or erase the selected dimension from the drawing view.

Number of Decimal Places

To modify the number of decimal values for the dimension text, choose **Format > Decimal Places** from the menu bar. You will be prompted to specify the number of decimal places in the **Message Input Window**. After you have specified the value, you will be prompted to select the dimensions to be modified. Select the dimension using the left mouse button and then press the middle mouse button to view the change in the decimal places.

Cleaning Up the Dimensions

Generally, the dimensions displayed in the drawing views are scattered and improperly placed. This in turn makes the drawing views look very confusing. To avoid this confusion, one method is to place every dimensions manually at the appropriate location in the view. The second and the more convenient method is to place these dimensions in proper order by choosing **Edit > Cleanup > Dimensions** from the menu bar, the **Clean Dimensions** dialog box is displayed and you are prompted to select dimensions or individual views to clean. When you select the dimension to clean, the options in this dialog box are enabled, as shown in Figure 11-8. These options are discussed next.

Placement Tab

The options under this tab of the **Clean Dimensions** dialog box are used for placing the dimensions. The distance between the two dimensions, the distance between the view boundary and the dimension, are set under this tab.

Space out dimensions

This option is used to maintain the distance between two dimensions. Most of the other options in this tab are available only when you select this option.

Dimensioning the Drawing Views

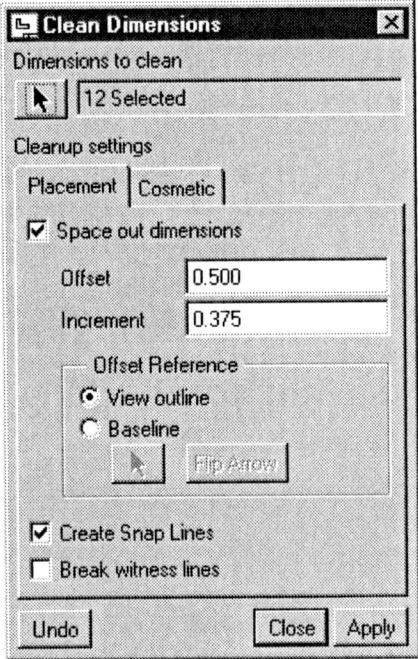

*Figure 11-8 The **Placement** tab of the **Clean Dimensions** dialog box*

Offset
The **Offset** edit box is used to specify the offset distance between the first dimension and the view boundary.

Increment
The **Increment** edit box is used to specify the offset distance between the two dimension lines.

Offset Reference Area
View outline. The **View outline** radio button is selected to specify the offset distances taking the outline of the view as reference.

Baseline. The **Baseline** radio button is selected to specify the offset distance taking a selected flat edge, datum plane, snap line, detail axis line, or view border as the reference. When you invoke this option, you will be prompted to select either of the above-mentioned entities as the reference.

Create Snap Lines
The **Create Snap Lines** check box is selected to create the snap lines using which the dimensions will be cleaned. The snap lines will be created at the distances specified using the **Offset** and the **Increment** edit boxes. If this check box is cleared, the snap lines will not be created.

Break witness lines

The **Break witness lines** check box is selected to break the witness lines if some dimensions are overlapping each other.

Cosmetic Tab

The options under the **Cosmetic** tab of the **Clean Dimensions** dialog box shown in Figure 11-9 are used for determining a dimension value's location with the dimension line. The arrows and the orientation of the dimension text can also be controlled using this tab.

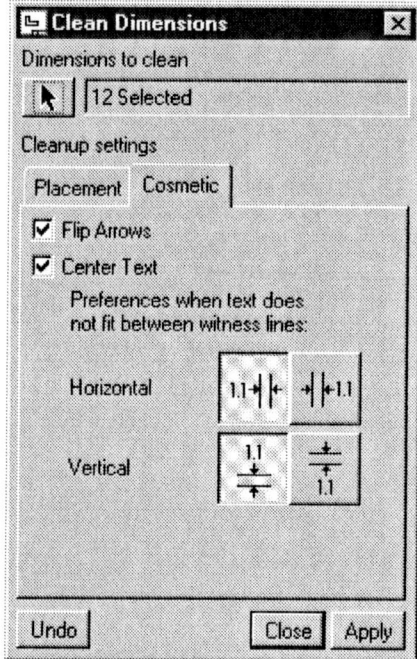

*Figure 11-9 The **Cosmetic** tab of the **Clean Dimensions** dialog box*

Flip Arrows

The **Flip Arrows** check box is selected to flip the arrows of the dimensions that are not easily adjusted inside the dimension lines.

Center Text

The **Center Text** check box is selected to place the dimensions at the center of the dimension lines.

Horizontal/Vertical

These buttons are used to specify the placement of the horizontal or vertical dimension text outside the dimension line if they do not fit.

Figures 11-10 and 11-11 explain the use of **Clean Dimensions** dialog box.

Dimensioning the Drawing Views

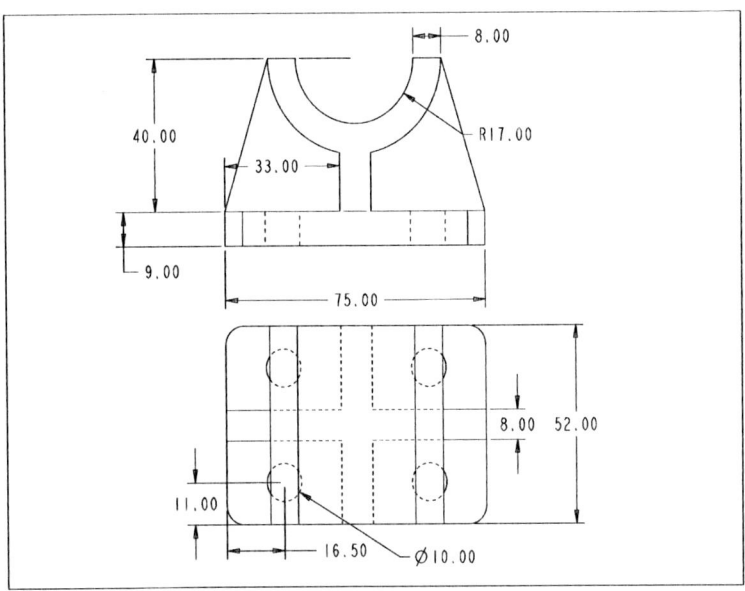

Figure 11-10 *Drawing views with dimensions before cleaning*

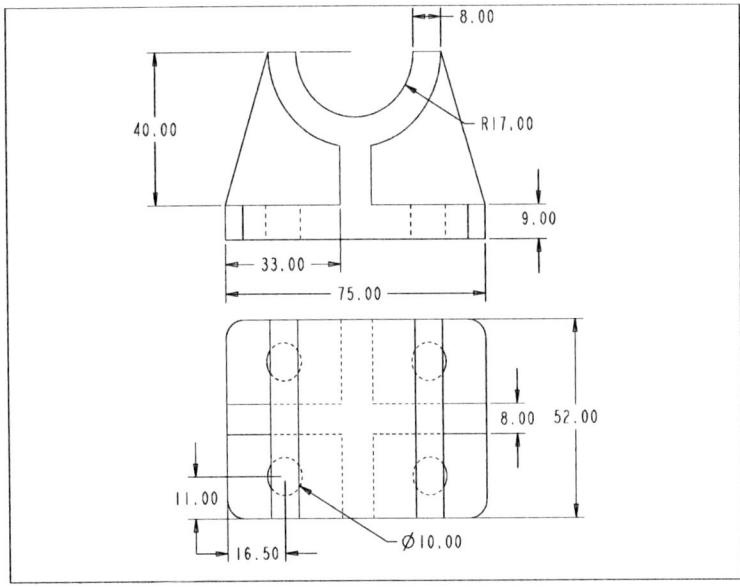

Figure 11-11 *Drawing views with dimensions after cleaning*

ADDING REFERENCE DATUMS TO THE DRAWING VIEWS

Reference datums are used as references for geometric tolerance. Datums are to be set before adding the geometric tolerance.

To create a reference datum, choose the **Datum Plane Tool** button from the **Right Toolchest** or choose **Insert > Model Datum > Plane** from the menu bar; the **Datum** dialog box is displayed, as shown in Figure 11-12. You can rename the selected datum and then choose the second button from the **Type** area to enclose the datum inside a feature control frame. You can place these reference datums freely in the drawing views or in a selected dimension. Figure 11-13 shows the drawing views with reference datums.

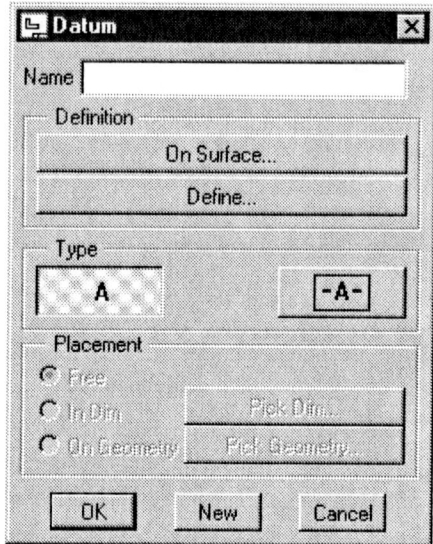

Figure 11-12 The **Datum** *dialog box*

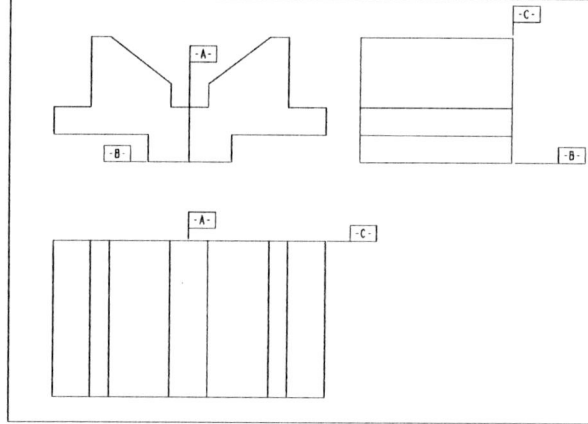

Figure 11-13 Drawing views with reference datums

Note

The system does not show a reference datum plane in a drawing view unless it is perpendicular to the graphics window.

ADDING TOLERANCES IN THE DRAWING VIEWS
Tolerance is defined as the difference between the maximum and minimum variations in the dimensions of the selected component. It is almost impossible to manufacture a component to the exact dimensions. In such cases, the tolerance value is added to dimensions to make sure that some variation that occurs during manufacturing can be taken care of. However, when you actually send a part for manufacturing, there are other parameters along with the dimension tolerances that may vary and that need some tolerances. Depending upon these factors, the tolerances are divided into two types: the dimensional tolerances and the geometric tolerances.

Dimensional Tolerances
These are the variation in the standard dimensional values. The dimensional tolerances can be easily displayed along with the dimensions in the drawing views by simply setting the value of **tol_display** to **yes** in the configuration file. You can select and modify the type of dimensional tolerances from the **Properties** tab of the **Dimension Properties** dialog box. Figure 11-14 shows the drawing views with the tolerance added to the dimensions.

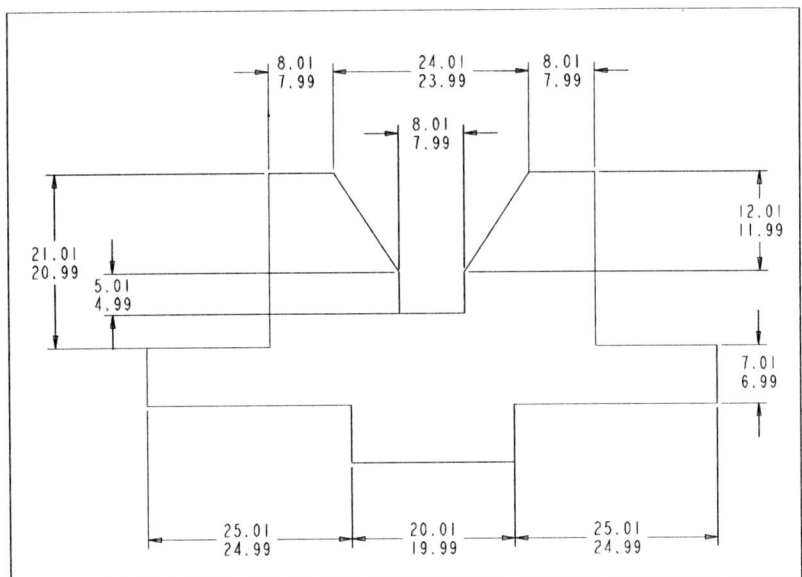

Figure 11-14 Dimensions with tolerances

Geometric Tolerances
When you send a component for manufacturing, you have to provide various other parameters

along with the dimensions and dimensional tolerances. These parameters can be the geometric condition, the surface profile, the material condition, and so on. All these parameters are defined using the geometric tolerances. The geometric tolerances (will be called gtol henceforth) can be added to the drawing using the **Geometric Tolerance** dialog box, shown in Figure 11-15. This dialog box is displayed when you choose **Insert > Geometric Tolerance** from the menu bar.

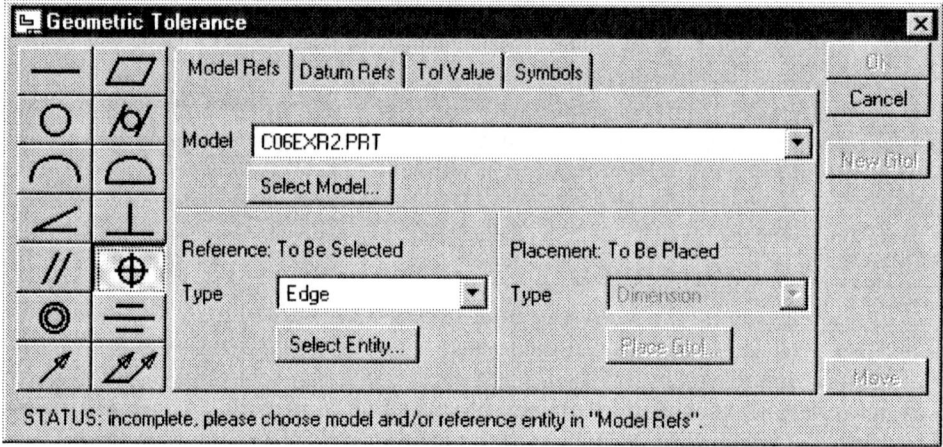

*Figure 11-15 The **Model Refs** tab of the **Geometric Tolerance** dialog box*

The options in the **Geometric Tolerance** dialog box are discussed next.

Model Refs Tab

The options under this tab are used to select the model or drawing where the gtol have to be placed and to specify the model references.

Model
This drop-down list is used to select the model on which you need to display the gtol. You can select the model from this drop-down list or by using the **Select Model** button.

Reference
The options under this area are used to select the reference for adding the gtol. The reference can be an edge, axis, surface, feature, datum, or an entity. You can select the reference type from the **Type** drop-down list or by using the **Select Entity** button.

Placement
The options under this area will be available only when you select a reference type for placing the gtol. These options are used to specify the placement type for the gtol. The placement type can be selected from the **Type** drop-down list.

Datum Refs Tab

The options under the **Datum Refs** tab, shown in Figure 11-16, are used to select the primary, secondary, or tertiary datum references for the gtol. You can also specify the composite tolerances to the primary, secondary, or tertiary datum references by using the **Composite**

Dimensioning the Drawing Views

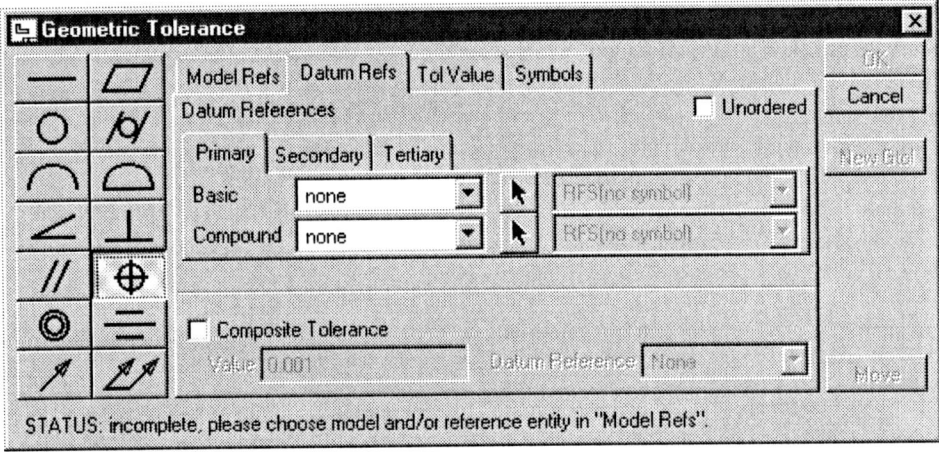

*Figure 11-16 The **Datum Refs** tab of the **Geometric Tolerance** dialog box*

Reference check box provided in this area. The **Composite Tolerance** check box is available only when the **Position** or the **Surface Profile** buttons from the types of gtol buttons available on the left side of the **Geometric Tolerance** dialog box are selected. The value of the composite reference can be specified in the **Value** edit box that is available only when you select the **Composite Reference** check box. The datum reference for the composite tolerance can be selected using the **Datum Reference** drop-down list that is available only when you select the **Composite Reference** check box.

Tol Value Tab

The options under the **Tol Value** tab, shown in Figure 11-17, are used to specify the tolerance value and the material condition for the gtol. You can specify the overall tolerance value or the per units tolerance value using the options under this tab. You can also specify the material condition using the **Material Condition** drop-down list.

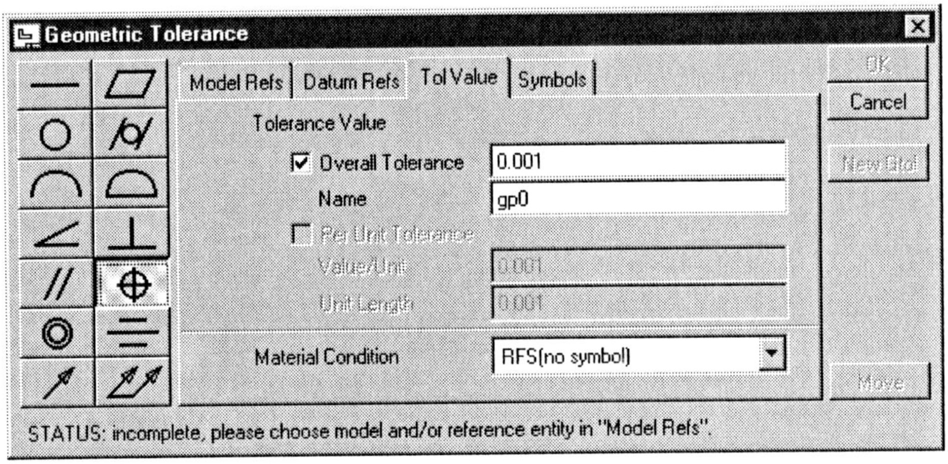

*Figure 11-17 The **Tol Value** tab of the **Geometric Tolerance** dialog box*

Symbols Tab

The options under the **Symbols** tab, shown in Figure 11-18, are used to specify the symbols for the gtol. The symbols and the modifiers for the gtol can be selected from the **Symbols and Modifiers** area and the projected tolerance zone symbol can be selected from the **Projected Tolerance Zone** area. The location of the projected tolerance zone symbol and its height can be specified using the options provided under this area. You can also specify the profile boundary symbol for the profile gtol from the **Profile Boundary** area that appears in the place of **Projected Tolerance Zone** area. The **Profile Boundary** area is available only if you choose the **Surface Profile or Line Profile** button.

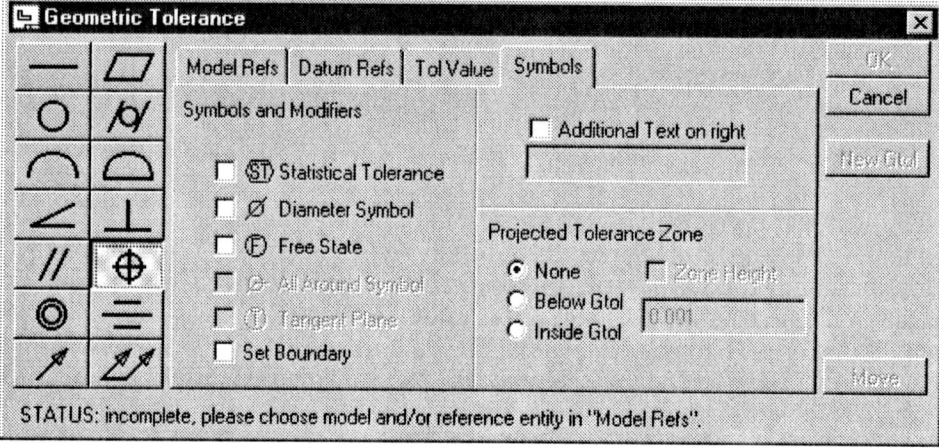

*Figure 11-18 The **Symbols** tab of the **Geometric Tolerance** dialog box*

New Gtol

The **New Gtol** button in the **Geometric Tolerance** dialog box is chosen to accept the current gtol and place a new gtol.

Move

The **Move** button is chosen to change the location of the current gtol.

Figure 11-19 shows the drawing view with reference datums and gtol.

EDITING THE GEOMETRIC TOLERANCES

The gtol added to the drawing views can be edited using the **Geometric Tolerance** dialog box. This dialog box is displayed when you select the text of the tolerance from the drawing view and then right-click to display the shortcut menu. Choose the **Properties** option from the shortcut menu to display the **Geometric Tolerance** dialog box. All options under the various tabs of the **Geometric Tolerance** dialog box are similar to those discussed in the **Geometric Tolerance** dialog box for adding the gtol.

Dimensioning the Drawing Views

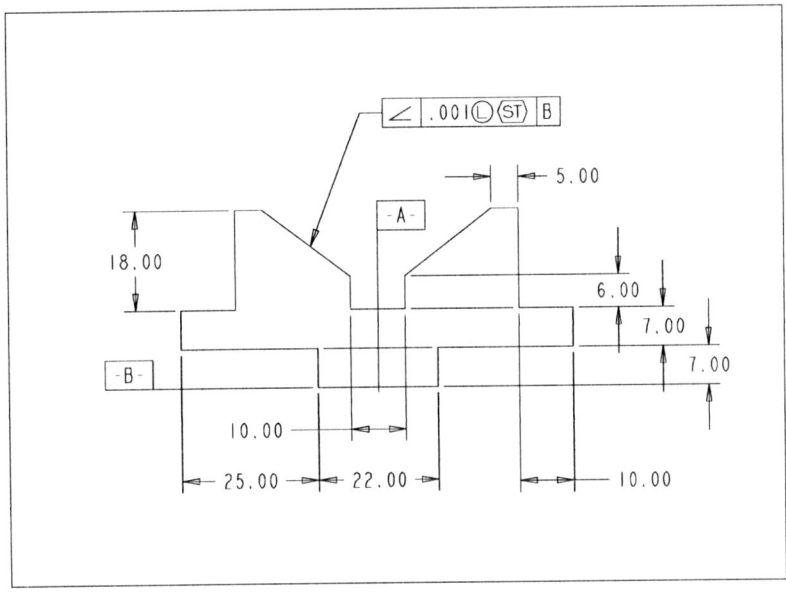

Figure 11-19 The drawing view with reference datums and gtol

ADDING NOTES TO THE DRAWING

The notes added to the model in the **Part** mode can be easily displayed in the drawing views using the **Show / Erase** dialog box. However, when you have not added notes to the model in the **Part** mode and still want to add them in the drawing views then you will have to create them. The notes can be created by choosing **Insert > Note** from the menu bar. When you choose this option, the **NOTE TYPES** menu is displayed, see Figure 11-20. In the **NOTE TYPES** menu, choose the type of leader to be used if required. Choose the **Make Note** option from this submenu, the **ATTACH TYPE** menu is displaced, choose the options from this menu .Now select the point on the drawing where you what to attach the note with left mouse button and then select note location with middle mouse button. Enter the text in the **Message Input Window** and choose the **Accept value** button. You are again prompted to enter the note in the Message Input Window, this time choose the Cancel input button. The note will be placed at the selected point.

 Note
You can also select the text symbol from the Text Symbol dialog box to attach as the note. Remember that ATTACH TYPE menu is displayed only when With Leader or ISO Leader option is selected in the NOTES TYPES menu

ADDING BALLOONS TO THE ASSEMBLY VIEWS

The balloons can be added to the assembly drawing views by choosing **Insert > Balloon** from the menu bar. The procedure to place balloons is similar to that used while adding notes to the drawing. See Figure 11-21.

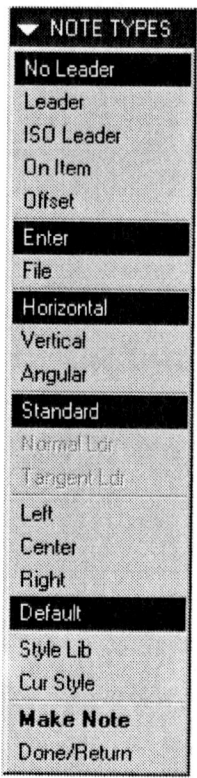

Figure 11-20 NOTE TYPES menu

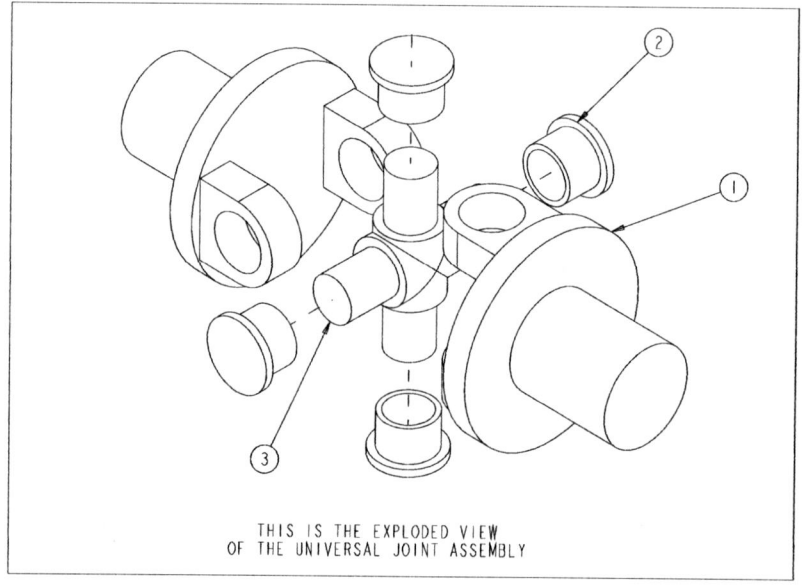

Figure 11-21 Drawing view with balloons and note

Dimensioning the Drawing Views   11-23

Tip: *The notes or balloons can be modified and edited using the options in the shortcut menu that is displayed when you select them using the left mouse button and hold down the right mouse button.*

TUTORIALS

Tutorial 1

In this tutorial you will generate the drawing views of the part shown in Figure 11-22. Generate the dimensions in the drawing views. The views to be generated are shown in Figure 11-23.

(Estimated time: 45 min)

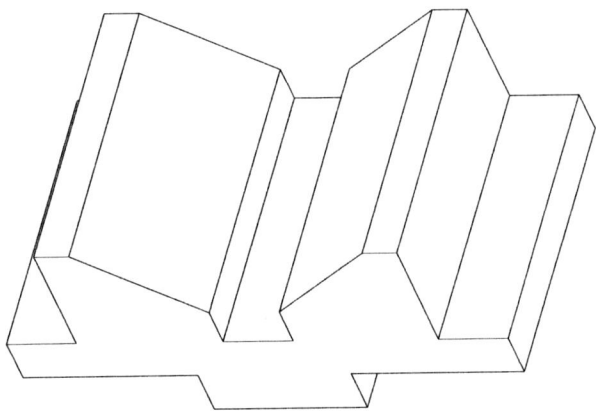

Figure 11-22 Model for generating the drawing views

The following steps are required to complete this tutorial:

a. Create the model in the **Part** mode using the given dimensions and save the file.
b. Open a new drawing file in the **Drawing** mode and generate the top view and the front view of the model on the drawing sheet, refer to Figure 11-24.
c. Generate the dimension to the views and clean them using the **Clean Dimensions** dialog box, refer to Figure 11-25.
d. Add geometric tolerance to the required entities, refer to Figure 11-26.

Before you start creating the model and the drawing views, set the working directory to *C:\ProE-WF-2.0\c11*. Make sure that the *c11* folder exists inside the *ProE-WF-2.0* folder. If these folders do not exist, create them using the **Select Working Directory** dialog box.

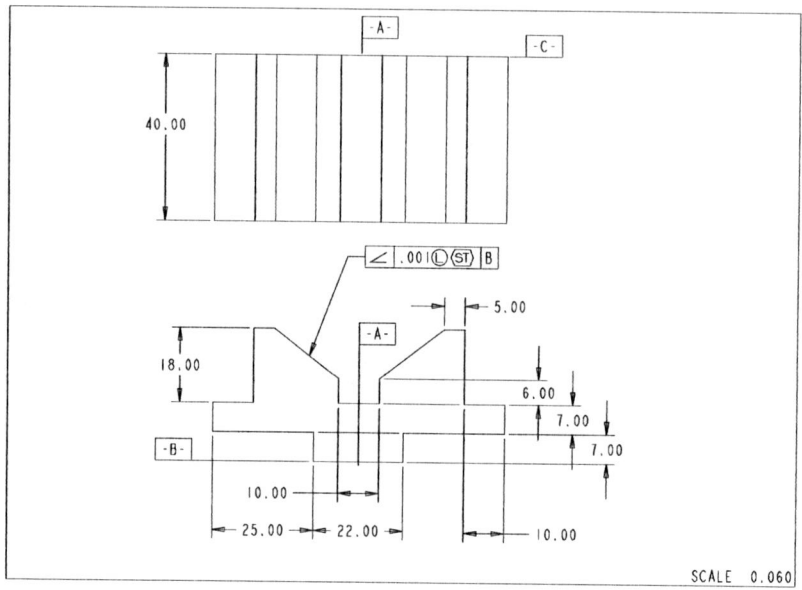

Figure 11-23 Drawing views to be created along with the dimensions

Creating the Model

First, create in the **Part** mode the given model whose views are to be generated later.

1. Open a new object file in the **Part** mode and name it *c11tut1*.

2. Create the model in the **Part** mode and save this model.

Starting a New Drawing File

To generate the drawing views, you need to start a new drawing file.

1. Choose the **Create a new object** button from the **File** toolbar; the **New** dialog box is displayed.

2. Select the **Drawing** radio button from the **Type** area and enter the name of the drawing as *c11tut1* in the **Name** edit box.

3. Choose **OK** from the **New** dialog box to proceed further; the **New Drawing** dialog box is displayed.

4. If the name of the model is not displayed in the **Default Model** area, choose the **Browse** button and select the model. Select the **Empty** radio button from the **Specify Template** area, **Landscape** button from the **Orientation** area, and the **A4** option from the **Standard Size** drop-down list.

Dimensioning the Drawing Views

5. Choose **OK** from the **New Drawing** dialog box to enter the **Drawing** mode. The **Drawing** mode is invoked and a new A4 size drawing sheet is displayed.

Generating the Drawing Views

You need to generate the top view and the front view of the model. Any of the two views can be generated first. However, you will generate the top view first.

1. Choose the **Create a general view** button from the **Top Toolchest**.

2. Select the center point for drawing view, as shown in Figure 11-24. As soon as you specify the point, the **Drawing View** dialog box is displayed

3. Select the **TOP** option from the **Models view names** list in the **View orientation** area.

4. Choose the **Apply** button. The default view is oriented as the top view.

5. Choose **Scale** option in the **Categories** area and edit the scale factor for the view to **0.06**. Exit the dialog box.

6. Choose **No Hidden** from the **Model Display** toolbar and repaint the screen using the **Redraw the current view** button from the **View** toolbar.

7. Choose **Insert > Drawing View > Projection** in the menu bar.

8. Select the center point for drawing view below the top view, as shown in Figure 11-24.

The sheet after generating the two views is shown in Figure 11-24.

Renaming the Datum Planes

The datum planes are renamed because the gtol values to the inclined surface will be applied with their reference. This is evident from Figure 11-23.

1. Choose the **Datum planes on/off** button from the **Datum Display** toolbar to display the datum planes and then repaint the screen.

2. Select the **RIGHT** datum plane from the top view and when it turns red in color hold down the right mouse button to display the shortcut menu.

3. Choose the **Properties** option from the shortcut menu; the **Datum** dialog box is displayed.

4. Enter **A** in the **Name** edit box and then select the second button from the **Type** area to enclose the datum in the reference box. Make sure the **Free** radio button is selected in the **Placement** area. Choose **OK** to exit the **Datum** dialog box.

5. Repeat steps 3 and 4 to rename **TOP** and **FRONT** datum planes as **B** and **C**, respectively.

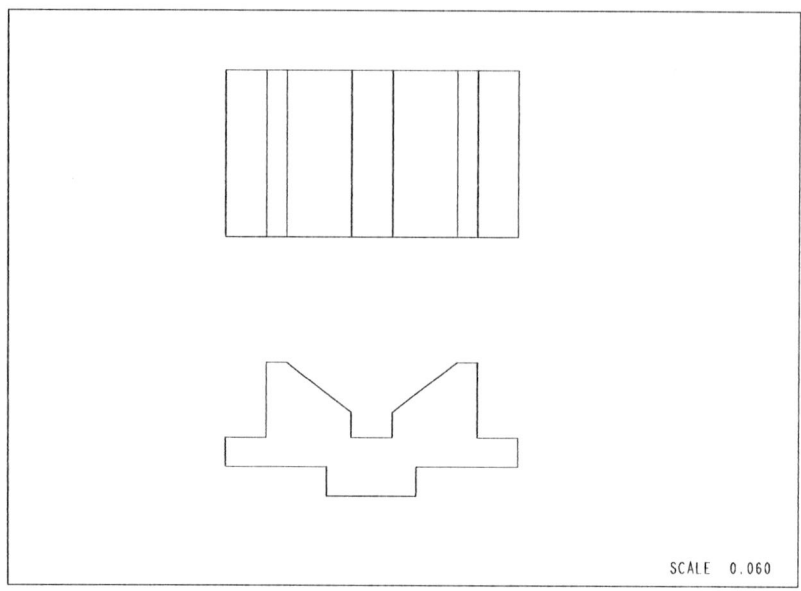

Figure 11-24 Top view and the front view of the model

Dimensioning the Drawing Views

1. Choose the **Open the Show/Erase dialog box** button from the **Top Toolchest**. The **Show / Erase** dialog box is displayed.

2. Choose the **Show** button, if it is not chosen by default, and then choose the **Dimension** button from the **Type** area in the **Show / Erase** dialog box.

3. Select the **View** radio button from the **Show By** area in the **Show / Erase** dialog box.

4. Select the front view (the view at the bottom) to display all dimensions of this view.

5. Press the middle mouse button.

6. Now, select the top view and then press the middle mouse button. All dimensions for this view are displayed. Exit the dialog box.

Placing the Dimensions in Order

The dimensions displayed in the drawing views are scattered and improperly placed. You need to place the dimensions in order using the **Clean Dimensions** dialog box.

1. Choose **Edit > Cleanup > Dimensions** from the menu bar to display the **Clean Dimensions** dialog box. You are prompted to select the dimensions to clean.

Dimensioning the Drawing Views 11-27

2. Select all dimensions from both the views by window selection method; the selected dimensions are highlighted. Press the middle mouse button.

3. Enter **0.5** in the **Offset** edit box and **0.5** in the **Increment** edit box.

4. Clear the **Create Snap Lines** check box. Choose **Apply** and then choose **Close** to exit the **Clean Dimensions** dialog box.

5. After cleaning the dimensions, the drawing views should look similar to the one shown in Figure 11-25. Select and move the dimensions, if necessary.

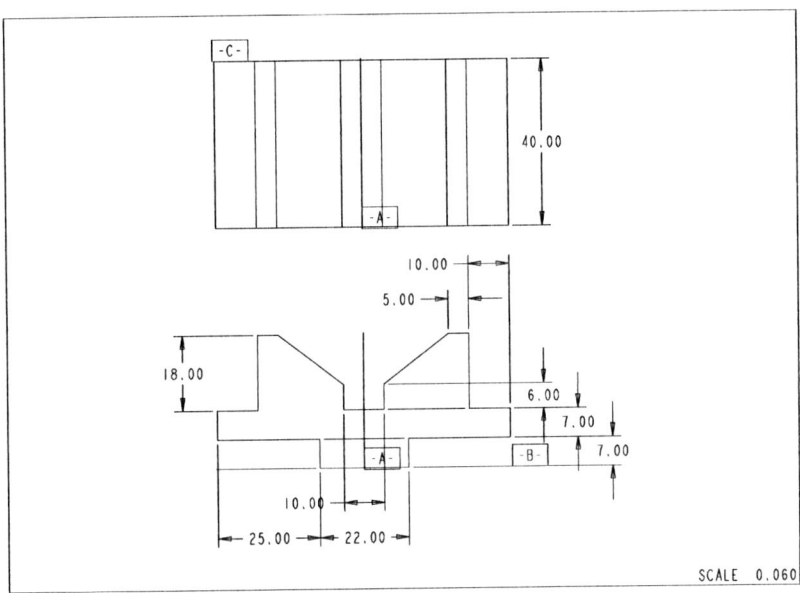

Figure 11-25 Top and the front views of the model with the dimensions and datum planes

Displaying the Geometric Tolerance

1. Choose **Insert > Geometric Tolerance** from the menu bar to display the **Geometric Tolerance** dialog box. You will notice that the name of the model *c11tut1.prt* is displayed in the **Model** drop-down list.

2. Choose the geometric condition as angularity using the **Angularity** button from the **Geometric Tolerance** dialog box.

3. Select the **Surface** option from the **Type** drop-down list in the **Reference** area and select the left inclined edge from the front view.

4. Select **With Leader** from the **Type** drop-down list in the **Placement** area. The **ATTACH TYPE** menu is displayed and you are prompted to select an edge for attaching the leader.

5. Select the left inclined edge from the front view and then middle-click at a position in the graphics window where you want the note to be placed. Even if the note is placed at an improper position in the graphics window, you can move its location later.

6. Choose the **Datum Refs** tab from the **Geometric Tolerance** dialog box.

7. Select **B** from the **Basic** drop-down list under the **Primary** tab of the **Datum References** area.

8. Select the **Tol Value** tab from the **Geometric Tolerance** dialog box.

9. Enter **0.001** in the **Overall Tolerance** edit box.

10. Select **LMC** from the **Material Condition** drop-down list.

11. Select the **Symbols** tab and then select the **Statistical Tolerance** check box from the **Symbols and Modifiers** area in the **Geometric Tolerance** dialog box.

12. Select the **Below Gtol** radio button from the **Projected Tolerance Zone** area and then select the **Zone Height** check box. Enter **0.001** in this edit box. Choose **OK** to exit the **Geometric Tolerance** dialog box.

13. The drawing views after adding the gtol should look similar to the one shown in Figure 11-26.

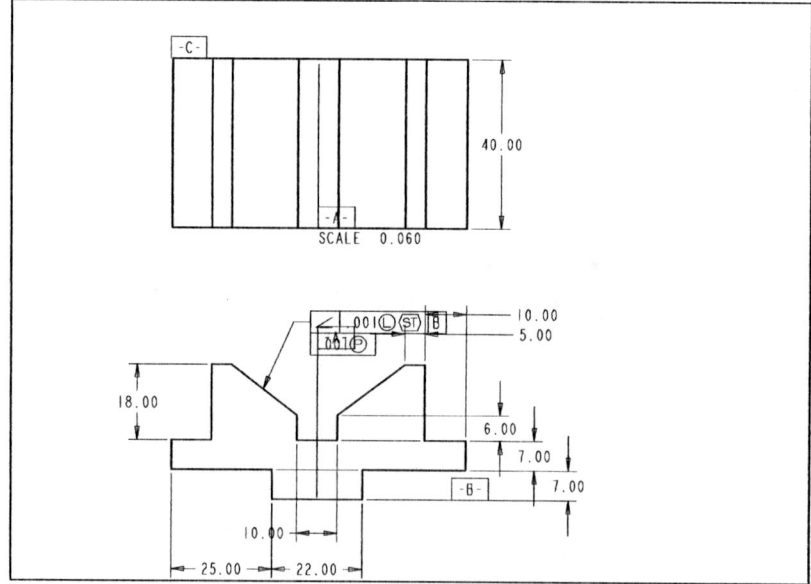

Figure 11-26 *Top and front views of the model with the dimensions, datum planes, and the geometric tolerances*

Dimensioning the Drawing Views

You need to move the note for gtol to a location shown in Figure 11-27 by selecting it and dragging it. Similarly, arrange all dimensions and datums on the drawing sheet, as shown in Figure 11-27.

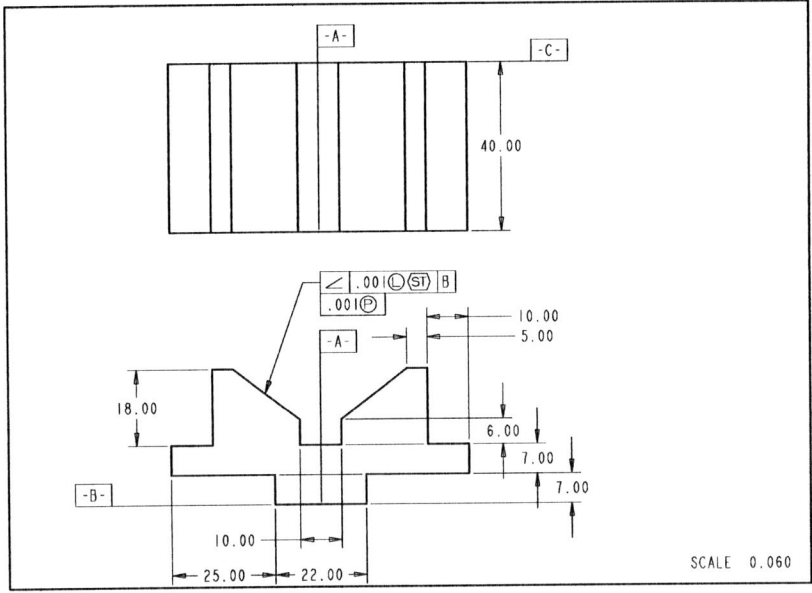

Figure 11-27 *The required drawing*

Saving the Drawing File

You need to save the drawing file that you have created as you might need it later.

1. Choose the **Save the active object** button from the **Top Toolchest**. The **Save Object** dialog box is displayed with the name of the drawing file that you had entered earlier.

2. Press ENTER to confirm the saving of the file.

Tutorial 2

In this tutorial, you will generate the drawing views of the model shown in Figure 11-28. The dimensioned drawing views are shown in Figure 11-29.   **(Estimated time: 45 min)**

The following steps are required to complete this tutorial:

a. Create the model in the **Part** mode using the given dimensions and save the file.
b. Open a new drawing file in the **Drawing** mode and generate the top view, front view, right-side view, and detailed view of the model on the drawing sheet, refer to Figure 11-30.
c. Dimension the views and clean the dimensions using the **Clean Dimensions** dialog box, refer to Figure 11-31.

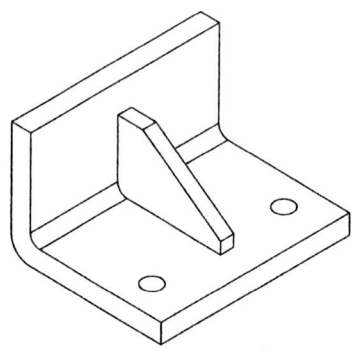

Figure 11-28 Model for generating the drawing views

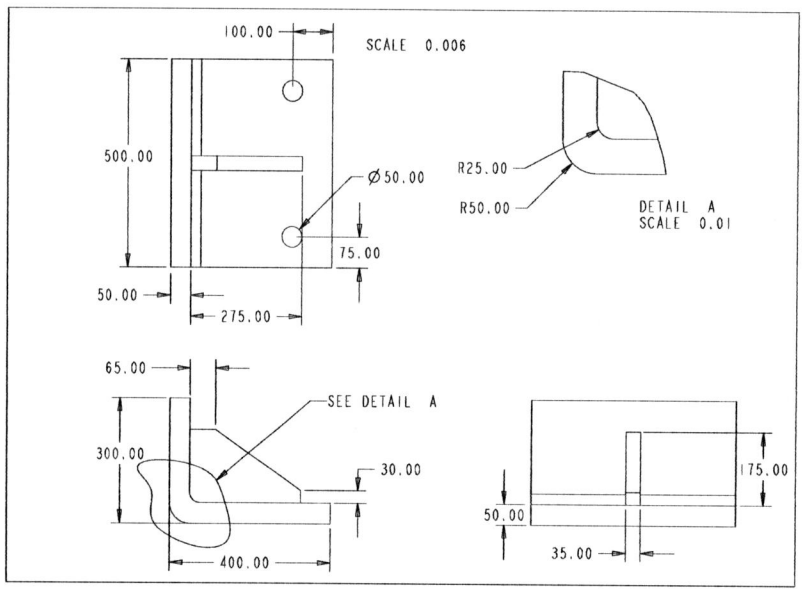

Figure 11-29 The drawing views to be generated in Tutorial 2

The working directory was selected in tutorial 1 of this chapter. But, if you still want to change the working directory set, it to *C:\ProE-WF-2.0\c11*.

Creating the Model

First, you need to create in the **Part** mode the model whose drawing views are to be generated.

1. Create the model in the **Part** mode using the given dimensions.

2. Save this model created in the **Part** mode as *c11tut2*.

Starting a New Drawing File

To generate the drawing views, you need to first start a new drawing file.

1. Choose the **Create a new object** button from the **File** toolbar; the **New** dialog box is displayed.

2. Select the **Drawing** radio button from the **Type** area and enter the name of the drawing as *c11tut2* in the **Name** edit box.

3. Choose **OK** from the **New** dialog box to display the **New Drawing** dialog box; the **New Drawing** dialog box is displayed with the name of the model in the **Default Model** area.

4. If the name of the model is not displayed in the **Default Model** area, choose the **Browse** button and select the model.

5. Select the **Empty** radio button from the **Specify Template**, the **Landscape** button from the **Orientation** area, and the **A4** option from the **Standard Size** drop-down list.

6. Choose **OK** from the **New Drawing** dialog box to enter the **Drawing** mode. A new drawing sheet of A4 size is displayed on the screen.

Creating the Drawing Views

First, the top view of the model needs to be generated and then the front view will be generated by projecting it from the top view. The right-side view will be projected from the front view. The detail view will be placed on the drawing sheet by specifying a scale for it.

1. Choose the **General** button from the **Top Toolchest**.

2. Select the center point for drawing view, as shown in Figure 11-30. As soon as you specify the point, the **Drawing View** dialog box is displayed.

3. Select the **TOP** option from the **Models view names** list in the **View orientation** area.

4. Choose the **Apply** button. The default view is oriented as the top view.

5. Choose **Scale** option in the **Categories** display area and edit the scale factor for the view as **0.006**. Exit the dialog box and then choose **No Hidden** from the **Model Display** toolbar. Repaint the screen using the **Redraw the current view** button from the **View** toolbar.

    The front view of the model is generated from the top view.

6. Choose the **Insert > Drawing View > Projection** in the menu bar. Select the center point for drawing view below the top view, as shown in Figure 11-30.

7. Similarly, place the right-side view and the detailed view on the drawing sheet, as shown in Figure 11-30.

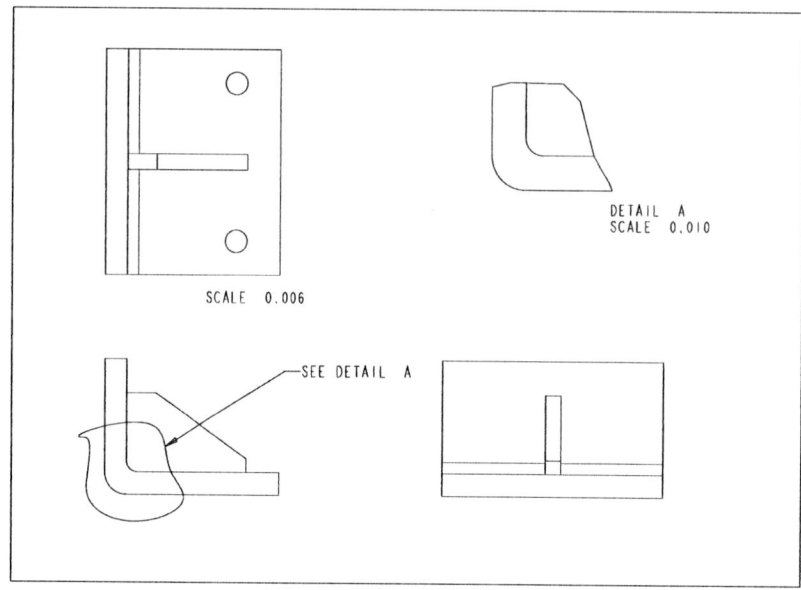

Figure 11-30 The drawing views before adding the dimensions

Dimensioning the Drawing Views

1. Choose the **Open the Show/Erase dialog box** button from the **Top Toolchest**. The **Show / Erase** dialog box is displayed.

2. Choose the **Dimension** button from the **Type** area and then select the **Feature** radio button from the **Show By** area. Select the feature in the detailed view. All dimensions related to the base feature are displayed in all views. Press the middle mouse button.

3. Select the **View** radio button from the **Show By** area and then select all three views one by one to display all dimensions in the views. Press the middle mouse button and then close the **Show / Erase** dialog box by choosing the **Close** button.

Placing the Dimensions in Order

The dimensions displayed in the drawing views are scattered and improperly placed. You need to use the **Clean Dimensions** dialog box to place the dimensions in order.

1. Choose **Edit > Cleanup > Dimensions** from the menu bar to display the **Clean Dimensions** dialog box. You are prompted to select the dimensions to clean.

2. Select all dimensions from both the views by dragging a rectangle around them. Press the middle mouse button so that the options in the **Clean Dimensions** dialog box are available.

When you select the dimensions, the radius dimensions in the drawing views are not selected. This means that the radius dimensions will remain unaffected by the values that you will enter in the **Clean Dimensions** dialog box.

Dimensioning the Drawing Views 11-33

3. Enter **0.4** in the **Offset** edit box and **0.3** in the **Increment** edit box.

4. Clear the **Create Snap Lines** check box. Choose **Apply** and then choose **Close** to exit the **Clean Dimensions** dialog box.

   After cleaning the dimensions, you will notice that the dimensions are not placed in the order that is required. Now, you need to manually place the dimensions in order.

5. Select the dimension and drag it to place it at the desired location on the drawing sheet.

   Some dimensions are repeated, for example, the diameter of the hole feature is displayed twice in the drawing view. So, you need to erase the dimensions that are repeated and that are not needed.

6. Select the dimension and hold down the right mouse button on the repeated dimensions to display the shortcut menu. Choose the **Erase** option from the shortcut menu.

   In some cases, the dimensions are displayed in the views in which you do not want them to be displayed. In such cases you can switch that dimension to the other views. The **Move Item to View** option is available in the shortcut menu that is displayed when you select a dimension and hold down the right mouse button.

   After manually placing the dimensions the drawing views should look similar to views shown in Figure 11-31.

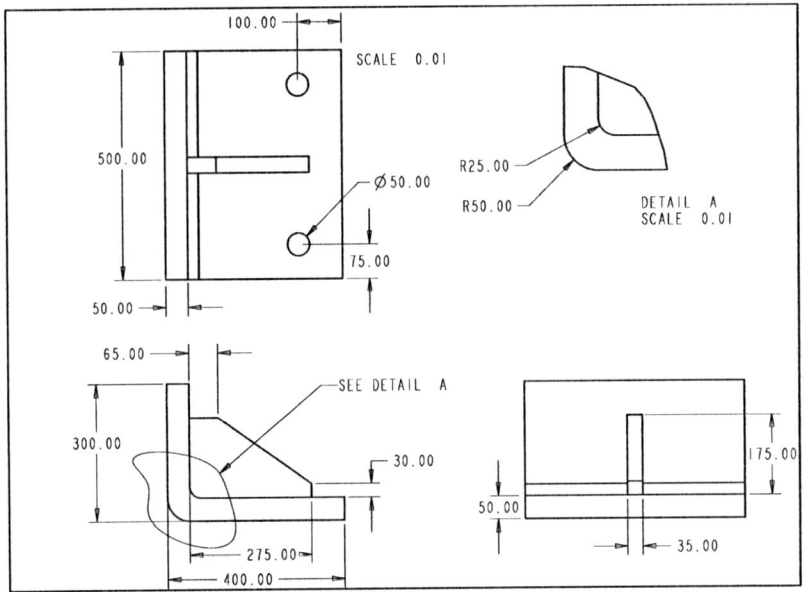

Figure 11-31 The required drawing views

Saving the Drawing File

1. Choose the **Save the active object** button from the **Top Toolchest**. The **Save Object** dialog box is displayed with the name of the drawing file that you had entered earlier.

2. Press ENTER to confirm the saving of the file.

Self-Evaluation Test

Answer the following questions and then compare your answers to the answers given at the end of this chapter.

1. The reference datum planes are used to display the geometric tolerances in a drawing view. (T/F)

2. If you modify a dimension value in the **Drawing** mode and then regenerate the drawing view, the modification is applied to the model in the **Part** mode also. (T/F)

3. The notes created in the **Part** mode on a model can also be displayed in a drawing view in the **Drawing** mode. (T/F)

4. When you select a drawing view to display dimensions, all dimensions are placed in proper order in the drawing view. (T/F)

5. Most of the editing operations in a drawing view can be done using the shortcut menu that is displayed when you hold down the right mouse button on an item to be modified. (T/F)

6. The _____ dialog box is used to display the dimensions in the drawing views.

7. The dimension values can be modified using the _____ option from the **Edit** menu in the menu bar.

8. The _____ dialog box is used to clean the dimensions in the drawing views.

9. You have to set the value of _____ to **Yes** in the configuration file to display the dimensional tolerance in the drawing views.

10. The configuration file can be opened from _____.

Review Questions

Answer the following questions:

1. Which of the following dialog boxes is used to display dimensions in a drawing view?

   (a) **Dimension Properties**   (b) **Clean Dimensions**
   (c) **Show / Erase**   (d) **None**

Dimensioning the Drawing Views

2. Which of the following dialog boxes is used to modify properties of text in a drawing view?

    (a) **Dimension Properties** (b) **Clean Dimensions**
    (c) **Show / Erase** (d) None

3. Which of the following dialog boxes is used to display the geometric tolerances in a drawing view?

    (a) **Geometric Tolerance** (b) **Clean Dimensions**
    (c) **Show / Erase** (d) None

4. Which of the following options available in the shortcut menu is used to switch a dimension from one view to another?

    (a) **Erase** (b) **Flip Arrows**
    (c) **Move Item to View** (d) None

5. Which of the following radio button in the **Type** area is selected in the **New** dialog box to open a new drawing file?

    (a) **Assembly** (b) **Sketch**
    (c) **Part** (d) None

6. The dimensions erased using the **Erase** option from the shortcut menu are deleted from the model in the **Part** mode also. (T/F)

7. The notes or balloons can be modified and edited using the options in the shortcut menu. (T/F)

8. The system does not show a reference datum plane in a drawing view unless it is perpendicular to the graphics window. (T/F)

9. Generally, the dimensions displayed in the drawing views are scattered and improperly placed. (T/F)

10. You cannot move text dynamically on the drawing sheet. (T/F)

Exercise

Exercise 1

In this exercise you will generate the drawing views of the model created in Exercise 2 of Chapter 6, shown in Figure 11-32, and add dimensions to the views, as shown in the figure.

(Estimated time: 45 min)

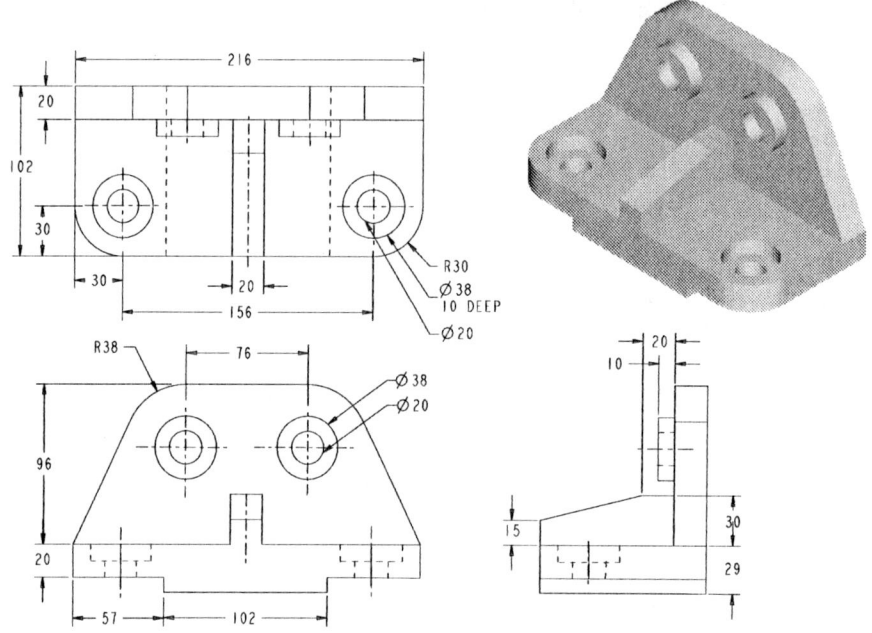

Figure 11-32 The required drawing views with the solid model

Note

In the given drawing views, the center lines are not generated but are created manually. After creating the center lines you need to define a line style and then apply it on the center lines.

Answers to the Self-Evaluation Test

1 - T, 2 - T, 3 - T, 4 - F, 5 - T, 6 - Show / Erase, 7 - Value, 8 - Clean Dimensions, 9 - tol_display, 10 - **File > Properties > Drawing Options**.

Chapter 12

Other Drawing Options

Learning Objectives

After completing this chapter you will be able to:
- *Sketch in the Drawing mode.*
- *Modify the sketched entities and other items in the drawing views.*
- *Create a user-defined drawing format for the drawing sheets.*
- *Add or remove the sheets from the current drawing.*
- *Create the tables in the current sheet.*
- *Generating associative Bill Of Material in the Drawing mode.*

SKETCHING IN THE DRAWING MODE

Sketching is one of the most important tools available in the **Drawing** mode. Sketching in the **Drawing** mode is called drafting. As discussed earlier, there are two types of drafting techniques in Pro/ENGINEER Wildfire 2.0: Generative drafting and Interactive drafting. Any item on the drawing sheet that is not generated from a model is called a draft entity or a draft item. Drafting is extensively used for creating user-defined formats, drawing tables, and also for drawing the title blocks in the formats. The sketching in the **Drawing** mode is almost similar to the sketching in the other modes of Pro/ENGINEER. The sketching can be done by using the **Sketch** menu in the menu bar or the tool buttons available in the **Drawing Sketcher Tools** toolbar from the **Right Toolchest**. The tools available in the **Drawing Sketcher Tools** toolbar and the options available in the **Sketch** menu that are used to sketch in the **Drawing** mode are discussed next.

Select Items

The **Select items** button in the **Right Toolchest** is used to select drawing views, sketched entities, dimensions, notes, and so on from the drawing views. This button is used extensively to move drawing views or the items in the drawing view.

Tip: *To delete an entity or an item from a drawing view, you can select the item and then press the DELETE key. You can also select the entity and use the **Delete** option from the shortcut menu that is displayed when you hold down the right mouse button.*

Line

The **Create 2 point lines** button from the **Right Toolchest** or the **Line** option in the **Sketch** menu is used to draw line segments. When you choose this option, the **Snapping References** dialog box is displayed, as shown in Figure 12-1. The options in the above dialog box are discussed next.

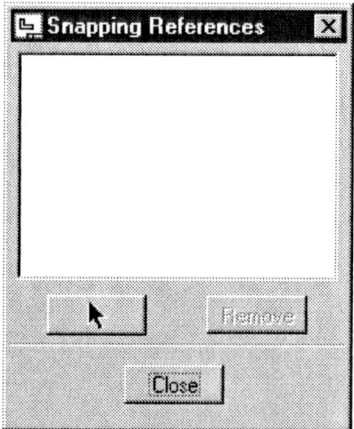

*Figure 12-1 The **Snapping References** dialog box*

Select references Button

 When you choose the **Select references** button from the **Snapping References** dialog box, you are prompted to select an entity to which you want the line to be snapped.

Remove

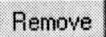 The **Remove** button is used to remove a selected entity from the **Snapping References** dialog box.

Circle

 The **Create circle** button is used to create circles. To create a circle you need to specify a center point and a point on the circumference of the circle.

Arc

 The **Create an arc by 3 points or tangent to an entity at its endpoint** button is used to create arcs.

The other geometric entities that you can draw using the tools available in the **Right Toolchest** are splines, construction circles and lines, ellipses, and points. You can also create a fillet and a chamfer between two entities.

Chain

 The **Enable sketching chain** button is used when the object you want to draw consists of more than one entity. For example, when you choose the **Create 2 points line** button to draw a line, you can draw a single line. But when you draw the line after choosing the **Enable sketching chain** button, you are able to draw a continuous chain of lines. The chain can be ended by pressing the middle mouse button.

Sketching Using References

 The **Remember parametric sketcher references** button is used to draw the sketches by automatically applying constraints.

MODIFYING THE SKETCHED ENTITIES

There are various options to modify the sketched entities that are termed as draft entities. The operations or the modifications that can be applied on the draft entities in Pro/ENGINEER Wildfire 2.0 are discussed next.

Trimming the Draft Entities

The sketched or draft entities can be trimmed by choosing **Edit > Trim** from the menu bar. When you choose this option, the cascading menu is displayed, as shown in Figure 12-2. The options in this menu are discussed next.

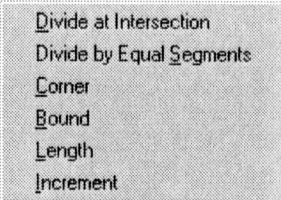

*Figure 12-2 Cascading menu of the **Trim** option in the **Edit** menu*

Divide at Intersection

The **Divide at Intersection** option is used to divide two selected entities at the point of intersection. When you choose this option you are prompted to select the two entities. Select the two entities to divide them at the intersection point. When you select an entity if a warning message window is displayed then you need to choose **OK** from it. This prompts you that this command will break all parametric sketching references.

Divide by Equal Segments

When you choose this option, you are prompted to select the entities you need to divide into equal segments. Select the entities to divide. After selecting the entity or entities, press the middle mouse button. The **Message Input Window** is displayed and you are prompted to enter the number of equal segments. Enter the number of segments into which you need to divide the entity.

Corner

This option is the same as that was discussed in the **Sketch** mode. This option trims the two selected entities so that a corner is formed after trimming.

Bound

This option is used for deleting or extending a portion of an entity by defining a bounding entity.

Length

This option is used to trim or extend an entity up to a specified length.

Increment

This option is used to extend or shorten an entity in specified increments. To extend the entity, enter a positive increment value and to shorten the entity, enter a negative increment value.

Transforming the Draft Entities

To translate or move the draft entities, choose **Edit > Transform**. The cascading menu is displayed, as shown in Figure 12-3. The options in this cascading menu are discussed next.

Translate

This option is used to move a selected entity or entities on the drawing sheet. To translate the

Other Drawing Options

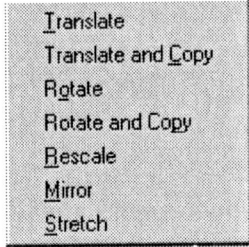

Figure 12-3 Cascading menu of
Translate *option in **Edit** menu*

entities invoke this tool and select the entities. After selecting the entities press the middle mouse button to confirm the selection. After you confirm the selection, the **GET VECTOR** menu along with the **GET POINT** submenu are displayed, as shown in Figure 12-4. The options in the **GET VECTOR** menu are used to specify the direction of translation.

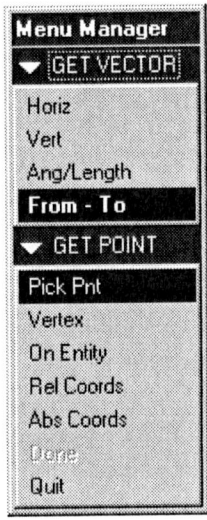

Figure 12-4 GET VECTOR
*menu with **GET POINT** submenu*

Translate and Copy
The **Translate and Copy** option is used to translate the selected entity and at the same time make a copy of the entity. After selecting the entities when you press the middle mouse button, the **GET VECTOR** menu and the **GET POINT** submenu are displayed. After defining the translation distance and direction you need to specify the number of copies.

Rotate
The **Rotate** option is used to rotate the selected entity from its actual location. When you choose this option, you are prompted to select the entity you want to rotate. After selecting the entity or entities, press the middle mouse button. You are prompted to select the center point for rotation. After you select the center point the **Message Input Window** is displayed

and you are prompted to enter the rotation angle in the counterclockwise direction. The default direction of rotation will be counterclockwise. However, you can rotate the selected entity in the clockwise direction by specifying a negative rotation angle.

Rotate and Copy

The **Rotate and Copy** option is used to create a pattern of the selected draft entity. When you choose the entity after choosing this option, press the middle mouse button. You are prompted to select the center point of rotation. Select the center point of rotation. The **Message Input Window** is displayed and you are prompted to enter the rotation angle. Press middle mouse button to accept the entered rotation angle value, now enter the number of copies in the **Message Input Window**. The rotated copies of the selected entity are placed on the drawing sheet.

Rescale

The **Rescale** option is used to rescale the selected entity. After choosing this option when you select the entity to rescale, press the middle mouse button. You are prompted to select the origin point for scaling. The new scale factor can be specified in the **Message Input Window**. A scale factor greater than 1 will increase the size of the selected entity and a scale factor less than 1 will reduce the size of the selected entity.

Mirror

The **Mirror** option is used to mirror the selected entity using an existing sketched line.

Stretch

The **Stretch** option is used to stretch a sketched entity by selecting it using a window. The vertices of the sketched entity that are included inside the window will be moved from their original location and stretched and the remaining vertices will remain stationary.

Grouping the Entities

A draft entity can be grouped with other draft entity, with a note, a geometric tolerance (gtol), or a dimension. In Pro/ENGINEER, the note, gtol, and dimensions are called detail items. After grouping the draft entities, a operation applied on any one of them is applied to the grouped entity also. The options to group can be invoked by choosing **Edit > Group**. The cascading menu is displayed, as shown in Figure 12-5. The options in this cascading menu are discussed next.

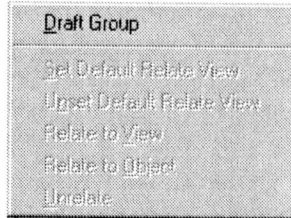

Figure 12-5 Cascading menu of the ***Group*** *option in the* ***Edit*** *menu*

Draft Group

This option is used to form a group among draft entities like lines, arcs, circles, and so on. The operation performed on any one of the grouped entity is performed on all entities forming that group. When you choose this option, the **DRAFT GROUP** menu is displayed, as shown in Figure 12-6. Use the option from this menu to create the draft group.

Figure 12-6 DRAFT GROUP menu

Set Default Relate View

This option is available only when the drawing view is selected. This option is used to set the draft view as the current view. The draft entities that are drawn after setting a default draft view are grouped with the current draft view. Now, if you move the draft view then the grouped entities will also move with it.

Note
When you move a view that forms a group with draft entities, the draft entities also move. But when you move the entities, the draft view does not move.

Unset Default Relate View

This option is used to unset the current draft view. This option is available only when a current draft view exists.

Relate to View

This option is used to relate a view with a draft entity or entities. This option is available in the cascading menu only when a draft entity is selected. When you choose this option after selecting a draft entity, you are prompted to select a view to which you want to relate the draft entity. Select the drawing view from the drawing sheet.

Relate to Object

This option is used to relate notes, dimensions, gtols, and other objects to a drawing view. This option is available in the cascading menu only when an object is selected. When you choose this option after selecting an object, you are prompted to select dim, dim arrow, gtol, note, or symbol which will serve as the host.

Unrelate

This option is used to unrelate the previously related objects. This option is available in the cascading menu only when a related object is selected.

Drafting Using the Drawing Views

You can use the generated drawing views to create an entity or the whole drawing view. To use a generated drawing view, choose **Sketch > Edge**. The cascading menu is displayed with two options; **Use** and **Offset**.

Use

The **Use** option is selected to draw a copy of the selected edge in the drawing view. This option will be available only when you have at least one drawing view in the current drawing sheet and if it is selected.

Offset

When you choose this option from the cascading menu, the **OFFSET OPER** menu is displayed, as shown in Figure 12-7. There are two options in this menu. The **Single Ent** option is used to select a single entity from the drawing view. The **Ent Chain** option is used to select a chain of entities from the drawing view. After choosing any one of the options from the **OFFSET OPER** menu, when you move the cursor over the drawing view, each edge of the model in drawing view is highlighted in cyan color separately. You need to define the offset distance.

Figure 12-7 OFFSET OPER menu

Converting a Generated View to a Draft View

To convert a generated view to a draft view, select the view to convert and choose **Edit > Convert to Draft Entities** from the menu bar. After converting a generated view to a draft view, all objects in the view are converted to individual entities. When you choose this option, the **SNAPSHOT** menu is displayed, as shown in Figure 12-8. The options in this menu are used to select either all views or the current view. After selecting the option from this menu, a **Confirm** dialog box is displayed. This prompts you that selected view(s) will be broken up into draft entities. Choose the **Yes** button from this dialog box.

Figure 12-8 SNAPSHOT menu

USER-DEFINED DRAWING FORMATS

Pro/ENGINEER provides you with some standard drawing formats for generating the drawing views. These standard formats have standard sheet sizes, tables, and title blocks. However,

Other Drawing Options 12-9

sometimes you may need to create a user-defined drawing format that is specifically designed as per your requirements. This format can include the sheet size, tables, and title block specified as per your requirements. Choose the **Create a new object** button to display the **New** dialog box. In this dialog box, select the **Format** radio button from the **Type** area and then specify the name of the format in the **Name** edit box to proceed for creating the user-defined format. When you choose **OK**, the **New Format** dialog box will be displayed, see Figure 12-9.

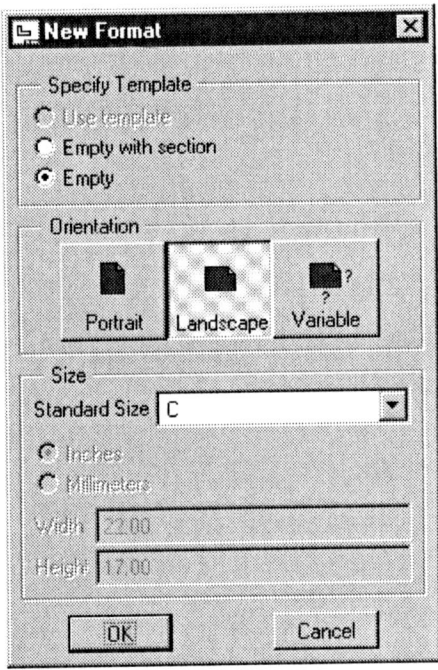

Figure 12-9 The New Format dialog box

You can set the size and the orientation of the format sheet using this dialog box. After setting the parameters choose **OK** from this dialog box to proceed to the **Format** mode.

You can create the desired entities in this format by using the **Drawing Sketcher** toolbar. You can also add the text material to the format by using the **Note** option from the **Insert** menu in the menu bar. Figure 12-10 shows a user-defined format created using the A4 size sheet. This format consists of the user-defined title block. Figure 12-11 shows the drawing views generated on the user-defined format.

RETRIEVING THE USER-DEFINED FORMATS IN THE DRAWINGS

For generating the drawing views you can also retrieve the user-defined format. Once you have created the user-defined formats, you can use them in the **Drawing** mode as sheets for generating the drawing views. To retrieve the user-defined format, select the **Empty with format** radio button and choose the **Browse** button from the **New Drawing** dialog box to

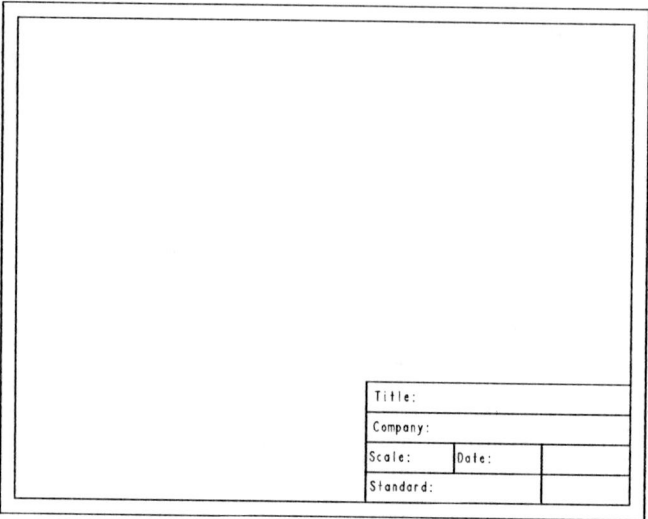

Figure 12-10 A user-defined format

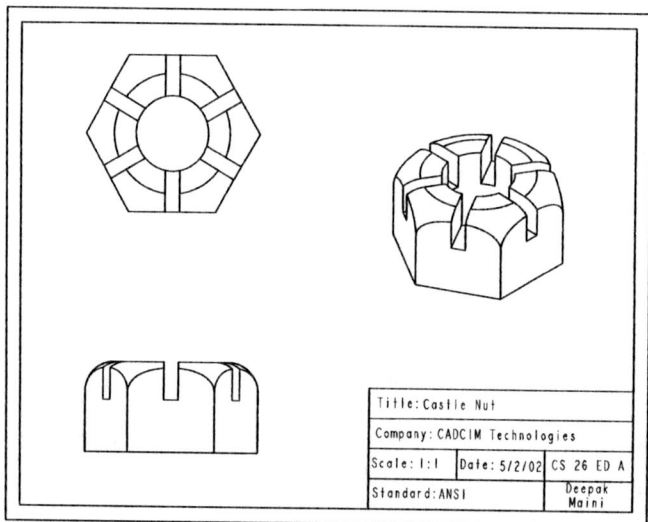

Figure 12-11 Drawing views generated using a user-defined format

invoke the **Open** dialog box. A folder named **System Formats** will be displayed in the **Open** dialog box. Some predefined formats are given in this folder. You can retrieve these predefined formats or browse the location where you have saved the user-defined format created earlier.

 **Tip**: *The user-defined format that you create is saved in the .frm format. The location of this format will be the working directory that was selected at the time when the format was created.*

 **Tip**: *You can add or replace the formats in drawing by using the options available in the **Page Setup** dialog box that is displayed when you hold down the right mouse button on the drawing sheet. You can add another sheet and then select the format from this dialog box.*

ADDING AND REMOVING SHEETS IN THE DRAWING

To add sheets in the current drawing, choose **Insert > Sheet** from the menu bar. A new sheet is displayed on the screen. At the bottom left of the screen, the sheet number is displayed. You can generate various views of a model on multiple sheets. All these drawing views on different sheets are contained in a single drawing file.

To remove sheets from a drawing, choose **Edit > Remove > Sheets** from the menu bar. The **Message Input Window** is displayed. Enter the sheet number that you want to remove.

CREATING TABLES IN DRAWING MODE

You can easily create any kind of tabular representation in **Drawing** mode by using the options available in the cascading menu that is displayed when you choose **Table > Insert** from the menu bar. The cascading menu is displayed, as shown in Figure 12-12 and the options in it are discussed next.

Figure 12-12 Cascading menu

Table Option

The **Table** option is used to create a table. When you choose this option from the cascading menu, the **TABLE CREATE** submenu is displayed, as shown in Figure 12-13. The table can be created in the ascending or descending order of the rows progress and in the right or the left direction. The length and width of the cells can be specified by picking the points on the screen or by specifying the value of their length and width.

Table From File Option

The **Table From File** option is used to insert a table that exists in the *.tbl* file format. When you choose this option, the **Open** dialog box is displayed. Select the file from this dialog box to open.

Column Option

The **Column** option is used to add a column to an existing table. When you choose this option from the cascading menu, you are prompted to select a point inside an existing table.

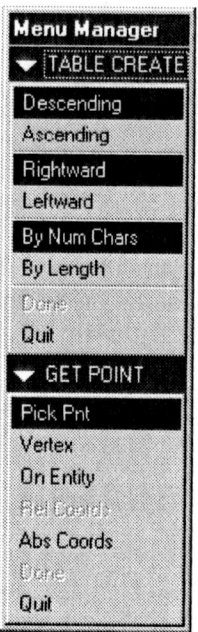

Figure 12-13 TABLE CREATE submenu

Tip: *To save a table with a .tbl file format, choose* **Table** > **Save Table** > **As Table File** *from the menu bar. This option is available only when a table exists on the drawing sheet.*

Row Option

The **Row** option is used to add a row to an existing table. When you choose this option from the cascading menu, you are prompted to select a point inside an existing table.

Following are the editing operations that can be performed on the tables.

1. To move a table, select it and then drag it by holding it from one of the points that are highlighted on the table.

2. To enter text in any of the cells of the table you need to double-click inside the cell to display the **Note Properties** dialog box.

3. To delete a table, choose **Table** > **Select** > **Table** from the menu bar to select the table. Hold down the right mouse button and choose **Delete** from the shortcut menu.

GENERATING THE BOM AND BALLOONS IN DRAWINGS

The Bill Of Material (BOM) is a representation of the components and their parameters that are used in the assembly. In the **Drawing** mode, the BOM is generated and is associative in nature. Therefore, any modification in the assembly, such as addition or removal of components, is automatically reflected in the BOM.

Other Drawing Options

Tip: *In the **Drawing** mode, most of the modifications and editing can be done using the shortcut menu that is displayed when you select the item to modify and then hold down the right mouse button. The options in this menu vary and depend on the item selected.*

*It is recommended that you use the options under the **Table** menu in the menu bar to create the title blocks in the formats. The text can be easily added to the title block by double-clicking in the cell.*

The procedure to generate the BOM and Balloons in the **Drawing** mode is discussed next.

Consider a case in which you have started a new drawing sheet of standard size A with a user-defined format. Now, follow the steps discussed next to generate BOM and balloons.

Creating the Table for BOM

Before proceeding to generate the BOM, you need to first create a table for the BOM. The items to be listed in BOM will be displayed in this table.

1. Choose **Table > Insert > Table** from the menu bar to display the **TABLE CREATE** menu.

2. Choose **Descending > Leftward > By Length** from the **TABLE CREATE** menu. Choose **Pick Pnt** from the **GET POINT** menu.

3. Specify a point close to the upper right corner of the inner format boundary. Now, you need to enter the dimensions of the BOM table.

4. Enter the width of the first column as **0.5** in the **Message Input Window** and press ENTER. Similarly, enter the width of the second column as **3** and the width for third column as **1**. Now, press a blank ENTER; you are prompted to enter the height for the first row.

5. Enter the height of the first row as **0.5** and of the second as **0.5** in the **Message Input Window**. Press ENTER twice.

Setting the Repeat Region

After creating the table for the BOM, you need to define the repeat region in this table. Repeat regions are smart cells that can expand depending on the amount of data inserted in the cell. The second row will be defined as the repeat region.

1. Choose **Table > Repeat Region** from the menu bar; the **TBL REGIONS** menu is displayed.

2. Choose the **Add** option; you are prompted to locate the corners of the region.

3. Select the first cell of the second row and then select the third cell of the second row. Press the middle mouse button twice to exit the menu.

Creating the Column Headers

After creating the table and setting the repeat region, you need to define the column headings in the table.

1. Double-click in the first cell of the first row; the **Note Properties** dialog box is displayed. Enter **Index** and choose **OK** to exit the dialog box.

2. Similarly, enter **Part Name** and **Qty** in the second and the third cells.

Assigning the Report Symbols in the Repeat Region

The information in the cells that are defined as repeat region is determined by the text that is written in the form of report symbols. These report symbols associate the information in the cells with the assembly file directly.

1. Double-click in the first cell of the second row; the **Report Symbol** dialog box is displayed. Select **rpt** from the dialog box; the **rpt** options are displayed in the dialog box.

2. Select the **index** option from the dialog box.

3. Double-click in the second cell of the second row; the **Report Symbol** dialog box is displayed. Select **asm** from the dialog box. The **asm** options are displayed.

4. Select the **mbr** option from the dialog box then select the **name** option from the dialog boxes that appear. This will ensure the names of the parts in the assembly are displayed in this column. The number of rows will be automatically modified depending on the number of parts in the assembly.

5. Double-click in the third cell of the second row; the **Report Symbol** dialog box is displayed. Select **rpt** from the dialog box; the **rpt** options are displayed. Select the **qty** option from the dialog box.

Generating the Exploded Drawing View

After setting all option, you need to generate the isometric view of the exploded state of the assembly. Therefore, you should have created an exploded state of the assembly. To generate the exploded drawing view follow the steps listed next.

1. Choose the **Create a general view** button from the **Top Toolchest**; you are prompted to select the center point for drawing view.

2. Select the point on the drawing sheet to place the view; the **Drawing View** dialog box is displayed. Select the **Isometric** option from the **Default orientation** drop-down list in the **Drawing View** dialog box. Choose the **Apply** button. Specify the scale factor, if required.

Other Drawing Options 12-15

4. Select the **View States** option from the **Categories** list box. Select the **Explode components in view** check box. Select the name of the exploded state from the **Assembly explode state** drop-down list. Choose the **Apply** button.

5. Close the **Drawing View** dialog box to complete placing the required view.

Note that the BOM automatically appears in the table by extending the repeat region. Also, note that the Qty column is empty. If the BOM is not displayed, choose the **Update the information displayed in tables** button from the **Top Toolchest**.

Setting the No Repeat Option
If the assembly has more than one quantity of a component, the components are repeated in the BOM. Therefore, you need to set the option to omit the repeated instances in the BOM.

1. Choose **Table > Repeat Region**; the **TBL REGIONS** menu is displayed.

2. Choose the **Attributes** option from the menu. You are prompted to select a region.

3. Move the cursor over the table below the first row. All rows, except the first row, are highlighted in cyan. The highlighted portion is called region. Select the region; the **REGION ATTR** submenu is displayed. Choose the **No Duplicates** option then choose the **Done/Return** button.

4. Press the middle mouse button twice and exit the menu.

5. Choose the **Update the information displayed in tables** button from the **Top Toolchest**, if the information is not updated in the table.

Generating Balloons
Next, you need to generate the Balloons. The balloons are generated using the repeat region. Before generating the balloons, you need to set the repeat region for balloon generation.

1. Choose **Table > BOM Balloons**; the **BOM BALLOONS** menu is displayed. The **BOM BAL TYPE** submenu is also displayed with the **Simple** option chosen by default. Also, you are prompted to select a region.

2. Select the region that was selected earlier. Press the middle mouse button to confirm the selection.

3. From the **BOM BALLOONS** menu, choose the **Create Balloon** option; the **BOM VIEW** submenu is displayed.

4. Choose the **Show All** option and exit the menu. Balloons are generated and are attached to the components of the assembly.

TUTORIALS

Tutorial 1

In this tutorial you will create a format of size A and add the title block in the format. Then you will retrieve the format in the **Drawing** mode and create the table for BOM and then generate the exploded isometric view of the **Pedestal Bearing** assembly created in Tutorial 2 of Chapter 9. Also, add the balloons to the drawing view, as shown in Figure 12-14.

(Estimated time: 45 min)

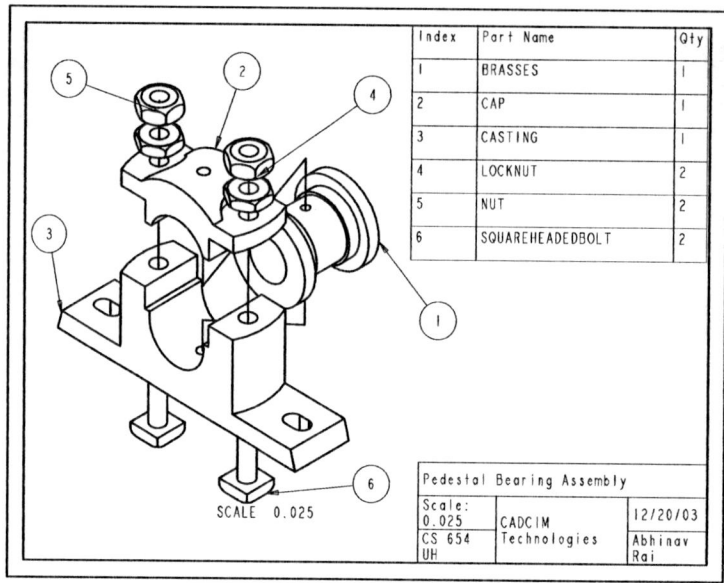

Figure 12-14 The drawing view of the assembly showing the BOM and balloons

The following steps are required to complete this tutorial:

a. Start a new file in the **Format** mode. Create the format of the drawing, and add the title block in the format, refer to Figures 12-15 and 12-16.
b. Save the format file and then close it.
c. Start a new drawing file in the **Drawing** mode. Select **Pedestal Bearing** as the model and retrieve the format that you have created.
d. Create the table and enter the headers.
e. Define the repeat region and the assign the report symbols, refer to Figure 12-19.
f. Generate the exploded isometric view of the assembly, refer to Figure 12-20.
g. Add balloons to the drawing, refer to Figure 12-22.

Other Drawing Options

Copy the *Pedestal Bearing* folder from the *c09* folder to the *c12* folder. Because this is the first tutorial of this chapter, you need to create the *c12* folder inside the *ProE-WF-2.0* folder. Before you start generating the drawing views, you need to set the working directory. Set the working directory to *C:/ProE-WF-2.0/c12/Pedestal Bearing*.

Starting the Format File

As evident from Figure 12-14, the exploded view of the Pedestal bearing is generated on a drawing sheet with a customized format. Therefore, you need to create a format before generating the drawing view. This format will then be retrieved later.

1. Choose the **Create a new object** button from the **Top Toolchest**, the **New** dialog box is displayed.

2. Select the **Format** radio button from the **Type** area in the **New** dialog box and name the file as *Format1*. Choose **OK** from the **New** dialog box.

   The **New Format** dialog box is displayed.

3. The **Empty** radio button in the **Specify Template** area and the **Landscape** button in the **Orientation** area of the **New Format** dialog box are selected by default. If they are not selected, you need to select them.

4. Select **A** from the **Standard Size** drop-down list in the **Size** area and then choose **OK** to proceed to the **Format** mode. Note, that a sheet of size **A** is displayed on the screen. This is evident from the text displayed below the sheet on the screen.

Creating Format

1. Choose **Sketch > Edge > Offset** from the menu bar; the **OFFSET OPER** menu is displayed. Choose the **Ent Chain** option and then select all four border lines of the format by drawing a window around them. Press the middle mouse button after you have made the selection.

   An arrow is displayed pointing outwards. The arrow displays the direction of offset. Because you need to offset the lines in the opposite direction, you need to specify a negative offset value.

2. Specify a value of **-0.25** in the **Message Input Window** and press ENTER. Press the middle mouse button to exit the **OFFSET OPER** menu.

3. Choose **Table > Insert > Table** from the menu bar to display the **TABLE CREATE** menu.

4. Choose **Ascending > Leftward > By Length** from the **TABLE CREATE** menu and then choose **Vertex** from the **GET POINT** submenu.

5. Select the lower right corner of the new rectangle created by offsetting the border lines. The corner of the new rectangle is selected because the title block will be created such

that it is attached to the new rectangle. Now, you need to enter the dimensions for the title block.

6. Enter the width of the first column as **1.2** in the **Message Input Window** and press ENTER. Similarly, enter the width of the second column as **2** and for the third column as **1.2** in the **Message Input Window** and press ENTER. You are prompted to enter the width for next column. Press ENTER.

   Now, you are prompted to enter the height for the first row.

7. Enter the height of the first row as **0.5**, the second as **0.5**, and the third as **0.5** in the **Message Input Window**. Press ENTER twice.

   The title block created will be similar to the one shown in Figure 12-15. Note that the title block created is not as required. Therefore, you need to modify it.

Figure 12-15 Format with the title block

8. Select the table by clicking on any cell. Choose **Table > Merge Cells** from the menu bar to display the **TABLE MERGE** menu. The **Rows & Cols** option is chosen by default in the **TABLE MERGE** menu.

9. Select the first column and the second column of the top row to merge. Similarly, merge the third column of the top row with the new bigger column.

10. Now, merge the second column of the second row with the second column of the third row. Press middle mouse button twice to exit this tool.

The table after merging the rows and columns should look similar to the one shown in Figure 12-16.

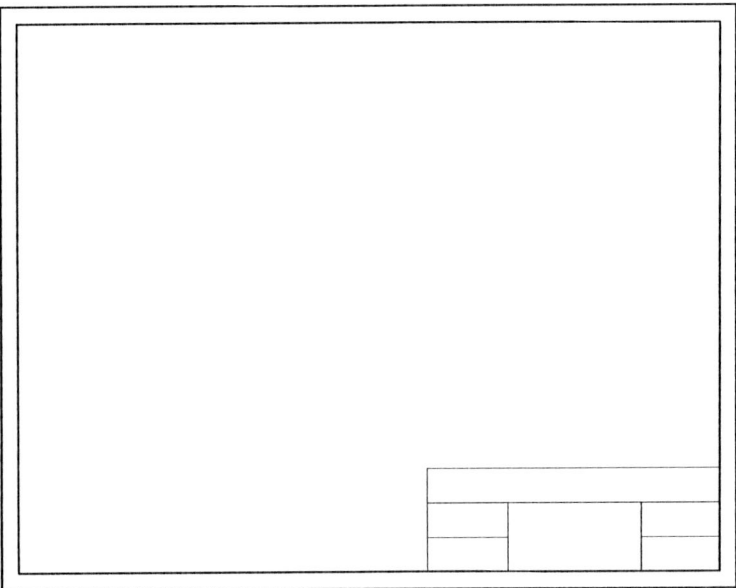

Figure 12-16 Modified title block

11. You need to set the alignment of the text that will be entered later in the cells of the table. Double-click in a cell to invoke the **Note Properties** dialog box.

12. In the dialog box, select the **Text Style** tab.

13. In the **Note/Dimension** area of this dialog box, from the **Vertical** drop-down list, select the **Middle** option. Exit the **Note Properties** dialog box by choosing the **OK** button.

14. Now, the alignment of the text is set to middle aligned vertically and left aligned horizontally. Similarly, individually set the alignment in all cells of the table.

Saving the Format File

You need to save the format file that you have created so that you can use it as a template in the **Drawing** mode where you will generate the exploded drawing view of the Pedestal Bearing. The file will be stored in the *.frm* file format.

1. Choose the **Save the active object** button from the **Top Toolchest**. The **Save Object** dialog box is displayed with the name of the file that you had entered earlier. Press ENTER.

2. Close the current window by choosing **Window > Close** from the menu bar.

Starting a New Drawing File

You need to start a new drawing file to generate the exploded drawing view of the Pedestal Bearing.

1. Choose the **Create a new object** button to display the **New** dialog box. Select the **Drawing** radio button and specify the name of the drawing as *c12tut1*. Choose **OK**; the **New Drawing** dialog box is displayed.

2. Choose the **Browse** button from the **Default Model** area and select the assembly file named *pedestalbearing.asm*.

3. Select the **Empty with format** radio button from the **Specify Template** area in the **New Drawing** dialog box.

4. Select *FORMAT1* from the **Format** drop-down list. If *FORMAT1* is not available in the drop-down list, choose the **Browse** button to locate *FORMAT1*. Choose **OK** to start a new drawing file with the selected format.

Creating the Table for BOM

1. Choose **Table > Insert > Table** from the menu bar to display the **TABLE CREATE** menu.

2. Choose **Descending > Leftward > By Length** from the **TABLE CREATE** menu. Choose **Pick Pnt** from the **GET POINT** menu.

3. Specify a point close to the upper right corner of the inner rectangle of the border. Now, you need to enter the dimensions of the BOM table.

4. Enter the width of the first column as **0.5** in the **Message Input Window** and press ENTER. Similarly, enter the width of the second column as **3** and press ENTER. You are prompted to enter the width for next column, enter **1** and press ENTER twice.

   Now, you are prompted to enter the height for the first row.

5. Enter the height of the first row as **0.5** and of the second as **0.5** in the **Message Input Window**. Press ENTER twice.

Setting the Repeat Region

Repeat regions are the smart cells in a table that can expand, depending on the amount of data inserted in the cell. The second row will be defined as the repeat region.

1. Choose **Table > Repeat Region** from the menu bar; the **TBL REGIONS** menu is displayed.

2. Choose the **Add** option; you are prompted to locate the corners of the region.

Other Drawing Options

3. Select the first cell of the second row and then select the third cell of the second row. Press the middle mouse button twice to exit the menu.

Creating the Column Headers

Next, the heading in the table will be entered.

1. Double-click in the first cell of the first row; the **Note Properties** dialog box is displayed. Enter **Index** and choose **OK** to exit the dialog box.

2. Similarly, enter **Part Name** and **Qty** in the second and the third cells.

Assigning the Report Symbols in the Repeat Region

The information in the cells that are defined as repeat region is determined by the text that is written in the form of report symbols. These report symbols associate the information in the cells with the assembly file directly.

1. Double-click in the first cell of the second row; the **Report Symbol** dialog box is displayed, as shown in Figure 12-17. Select **rpt** from the dialog box; the **rpt** options are displayed, as shown in Figure 12-18.

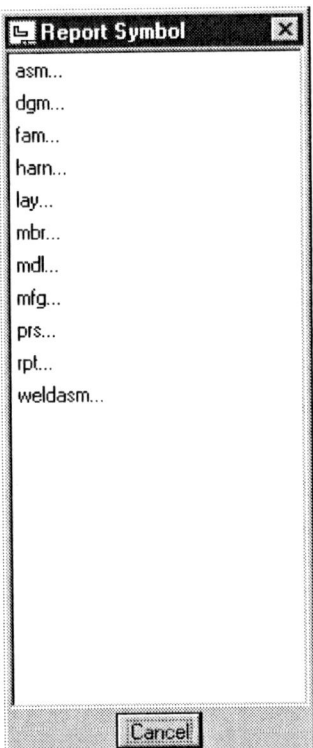

Figure 12-17 The **Report Symbols** dialog box   *Figure 12-18* The **Report Symbols** dialog box

2. Select the **index** option from the dialog box.

3. Double-click in the second cell of the second row; the **Report Symbol** dialog box is displayed. Select **asm** from the dialog box. The **asm** options are displayed.

4. Select the **mbr** option from the dialog box then select the **name** option from the dialog boxes that appear.

5. Double-click in the third cell of the second row; the **Report Symbol** dialog box is displayed. Select **rpt** from the dialog box. The **rpt** options are displayed.

6. Select the **qty** option from the dialog box.

If the text you have selected to enter in the cells is overlapping the other text or is extending out of the cell, ignore it. This is because the BOM will be created in the repeat region where this text exists and after the BOM is created the text of the BOM will fit inside the cells. The drawing sheet after creating the column headers and entering the report symbols is shown in Figure 12-19.

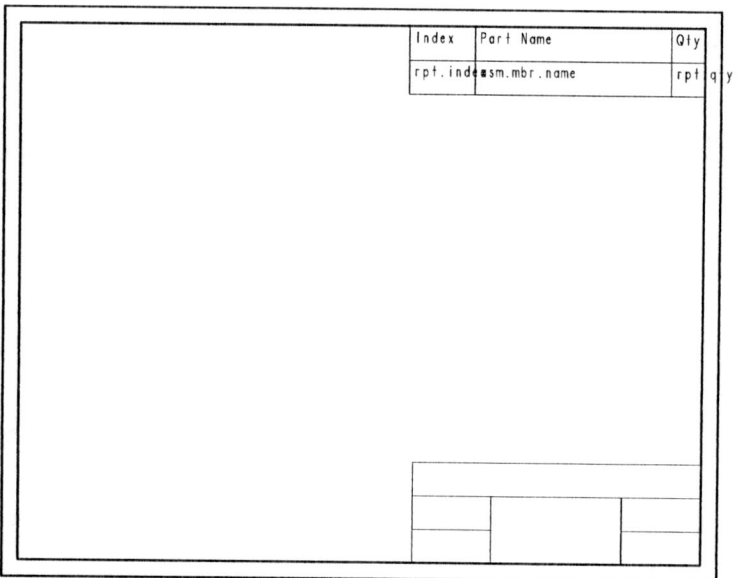

Figure 12-19 Drawing sheet after adding the report symbols

Generating the Exploded Drawing View

In Tutorial 2 of Chapter 9, you created the exploded view of the Pedestal Bearing in the **Assembly** mode. The name of the exploded view that you created was **EXP1**. Hence, the exploded view of the Pedestal Bearing is integrated with the assembly file that you copied from the *c09* folder. Now, you will use this exploded state to generate the exploded drawing view.

Other Drawing Options 12-23

1. Choose the **Create a general view** button from the **Top Toolchest**, you are prompted to select the center point for drawing view.

2. Select the point on the left side of the drawing sheet, refer to Figure 12-20. The **Drawing View** dialog box is displayed. Select the **Isometric** option from the **Default orientation** drop-down list in the **Drawing View** dialog box. Choose the **Apply** button.

3. Choose the **Scale** option from the **Categories** list box. Choose the **Custom scale** radio button from the **Scale and perspective options** area. Enter a value of **.025** in the edit box. Choose the **Apply** button.

4. Select the **View States** option from the **Categories** list box. Select the **Explode components in view** check box. Select the **EXP1** from the **Assembly explode state** drop-down list. Choose the **Apply** button.

5. Close the **Drawing View** dialog box to complete placing the required view.

If the model is not placed properly inside the boundaries of the drawing sheet, unlock it and then drag the drawing view and place it, as shown in Figure 12-20. Note that the BOM automatically appears in the table by extending the repeat region and the Qty column is empty.

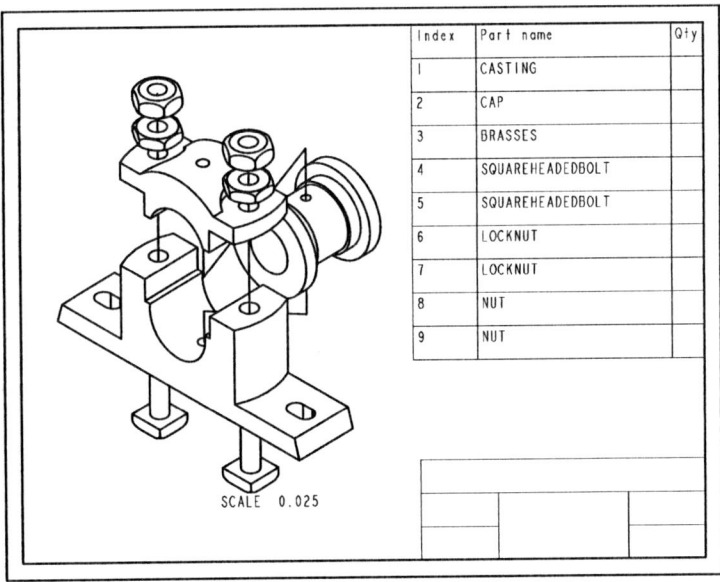

Figure 12-20 Exploded isometric view of the assembly

Note
If the BOM is not displayed, then choose the **Update the information displayed in tables** *button from the* **Top Toolchest**.

Setting the No Repeat Option

In the assembly, some components are repeated. Therefore, in the BOM also they are repeated. The BOM need to be set such that no part name is repeated.

1. Choose **Table > Repeat Region** from the menu bar; the **TBL REGIONS** menu is displayed.

2. Choose the **Attributes** option from the menu; you are prompted to select a region.

3. Move the cursor over the table below the first row. All rows, except the first row, are highlighted in cyan. The highlighted portion is called region. Select the region; the **REGION ATTR** submenu is displayed. Select the **No Duplicates** option then select the **Done/Return** button.

4. Press the middle mouse button twice and exit the menu. The drawing sheet appears, as shown in Figure 12-21.

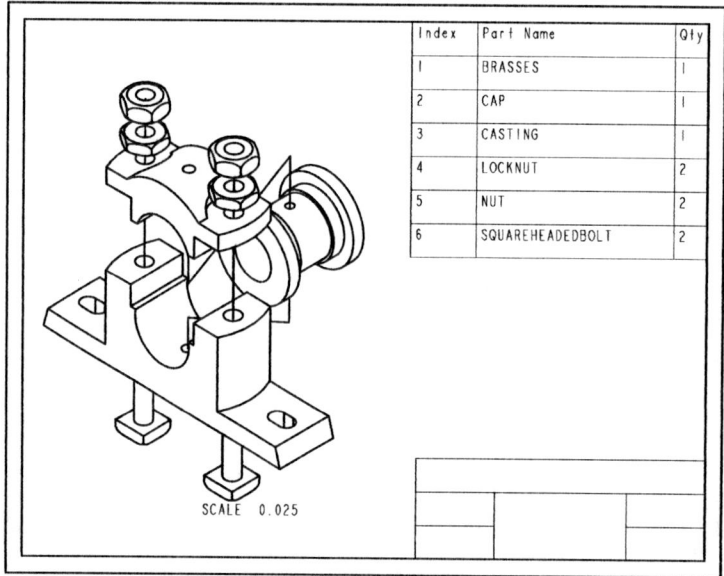

Figure 12-21 The drawing view showing BOM with no duplicates

5. Choose the **Update the information displayed in tables** button from the **Top Toolchest**, if the information is not updated in the table.

Generating Balloons

The balloons are generated by using the repeat region. Before generating the balloons, you need to set the repeat region for balloon generation.

1. Choose **Table > BOM Balloons**; the **BOM BALLOONS** menu is displayed. The **BOM**

Other Drawing Options

BAL TYPE submenu is also displayed with the **Simple** option chosen by default. Also, you are prompted to select a region.

2. Select the region that was selected earlier. Press the middle mouse button

3. From the **BOM BALLOONS** menu, choose the **Create Balloon** option. The **BOM VIEW** submenu is displayed.

4. Choose the **Show All** option. Balloons are attached with all parts in the exploded view. The balloons contain the index number so that the parts in the assembly can be referred to the BOM.

5. Choose **Done** to exit this menu.

6. Drag the balloons to place them appropriately on the drawing sheet, as shown in Figure 12-22. You can select the balloon and then hold down the right mouse button to invoke the shortcut menu. From this menu, you can select the options to modify the attachment points of the balloons.

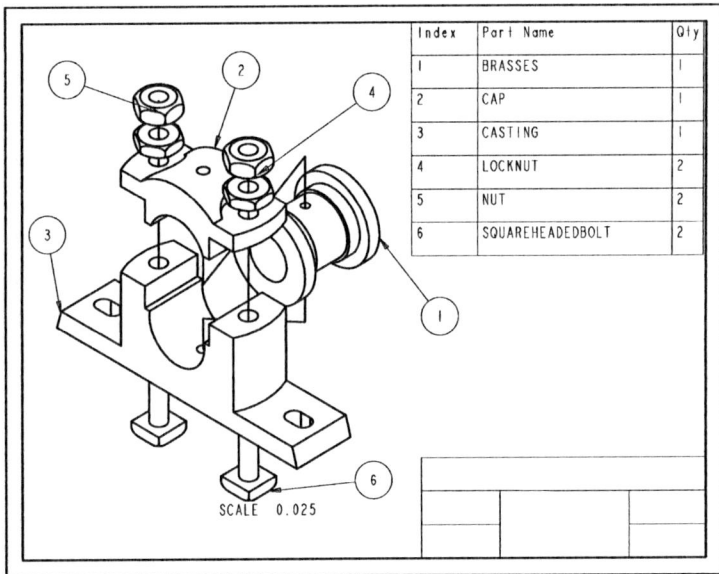

Figure 12-22 The drawing view of the assembly showing the BOM and balloons

Entering Text in the Title Block

The text alignment in various cells of the title block was defined while creating the format. As a result, when you enter the text, it will be automatically aligned middle left.

1. Double-click the top cell in the table. The **Note Properties** dialog box is displayed.

2. Enter **Pedestal Bearing Assembly** and choose **OK**.

3. Double-click on the cell that is in the first column and the second row. The **Note Properties** dialog box is displayed.

4. Enter **Scale:** and press ENTER. Now, enter **0.025** and choose **OK**. The text Scale: 0.025 is entered in two lines in the title block.

5. Similarly, enter the text in all remaining cells one by one, as shown in Figure 12-23.

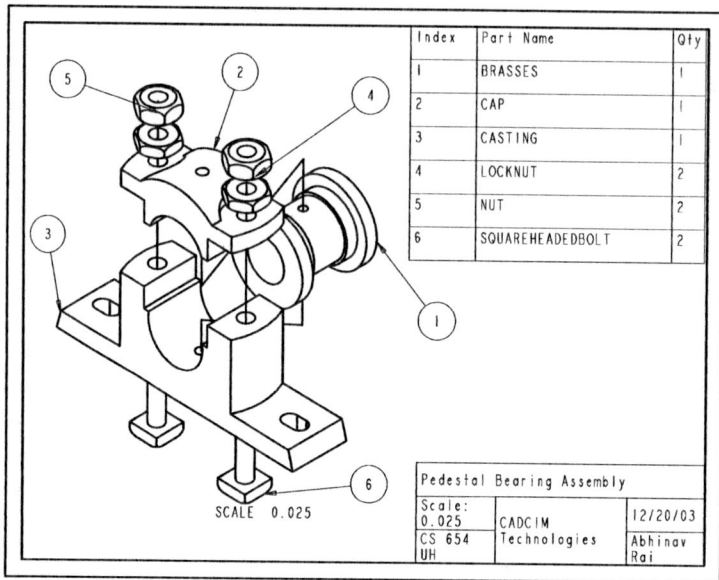

Figure 12-23 The drawing view of the assembly showing the title block, balloons, and the BOM

Saving the Drawing File

1. Choose the **Save the active object** button from the **Top Toolchest** to save the drawing file.

2. The **Save Object** dialog box is displayed. Press ENTER.

Closing the Window

The drawing file is saved and now you can exit the **Drawing** mode.

1. Choose **Window > Close** from the menu bar.

Note
The occurrence of the components in BOM depends on the order in which they were assembled in the assembly. In other words, the component placed first will occur first in the BOM.

Other Drawing Options 12-27

Tutorial 2

In this tutorial you will create a format of size A and add the title block in the format. You will retrieve the format in the **Drawing** mode and generate the exploded isometric view of the **Shock** assembly created in Tutorial 1 of Chapter 9. Also, add the associative Bill of Material and the Balloons to the drawing view, as shown in Figure 12-24. **(Estimated time: 45 min)**

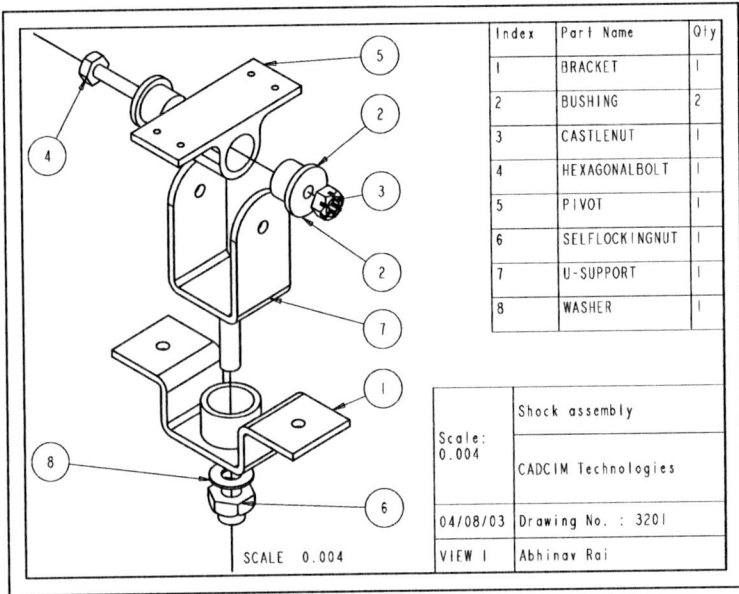

Figure 12-24 The drawing view of the exploded assembly with BOM

The following steps are required to complete this tutorial:

a. Start a new file in the **Format** mode. Create the format of the drawing and add the title block in the format, refer to Figure 12-26.
b. Save the format file and then close it.
c. Start a new drawing file in the **Drawing** mode, select *shockassembly.asm* as the model, and retrieve the format that you had created.
d. Create the table for BOM and enter the headers.
e. Define the repeat region and then assign the report symbols, see Figure 12-27.
f. Generate the exploded isometric view of the assembly and generate balloons in the drawing view, refer to Figure 12-29.
g. Add text in the title block, refer to Figure 12-30.

Copy the *Shock Assembly* folder from the *c09* folder to *c12* folder. Set the *Shock Assembly* folder as the working directory.

Starting the Format File

As mentioned in the tutorial description, you need to create a format of size A that will be retrieved later to create the drawing view of the Shock assembly.

1. Choose the **Create a new object** button from the **Top Toolchest**, the **New** dialog box is displayed.

2. Select the **Format** radio button from the **Type** area in the **New** dialog box and name the file as *Format2*. Choose **OK** from the **New** dialog box.

   The **New Format** dialog box is displayed.

3. Select the **Empty** radio button from the **Specify Template** area and the **Landscape** button from the **Orientation** area of the **New Format** dialog box, if they are not selected.

4. Select **A** from the **Standard Size** drop-down list in the **Size** area and then choose **OK** to proceed to the **Format** mode. Note, that a sheet of size A is displayed on the screen. This is evident from the text displayed below the sheet on the screen.

Creating Format

1. Choose **Sketch > Edge > Offset** from the menu bar; the **OFFSET OPER** menu is displayed. Choose the **Ent Chain** option and then select all four border lines of the format by drawing a window around them. Press the middle mouse button after you have made the selection.

   An arrow is displayed pointing outwards. The arrow displays the direction of offset. Since you need to offset the lines in the opposite direction, therefore, you will specify a negative offset distance.

2. Specify a value of **-0.25** in the **Message Input Window** and press ENTER. Press the middle mouse button twice to exit the **OFFSET OPER** menu.

3. Choose **Table > Insert > Table** from the menu bar to display the **TABLE CREATE** menu.

4. Choose **Ascending > Leftward > By Length** from the **TABLE CREATE** menu and then choose **Vertex** from the **GET POINT** submenu.

5. Select the lower right corner of the new rectangle created by offsetting the border lines. The corner of the new rectangle is selected because the title block will be created such that it is attached to the new rectangle. Now, you need to enter the dimensions for the title block.

6. Enter the width of the first column as **3.2** in the **Message Input Window** and press ENTER. Similarly, enter the width of the second column as **1.2** and press ENTER. You are prompted to enter the width for next column. Press ENTER.

Other Drawing Options 12-29

Now, you are prompted to enter the height for the first row.

7. Enter the height of the first row as **0.5**, the second as **0.5,** the third as **1**, and the fourth as **0.7** in the **Message Input Window**. Press ENTER twice.

   The title block created will be similar to the one shown in Figure 12-25. Note the difference between the title block shown in Figure 12-25 and that shown in Figure 12-26. You need to modify the title block that you have created so that it is similar to the one shown in Figure 12-26.

Figure 12-25 Format with the title block

8. Select a cell in the table and choose **Table > Merge Cells** to display the **TABLE MERGE** menu.

   The **Rows & Cols** option is chosen by default in the **TABLE MERGE** menu.

9. Select the cell in the first row and the cell in the second row of the first column to merge. Exit the **TABLE MERGE** menu.

   The table after merging the rows and columns should look similar to the one shown in Figure 12-26.

   You need to set the alignment of the text that will be entered later in the cells of the table.

10. Double-click in a cell to invoke the **Note Properties** dialog box.

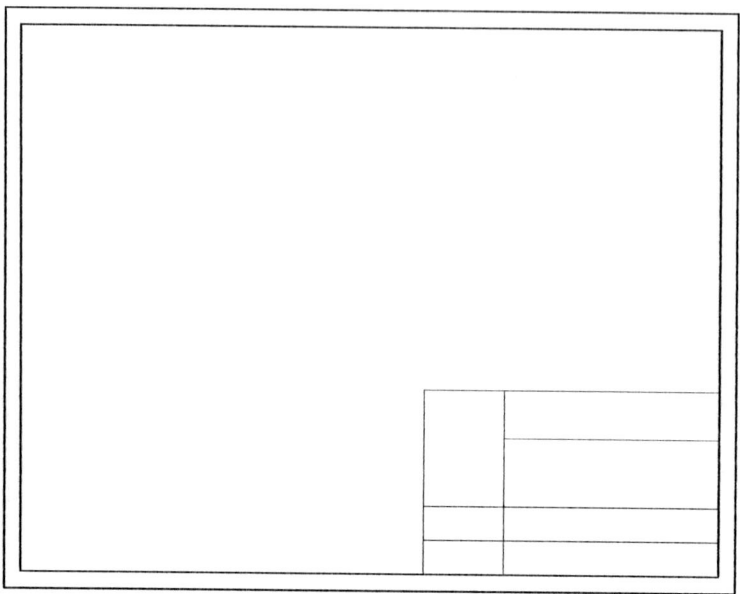

Figure 12-26 Modified title block

11. In the dialog box, select the **Text Style** tab.

12. In the **Note/Dimension** area of this dialog box, from the **Vertical** drop-down list, select the **Middle** option. Exit the **Note Properties** dialog box by choosing the **OK** button.

   Now, the alignment of the text is set to middle aligned vertically and left aligned horizontally. Similarly, set the alignment in all cells of the table individually.

Saving the Format File

You need to save the format file that you have created so that you can use it as a template in the **Drawing** mode where you will generate the exploded drawing view of the Shock assembly. The file will be stored in the *.frm* file format.

1. Choose the **Save the active object** button from the **Top Toolchest**. The **Save Object** dialog box is displayed with the name of the file that you had entered earlier. Press ENTER.

2. Close the current window by choosing **Window > Close** from the menu bar.

Starting a New Drawing File

You need to start a new drawing file for generating the exploded drawing view of the Shock assembly and for generating the BOM

1. Choose the **Create a new object** button to display the **New** dialog box. Select the

Other Drawing Options 12-31

Drawing radio button and specify the name of the drawing as *c12tut2* and then choose **OK**. The **New Drawing** dialog box is displayed.

2. Choose the **Browse** button and select the assembly file named *shockassembly.asm*.

3. Select the **Empty with format** radio button from the **Specify Template** area in the **New Drawing** dialog box.

4. Select **Format2** from the **Format** drop-down list. If **Format2** is not available in the drop-down list, choose the **Browse** button to locate **Format2**. Choose **OK** to exit the **New Drawing** dialog box.

Creating the Table for BOM

1. Choose **Table > Insert > Table** from the menu bar to display the **TABLE CREATE** menu.

2. Choose **Descending > Leftward > By Length** from the **TABLE CREATE** menu and choose **Pick Pnt** from the **GET POINT** menu.

3. Select a point close to the upper right corner of the inner border rectangle. Now, you need to enter the dimensions of the BOM table.

4. Enter the width of the first column as **0.5** in the **Message Input Window** and press ENTER. Similarly, enter the width of the second column as **3** and press ENTER. You are prompted to enter the width for next column, enter **1** and press ENTER twice.

    Now, you are prompted to enter the height for the first row.

5. Enter the height of the first row as **0.5** and of the second as **0.5** in the **Message Input Window**. Press ENTER twice.

Setting the Repeat Region

Repeat regions are the smart cells in a table that can expand depending on the amount of data inserted in the cell. The second row will be defined as the repeat region.

1. Choose **Table > Repeat Region** from the menu bar. The **TBL REGIONS** menu is displayed.

2. Choose the **Add** option; you are prompted to locate the corners of the region.

3. Select the first cell of the second row and then select the third cell of the second row. Press the middle mouse button twice to exit the menu.

Creating the Column Headers

Next, the heading in the table will be entered.

1. Double-click in the first cell of the first row; the **Note Properties** dialog box is displayed. Enter **Index** and choose **OK** to exit the dialog box.

2. Similarly, enter **Part Name** and **Qty** in the second and the third cells.

Entering the Report Symbols in the Repeat Region

As mentioned earlier, the information in the cells that are defined as repeat region is determined by the text that is written in the form of report symbols. These report symbols associate the information in the cells with the assembly file directly.

1. Double-click in the first cell of the second row; the **Report Symbol** dialog box is displayed. Select **rpt** from the dialog box; the **rpt** options are displayed.

2. Select the **index** option from the dialog box.

3. Double-click in the second cell of the second row; the **Report Symbol** dialog box is displayed. Select **asm** from the dialog box; the **asm** options are displayed.

4. Select the **mbr** option from the dialog box then select the **name** option from the options that appear.

5. Double-click in the third cell of the second row; the **Report Symbol** dialog box is displayed. Select **rpt** from the dialog box; the **rpt** options are displayed.

6. Select the **qty** option from the dialog box. The drawing sheet after adding the report symbols is shown in Figure 12-27.

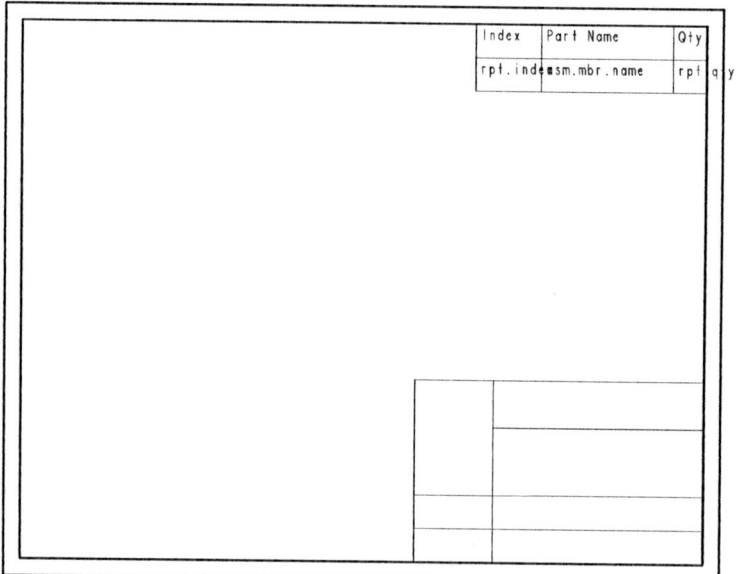

Figure 12-27 *Drawing sheet after adding the report symbols*

Generating the Drawing View

In Tutorial 1 of Chapter 9 you have created the exploded view of the Shock assembly in the **Assembly** mode. The name of the exploded view that you specified was **EXP1**. Hence, the exploded view of the Shock assembly is integrated with the assembly file that you copied from the *c09* folder. Now, you will use this exploded state to generate the exploded drawing view in the **Drawing** mode.

1. Choose the **Create a general view** button from the **Top Toolchest**; you are prompted to select the center point for drawing view.

2. Select the point on the left side on the drawing sheet; the **Drawing View** dialog box is displayed. Select the **Isometric** option from the **Default orientation** drop-down list in the **Drawing View** dialog box. Choose the **Apply** button.

3. Choose the **Scale** option from the **Categories** list box. Choose the **Custom scale** radio button from the **Scale and perspective options** area. Enter a value of **.004** in the edit box and choose the **Apply** button.

4. Choose the **View States** option from the **Categories** list box. Select the **Explode components in view** check box. Select the **EXP1** from the **Assembly explode state** drop-down list. Choose the **Apply** button.

5. Close the **Drawing View** dialog box to complete placing the required view.

If the model is not placed properly inside the boundaries of the drawing sheet, unlock it and then drag the drawing view and place it, as shown in Figure 12-28. If the BOM is not

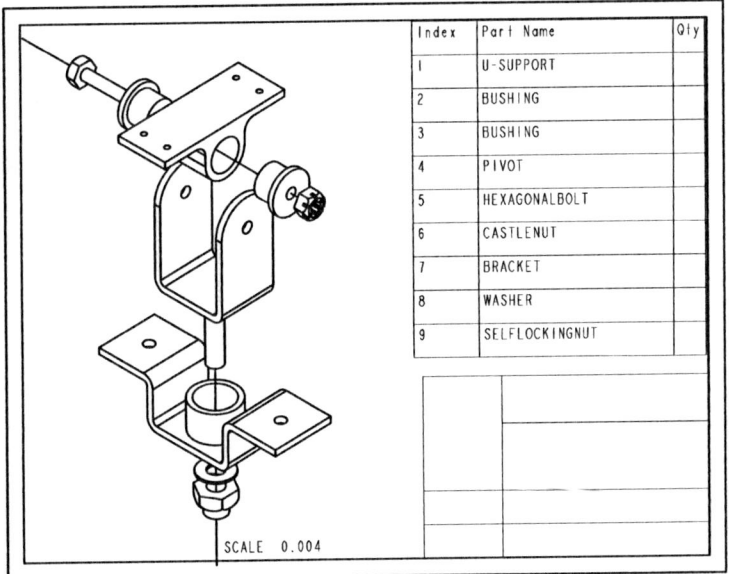

Figure 12-28 Exploded isometric view of the assembly

displayed, then choose the **Update the information displayed in tables** button from the **Top Toolchest**.

Setting the No Repeat Option

The Shock assembly consists of two instances of Bushing. Therefore, in the BOM, it is also repeated. But the BOM needs to be set such that no part name is repeated.

1. Choose **Table > Repeat Region**; the **TBL REGIONS** menu is displayed.

2. Choose the **Attributes** option from the menu; you are prompted to select a region.

3. Move the cursor over the table below the first row. All rows, except the first row, are highlighted in cyan. Select the region, the **REGION ATTR** submenu is displayed. Choose the **No Duplicates** option then choose **Done/Return** button.

4. Press the middle mouse button twice to exit the menu. The drawing sheet appears, as shown in Figure 12-29.

5. Choose the **Update the information displayed in tables** button from the **Top Toolchest**, if the information is not updated in the table.

Generating Balloons

The balloons are generated by setting the repeat region for generating the balloon. Therefore, before generating the balloons you need to set the repeat region.

1. Choose **Table > BOM Balloons**; the **BOM BALLOONS** menu is displayed.

   The **BOM BAL TYPE** submenu is also displayed with the **Simple** option chosen by default and you are prompted to select a region.

2. Select the repeat region that was created earlier. Press the middle mouse button

3. Choose the **Create Balloon** option from the **BOM BALLOONS** menu; the **BOM VIEW** submenu is displayed.

4. Choose the **Show All** option. Exit the menu.

   Balloons are attached with all parts in the exploded view. The balloons contain the index number so that the parts in the assembly can be referred to the BOM.

5. Drag the balloons to place them appropriately on the drawing sheet, as shown in Figure 12-29. You can select the balloon and then hold down the right mouse button to invoke the shortcut menu. From this menu you can select the options to modify the attachment points of the balloons.

Other Drawing Options

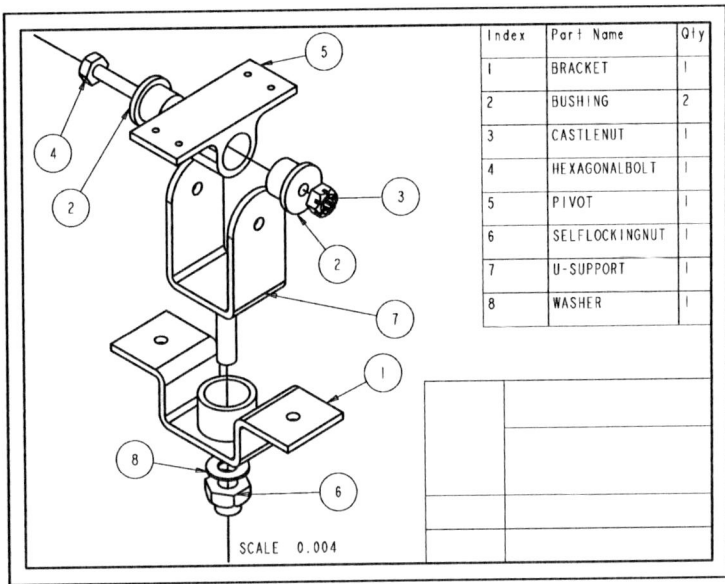

Figure 12-29 The drawing view of the assembly showing the BOM and balloons

Entering Text in the Title Block

1. Double-click in the cell formed by the first row and the second column. The **Note Properties** dialog box is displayed.

2. Enter **Shock assembly** and choose the **OK** button.

3. Now, select the cell from the second row and the second column. Enter **CADCIM Technologies** in the **Enter Text** dialog box and choose the **OK** button.

4. Similarly, enter the text in all remaining cells one by one, as shown in Figure 12-30.

Saving the Drawing File

1. Choose the **Save the active object** button from the **Top Toolchest** to save the drawing file.

2. The **Save Object** dialog box is displayed. Press ENTER.

Closing the Window

The drawing file is saved and now you can exit the **Drawing** mode.

1. Choose **Window > Close** from the menu bar.

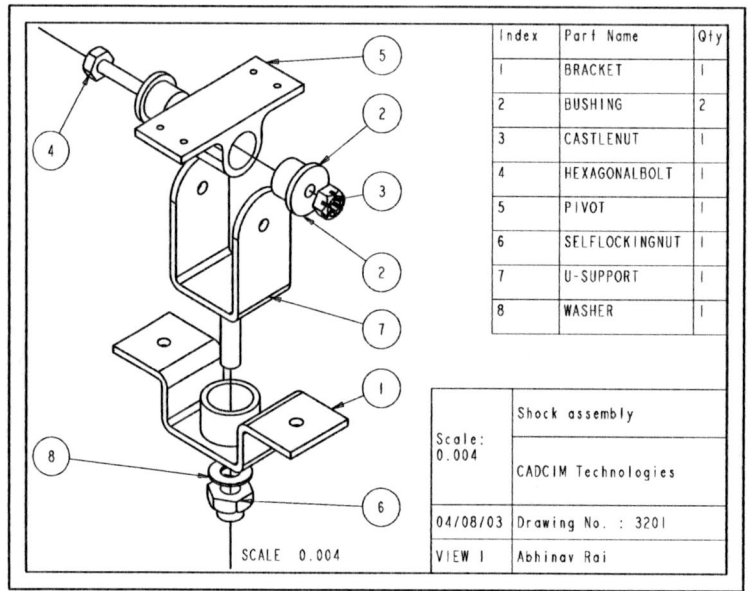

Figure 12-30 The drawing view of the assembly showing the title block, balloons, and the BOM

Self-Evaluation Test

Answer the following questions and then compare your answers to the answers given at the end of this chapter.

1. The **Select one item at a time - shift to gather more than one item** button in the **Right Toolchest** is used to select any item from the drawing view or the drawing view itself to modify. (T/F)

2. Pro/ENGINEER Wildfire 2.0 provides you with some standard drawing formats for generating the drawing views. (T/F)

3. Pro/ENGINEER Wildfire 2.0 allows you to create the user-defined formats. (T/F)

4. The BOM generated using repeat region is associative. (T/F)

5. Only one drawing sheet is available in one drawing file. (T/F)

6. The _____ option of the **Table** menu in the menu bar is used to generate balloons.

7. The _____ dialog box is displayed when you double-click in the cell of a table.

Other Drawing Options

8. The _____ dialog box is used to edit the text.

9. The _____ key from the keyboard can also be used to delete any draft entity.

10. The _____ menu in the menu bar is used to draw tables in the **Drawing** mode.

Review Questions

Answer the following questions:

1. Which of the following modes in Pro/ENGINEER Wildfire 2.0 is used to create an associative BOM?

   (a) Part (b) Format
   (c) Drawing (d) None

2. Which of the following dialog box is displayed, when you double click in the repeat region cell.

   (a) **Drawing View** (b) **Note Properties**
   (c) **Report Symbol** (d) None

3. Which of the following buttons on the **Right Toolchest** enables you to draw the geometric entities continuously?

   (a) **Create 2 point lines** (b) **Create circle**
   (c) **Enable sketching chain** (d) None

4. Which of the following radio buttons is selected to create a user-defined format for the drawing sheet?

   (a) **Part** (b) **Format**
   (c) **Drawing** (d) None

5. Which of the following menus in the menu bar has the **Close** option to close the current window?

   (a) **Info** (b) **Window**
   (c) **View** (d) None

6. You can open more than one windows in Pro/ENGINEER Wildfire 2.0. (T/F)

7. It is not possible to draw splines in the **Drawing** mode. (T/F)

8. There are two types of drafting in Pro/ENGINEER Wildfire 2.0: Generative drafting and Interactive drafting. (T/F)

9. You can import a text file in the **Drawing** mode. (T/F)

10. The **Draft Group** option in the cascading menu is used to add or remove notes, surface symbols, or gtols from the selected dimension. (T/F)

Exercise

Exercise 1

Create a format of size A and then retrieve it in the **Drawing** mode to place the top, front, and an exploded isometric view of the **Cross Head** assembly created in Exercise 1 of Chapter 9. Add the associative assembly BOM and the balloons to the drawing view, as shown in Figure 12-31. **(Estimated time: 45 min)**

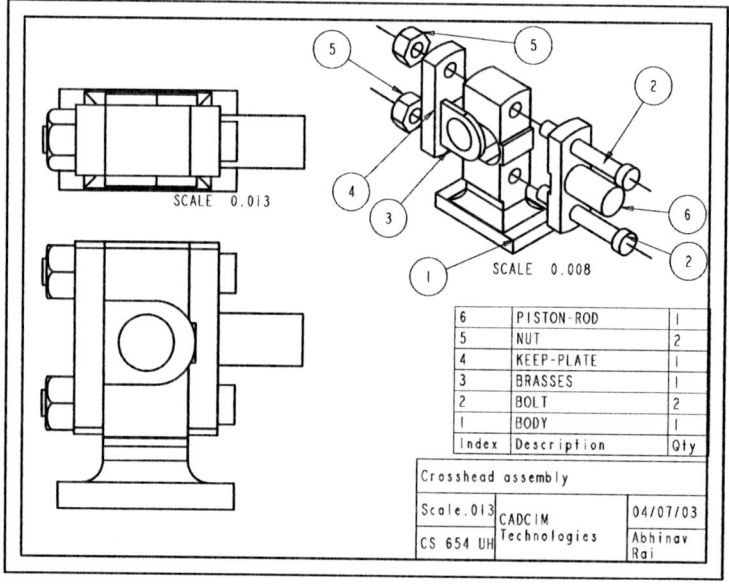

Figure 12-31 Drawing for Exercise 1

Answers to the Self-Evaluation Test
1 - T, 2 - T, 3 - T, 4 - T, 5 - F, 6 - Bom Balloons, 7- Note Properties, 8 - Note Properties, 9 - DELETE, 10 - Table.

Index

A

Add Inn Fcs option 7-3
Adding and Removing Sheets in the Drawing 12-11
Adding Balloons to the assembly views 11-21
Adding New Parts or Assemblies to the Current Drawing 10-21
Adding Notes to the Drawing 11-21
Adding or Removing the Cross-section Arrows 10-23
Adding reference datums to the drawing views 11-16
Adding Tolerances In The Drawing Views 11-17
Advantages of Nesting the Sketches 3-24
Align 9-7
Align Coincident 9-7
Align Offset 9-7
Align Oriented 9-7
Angular Dimensioning an Arc 1-14
Arc 1-10
Assembling Repeated Copies Of A Component 9-23
Assembling The Components 9-10
Assembly Datum Planes 9-10
Assembly Modeling 9-2
Assigning the Report Symbols in the Repeat Region 12-14
ATTRIBUTES menu 7-3
Auxiliary View 10-7
Axis 11-3
Axis Patterns 6-4

B

Balloon 11-3
Base features 3-2
Base Features toolbar 3-5
Basic Rounds 5-11
Blend Between Surfaces 8-11
Blend features 7-8
BLEND OPTS submenu 7-8
Blend Section To Surfaces 8-11
Blind 3-9
Bndry Chain 7-6
BOM dialog box 9-30
Bottom-Up Approach 9-3
Broken View 10-11
Build feature 3-12

C

Centerline 1-7
CHAMFER (CORNER) dialog box 5-21
Chamfer dashboard 5-22
Chamfers 5-21
Changing the View Scale 10-22
Changing the View Type 10-21
Circle 1-7
Circular fillets 2-3
Clean Dimensions dialog box 11-13
Cleaning Up the Dimensions 11-12
Collection 4-5
Component Create dialog box 9-5
Component Reference Area 9-12
Constant Normal Direction 8-5
Constant section 8-6
Constant Thickness Shell 7-13
Constraints dialog box 1-17
Convert 9-10
Converting a Generated View to a Draft View 12-8
Converting a Weak Constraint into a Strong Constraint 1-17

Coord Sys 9-9
Coordinate System 2-7
Copy and Align View 10-19
COPY FEATURE submenu 6-15
Copying features 6-13
Copying sketched entities in Sketch mode 2-13
Corner Chamfer 5-21
Cosmetic Feature 11-3
Create defining dimension 1-13
Creating a Protrusion 3-5
Creating A Reference Coordinate System 2-7
Creating a Spline 2-7
Creating a Sweep Cut 7-7
Creating base features 3-2
Creating bottom-up assemblies 9-5
Creating Circular Fillets 2-3
Creating Components in the Assembly Mode 9-4
Creating Cuts 4-28
Creating Datum Curve by Sketching 7-18
Creating Datum Planes 4-8
Creating draft features 7-24
Creating Elliptical Fillets 2-6
Creating Feature Patterns 6-2
Creating Fillets 2-3
Creating Ribs 5-27
Creating Simplified Representations 9-17
Creating Sweep Feature by Selecting a Trajectory 7-5
Creating Swept Protrusions 7-2
Creating Tables in Drawing mode 12-11
Creating the Column Headers 12-14
Creating The Exploded State 9-25
Creating the Offset Lines 9-29
Creating the Table for BOM 12-13
Creating Thin Sweep Protrusion 7-7
Creating top-down assemblies 9-4
Creating X-section of the Assembly 9-22
Curve Chain 7-6

D

Datum Axes 4-15
DATUM AXIS dialog box 4-16
Datum Axis Normal to a Plane 4-19
Datum Axis Passing Through a Point and Normal to plane 4-19
Datum Axis Passing Through an Edge 4-18
Datum Axis Passing Through the Center of a Round Surface 4-20
Datum curves 7-15
Datum dialog box 11-16
Datum Options 4-5
Datum Plane 11-3
DATUM PLANE dialog box 4-7
Datum Planes 4-5
Datum Planes Created On-The-Fly 4-14
Datum Point by Clicking 4-27
DATUM POINT dialog box 4-23
Datum Points 4-21
Datum Target 11-3
Datum toolbar 4-5
Datums 4-2
Default Datum Planes 4-2
default datum planes 3-3
Delete 1-20
Deleting a feature 5-33
Deleting a Pattern 6-13
Deleting the Drawing Views 10-21
Deleting the sketcher entities 1-20
Detailed View 10-7
Diameter Dimensioning 1-14
Dim Orientation dialog box 2-2
Dimension Patterns 6-3
Dimension Properties dialog box 11-8
Dimensional Tolerances 11-17
Dimensioning 2-2
Dimensioning a Sketch Using the Baseline Option 2-2
Dimensioning of Splines 2-8
Dimensioning options 1-12
Dimensioning Revolved Sections 1-15
Dimensioning the basic sketched entities 1-14
Dimensioning the drawing views 11-2
Dimensioning the sketch 1-13, 2-2
Direction Patterns 6-4
Disabling the Constraints 1-17
Disadvantages of Nesting the Sketches 3-24
Displaying the Components in a Separate Window 9-10
Displaying the Components in the Same Window 9-11
Divide an entity 1-22
Draft dashboard 7-25
Draft features 7-24

Draft Group 12-7
Drafting Using the Drawing Views 12-8
Drawing a Centerline 1-7
Drawing a Circle 1-7
Drawing a Construction Circle 1-8
Drawing a Line 1-5
Drawing a Rectangle 1-7
Drawing a sketch using the sketcher tools toolbar 1-5
Drawing an Arc 1-10
Drawing an Ellipse 1-10
Drawing display options 1-22
Drawing Mode 10-2
Drawing View dialog box 10-7
Drawing View Dialog Box Options 10-9
Dynamically Modifying the Sketch of a Feature 5-33
Dynamically trim section entities tool 1-21

E

Edge Chamfer 5-22
Edge On Srf 9-9
Editing Definition or Redefining Features 5-29
Editing features of a model 5-29
Editing the Cross-section Hatching 10-24
Editing the drawing views 10-20
Editing The Geometric Tolerances 11-20
Ellipse 1-10
Ellipse Radius dialog box 2-7
Elliptical Fillets 2-6
Enable sketching chain 12-3
Erase 11-5
Erasing the Drawing View 10-20
Explicit Relationship 3-23
Explode Position dialog box 9-26
Extrude as solid 3-10
Extrude as surface 3-10
Extrude dashboard 3-6, 3-8
Extrude Tool 3-5
Extruding a Sketch 3-5

F

Feature Operations 6-13
Field Datum Poin 4-26
File formats 2-14
Fill Patterns 6-6

Filter drop-down list 4-4
Fix 9-10
Free Ends 7-5
Full (Aligned) view 10-18
Full section view 10-13
Full(Unfold) view 10-18

G

General Blend 7-12
General Pattern 6-12
General View 10-5
Generating Balloons 12-15
Generating BOM and Balloons in the Drawings 12-12
Generating Drawing Views 10-5
Generating the Auxiliary View 10-7
Generating the Copy and Align View 10-19
Generating the Detailed View 10-7
Generating the Exploded Drawing View 12-14
Generating the General View 10-5
Generating the Projection View 10-6
Generating the Revolved Section View 10-8
Geometric Tolerance 11-3
Geometric Tolerance dialog box 11-18
Geometric Tolerances 11-17
Grouping the Entities 12-6

H

Half section view 10-16
Half View 10-10
Helical Sweep 8-8
Hole Dashboard 5-2
Holes 5-2

I

Identical Pattern 6-11
imensioning the basic sketched entities 1-14
Implicit Relationship 3-23
Importing 2D Drawings in the sketch mode 2-13
Insert 9-9
Inserting Components in the Assembly 9-5
Intent Chain 7-7
Intersect Option 7-18
Invoking the Sketch Mode 1-3

L

Line 1-5
Line tangent to two entities 1-6
Linear Dimensioning a Line 1-14
Lines using two point 1-6
Local section view 10-16

M

Mate 9-6
Mate Coincident 9-6
Mate Offset 9-6
Mate Oriented 9-7
Merge Ends 7-5
Mirror 6-14, 12-6
Mirroring a geometry 6-16
Mirroring the sketcher entities 1-22
Model edge visibility 10-16
Modify Attach 11-11
Modify Dimensions dialog box 1-19
Modifying a Dimension by Double-Clicking 1-19
Modifying a feature 5-33
Modifying a Spline 2-9
Modifying And Editing Dimensions 11-6
Modifying Boundaries of Views 10-23
Modifying Dimensions Dynamically 1-19
Modifying other Parameters 10-23
Modifying The Components Of The Assembly 9-25
Modifying the Cross-sections 10-23
Modifying the Dimensions of a Feature of a Component 9-25
Modifying The dimensions of a sketch 1-18
Modifying The Drawing views 10-21
Modifying the Offset Lines 9-29
Modifying the Perspective Views 10-23
Modifying the Sketched entities 12-3
Move 6-15
Move dialog box 9-17
Move Text 11-11
Moving the Drawing View 10-20

N

Need for datums in modeling 4-2
Nesting of Sketches 3-23
New dialog box 1-3
New File Options dialog box 3-3
New Format dialog box 12-9
No Inn Fcs option 7-4
Norm To Traj 8-7
Normal Constraint 4-10
Normal To Projection 8-5
Normal To Trajectory 8-4
NormalToOriginTraj 8-7
Note 11-3
NOTE TYPES menu 11-22

O

Offset Constraint 4-12
Offset Lines 9-28
On-The-Fly 4-14
One By One 7-6
Open dialog box 2-15
Options aiding construction of parts 5-2
Options tab slide-up panel 3-9
Orientation dialog box 3-15
Other protrusion options 7-2

P

Package 9-4
Packaging the Components 9-16
Parallel blend 7-8
Parallel Constraint 4-11
Parent-child relationship 3-23
Part mode 3-2
PART SETUP menu 4-2
Partial View 10-10
Patterns 6-2
Pivot Dir 8-7
PLACEMENT Constraints 9-5
Placement Constraints 9-4
Placement tab slide-up panel 3-9
Placing a Point 1-5
Pnt On Line 9-9
Pnt On Srf 9-9
Point 1-5
Pro/TABLE 6-10
Project Option 7-19
Projection View 10-6
Properties Tab 3-9
Protrusion 3-5

Index

R

Radial Dimensioning 1-15
Rectangle 1-7
Redefining a Feature of a Component 9-25
Redefining The Components Of The Assembly 9-22
Redraw the current view 1-23
Reference Patterns 6-8
References Dialog Box 3-15
Refit object 1-23
Remember parametric sketcher references 12-3
Remove Material 3-10
Removing Material by Extruding a Sketch 4-28
Removing Material by Revolving a Sketch 4-29
Reordering Features 5-30
Reordering The Components 9-22
Reorienting the Views 10-22
Replace dialog box 9-22
Replacing Components 9-23
Replacing the Dimensions 2-3
Rerouting Features 5-31
Rescale 12-6
Resolve Sketch dialog box 1-20
Resolve sketch Dialog box 1-19
Retrieving The user-defined formats in the drawing 12-9
Revolve dashboard 3-16
Revolved Section View 10-8
Revolving a Sketch 3-16
Revolving a Sketch as a Solid 3-16
Revolving a Sketch with Thickness 3-17
Rib dashboard 5-26
Ribs 5-27
Rotate 12-5
Rotate and Copy 12-6
Rotating entities 2-12
Rotating the features 6-16
Rotational Blend 7-10
Round dashboard 5-12
Rounds 5-11

S

Saved view list 3-14
Scale 10-12
Scale Rotate dialog box 2-13
Scaling and Rotating entities 2-12
Section of the Solid Model 6-17
Section views 10-13
Select Items 12-2
Select Sec 8-7
Select Traj 7-4
Selected Items dialog box 4-4
Selection 4-3, 4-4
Setting the No Repeat Option 12-15
Setting the Repeat Region 12-13
Shape slide-up panel 5-6
Shell 7-12
Show / Erase dialog box 11-2
Sketch Dialog Box 3-12
Sketch dialog box 3-6
Sketch mode 1-2, 1-4
Sketch placement point collector 8-6
Sketch Sec 8-7
Sketch Traj Option 7-2
Sketched Datum Point 4-26
Sketched Hole 5-6
Sketcher environment 1-3
sketcher environment 1-3
Sketcher Tools toolbar 1-5
Snapping References dialog box 12-2
Spline 2-7
Standard Hole 5-8
Straight holes 5-2
Stretch 12-6
Suppressing Features 5-32
Suppressing/Resuming The Components 9-22
Surf Chain 7-7
Surface Finish 11-3
Sweep Cut 7-7
Sweep features 7-2
SWEEP TRAJ menu 7-2
Swept Blend 8-7
Symbol 11-3
Symmetric 3-9

T

Table Patterns 6-9
Tangent 9-9
Tangent Constraint 4-14
Tangnt Chain 7-6
Text dialog 2-11
Text Symbol 11-11

The Bill Of Material 9-30
The Sketch mode 1-2
Thicken Sketch 3-11
Through Constraint 4-8
To Selected 3-9
Top-Down Approach 9-2
Toroidal Bend 8-12
Trajectories collector 8-3
Transforming the Draft Entities 12-4
Translate 12-4
Translate and Copy 12-5
Translating the Features 6-15
Trimming the Draft Entities 12-3
Trimming the lines 1-20
Trimming the sketcher entities 1-21

U

Understanding the orientation of datum planes
    3-17
Undo 1-20
User-defined drawing formats 12-8
Using Blend Vertex 7-12
Using the Edit Menu 1-19
Using the Sketch Mode 1-2
Using the View Manager 9-19

V

Variable Pattern 6-12
Variable Radius Round 5-19

List of other Publications by CADCIM Technologies

The following is the list of some of the publications by Prof. Sham Tickoo and CADCIM Technologies.

CATIA Textbooks
- CATIA for Designers, V5R13
  CADCIM Technologies, USA
- CATIA for Engineers and Designers, V5R13
  dreamtech Press, India

Solid Edge Textbooks
- Solid Edge for Designers, Version 15
  CADCIM Technologies, USA
- Solid Edge for Engineers and Designers, Version 15
  dreamtech Press, India

SolidWorks Textbooks
- SolidWorks for Designers, Release 2004
  CADCIM Technologies, USA
- SolidWorks for Designers, Release 2004
  Piter Publishing Press, Russia
- SolidWorks for Designers, Release 2003
  CADCIM Technologies, USA
- SolidWorks for Engineers and Designers, Release 2003
  dreamtech Press, India

Autodesk Inventor Textbooks
- Autodesk Inventor for Designers, Release 9
  CADCIM Technologies, USA
- Autodesk Inventor for Designers Release 6 with Release 7 Update Guide
  CADCIM Technologies, USA
- Autodesk Inventor for Designers, Release 6
  CADCIM Technologies, USA
- Autodesk Inventor for Designers: Update Guide Release 6
  CADCIM Technologies, USA
- Autodesk Inventor for Engineers and Designers, Release 6
  dreamtech Press, India
- Autodesk Inventor for Designers, Release 5
  CADCIM Technologies, USA

Pro/ENGINEER Textbooks
- Pro/ENGINEER Wildfire for Designers
  CADCIM Technologies, USA
- Pro/ENGINEER for Designers, Release 2001
  CADCIM Technologies, USA

- Pro/ENGINEER Wildfire for Engineers and Designers
  dreamtech Press, India
- Pro/ENGINEER for Engineers and Designers, Release 2001
  dreamtech Press, India
- Designing with Pro/ENGINEER, Release 2001
  dreamtech Press, India

Mechanical Desktop Textbook
- Mechanical Desktop Instructor, Release 5
  McGraw Hill Publishing Company, USA

AutoCAD Textbooks (US Edition)
- AutoCAD 2005: A Problem-Solving Approach
  Autodesk Press
- AutoCAD LT 2005: A Problem-Solving Approach
  Autodesk Press
- Customizing AutoCAD 2004 and 2005
  Autodesk Press
- AutoCAD 2004: A Problem-Solving Approach
  Autodesk Press
- AutoCAD LT 2004: A Problem-Solving Approach
  Autodesk Press
- Customizing AutoCAD 2004
  Autodesk Press
- AutoCAD 2002: A Problem-Solving Approach
  Autodesk Press
- AutoCAD LT 2002: A Problem-Solving Approach
  Autodesk Press
- Customizing AutoCAD 2002
  Autodesk Press
- AutoCAD 2000: A Problem Solving Approach
  Autodesk Press
- Customizing AutoCAD 2000
  Autodesk Press
- AutoCAD LT 2000: A Problem-Solving Approach
  Autodesk Press

AutoCAD Textbooks (Russian Edition)
- AutoCAD 2005
  Piter Publishing Press, Russia
- AutoCAD 2004
  Piter Publishing Press, Russia
- AutoCAD 2002
  Piter Publishing Press, Russia
- AutoCAD 2000
  Piter Publishing Press, Russia

AutoCAD Textbooks (Italian Edition)
- AutoCAD 2000 Fondamenti
- AutoCAD 2000 Tecniche Avanzate

AutoCAD Textbook (Chinese Edition)
- AutoCAD 2000

AutoCAD Textbooks (Indian Edition)
- AutoCAD 2005 for Engineers and Designers
  dreamtech Press, India
- AutoCAD 2004 for Engineers and Designers
  dreamtech Press, India
- Understanding AutoCAD 2004: A Beginner's Guide
  dreamtech Press, India
- Customizing AutoCAD 200
  dreamtech Press, India
- AutoCAD 2002 with Applications
  Tata McGraw Hill Publishers
- Understanding AutoCAD 2002
  Tata McGraw Hill Publishers
- Advanced Techniques in AutoCAD 2002
  Tata McGraw Hill Publishers
- AutoCAD 2000 with Applications
  Galgotia Publishers
- Understanding AutoCAD 2000
  Galgotia Publishers
- Advanced Techniques in AutoCAD 2000
  Galgotia Publishers

3D Studio MAX and VIZ Textbooks
- Leaning 3ds max5: A Tutorial Approach
  (Complete manuscript available for free download on *www.cadcim.com*)
- Learning 3DS Max: A Tutorial Approach, Release 4
  Goodheart-Willcox Publishers (USA)
- Learning 3D Studio VIZ: A Tutorial Approach
  Goodheart-Willcox Publishers (USA)
- Learning 3D Studio R4: A Tutorial Approach
  Goodheart-Willcox Publishers (USA)
- Learning 3D Studio MAX/VIZ 3.0: A Tutorial Approach
  BPB Publishers (India)

Paper Craft Book
- Constructing 3-Dimensional Models: A Paper-Craft Workbook
  CADCIM Technologies

Coming Soon: New Textbooks from CADCIM Technologies
- SolidWorks for Designers Release 2005 (September 2004)
- Solid Edge for Designers Version 16 (October 2004)
- Autodesk Revit for Architects and Designers (November 2004)